ALGEBRA AND TRIGONOMETRY WITH CALCULATORS

ALGEBRA AND TRIGONOMETRY WITH CALCULATORS

Marshall D. Hestenes
Michigan State University

Richard O. Hill, Jr.
Michigan State University

Prentice-Hall, Inc., Englewood Cliffs, New Jersey 07632

Library of Congress Cataloging in Publication Data

HESTENES, MARSHALL D
    Algebra and trigonometry with calculators.

    Includes index.
    1. Algebra—Data processing.   2. Trigonometry
—Data processing.   3. Calculating-machines.
I. Hill, Richard O., joint author.   II. Title.
QA155.7.E4H47   512′.13′02854   80-39723
ISBN 0-13-021857-X

© 1981 by Marshall D. Hestenes and Richard O. Hill, Jr.

Published by Prentice-Hall, Inc., Englewood Cliffs, N.J. 07632.

All rights reserved. No part of this book may be reproduced in any form or by any means without permission in writing from the publisher.

Printed in the United States of America

10 9 8 7 6 5 4 3 2 1

Editorial/production supervision by Paul Spencer
Cover and interior design by Janet Schmid
Manufacturing buyer: John Hall

Prentice-Hall International, Inc., *London*
Prentice-Hall of Australia Pty. Limited, *Sydney*
Prentice-Hall of Canada, Ltd., *Toronto*
Prentice-Hall of India Private Limited, *New Delhi*
Prentice-Hall of Japan, Inc., *Tokyo*
Prentice-Hall of Southeast Asia Pte. Ltd., *Singapore*
Whitehall Books Limited, *Wellington, New Zealand*

CONTENTS

PREFACE — xi

CHAPTER ZERO Calculators — 1

    0.1. Accuracy and Estimation — 2
    0.2. Problem Solving with a Calculator — 4
    0.3. Calculators for This Text — 5
    0.4. Diagnostic Test — 6

CHAPTER ONE Fundamental Concepts of Algebra — 8

    1.1. Real Numbers, Rational Numbers, and Integers — 9
    1.2. Integral Exponents — 14
    1.3. Scientific Notation and Approximations — 20
    1.4. Algebraic Expressions — 25
    1.5. Factoring — 29
    1.6. Rational Expressions — 33
    1.7. Rational Exponents and Radicals — 38
    Review Exercises — 45

CHAPTER TWO Equations and Inequalities — 48

- 2.1. Basic Algebraic Equations — 49
- 2.2. Applications — 55
- 2.3. Quadratic Equations and Applications — 63
- 2.4. Equations with Radicals and Equations Quadratic in Form — 72
- 2.5. Inequalities and Sets — 79
- 2.6. Expressions, Equations, and Inequalities with Absolute Value — 88
- 2.7. Quadratic and Other Inequalities — 93
- Review Exercises — 98

CHAPTER THREE Relations, Functions, and Graphs — 100

- 3.1. The Coordinate System — 101
- 3.2. Relations, Graphs, and Symmetry — 109
- 3.3. Functions — 121
- 3.4. Graphing Techniques — 129
- 3.5. Linear Functions and Lines — 143
- 3.6. Variation — 152
- 3.7. Composite and Inverse Functions — 157
- Review Exercises — 170

CHAPTER FOUR Graphs of Polynomial and Rational Functions and Conic Sections — 172

- 4.1. Quadratic Functions — 173
- 4.2. Graphs of Polynomial Functions — 180
- 4.3. Rational Functions — 184
- 4.4. Conic Sections — 188
- Review Exercises — 198

CHAPTER FIVE Exponential and Logarithmic Functions — 199

- 5.1. Exponential Functions — 200
- 5.2. Logarithms — 207
- 5.3. Graphs and Properties of Logarithms — 213

5.4.	Applications	221
5.5.	Two Important Formulas and Their Applications	228
5.6.	Exponential and Logarithmic Equations	232
5.7.	Linear Interpolation	235
	Review Exercises	238

CHAPTER SIX Systems of Equations and Inequalities   240

6.1.	Systems of Two Equations in Two Variables	241
6.2.	Linear Equations in Two Variables	246
6.3.	Systems of Linear Equations in Three or More Variables	253
6.4.	Matrix Solutions of Linear Systems	260
6.5.	Determinants and Cramer's Rule	264
6.6.	Operations on Matrices	269
6.7.	Systems of Inequalities	276
6.8.	Linear Programming	283
	Review Exercises	290

CHAPTER SEVEN Right Triangle Trigonometry   293

7.1.	Angles	294
7.2.	The Trigonometric Functions	302
7.3.	Evaluation of Trigonometric Functions	309
7.4.	Inverse Trigonometric Functions	314
7.5.	Applications of Right Triangles	319
	Review Exercises	324

CHAPTER EIGHT General Trigonometry   327

8.1.	The Trigonometric Functions	328
8.2.	Evaluating Trigonometric Functions	333
8.3.	Properties of the Trigonometric Functions	337
8.4.	Reference Angles	341
8.5.	Graphs of the Trigonometric Functions	346
8.6.	The Graphs of $y = a \sin(bx - c)$ and $y = a \cos(bx - c)$	354
8.7.	Inverse Trigonometric Functions	359
8.8.	Applications	366
	Review Exercises	371

CHAPTER NINE Analytic and Geometric Trigonometry — 373

- 9.1. Trigonometric Identities — 374
- 9.2. Conditional Trigonometric Equations — 378
- 9.3. The Addition Formulas — 381
- 9.4. The Multiple-Angle Formulas — 388
- 9.5. Oblique Triangles; The Law of Sines — 391
- 9.6. The Law of Cosines — 396
- 9.7. Vectors — 401
- 9.8. Polar Coordinates — 408
- Review Exercises — 414

CHAPTER TEN Complex Numbers — 416

- 10.1. The Complex Numbers — 417
- 10.2. Complex Roots of Equations — 422
- 10.3. Trigonometric Form of Complex Numbers — 427
- 10.4. Powers and Roots of Complex Numbers — 433
- Review Exercises — 439

CHAPTER ELEVEN Polynomials — 440

- 11.1. Polynomials; Remainder and Factor Theorems — 441
- 11.2. Synthetic Division — 446
- 11.3. Zeros of Polynomials — 452
- 11.4. The Rational Root Theorem — 456
- Review Exercises — 460

CHAPTER TWELVE Sequences and Enumeration — 461

- 12.1. Sequences and Arithmetic Progressions — 462
- 12.2. Geometric Progressions and Geometric Series — 467
- 12.3. Enumeration; Permutations — 474
- 12.4. Distinguishable Permutations; Combinations — 481
- 12.5. Probability — 487
- 12.6. The Binomial Theorem — 490
- 12.7. Mathematical Induction — 494
- Review Exercises — 500

APPENDIX
A Reference Material — 503

APPENDIX
B Using a Calculator — 506

  B.1 Basic Information — 507
  B.2 Memory and Parentheses — 509
  B.3 The Square, Square Root, and Reciprocal Buttons — 511

ANSWERS TO ODD-NUMBERED EXERCISES — 514

INDEX — 547

PREFACE

The advent of inexpensive scientific calculators brought the promise that the traditional subject matter of college algebra and trigonometry could benefit from this technical innovation. This textbook fulfills that promise. Unlike early attempts in this direction (which consisted mostly of tacking on a few exercises with messy numbers), it fully integrates the calculator into the subject matter and uses it as a pedagogical device as well as a computational aid. The traditional logarithm and trigonometric tables are discarded (though an optional section on using tables and interpolation is included); the liberated time is used for a deeper investigation into topics such as the number e, and the use of exponentials and logarithms to solve mathematical, scientific, and business applications.

There are two very important consequences of the calculator-oriented approach used in this text. The first is that students are delighted at being able to use calculators, and that in itself helps to improve their attitude. The second is that calculators eliminate tedious computations, giving students more time to think about the mathematical concepts and processes. As a result of both factors, the whole learning process is usually enhanced.

In developing the materials involving calculators for this book, we had two goals in mind: (a) *The book should be easily used with any scientific calculator on the market.* (b) Calculators should be used whenever possible to help simplify, illustrate, and explain the material, but calculator usage should not encumber the mathematics. Towards these goals, a Chapter 0 was written to set the stage for using calculators in the text. It includes a discussion of round-off error and estimation together with a diagnostic test on calculator usage. This test has references to Appendix B, which provides a brief but thorough explanation of the use of calculators, sufficient to prepare anyone to begin the text. However, we must emphasize that this is a mathematics textbook and not a manual which employs calculators for all topics at any cost. There are several topics, such as systems of linear equations, which require a facility with manipulating buttons that is greater than that developed by the average student at this level. Hence calculator usage for these topics is minimal and our approach to them is quite traditional.

This text was written to provide a highly readable, student-oriented presentation of algebra and trigonometry. The level assumes the student has had the equivalent of one and a half years of high school algebra or geometry.

The following are some of the features of our approach:

Word problems and applications. Wherever possible, we have included word problems and/or applications to illustrate the usefulness of the mathematics presented. The application sections in trigonometry (Sec. 8.8) and exponentials (Secs. 5.4 and 5.5) are particularly unique examples of this. This is also exemplified by the fact that there are 562 word problems among the 3526 exercises in the text. In fact, 40 out of 76 sections have word problems in their exercise sets. Since calculators are assumed, the problems have more realistic numbers and are more interesting and varied.

Graphing. There is a heavy emphasis on graphing throughout the text.

Estimation. Approximations are used throughout the text, both to counteract mindless button-pushing and to emphasize mathematical relationships.

Functional approach. A functional approach is emphasized, both because it facilitates explaining many relationships and because it is necessary for those students continuing into calculus. This is nicely illustrated by calculators playing the role of the traditional "black box."

Informal approach. Our approach is highly readable and informal, containing 634 worked-out examples. (This averages better than 8 examples per section.)

Right triangle trigonometry. The right triangle approach to trigonometry is used, since we have found this to be the most understandable when using calculators.

Another feature of the text which enhances student interest is that each chapter begins with a *historical note*. These introductions are designed to increase interest by providing some historical perspective to the ensuing material. The main sources of information for these notes were Carl B. Boyer, *A History of Mathematics* (John Wiley and Sons, Inc., New York, 1968); W. W. Rouse Ball, *A Short Account of the History of Mathematics* (Dover Publications, Inc., New York, 1960); and *Historical Topics for the Mathematics Classroom* (NCTM, Washington, D.C., 1969).

Most of the material in this textbook has been developed over a period of three years and class-tested both at Michigan State University and at Lansing Community College. We would like to express appreciation to our colleagues, teaching assistants, and students for their many valuable suggestions. We are especially grateful to the following reviewers for their helpful criticism and the time they took in sharing their ideas and often their enthusiasm:

John Ewing, Indiana University
Paul R. Fallone, The University of Connecticut—Hartford
George F. Feissner, SUNY College at Cortland
Mark Hale, University of Florida
James E. Hall, University of Wisconsin—Madison
Janice B. Koop, San Diego State University
Norman Levine, Ohio State University
Bernard Madison, Louisiana State University
Richard Marshall, Eastern Michigan University
Helen Medley, Kent State University
Louis J. Nachman, Oakland University
Carol W. Penney, The University of Georgia
Jack R. Porter, The University of Kansas
M. James Stewart, Lansing Community College
Clifford Weil, Michigan State University
Ronald H. Wenger, University of Delaware
Bostwick F. Wyman, Ohio State University

In addition, Professors Hall, Nachman, Porter, and Stewart met with the authors in a conference discussion of the book.

Our heartfelt thanks go to Mary Reynolds for converting our handwritten notes with multiple changes into a beautifully typewritten manuscript. We also thank the editorial and production staff at Prentice-Hall for their contribution.

Most importantly, we express our extreme gratitude to our families, particularly our wives, for their support, encouragement, and tolerance during the preparation of this textbook.

Marshall D. Hestenes
Richard O. Hill, Jr.

Michigan State University
East Lansing, MI 48824

ALGEBRA AND TRIGONOMETRY WITH CALCULATORS

Calculators

CHAPTER ZERO

This book has been written with the assumption that you have a scientific calculator. This approach was used because it was felt that not only can the calculator help enormously with routine calculations, but it can also assist you in learning and applying the mathematics. Consequently, a calculator is required, and it is extremely important that you know how to use your calculator and understand how it works. Although this is not a text on how to use a calculator, there are instructions regarding calculators throughout the text. These instructions are guidelines and aids to help insure that you are using your machine correctly, particularly when new functions are introduced. Of necessity the instructions are general, because calculators vary considerably.

    The purpose of this preliminary chapter is to set the stage for using calculators with this text. We discuss accuracy and estimation in Sec. 0.1 and problem solving in Sec. 0.2. In Sec. 0.3, we discuss what properties your calculator must have to be used with this textbook, some things to be aware of if you are about to buy a calculator, and using a calculator on examinations. In Sec. 0.4, there is a diagnostic test to determine your proficiency at using the calculator for basic calculations. If you cannot do all the problems, there are references to Appendix B, which contains detailed instructions and exercises to help you learn to use your calculator. In addition, you may wish to refer to Appendix B later if you are having difficulty.

SECTION 0.1. ACCURACY AND ESTIMATION

While calculators usually do calculations with far greater accuracy than humans, the user should be aware that they have some inaccuracies. Most machine inaccuracies are due to either round off [e.g., $1 \div 3$ is rounded to .33333333, so $(1 \div 3) \times 3 = .99999999$*] or the algorithm used by the calculator to compute the special functions (e.g., on one calculator if 2^3 is computed using the y^x button, the answer is given as 7.999998). Moreover, different calculators may give slightly different answers to the same problem because they treat round off differently or use different algorithms to compute the functions.

While the preceding inaccuracies do not usually have a great effect on the answer, there are circumstances where they can, such as when the problem has numbers which differ greatly in magnitude. For example, clearly $1/(10^{15} + 10^{-2} - 10^{15}) = 1/10^{-2} = 100$. But if you used your calculator and first computed $10^{15} + 10^{-2}$, this rounds to 10^{15}, so $(10^{15} + 10^{-2}) - 10^{15}$ would be computed as 0. Then trying to compute its reciprocal would produce an error message.

There is another question which arises concerning accuracy. In a problem that states "the length is 5 meters," how accurate is the number 5? After all, if the length were measured by some means, there would be some error or *degree of accuracy*. The accuracy of the numbers given is important because the result of any calculation is generally no more accurate than the least accurate number

---

Note: Some calculators are tricky, here, in that they will round off this number to 1. However, internally the machine thinks the number is not 1. You can see this by now subtracting 1, for the answer will not be zero.

used in the calculation. We shall make the following assumption:

> **ASSUMPTION**
> Unless otherwise indicated in a particular situation, all numerical data will be assumed to be accurate to at least eight significant figures.

Note: Significant figures are discussed fully in Sec. 1.3.

When this assumption applies, you should not round off any data or intermediate results to fewer than eight places. Then usually the answers will be correct to at least six places (even with most machine inaccuracies), and **we shall give the answers to six places.** For problems in which the accuracy of the data is given, you should still *not round off any intermediate results to fewer than eight places*, but the answer should be given to only the number of places in the least accurate number used. The reason for the prohibition against unnecessary rounding off is that accuracy can be decreased even under the simple operations of addition and multiplication. For example, if we round 21.4 to 21 and 32.3 to 32, then 21 + 32 = 53, but 21.4 + 32.3 = 53.7, which rounds to 54. Multiplication is worse: 21 × 32 = 672, but 21.4 × 32.3 = 691.22. Thus it pays to keep as many significant digits as possible (preferably in your calculator) throughout a calculation.

Notation: When rounding off, we shall use $\approx$ for *approximately equals*. For example, $\frac{2}{3} \approx .666667$.

We shall be rounding off numbers for two different reasons. One is to give the answer to an appropriate number of places, as we have just discussed. The other is to give a very rough approximation to a computation, correct to only one or two significant figures. This is usually done in your head in order to provide a rough check to that computation (for example, to help check that you have not pushed a wrong button). We have found that it is easier for students if we distinguish between these two uses, so we shall use the following terminology:

> ***Estimate:*** a rough educated guess as to the size of a number or computation.
> ***Approximation:*** a finite decimal round off of a number or computation.

For example, an estimate of $\sqrt{5}$ done in your head, $\sqrt{5} \approx 2.2$, is a check to an approximation to $\sqrt{5}$ done on your calculator, $\sqrt{5} \approx 2.23607$. Of course, most answers in the back of the text which are given to six significant figures are approximations rather than exact answers.

SECTION 0.2. PROBLEM SOLVING WITH A CALCULATOR

There are two things to keep in mind when using a calculator to help you solve problems.

> 1. Whenever you are about to use your calculator, always estimate the answer if it is at all reasonable to do so.
> 2. Try to save the calculator for one computation rather than using it for several intermediate computations.

Before illustrating this, recall that when solving problems with "nice" numbers, you do all computations as you go along. For example,

$$2x + 5 = 7x + 8$$
$$2x - 7x = 8 - 5$$
$$-5x = 3$$
$$x = -\tfrac{3}{5} = -.6$$

Now suppose you have a similar problem in which numbers are "messy," so that you might want to use a calculator. While it is possible to do the computations as you go along, it is much easier in the long run to do all the algebra first, and *save the calculator for one computation at the end*. For example,

$$2.187x + 4.918 = 7.419x + 8.213$$
$$2.187x - 7.419x = 8.213 - 4.918$$
$$(2.187 - 7.419)x = 8.213 - 4.918$$

(1) $$x = \frac{8.213 - 4.918}{2.187 - 7.419}$$

Now *estimate first*:

$$x \approx \frac{8 - 5}{2 - 7} = \frac{3}{-5} = -.6$$

Using a calculator, $x \approx -.629778$.

Note that when the numbers are "messy" the algebra is very similar to the case where the constants are letters, and this is why it is important to understand how to do the problems in this form. For example, solve for x:

$$ax + b = cx + d, \quad a \neq c$$
$$ax - cx = d - b$$
$$(a - c)x = d - b$$
$$x = \frac{d - b}{a - c}$$

Compare this with equation (1).

SECTION 0.3. CALCULATORS FOR THIS TEXT

Virtually any calculator that is presently being sold as a *scientific calculator* is adequate for this text. However, some older models have deficiencies. Your calculator is satisfactory for this text if it has the following capabilities:

1. Add [+], subtract [−], multiply [×], divide [÷]
2. Natural logarithm [ln x]
3. Exponentiation [y^x] or [a^x] and [e^x] (e^x may be [inv] [ln x])
4. Scientific notation and a display of at least either eight significant digits or five significant digits plus the exponent
5. Trigonometric functions [sin], [cos], [tan]
6. Pi [π]
7. Inverse trigonometric functions [inv] or [$\sin^{-1}$], [$\cos^{-1}$], [$\tan^{-1}$]

The following are optional but useful features:

1. On algebraic logic machines, parentheses. These can be highly useful.
2. (Trigonometry) Calculators vary as to the values of x for which they will compute $\sin x$, $\cos x$, and $\tan x$. It is strongly recommended that your calculator *should* compute these functions *at least* for $0° \leq x \leq 360°$ (x in degrees) or $0 \leq x \leq 2\pi$ (x in radians).
3. A factorial button [$x!$].
4. A noticeable click when an entry is made.

If you are about to buy a calculator, here are some things to consider.

1. **Logic:** There are two basic kinds of machine logic, *algebraic* and *reverse Polish notation* (RPN). In addition, there are two kinds of algebraic calculators, standard and algebraic hierarchy. See Sec. B.1 in Appendix B for a discussion of these. Our experience seems to indicate that if you have trouble with algebra, you probably should get an algebraic calculator and that a calculator with algebraic hierarchy is more advantageous than a standard algebraic calculator. On the other hand, if you are technically oriented, you might also consider an RPN calculator. Using such a calculator requires keeping track of a *stack* but requires slightly fewer key strokes for many computations.
2. **Batteries:** Calculators have either rechargeable batteries, replaceable *watch* batteries which typically last 1000 or more hours of use, or replaceable AA, AAA, or the like batteries which typically last 10 to 20 hours. Although in general the last type is cheaper than the first two, in the long run the frequency of replacing batteries makes them more expensive.
3. The following is a list of some of the calculators which are adequate. By no means is this list all-inclusive, and new models are being

introduced all the time: TI30, SR40, TI25, TI50 (or any higher number); HP31E (or any higher number); Casio fx-21, fx-31; Sharp 5806; and many others. Any local dealer should be able to help you decide among the various models available.

Finally, if you are taking a course which requires calculators for examinations, you probably should follow these guidelines:

1. If your machine has replaceable batteries, you should probably carry an extra set to tests. If it has rechargeable batteries, make sure you fully charge it the night before a test.
2. If you have a new calculator, make sure it has had some use before walking into a test with it. (If something is wrong with a machine, it usually appears early in the calculator's life.)
3. Different models work differently. If you have to suddenly use a different machine (for example, if you have to borrow a friend's for a test), make sure you understand its idiosyncrasies.

SECTION 0.4. DIAGNOSTIC TEST

The following diagnostic test is designed to determine if you are sufficiently proficient at using your calculator to begin this text. As you do the problems, remember to *estimate first*. If you have no trouble, you are ready. Otherwise, refer to the appropriate parts of Appendix B:

Go to Sec. B.1 if you have difficulty with problems 1–6.
Go to Sec. B.2 if you have difficulty with problems 7–9.
Go to Sec. B.3 if you have difficulty with problems 10–13.

PROBLEM	ESTIMATE (DONE IN YOUR HEAD AS MUCH AS POSSIBLE)	ANSWER (TO SIX SIGNIFICANT FIGURES)
1. $672.81 + 919.07$	$\approx 700 + 900 = 1600$	1591.88
2. $.0013412 - .0079492$	$\approx .001 - .008 = -.007$	$-.0066080$
3. $5.932 \times (-411.8)$	$\approx 6 \times (-400) = -2400$	-2442.80
4. $.7735 \div (-.02134)$	$\approx .8 \div (-.02) = 80 \div (-2) = -40$	-36.2465
5. $5.813 + 2.114(7.812)$	$\approx 6 + 2(8) = 22$	22.3276
6. $4.819 - 20.81 \div (2.119)$	$\approx 5 - 20 \div (2) = -5$	-5.00167
7. $2.134 \times 4.691 + 3.814 \times 4.192 + 1.992 \times 6.874$	$\approx 2(5) + 4(4) + 2(7) = 40$	39.6919
8. $(2.134 + 4.691)(3.814 + 4.192)(1.992 + 6.874)$	$\approx 7(8)(9) \approx 500$	484.447

PROBLEM	ESTIMATE (DONE IN YOUR HEAD AS MUCH AS POSSIBLE)	ANSWER (TO SIX SIGNIFICANT FIGURES)
9. $\dfrac{2.819 - 1.414}{7.196 + 3.192}$	$\approx \dfrac{3-1}{7+3} = \dfrac{2}{10} = .2$	.135252
10. $5.213 + \sqrt{4.115}$	$\approx 5 + \sqrt{4} = 7$	7.24155
11. $\sqrt{5.213 + 4.115}$	$\approx \sqrt{5+4} = 3$	3.05418
12. $(1 + \dfrac{4.1}{8})^2$	$\approx (1 + .5)^2 \approx 2$	2.28766
13. $1/(2.19\sqrt{4.12} - 1/9)$	$\approx 1/[2(2) - .1] \approx 1/4 = .25$	.230728

Fundamental Concepts of Algebra

CHAPTER ONE

Algebra had its roots in Babylonian, Greek, and Hindu mathematics. Then, as followers of Muhammad conquered parts of three continents in religious zeal, Arab scholars went in and absorbed learning from the neighbors who were conquered. Mathematics flourished under Arab hands, and they introduced it into Western Europe, where it evolved into what we have today.

The name *algebra* comes from the title of an Arab textbook, *Al-jabr wa'l muqabalah* (rough translation, *Balancing Equations*), written in the early 800s by al-Khowarizmi (from whose name we derive the word *algorithm*). It was from this text that Europe, coming out of the dark ages, learned algebra.

As mathematics evolved, so did symbolism, but it developed at a painfully slow pace, and it often took the geniuses of the time to concoct it. The following illustrates some of the steps along the way as symbolism progressed from no symbols (in *Al-jabr*, equations were written out in words) to the extensive symbolism used today.

Oresme (1323–1382) wrote

p	1
1	2

for $x^{1\frac{1}{2}}$.

Chuquet (ca. 1500) wrote 72^1 (divided by) 8^3 (is) $9^{2.m}$ for $72x \div 8x^3 = 9x^{-2}$.

Bombellie (1526–1573) wrote $1^2 p.5^1 m.4$ for $x^2 + 5x - 4$.

Viete (1540–1603) first used letters to represent "the unknown" and wrote A cu 7 for $7A^3$.

Descartes (1596–1650) first used $+$, $-$, and integer exponents with letters representing "the unknown." (Descartes was the founder of analytic geometry.)

Newton (1642–1727) first used rational and negative exponents. (Newton was a co-inventor of calculus.)

As you read this text, occasionally think about how much harder it would be if you had to use some of the preceding notations.

SECTION 1.1. REAL NUMBERS, RATIONAL NUMBERS, AND INTEGERS

Numbers of one kind or another occur in countless situations in everyday life. They arise frequently in the natural sciences, social sciences, engineering, and business. Hence, you are undoubtedly well acquainted with some of the various properties that numbers have. But to ensure that everyone has a solid foundation, we shall review the systems of real numbers, rational numbers, and integers and their basic properties. Since this text is oriented toward the calculator, we shall discuss the strengths and limitations of calculators when dealing with these numbers.

We begin by reviewing the real numbers and their basic properties. For our purposes, a **real number** is a number which can be expressed as a (possibly infinite) decimal, for example,

$$3\tfrac{1}{2} = 3.5, \qquad -\tfrac{7}{8} = -.875, \qquad \tfrac{2}{3} = .6666\ldots, \qquad \sqrt{2} = 1.414\ldots$$

The collection of all real numbers is denoted by $\mathbb{R}$. There are two (binary) operations defined on real numbers, addition (denoted by $+$) and multiplication (denoted by $\cdot$). The real numbers are **closed** with respect to these operations, which means that to every pair of real numbers a, b there corresponds a unique real number $a + b$, called the **sum** of a and b, and another unique real number $a \cdot b$ (or simply ab), called the **product** of a and b. These operations satisfy the following properties:

Commutative properties:

$$a + b = b + a, \qquad ab = ba$$

Associative properties:

$$a+(b+c) = (a+b)+c, \qquad a(bc) = (ab)c$$

Identities: There are special real numbers 0, called the **additive identity**, and 1, called the **multiplicative identity**, such that

$$a+0 = a = 0+a, \qquad a\cdot 1 = a = 1\cdot a$$

for every real number a.

Inverses: For every real number a, there is a real number $-a$, called the **negative** or **additive inverse** of a, such that

$$a+(-a) = 0 = -a+a$$

For every real number $a \neq 0$, there is a real number a^{-1} or $1/a$, called the **(multiplicative) inverse** of a, such that

$$a\cdot a^{-1} = 1 = a^{-1}\cdot a$$

The first four properties hold for each operation individually. The last property connects the two operations.

Distributive properties:

$$a(b+c) = ab+ac, \qquad (a+b)c = ac+bc$$

From addition and multiplication, we easily obtain two other familiar operations on the real numbers: subtraction and division. The operation of **subtraction** is defined by

$$a-b = a+(-b)$$

while for $b \neq 0$ **division** is defined by

$$a \div b = ab^{-1} = a\left(\frac{1}{b}\right)$$

We usually write $\dfrac{a}{b}$ for $a \div b$.

A few of the more important equations relating the operations of addition, subtraction, multiplication, and division are the following:

$$\frac{ac}{bc} = \frac{a}{b}$$

$$\frac{a}{c} \pm \frac{b}{c} = \frac{a \pm b}{c}$$

$$\frac{a}{c} \pm \frac{b}{d} = \frac{ad \pm bc}{cd}$$

$$\frac{a}{c} \cdot \frac{b}{d} = \frac{ab}{cd}$$

$$\frac{a}{c} \div \frac{b}{d} = \frac{a}{c} \cdot \frac{d}{b} = \frac{ad}{cb}$$

The last equation says the way to divide $\frac{a}{c}$ by $\frac{b}{d}$ is to invert the $\frac{b}{d}$ and multiply.

EXAMPLE 1 Simplify

$$\frac{\frac{2}{3}}{\frac{8}{5}} \quad \text{and} \quad \frac{\frac{5}{9}}{20}.$$

SOLUTION

$$\frac{\frac{2}{3}}{\frac{8}{5}} = \frac{2}{3} \div \frac{8}{5} = \frac{\cancel{2}^{1}}{3} \cdot \frac{5}{\cancel{8}_{4}} = \frac{5}{12}$$

while

$$\frac{\frac{5}{9}}{20} = \frac{5}{9} \div 20 = \frac{5}{9} \div \frac{20}{1} = \frac{5}{9} \cdot \frac{1}{20} = \frac{\cancel{5}}{9} \cdot \frac{1}{4 \cdot \cancel{5}} = \frac{1}{36}.$$

Note: Always cancel common factors before multiplying out.

There are several kinds of numbers which are very important. The **positive integers** or **natural numbers** are the numbers $1, 2, 3, 4, \ldots$. They may be obtained by beginning at 1 and successively adding 1. Their negatives $-1, -2, -3, -4, \ldots$ are called the **negative integers**. The **integers** are the positive integers and the negative integers together with the number zero. Calculators can handle integers easily, within the limits of the machine. (That is, if the integers are not too large or take too many digits to write.)

A number is a **rational** number if it can be expressed as the ratio of two integers. For example,

$$3\frac{1}{2} = \frac{7}{2}, \quad -5 = \frac{-5}{1}, \quad 23.41 = \frac{2341}{100}, \quad .666\ldots = \frac{2}{3}$$

are all rational numbers. Any integer n is a rational number, since $n = \frac{n}{1}$.

Suppose that given a rational number $\frac{m}{n}$, we divide m by n to obtain the decimal expansion of the number. For example,

$$\frac{17}{4} = 4.2500\ldots, \quad -\frac{23}{11} = -2.0909\ldots, \quad \text{and} \quad \frac{9}{7} = 1.285714285714\ldots$$

We call these decimal numbers **infinite repeating decimals**, since in each case eventually there is a block of numbers which is repeated over and over (in the above examples, 0, 09, and 285714, respectively). Often a bar is placed over the repeating blocks instead of writing the dots, so that $\frac{17}{4} = 4.25\overline{0}$, $-\frac{23}{11} = -2.\overline{09}$ and $\frac{9}{7} = 1.\overline{285714}$. (When the repeating block is all zeros, it is usually omitted; thus in the first case we simply write $\frac{17}{4} = 4.25$.)

The property of having repeating blocks completely characterizes the rational numbers.

> **(1) THEOREM**
>
> A real number is a rational number if and only if it has a repeating decimal expansion.

A real number which is not a rational number is an **irrational number**. For example, the number .1010010001... (where after each 1 there is one more zero than before) is an irrational number, since it does not repeat. Other more familiar examples of irrational numbers are $\sqrt{2}$, $\sqrt{3}$, and π.

If we restrict our attention to the rational numbers by themselves, then all the properties for addition and multiplication mentioned earlier are still satisfied. If we restrict still further to the integers, then all the properties remain satisfied except for multiplicative inverses. However, if we restrict to the irrationals, then addition and multiplication are not even closed. For example, if x is an irrational number, then $-x$ is also irrational, but $x + (-x) = 0$, which is rational. Of course, the sum or product of two irrational numbers is always a real number.

Now let us consider real numbers in relation to calculators. A calculator usually handles a number in decimal form. (There are a few calculators which can manipulate small fractions, but we will ignore this feature.) A calculator can handle only somewhere between the first 8 and 15 digits of a number, depending on the calculator. Consequently, if the decimal form of the number does not start repeating zero by the eighth (or so) place, then that number does not fit into the calculator exactly. Thus:

> All irrational numbers and most rational numbers (e.g., $1/3, -9/7, 289/113$) cannot be contained exactly in a calculator.

Consequently, if you press the $[\pi]$ button or take the square root of 2 or divide 1 by 3, etc., the calculator gives you only an *approximation* to the actual number which is correct to the number of digits carried by the calculator (8 to 15). The remaining digits are discarded.

This property of calculators, storing only the first 8 to 15 digits of a number, will occasionally cause significant inaccuracies. For example, by the associative law

$$(1 + 10^{18}) - 10^{18} = 1 + (10^{18} - 10^{18}) = 1.$$

On a calculator, $1 + (10^{18} - 10^{18}) = 1$. However, notice that $1 + 10^{18} = 1{,}000{,}000{,}000{,}000{,}000{,}001$ which is too many digits for a calculator. So, computing $1 + 10^{18}$ on a calculator would yield just 10^{18}. Thus, *on a calculator*,

$$1 + (10^{18} - 10^{18}) = 1 \quad \text{but} \quad (1 + 10^{18}) - 10^{18} = 10^{18} - 10^{18} = 0.$$

This would cause extreme difficulty if we were careless when computing $1 \div (1 + 10^{18} - 10^{18})$. You should be aware that such difficulties may arise, but usually they are only important when the numbers differ greatly in magnitude.

EXERCISES

In Exercises 1–12, state which property discussed in this section is being used.

1. $2 + 5 = 5 + 2$.
2. $(4 \cdot 7) \cdot 9 = 4 \cdot (7 \cdot 9)$.
3. $-3 + 0 = -3$.
4. $(-4) \cdot 6 = 6 \cdot (-4)$.
5. $(-2) \cdot (3 + 4) = -2 \cdot 3 + (-2) \cdot 4$.
6. $5 \cdot 5^{-1} = 1$.
7. $3 \cdot 5$ is a real number.
8. $4 \cdot 1 = 4$.
9. $7 + (-7) = 0$.
10. $5 + (-2)$ is a real number.
11. $(8 + 2) + 3 = 8 + (2 + 3)$.
12. $(3 + 2) \cdot 4 = 3 \cdot 4 + 2 \cdot 4$.

In Exercises 13–24, state whether the number is rational or irrational. If the number is rational, express it as a ratio of two integers.

13. 3
14. -4
15. 18.3
16. 9.17
17. $\sqrt{2}$
18. $1 + \sqrt{2}$
19. $\frac{1}{2}\sqrt{3}$
20. $\sqrt{6}$
21. $\frac{1}{3}\sqrt{4}$
22. $\frac{1}{2}\sqrt{9}$
23. $\frac{1}{2} + \frac{1}{3}$
24. $\frac{1}{2} + \frac{2}{3} + \frac{3}{4}$

25. Which of the following numbers can be entered exactly on a calculator?
    a. $\frac{3}{2}$
    b. $\frac{2}{3}$
    c. $\sqrt{2}$
    d. $3 + \sqrt{2}$
    e. π
    f. $.33\bar{3}^{-1}$
    g. $.625^{-1}$

26. Which of the following numbers can be entered exactly on a calculator?
    a. $\frac{3}{4}$
    b. $\frac{4}{3}$
    c. $\sqrt{3}$
    d. $\sqrt{4}$
    e. 2π
    f. $.16^{-1}$
    g. $.15^{-1}$

In Exercises 27–34, express the given rational number as an infinite repeating decimal.

27. $\frac{1}{2}$
28. $\frac{3}{4}$
29. $\frac{5}{6}$
30. $\frac{5}{9}$
31. $\frac{1}{11}$
32. $\frac{3}{22}$
33. $\frac{6}{7}$
34. $\frac{1}{13}$

In Exercises 35–40, simplify.

35. $\dfrac{\frac{2}{5}}{\frac{6}{25}}$
36. $\dfrac{14}{\frac{21}{4}}$
37. $\dfrac{\frac{14}{21}}{4}$
38. $\dfrac{\frac{1}{2} + \frac{1}{3}}{10}$
39. $\dfrac{\frac{9}{2}}{\frac{3}{5} + \frac{3}{4}}$
40. $\dfrac{\frac{168}{25}}{\frac{27}{10}}$

SECTION 1.2. INTEGRAL EXPONENTS

Positive integers are used as exponents to simplify repeated products. For example,
$$x^2 = x \cdot x, \qquad x^3 = x \cdot x \cdot x, \qquad x^4 = x \cdot x \cdot x \cdot x$$
and, in general, if n is a positive integer,

---

(2) **DEFINITION**

$$x^n = \underbrace{x \cdot x \cdot \cdots \cdot x}_{n \text{ factors}}$$

---

where n factors, all equal to x, appear on the right-hand side of the equals sign. The number x is called the **base**, and the positive integer n is called the **exponent**. The expression x^n is read "x to the nth power" or simply "x to the nth." Note that $x^1 = x$. Some examples are

$$\left(\frac{1}{3}\right)^5 = \frac{1}{3} \cdot \frac{1}{3} \cdot \frac{1}{3} \cdot \frac{1}{3} \cdot \frac{1}{3} = \frac{1}{243}$$
$$(-4)^3 = (-4)(-4)(-4) = -64$$
$$(\sqrt{3})^4 = \sqrt{3}\,\sqrt{3}\,\sqrt{3}\,\sqrt{3} = 3 \cdot 3 = 9$$

It is important to remember that ax^n means $(a)(x^n)$, not $(ax)^n$. For example, $2x^3$ means $(2)(x^3)$, not $(2x)(2x)(2x) = 8x^3$; $-3x^4$ means $(-3)(x^4)$, not $(-3x)^4 = 81x^4$; and, especially, $-x^6$ means $(-1)(x^6)$, not $(-x)^6 = x^6$. Thus if $x = 2$, $2x^3 = 16$, $-3x^4 = -48$, and $-x^6 = -64$.

We now state the laws of exponents.

---

(3) **LAWS OF EXPONENTS**

If x and y are real numbers, then

i. $x^m x^n = x^{m+n}$.
ii. $(x^m)^n = x^{mn}$.
iii. $(xy)^m = x^m y^m$.
iv. $\left(\dfrac{x}{y}\right)^m = \dfrac{x^m}{y^m}$ if $y \neq 0$.
v. If $x \neq 0$, then $\dfrac{x^m}{x^n} = \begin{cases} x^{m-n} & \text{if } m > n, \\ 1 & \text{if } m = n, \\ \dfrac{1}{x^{n-m}} & \text{if } n > m. \end{cases}$

---

(4) **THEOREM**

The laws of exponents (3) hold if m and n are positive integers.

We shall see that the laws of exponents hold for any integers m and n later in this section, that they hold for m and n rational numbers (if x and y are positive) in Sec. 1.7, and finally that they hold for any real numbers m and n (if x and y are positive) in Chapter 5. One of the first problems is to define what x^n means in these various situations.

A formal proof of (4) requires induction, but it can easily be seen intuitively. For example, to see that law (3i) holds,

$$x^m \cdot x^n = \underbrace{(x \cdot x \cdots x)}_{m \text{ factors}} \cdot \underbrace{(x \cdot x \cdots x)}_{n \text{ factors}}$$

$$= \underbrace{x \cdot x \cdots x}_{m + n \text{ factors}}$$

$$= x^{m+n}$$

The other laws can be seen similarly. These laws can be extended to such laws as $x^m \cdot x^n \cdot x^p = x^{m+n+p}$. The laws of exponents can be used to simplify many expressions.

EXAMPLE 1 Simplify

$$\frac{2^5 \cdot 3^2}{2^3 \cdot 3^7}$$

SOLUTION

$$\frac{2^5 \cdot 3^2}{2^3 \cdot 3^7} = \frac{2^5}{2^3} \cdot \frac{3^2}{3^7} = 2^{5-3} \cdot \frac{1}{3^{7-2}} = \frac{2^2}{3^5} = \frac{4}{243}$$

EXAMPLE 2 Simplify

$$\frac{4^3 \cdot 16^4 \cdot 2^5}{8^6 \cdot 32^2}$$

SOLUTION

$$\frac{4^3 \cdot 16^4 \cdot 2^5}{8^6 \cdot 32^2} = \frac{(2^2)^3 (2^4)^4 2^5}{(2^3)^6 (2^5)^2} = \frac{2^6 2^{16} 2^5}{2^{18} 2^{10}} = \frac{2^{27}}{2^{28}} = \frac{1}{2}$$

EXAMPLE 3 Simplify $(2a^2bc^3)^4$.

SOLUTION

$$(2a^2bc^3)^4 = 2^4(a^2)^4 b^4 (c^3)^4 = 16a^8 b^4 c^{12}$$

EXAMPLE 4 Simplify

$$\left(\frac{2r^3}{3s^2}\right)^3 \left(\frac{9s}{4r^2}\right)^2$$

SOLUTION

$$\left(\frac{2r^3}{3s^2}\right)^3\left(\frac{9s}{4r^2}\right)^2 = \frac{2^3(r^3)^3}{3^3(s^2)^3} \cdot \frac{(3^2)^2 s^2}{(2^2)^2(r^2)^2} = \frac{2^3 r^9 3^4 s^2}{3^3 s^6 2^4 r^4} = \frac{3r^5}{2s^4}$$

It is possible to extend the use of exponents to include negative integers and zero in such a way that the laws of exponents (3) apply. For example, if we want law (3i) to be true for $n = 0$, then $x^m \cdot x^0 = x^{m+0} = x^m$. If $x \neq 0$, then dividing by x^m leads to $x^0 = 1$. Thus we define x^0 as follows:

DEFINITION

If x is any nonzero real number, then $x^0 = 1$.

For example, $2^0 = 1$, $(-14.2)^0 = 1$, $\pi^0 = 1$, $(4 - \sqrt{5})^0 = 1$, etc. It is easy to check that all the laws of exponents (3) hold if we use this definition for x^0. The symbol 0^0 is left *undefined*. Whenever x^0 is written, it is understood that $x \neq 0$.

We next consider negative exponents. Suppose $m > 0$ so that $-m < 0$. For law (3i) to hold for $n = -m$, then $x^m \cdot x^{-m} = x^{m+(-m)} = x^0 = 1$. Dividing by x^m gives $x^{-m} = 1/x^m$. Thus we define x^{-m} as follows:

DEFINITION

If m is a positive integer and $x \neq 0$, then $x^{-m} = 1/x^m$.

For example,

$$3^{-2} = \frac{1}{3^2} = \frac{1}{9}, \qquad (ab)^{-4} = \frac{1}{(ab)^4} = \frac{1}{a^4 b^4}$$

$$\left(\frac{3}{4}\right)^{-2} = \frac{1}{\left(\frac{3}{4}\right)^2} = \frac{1}{\frac{9}{16}} = \frac{16}{9}$$

$$(-4)^{-1} = \frac{1}{-4} = -\frac{1}{4}$$

A special case is the following:

$x^{-1} = 1/x$, the reciprocal of x.

It is straightforward to check that all the laws of exponents (3) remain true with this definition of negative exponents. Hence we conclude the following:

---
(5)   **THEOREM**

The laws of exponents (3) hold if m and n are any integers (provided $x \neq 0$ if $m \leq 0$ or $n \leq 0$ and $y \neq 0$ if $m \leq 0$).

FUNDAMENTAL CONCEPTS OF ALGEBRA

Note: Negative exponents allow us to simplify law (3v) to read

$$\frac{x^m}{x^n} = x^{m-n}, \quad x \neq 0$$

where m and n are any integers.

EXAMPLE 5 Eliminate negative exponents and simplify (m and n are positive integers).

a. $\dfrac{1}{x^{-n}}$ b. $(x^{-2}y^3z)^{-3}$ c. $\dfrac{4a^2b^{-3}}{8a^{-5}b^2}$

d. $\dfrac{(9 \times 10^2)^3(2 \times 10^3)^{-5}}{(3 \times 10^{-7})^2(4 \times 10^{-2})^{-3}}$

SOLUTION

a. $\dfrac{1}{x^{-n}} = \dfrac{1}{1/x^n} = 1 \cdot \dfrac{x^n}{1} = x^n$

b. $(x^{-2}y^3z)^{-3} = (x^{-2})^{-3}(y^3)^{-3}z^{-3} = x^6 y^{-9} z^{-3} = x^6 \dfrac{1}{y^9}\dfrac{1}{z^3} = \dfrac{x^6}{y^9 z^3}$

c. $\dfrac{4a^2b^{-3}}{8a^{-5}b^2} = \dfrac{4}{8} \cdot a^2 \cdot \dfrac{1}{a^{-5}} \cdot b^{-3} \cdot \dfrac{1}{b^2} = \dfrac{1}{2} a^2 \cdot a^5 \cdot \dfrac{1}{b^3} \cdot \dfrac{1}{b^2} = \dfrac{1}{2}\dfrac{a^7}{b^5}$

d. $\dfrac{(9 \times 10^2)^3(2 \times 10^3)^{-5}}{(3 \times 10^{-7})^2(4 \times 10^{-2})^{-3}} = \dfrac{9^3 \times (10^2)^3 \times 2^{-5} \times (10^3)^{-5}}{3^2 \times (10^{-7})^2 \times 4^{-3} \times (10^{-2})^{-3}}$

$= \dfrac{9^3 \times 10^6 \times 2^{-5} \times 10^{-15}}{9 \times 10^{-14} \times (2^2)^{-3} \times 10^6}$

$= 9^2 \times \dfrac{2^6}{2^5} \times \dfrac{10^6}{10^6} \times \dfrac{10^{14}}{10^{15}}$

$= 9^2 \times 2 \times 10^{-1} = 162 \times 10^{-1} = 16.2$

Exponents on a Calculator

The computations x^2 and x^{-1} arise so often that most calculators have separate buttons for them. See Appendix B. Your calculator has a button for calculating general exponents that is commonly labeled y^x (but other variations are a^x and x^y). This button computes any *positive* number y raised to the power x within the limits of the calculator. (The reason for this will be explained in Chapter 5.) For example, to compute 1.51^7, press

Alg: [1.51][y^x][7][=];   RPN: [1.51][ENT][7][y^x]

You should obtain 17.899406.

It is important to remember that your calculator *cannot* compute *negative* numbers raised to a power x. (It may or may not be able to compute 0^x, but of course this should cause no trouble either way.) When you want to compute negative numbers raised to an integer power, you must compute the signs yourself and use your calculator for the positive part.

EXAMPLE 6 Compute $(-2.1)^{-4}$.

SOLUTION

$$(-2.1)^{-4} = [(-1)(2.1)]^{-4} = (-1)^{-4}(2.1)^{-4} = (+1)(2.1)^{-4}$$

Using a calculator, $2.1^{-4} \approx .0514189$, so $(-2.1)^{-4} \approx .0514189$.

If you are in the middle of a computation using the $[y^x]$ button and get an error message from your calculator, check to see if you have tried to raise a negative number to a power.

Warning: There are a few models of scientific calculators which do not have a $[y^x]$ button. These are older models and are no longer produced. Your instructions will explain how to compute y^x using the $[e^x]$ and $[\ln x]$ buttons.

EXAMPLE 7 Use your calculator to evaluate $19.2^3(19.2)^{-4}/19.2^5$. Then simplify the expression, using the laws of exponents, and finally evaluate the simplified expression.

SOLUTION *RPN:* [19.2] [ENT] [3] [y^x] [ENT] [19.2] [ENT] [−4] [y^x] [×] [ENT] [19.2][ENT][5][y^x][÷]. *Note:* The second and fourth [ENT]s were not necessary.
Alg. with parentheses: [19.2][y^x][3][=][×][(][19.2][y^x][−4][=][ )][÷][(][19.2][y^x][5][)][=]. Simplified, this is $19.2^{-6} \approx 1.99615 \times 10^{-8}$.

EXERCISES

In Exercises 1–4, use your calculator to evaluate the expression as it stands. Then simplify the expression, using the laws of exponents, and finally evaluate the simplified expression.

1. $91.4^4 \; 91.4^{-3} \; 91.4^2$
2. $\left(\dfrac{4.17^3}{4.17^2}\right)^{-5}$
3. $[(.515^2)^{-3}]^{-4}$
4. $(24.1^3 \; 24.1^{-4})^{-3}$

In Exercises 5–54, simplify the expression; m, n, and p are positive integers. Here **simplify** means replace the given expression by one in which letters representing real numbers appear at most once in each term and no negative exponents occur for letters.

5. $\left(-\dfrac{2}{5}\right)^3$
6. $(-2)^4$
7. $2^3 + (2 \cdot 3)$
8. $\left(\dfrac{2}{5}\right)^{-3}\left(\dfrac{4}{5}\right)^2$
9. $\dfrac{2^{-3}}{3^{-2}}$
10. $4^0 + 0^4$

18 FUNDAMENTAL CONCEPTS OF ALGEBRA

11. $(4 \times 10^5)^2$

12. $\dfrac{(2 \times 10^{-3})^4}{(4 \times 10^{-4})^{-2}}$

13. $\dfrac{(5 \times 10^{-4})^{-3}(3 \times 10^{-4})^3}{(25 \times 10^4)^{-1}(9 \times 10^6)^2}$

14. $(2x^2)(3x^3)(4x^4)$

15. $(x^2 x^3 x^4)^2$

16. $aa^5(-a)^3$

17. $(2s^3 t^4)^5$

18. $(3st^{-2}a^3)^{-4}$

19. $\dfrac{(3^0 k^2 l^3)(-3kl^{-2})^{-3}}{(3^{-1} k^{-4} l)^4}$

20. $(c^{-2} d^{-3})^{-1}$

21. $(x+y)^{-1}(x+y)$

22. $8c^2 d^3 (2c^4 d^{-5})\left(\dfrac{1}{4} c^{-3} d\right)$

23. $xy(x^{-1} + y^{-2})$

24. $\dfrac{9u^3 v^{-1}}{3uv^{-2}} \div \dfrac{u^2}{v^3}$

25. $(2x - 3y)^2 (2x - 3y)^{-3}$

26. $(18a^2 b^{-2} - 7a^3 b^{-5})^0$

27. $[(c^2 d^{-3})^2]^{-3}$

28. $(2^{-1} + 3^{-1})^{-1}$

29. $(3^{-2} + 2^{-3})^{-1}$

30. $[(x^{-1})^{-1}]^{-1}$

31. $\dfrac{(ab)^{-1}}{a^{-1} b^{-1}}$

32. $\dfrac{2^{-1} + 3^{-1}}{2^{-1} 3^{-1}}$

33. $\dfrac{x^{-1} + y^{-1}}{x^{-1} y^{-1}}$

34. $(8s^2 t^4)(2^{-1} st^{-2})$

35. $(-6a^3)^2 (12a^4)^{-1}$

36. $x^4 (-9x^3)^2 (6x^{-2})^{-2}$

37. $9a^2 b^3 (3a^5 b^{-3})^{-2} \left(\dfrac{1}{6} a^{-4} b\right)^{-1}$

38. $\left(\dfrac{4m^2 p^3}{3m^3 p^2}\right)^{-1} \left(\dfrac{6mp^4}{2m^5}\right)^{-2}$

39. $(-3xy^2)^3 \left(\dfrac{x^4}{6y}\right)^2$

40. $\dfrac{(10c^{-2} d)^{-3}}{(50c^3 d^{-1})^{-2}}$

41. $\dfrac{(3a^2 b)^{-1}}{(2ab^{-1})^2} \div \dfrac{(3a^{-1} b^3)^{-2}}{(4a^{-2} b)^{-3}}$

42. $\left(\dfrac{16a^2 b^{-2} c}{7a^3 b^{-2} c^3}\right)^0$

43. $4x^2 y^{-1}(2x^{-1} y^3 - 3x^{-2} y^{-1})$

44. $3c^{-1} d^{-1}(2cd - 5c^2 d^2)$

45. $[(a^2 y^{-1})^{-2}]^3$

46. $c[(c^{-1} d^3)^{-1}]^{-5}$

47. $(a^2 b)^{-1}(a^{-2} + b^{-2})^{-1}$

48. $(xy)^{-1}(x^{-1} + y^{-1})^{-1}$

49. $\dfrac{(-3ux^2 y^{-3} z^{-1})^3}{(6u^{-4} x^4 y^{-2} z^{-1})^{-2}}$

50. $\dfrac{(-2a^4 bc^{-6} d^{-1})^{-3}}{(4a^3 b^2 c^{-1} d^{-3})^{-2}}$

51. $\left[\dfrac{(3a^5 bc^{-2} d)^{-2}}{(5ab^2 c^{-2} d)^{-3}}\right]^0$

52. $\dfrac{3(3^n)}{2^m 2^p}$

53. $\dfrac{(x^m)^{p+m}}{(x^p)^m}$

54. $\left(\dfrac{x^m}{x^n}\right)^{m+n} \left(\dfrac{x^n}{x^p}\right)^{n+p} \left(\dfrac{x^p}{x^m}\right)^{p+m}$

Section 1.2 Integral Exponents

SECTION 1.3.   SCIENTIFIC NOTATION AND APPROXIMATIONS

Scientific Notation

Most of the numbers we deal with are of a reasonable size. However, when dealing with figures like the size of the U.S. budget or the diameter of an atom, numbers can get very large or very small. They require a lot of zeros to write, and this can be cumbersome to handle. Scientific notation was developed to deal with this difficulty.

DEFINITION

A number is written in **scientific notation** if it is expressed in the form

$$\pm c \times 10^n$$

where n is an integer, c is in decimal form, and $1 \leq c < 10$.

Suppose $\pm c \times 10^n$ is in scientific notation and we wish to write it in decimal form. Since $1 \leq c < 10$, the decimal point is to the right of the first non-zero digit of c (e.g., $c = 3.6418$). When we multiply a decimal number by 10, we just move the decimal point one place to the right, while when we divide a decimal by 10, we move the decimal point one place to the left. Thus to express a number $c \times 10^n$ in decimal notation, just move the decimal point n places, to the right if n is positive, to the left if n is negative.

EXAMPLE 1   Express the following numbers in decimal notation:

(a) 3.876×10^4 (b) -9.919×10^{-9}

SOLUTION

(a) $3.876 \times 10^4 = 38{,}760$ (b) $-9.919 \times 10^{-9} = -.000000009919$

In order to write a decimal number in scientific notation, we just reverse the above. First write c by moving the decimal point to the right of the first non-zero digit. The number of places the decimal point is moved is n. The sign of n is positive or negative according to which direction the point would be moved if the number were being converted back to decimal form, positive for right, negative for left.

EXAMPLE 2   Express the following numbers in scientific notation:

(a) 234.16007   (b) .00000021874
(c) $500,000,000,000   (d) .000000000053 meters

SOLUTION

(a) $234.16007 = 2.3416007 \times 10^2$ (b) $.00000021874 = 2.1874 \times 10^{-7}$
(c) $\$500{,}000{,}000{,}000 = \$5 \times 10^{11}$ (d) $.000000000053\text{m} = 5.3 \times 10^{-11}\text{m}$

20   FUNDAMENTAL CONCEPTS OF ALGEBRA

Note: The 1979 U.S. budget was approximately (c), while the diameter of a hydrogen atom is roughly (d).

Scientific Notation on a Calculator

Most scientific calculators are equipped to handle scientific notation, within limits. They have a button which allows you to enter the exponent of a power of 10 up to two digits. (The two digits are plenty for virtually any practical application.) On most machines, the button looks like [EXP], [EEX], or [EE]. To enter 3.29×10^7, press [3][·][2][9][EXP][7]. The display will read [3.29 07]. Of course, you could have entered [3][2][9][0][0][0][0][0], but using scientific notation to enter this number is quicker and easier. To enter .0000000000009123, you must use scientific notation and enter 9.123×10^{-11}. Press [9][·][1][2][3][EXP][1][2][+/−]. The [+/−] may be pressed anytime after pressing [EXP]. The display will read [9.123 −12]. Alternatively, the same number could be entered as $.09123 \times 10^{-10}$, $.9123 \times 10^{-11}$, 91.23×10^{-13}, etc.

EXAMPLE 3   Enter the following numbers into your calculator:

(a) 7.654×10^{12};   (b) 7.654×10^{-12};   (c) -7.654×10^{12};   (d) -7.654×10^{-12}.

SOLUTION

(a) [7.654][EXP][12];   (b) [7.654][EXP][12][+/−];
(c) [7.654][+/−][EXP][12];   (d) [7.654][+/−][EXP][12][+/−].

Be careful with the minus signs; on many machines they will not register if you press [+/−] *before* you enter the number.

Scientific notation can help make approximating easier.

EXAMPLE 4   For each of the following approximate first and then use your calculator to compute.

(a) 4239(78,411)(.00002991)
(b) $1.892 \times 10^{13}(7.196 \times 10^{12})(8.719 \times 10^{-9})$
(c) $894,100 + 912,812 - 42,371$.

SOLUTION   Estimations:

(a) $4 \times 10^3 \times 8 \times 10^4 \times 3 \times 10^{-5} \approx 90 \times 10^2 = 9000$
(b) $2 \times 10^{13} \times 7 \times 10^{12} \times 9 \times 10^{-9} \approx 140 \times 10^{16} \approx 1 \times 10^{18}$
(c) $9 \times 10^5 + 9 \times 10^5 - 4 \times 10^4 \approx 18 \times 10^5$.

With the calculator: **(a)** 9941.61; **(b)** 1.18708×10^{18}; **(c)** 1,764,541.

Significant Figures

When dealing with actual measurements or when rounding off numbers, it can be very important to know how accurate your numbers are. For example, suppose you are trying to determine the area of a rectangle. You measured its length as 3.31 inches and its width as 2.71 inches. Multiplying this on your calculator gives 8.9701. However, the first two measurements were correct only to *three significant figures*, i.e., the number of digits used to describe the accuracy of the data. The actual length might be anywhere from 3.3050 to 3.3150 inches, and the actual width might be anywhere from 2.7050 to 2.7150 inches. Consequently the product also can be accurate to at most three significant figures, so the answer must be rounded off to 8.97. In general, a computation is no more accurate than the least accurate of the numbers involved.

Now suppose the length of that rectangle was 3.32 inches and its width 2.71 inches. Then the product is 8.9972. Rounding this off to three significant figures gives 9.00, not just 9. The last two zeros indicate the answer is accurate to three significant figures, not just one.

To give a rigorous definition, a digit of a number is **significant** unless it is used *only* to place the decimal point.

EXAMPLE 5 How many significant figures do the following numbers have?

(a) 15.89; (b) .001589; (c) 1.589×10^{-3}; (d) .003600; (e) 1.589×10^6; (f) 3.600×10^6; (g) 1,589,217; (h) 1.589000×10^6; (i) 1,589,000.

SOLUTION

(a)–(f) all have four significant figures. The first two zeros in (b) and (d) are used only to place the decimal point and hence are not significant. (g) and (h) have 7. (i) is ambiguous. It probably is a number like (g) rounded off, but it might be like (h). When there is a real question concerning significant figures, (i) would be written using scientific notation, like (e) or (h).

If you state that all numbers are accurate to three significant figures, then you may write 9 instead of 9.00, etc. We are using this convention in this text and are assuming all numbers to be accurate to eight significant figures (see Sec. 0.1).

One type of problem that calculators make easy is converting measurements from one unit to another (for example, converting distance in miles to kilometers). Scientific notation can simplify conversions where the numbers are large. We explain how to convert units in the next two examples. Tables of equivalences are given in Appendix A.

EXAMPLE 6

(a) An inchworm can travel about 18 inches in a minute. How many centimeters is this?
(b) A butterfly flies about 1500 centimeters in a minute. How many inches is this?

SOLUTION   By Appendix A, 1 inch = 2.54 cm. We write this in the form
$$\frac{1 \text{ in}}{2.54 \text{ cm}} = 1 \quad \text{and} \quad \frac{2.54 \text{ cm}}{1 \text{ in}} = 1.$$

To change inches to centimeters or centimeters to inches, multiply by the fraction which enables the appropriate units to cancel. In this example,

(a) $18 \text{ in} = 18 \cancel{\text{in}} \cdot \frac{2.54 \text{ cm}}{1 \cancel{\text{in}}}$

$\quad = 18(2.54) \text{ cm}$

$\quad = 45.72 \text{ cm}$

(b) $1500 \text{ cm} = 1500 \cancel{\text{cm}} \cdot \frac{1 \text{ in}}{2.54 \cancel{\text{cm}}}$

$\quad = \frac{1500}{2.54} \text{ in}$

$\quad \approx 590.551 \text{ in}$

This method is easily adapted to two or more units.

EXAMPLE 7   The speed of light is 186,000 miles per second to three significant figures. How many inches per minute is this?

SOLUTION   Using 1 mile = 5280 feet, we have

$$1.86 \times 10^5 \frac{\text{mi}}{\text{sec}} = 1.86 \times 10^5 \frac{\cancel{\text{mi}}}{\text{sec}} \cdot \frac{5.28 \times 10^3}{1} \frac{\text{ft}}{\cancel{\text{mi}}}$$

$$= 1.86 \times 5.28 \times 10^8 \frac{\cancel{\text{ft}}}{\text{sec}} \times \frac{12 \text{ in}}{1 \cancel{\text{ft}}}$$

$$= 1.86 \times 5.28 \times 12 \times 10^8 \frac{\text{in}}{\cancel{\text{sec}}} \times \frac{60 \cancel{\text{sec}}}{1 \text{ min}}$$

$$= 1.86 \times 5.28 \times 12 \times 6 \times 10^9 \frac{\text{in}}{\text{min}}$$

First estimate: $\approx 2 \times 5 \times 10 \times 6 \times 10^9 = 600 \times 10^9 = 6 \times 10^{11}$.

Using a calculator, the answer is 7.07×10^{11}. This answer is rounded off to three significant figures, since the given speed of light is accurate only to three significant figures.

EXERCISES

1. How many significant figures do the following numbers have?

   a. 23.1   **b.** .00231
   c. 4.8125×10^{-3} **d.** 1.810×10^{-13}
   e. 1.438×10^6 **f.** 1.438000×10^6
   g. 1,438,000

2. Enter the following numbers using scientific notation:

   a. 7.184×10^{18}
   b. -7.184×10^{18}
   c. 7.184×10^{-18}
   d. -7.184×10^{-18}

In Exercises 3–10, write the given number in scientific notation.

3. 4761   **4.** .00278
5. 271000   **6.** .0003819

7. 67,100,000,000,000
8. .000000000000421
9. 10
10. .01

In Exercises 11–16, write the given number in decimal notation.

11. 4.102×10^2
12. 5.119×10^{-3}
13. 4×10^{-1}
14. 2×10
15. 8.149×10^9
16. 8.149×10^{-9}

In Exercises 17–24, (1) approximate the given number first and then (2) use your calculator to compute the number.

17. $(.0000124)(1,892,000)(-491)$
18. $(87,124)(-.002916)(-412.7)$
19. $(43.01)^3(.0219)^{-2}(8691)^2$
20. $8.12 \times 10^{-4} - 18 \times 10^{-5} + .02 \times 10^{-3}$
21. $143.2 + 18.11 + 8.71 - 900.01$
22. $(1.213 \times 10^2)^3(4.912 \times 10^{-4})^2$
23. $\dfrac{(1.89 \times 10^2)^3(-2.11 \times 10^{-4})^2}{(-7.19 \times 10^{-2})^2(1.09 \times 10^5)^{-3}}$
24. $\dfrac{(9.12 \times 10^3)^{-2}(8.19 \times 10^5)^2}{(7.91 \times 10^{-2})^3(2.91 \times 10^5)^{-4}}$

In Exercises 25–28, solve the equation. Whenever it is necessary to use a calculator, save it for one big final computation, but first approximate the answer.

25. $(4.12 \times 10^{-11})x + 2.91 \times 10^{-7}$
$= (6.81 \times 10^{-11})x$
26. $ax + b = cx,\ a \neq c$
27. $(1.291 \times 10^8)x - 4.681 \times 10^{-3}$
$= (4.371 \times 10^7)x - 9.117 \times 10^{-2}$
28. $mx + p = nx + q,\ m \neq n$

In Exercises 29–32, evaluate the given expression for $x = 2.117,\ -9.812,\ 3.717 \times 10^{-8},\ -5.616 \times 10^8$.

29. $3x^5$
30. $8x^{-4}$
31. $-3x^4$
32. $-8x^{-4}$

In Exercises 33–40, solve the given problem by first changing all large or small numbers into scientific notation.

33. If the budget of the U.S. government is $500,000,000,000 and there are 210,000,000 citizens of the United States, how much money per citizen does the U.S. government spend?

34. If the speed of light is 186,000 miles per second and the distance from the sun to the earth is 93,000,000 miles, how many minutes does it take for light to travel from the sun to the earth?

For Exercises 35 and 36, assume that it is 25,000 miles around the equator of the earth.

35. How many inches is it around the equator of the earth?
36. How many meters is it around the equator of the earth?
37. If the speed of light is 186,000 miles per second, how many miles is 1 light year? (One light year is the distance light travels in 1 year.)
38. How many kilometers is 1 light year? (See Exercise 37.)
39. Assume the earth is a sphere with a radius of 4000 miles. If the surface area of a sphere is $S = 4\pi r^2$, how many square inches of surface area does the earth have?
40. How many square meters of surface area does the earth have?
(See Exercise 39.)

FUNDAMENTAL CONCEPTS OF ALGEBRA

SECTION 1.4. ALGEBRAIC EXPRESSIONS

A symbol is a **variable** if it represents a typical member of some collection under discussion. For our purposes, that collection will normally be a collection of real numbers. If we begin with a collection of variables and real numbers and apply operations of addition, subtraction, multiplication, division, and taking roots, then the result is an **algebraic expression**. Examples of algebraic expressions are

$$7, \quad 2x^{12} - 8 + \sqrt{x-1}, \quad \frac{x^5 - 4y^{-1}z}{\sqrt{y^2 - 8xz}}$$

If specific numbers are substituted for the variables in an algebraic expression, the resulting number is called the **value** of the expression for these numbers.

EXAMPLE 1 Find the value of the third expression above if

(a) $x = 2, y = 3, z = -1$; (b) $x = 2.191, y = 2.917, z = -.8791$.

SOLUTION

(a) $$\frac{x^5 - 4y^{-1}z}{\sqrt{y^2 - 8xz}} = \frac{2^5 - 4(3)^{-1}(-1)}{\sqrt{3^2 - 8(2)(-1)}} = \frac{32 + \frac{4}{3}}{\sqrt{9 + 16}} = \frac{[3(32) + 4]/3}{\sqrt{25}} = \frac{100}{3(5)}$$

$$= \frac{20}{3} = 6\frac{2}{3}$$

(b) $$\frac{x^5 - 4y^{-1}z}{\sqrt{y^2 - 8xz}} = \frac{2.191^5 - 4(2.917)^{-1}(-.8791)}{\sqrt{2.917^2 - 8(2.191)(-.8791)}}$$

These numbers will probably make you want to use your calculator, so you should *estimate first*. The estimates are $x \approx 2, y \approx 3$, and $z \approx -1$, so (b)

$$\approx \frac{2^5 - 4(3)^{-1}(-1)}{\sqrt{3^2 - 8(2)(-1)}} \approx \frac{32 + 1}{\sqrt{25}} \approx \frac{35}{5} = 7$$

Compare this with (a). There are several ways to do (b) using your calculator, but all of them require a *little thought* and *planning ahead*. To do it as written requires several levels of parentheses on your calculator. If you have parentheses, do it that way. The following is a method which requires a memory but no parentheses. You compute the numerator first using the memory during the computation. Then store the numerator in the memory. Next compute the denominator by computing the product of three numbers first. Then take the reciprocal of the denominator and multiply that by the numerator stored in the memory:

$[2.191][x^y][5][=][M+][2.917][1/x][\times][-.8791][\times][-4][=]$
$[M+][-8][\times][2.191][\times][-.8791][=][+][2.917][x^y][=][\sqrt{\ }][1/x]$
$[\times][MR][=]$

The answer to (b) is ≈ 10.5706.

Often we are presented with algebraic expressions in a complicated form, so to deal with them they need to be simplified or manipulated into a more manageable form.

EXAMPLE 2 Simplify $-(x-3)-[x-(3-2x)]$.

SOLUTION This should be thought of as $(-1)\cdot[x+(-3)]+(-1)\cdot\{x+(-1)\cdot[3+(-2x)]\}$. It is usually, though not always, easier to clear the parentheses and brackets from the outside in. Hence the steps *might* be written as

$$(-1)x+(-1)(-3)+(-1)x+(-1)(-1)[3+(-2x)]$$
$$=(-1)x+3+(-1)x+[3+(-2x)]$$
$$=(-1)x+3+(-1)x+3+(-2x)$$
$$=6+(-4x)=6-4x$$

In practice, however, all those parentheses and addition signs are *not* necessary. Although these details should be included the first time or two that problems like these are attempted, the student should work to get algebraic proficiency to the point where the following would be written:

$$-(x-3)-[x-(3-2x)]=-x+3-x+(3-2x)=6-4x$$

EXAMPLE 3 Multiply and simplify $(x-2y)(3x^2+2xy+y^2)$.

SOLUTION This can be *thought* of as follows:

$$(x-2y)(3x^2+2xy+y^2)=[x+(-2y)](3x^2+2xy+y^2)$$
$$=x(3x^2+2xy+y^2)+(-2y)(3x^2+2xy+y^2)$$
$$=x(3x^2)+x(2xy)+x(y^2)+(-2y)(3x^2)$$
$$\quad+(-2y)(2xy)+(-2y)(y^2)$$
$$=3x^3+2x^2y+xy^2-6x^2y-4xy^2-2y^3$$
$$=3x^3-4x^2y-3xy^2-2y^3$$

In practice, what is usually written is

$$(x-2y)(3x^2+2xy+y^2)=3x^2+2x^2y+xy^2-6x^2y-4xy^2-2y^3$$
$$=3x^2-4x^2y-3xy^2-2y^3$$

There are some types of products that occur so frequently that it is best to memorize them. If you forget, you can always multiply them out as above.

> **(6)**
> i. $a(u + v) = au + av$
> ii. $(u + v)(u - v) = u^2 - v^2$
> iii. $(u + v)^2 = u^2 + 2uv + v^2$
> iv. $(u - v)^2 = u^2 - 2uv + v^2$
> v. $(u + c)(u + d) = u^2 + (c + d)u + cd$
> vi. $(au + c)(bu + d) = abu^2 + (ad + bc)u + cd$
> vii. $(u + v)(u^2 - uv + v^2) = u^3 + v^3$
> viii. $(u - v)(u^2 + uv + v^2) = u^3 - v^3$

It is extremely important to realize that the letters in these equations may be any quantities whatsoever. It is the *form* that is important. This generality is what makes mathematics so useful. For instance, $(5x^3 + 9a^4)(5x^3 - 9a^4)$ is of the form (6ii) with $u = 5x^3$ and $v = 9a^4$.

In (6), formulas (6ii), (6iii), (6iv), and (6v) are all special cases of (6vi), so you would not have to memorize them separately. However, they turn up so often that it is extremely helpful to do so.

We now give examples using some of the formulas.

Formula (6ii): $\qquad (u + v)(u - v) = u^2 - v^2$

> The product of the sum and difference of two expressions is the square of the first expression minus the square of the second.

EXAMPLE 4 Find the product $(5s - 2t)(5s + 2t)$.

SOLUTION This is of the form (6ii) with $u = 5s$, $v = 2t$. Thus

$$(5s - 2t)(5s + 2t) = (5s)^2 - (2t)^2 = 25s^2 - 4t^2$$

The second step should be done in your head.

Formula (6iii): $\qquad (u + v)^2 = u^2 + 2uv + v^2$

> The square of the sum of two quantities is the square of the first plus two times the first times the second plus the square of the second.

EXAMPLE 5 Multiply $(6x^2 + 4y)^2$.

SOLUTION Here $u = 6x^2$ and $v = 4y$. Thus

$$(6x^2 + 4y)^2 = (6x^2)^2 + 2(6x^2)(4y) + (4y)^2 = 36x^4 + 48x^2y + 16y^2$$

Again, you should learn to do the second step in your head.

Formula (6v): $\qquad (u+c)(u+d) = u^2 + (c+d)u + cd$

Formula (6vi): $\qquad (au+c)(bu+d) = abu^2 + (ad+bc)u + cd$

Rather then memorize these formulas, it is easier to remember to first multiply the first terms, then multiply and collect the two middle or crossed terms, and finally multiply the last terms, as indicated in Example 6. As with the above examples, you should try to do the second step in your head.

EXAMPLE 6 Multiply and simplify $(2x+3y)(5x+4y)$.

SOLUTION

$$(2x+3y)(5x+4y) = 10x^2 + (8+15)xy + 12y^2$$
$$= 10x^2 + 23xy + 12y^2$$

EXAMPLE 7 Multiply and simplify $(x+7)(x-4)$.

SOLUTION

$$(x+7)(x-4) = x^2 + (7-4)x + 7(-4)$$
$$= x^2 + 3x - 28$$

EXAMPLE 8 Multiply and simplify $(.2c + \sqrt{d})(.3c - 2\sqrt{d})$.

SOLUTION

$$(.2c + \sqrt{d})(.3c - 2\sqrt{d}) = .06c^2 + (-.4 + .3)c\sqrt{d} - 2(\sqrt{d})^2$$
$$= .06c^2 - .1c\sqrt{d} - 2d$$

EXERCISES

In Exercises 1–4, evaluate the expressions at $x=4, y=3, z=-2$ and at $x=180.3, y=-.0181, z=11.19$ (approximate first).

1. $3x^2 + 4xyz$
2. $(z^3 - 20y)/xy$
3. $(x^2 - z^2 - y^{-2})/2xyz$
4. $\sqrt{4 + z^2}/(z^2 + y^2)$

In Exercises 5–44, perform the indicated operations and simplify.

5. $(3x^5 - x^3 + x^2 + 5) + (2x^5 - x^4 - x^3 + 7)$
6. $(7x^4 - 8x^2 + x + 2) - (x^5 + 2x^4 - x^2 - 4x + 1)$
7. $2x(x^2 - x - 1) - (x^2 + 2x + 1)$
8. $(a + b - c) - (a - b - c)$
9. $4ab(a^3 - a^2b + ab^2 - b^3)$
10. $x - [1 - (x - 1)]$
11. $-(2x - 3) - 3[x - 2(x - 1)]$
12. $-(1 - \{1 - [1 - (1 - x)]\})$
13. $x^2 y^3 (x^{-2} + y^{-2})$
14. $x^3(x^2 + 5x)$
15. $x^9 y^{-7}(x^2 y + x^{-1} y^3)$
16. $(x - 1)(x^2 - 1)$
17. $(x + 2y)(x^2 - xy + 2y^2)$
18. $(a - 3b)(4a^2 - 2ab + 3b^2)$

28 FUNDAMENTAL CONCEPTS OF ALGEBRA

19. $(x-y)(x+y)(x^2+y^2)$
20. $(2a-5b)^2$
21. $(.2c+.3d)^2$
22. $(xy^2-z^3)^2$
23. $(3x^2-2x^{-2})^2$
24. $(2x-3)(2x+3)$
25. $(6m+7n)(6m-7n)$
26. $(2\sqrt{x}-y)(2\sqrt{x}+y)$
27. $(x+4)(x+9)$
28. $(y+3)(y+8)$
29. $(2z-5)(2z+7)$
30. $(3a-b)(2a+b)$
31. $(8c^2-3)(3c^2-2)$
32. $(3\sqrt{m}-4)(4\sqrt{m}-3)$
33. $(6p+q)(7p-3q)$
34. $(11x^3-5y)(3x^3-8y)$
35. $(9u+10v^5)(7u+9v^5)$
36. $(12s-7s^{-1})(10s+s^{-1})$
37. $(7x-4y)^2-(7x+4y)^2$
38. $(x^m-x^n)(x^m+x^n)$
39. $(x^m y^n - z^p)^2$
40. $xy^{-2}(x^{-2}+y^{-1})^2$
41. $(x+2)(x^2-2x+4)$
42. $(2x-3)(4x^2+6x+9)$
43. $(7-ab)(49+7ab+a^2b^2)$
44. $(2u+1)(4u^2-2u+1)(8u^3-1)$

SECTION 1.5. FACTORING

Factoring is the reverse of the process of multiplying out products described in the previous section. Thus, factoring is the process of converting expressions as sums into equivalent expressions as products. This process will be very useful later on when we solve equations.

The equations in (6) of Sec. 1.4 are the main tools you use in factoring, so it is extremely important that you be able to recognize when an expression is of the *form* of the right side of one of these equations. The following examples illustrate this.

1. *Terms with common factors.* Always factor out common factors first, using $au + av = a(u + v)$.

EXAMPLE 1 Factor the following: **(a)** $3x + 6$; **(b)** $12x^4 + 8x^3y + 4x^2y^2$; **(c)** $(x+y)(a+b) + (x+y)(a-b)$; **(d)** $x^3 + 2x^2 - 3x - 6$.

SOLUTION (a) 3 is common to both terms, so $3x + 6 = 3(x + 2)$.
(b) $4x^2$ is common to each term, so $12x^4 + 8x^3y + 4x^2y^2 = 4x^2(3x^2 + 2xy + y^2)$.
(c) The factor $x + y$ is common to both terms, so
$$(x+y)(a+b) + (x+y)(a-b) = (x+y)[(a+b)+(a-b)]$$
$$= (x+y)(2a) = 2a(x+y)$$

(d) Occasionally, regrouping to factor out common terms leads to the desired result:
$$x^3 + 2x^2 - 3x - 6 = x^2(x+2) - 3(x+2) = (x+2)(x^2-3)$$

2. *Difference of two squares.* This simply requires recognizing that
$$u^2 - v^2 = (u-v)(u+v).$$

EXAMPLE 2 Factor the following: (a) $4s^2 - 9t^2$; (b) $81a^4 - b^4$; (c) $(a - 5)^2 - (a + 5)^2$.

SOLUTION (a) Applying the above formula with $u = 2s$, $v = 3t$ gives
$$4s^2 - 9t^2 = (2s)^2 - (3t)^2 = (2s - 3t)(2s + 3t)$$

(b) Applying the formula twice gives
$$81a^4 - b^4 = (9a^2)^2 - (b^2)^2 = (9a^2 - b^2)(9a^2 + b^2)$$
$$= [(3a)^2 - b^2](9a^2 + b^2) = (3a - b)(3a + b)(9a^2 + b^2)$$

(c) Let $u = a - 5$, $v = a + 5$. Then
$$(a - 5)^2 - (a + 5)^2 = u^2 - v^2 = (u - v)(u + v)$$
$$= [(a - 5) - (a + 5)][(a - 5) + (a + 5)]$$
$$= (-5 - 5)(a + a) = (-10)(2a) = -20a$$

In these problems, whenever possible, intermediate steps should be done in your head.

Note that $x^2 - 2$ can be factored as $(x - \sqrt{2})(x + \sqrt{2})$. However, in this section we shall assume that if the coefficients of the original problem are integers or fractions, then the coefficients in the answers must be also.

3. *Perfect squares.* This requires recognizing when a trinomial (a sum of three terms) is of the form
$$u^2 + 2uv + v^2 = (u + v)^2$$
or
$$u^2 - 2uv + v^2 = (u - v)^2$$

This should be looked for whenever the first and last terms are perfect squares.

EXAMPLE 3 Factor (a) $x^2 - 6x + 9$; (b) $16x^2 + 40xy^3 + 25y^6$.

SOLUTION (a) If we let $u = x$ and $v = 3$, then the first term is u^2, and the last term is v^2. We examine the middle term to see if it is $\pm 2uv$. Since $-2uv = -6x$,
$$x^2 - 6x + 9 = (x - 3)^2$$

(b) Let $u = 4x$, $v = 5y^3$. Then the first and last terms are u^2 and v^2, respectively. Compute $2uv$ to see if it is the middle term. It is, so
$$16x^2 + 40xy^3 + 25y^6 = (4x + 5y^3)^2$$

4. *Trinomials that are not perfect squares.* In such cases, it may be possible to apply

(6v) $u^2 + (c + d)u + cd = (u + c)(u + d)$

or the more general form

(6vi) $$abu^2 + (ad + bc)u + cd = (au + c)(bu + d).$$

EXAMPLE 4 Factor $x^2 - 6x + 8$.

SOLUTION Since the coefficient of x is 1, we use the simpler formula (6v) with $u = x$. We are looking for integers c, d such that $c + d = -6$ and $cd = 8$. Since their product is positive and their sum is negative, they must both be negative. We write $(x -)(x -)$. We now look for two positive integers whose product is 8 and whose sum is 6. The only possible ways to factor 8 as a product of two positive integers are $8 = 1 \cdot 8$ and $8 = 2 \cdot 4$. Since $1 + 8 \neq 6$, but $2 + 4 = 6$, we have

$$x^2 - 6x + 8 = (x - 2)(x - 4)$$

EXAMPLE 5 Factor $2x^2 - 11x - 6$.

SOLUTION Since the coefficient 2 of x^2 factors only as $2 = 2 \cdot 1$, we write $(2x + c)(x + d)$. We must then choose c and d so that $c + 2d = -11$ and $cd = -6$. We see that c and d must have opposite signs, since their product is negative. Since $c + 2d = -11$, it is likely that d is negative. Since $6 = 1 \cdot 6 = 2 \cdot 3$, we see that $c = 1$ and $d = -6$ works:

$$2x^2 - 11x - 6 = (2x + 1)(x - 6)$$

EXAMPLE 6 Factor $4x^2 + 19xy + 12y^2$.

SOLUTION For this problem we write down the letters and signs first: $(x + y)(x + y)$. Then we look for factors of 4, $4 = 1 \times 4 = 2 \times 2$, and of 12, $12 = 1 \times 12 = 2 \times 6 = 3 \times 4$, and match up where the sum is 19. Guessing, or trying them, we see that

$$4x^2 + 19xy + 12y^2 = (4x + 3y)(x + 4y)$$

Hint: When guessing factors, when the coefficient of the middle term is large, guess the factors where one is large (see Example 5). When the middle term is relatively medium or small, guess the factors of about the same size (as in Example 6).

5. *The sum and difference of two cubes.* The two formulas

$$u^3 - v^3 = (u - v)(u^2 + uv + v^2)$$
$$u^3 + v^3 = (u + v)(u^2 - uv + v^2)$$

can be verified by simply multiplying out. They should be memorized.

EXAMPLE 7 Factor **(a)** $x^3 - 8$; **(b)** $27a^3 + 64b^3$; **(c)** $8c^3 - \frac{1}{125}d^6$.

SOLUTION **(a)** Let $u = x$, $v = 2$. Then
$$x^3 - 8 = x^3 - 2^3 = (x - 2)\left[x^2 + x(2) + 2^2\right] = (x - 2)(x^2 + 2x + 4)$$

(b) Let $u = 3a$, $v = 4b$. Then
$$27a^3 + 64b^3 = (3a)^3 + (4b)^3 = (3a + 4b)\left[(3a)^2 - (3a)(4b) + (4b)^2\right]$$
$$= (3a + 4b)(9a^2 - 12ab + 16b^2)$$

(c) Let $u = 2c$, $v = \frac{1}{5}d^2$. Then
$$8c^3 - \frac{1}{125}d^6 = (2c)^3 - \left(\frac{1}{5}d^2\right)^3$$
$$= \left(2c - \frac{1}{5}d^2\right)\left[(2c)^2 + (2c)\left(\frac{1}{5}d^2\right) + \left(\frac{1}{5}d^2\right)^2\right]$$
$$= \left(2c - \frac{1}{5}d^2\right)\left(4c^2 + \frac{2}{5}cd^2 + \frac{1}{25}d^4\right)$$

6. *General situation.* Usually when faced with factoring problems, several of the above apply. Always factor out common factors first. Then look for expressions which are in one of the preceding forms.

EXAMPLE 8 Factor $2x^2 - 18x - 20$.

SOLUTION $2x^2 - 18x - 20 = 2(x^2 - 9x - 10) = 2(x - 10)(x + 1)$.

EXAMPLE 9 Factor $3x^6 - 192$.

SOLUTION
$$3x^6 - 192 = 3(x^6 - 64) = 3(x^3 - 8)(x^3 + 8)$$
$$= 3(x - 2)(x^2 + 2x + 4)(x + 2)(x^2 - 2x + 4)$$

EXAMPLE 10 Factor $5x^4 - 5x^3 - 20x^2 + 20x$.

SOLUTION
$$5x^4 - 5x^3 - 20x^2 + 20x = 5x(x^3 - x^2 - 4x + 4)$$
$$= 5x\left[x^2(x - 1) - 4(x - 1)\right]$$
$$= 5x\left[(x - 1)(x^2 - 4)\right]$$
$$= 5x(x - 1)(x - 2)(x + 2)$$

EXAMPLE 11 Factor $50x^7y^2 - 80x^4y^3 + 32xy^4$.

SOLUTION
$$50x^7y^2 - 80x^4y^3 + 32xy^4 = 2xy^2(25x^6 - 40x^3y + 16y^2)$$
$$= 2xy^2(5x^3 - 4y)^2$$

EXERCISES

Factor the following:

1. $3x^3 - 6x^2 + 3x$
2. $y^8 - 16y^4$
3. $11z^3 - 11$
4. $a^2 - 169$
5. $a^3 - 6a^2 + 2a - 12$
6. $\frac{1}{4}a^2 - \frac{1}{9}b^2$
7. $\frac{1}{3}c^2 + \frac{10}{3}cd + \frac{25}{3}d^2$
8. $m^2 - m - 6$
9. $n^2 + 7n + 12$
10. $2b^3 - 8b^2 - 42b$
11. $2s^4t + 7s^2t^2 + 3t^3$
12. $2x^2 + 5xy + 3y^2$
13. $x^6 - 1$
14. $y^4 - 1$
15. $z^3 + 4z^2 + 4z$
16. $4x^2 - 49y^2$
17. $(a - b)^2 - (a + b)^2$
18. $b^2 - \frac{1}{9}b^4$
19. $a(x + y) + b(x + y)$
20. $x^2 + 11x + 24$
21. $x^3 + 5x^2 - 3x - 15$
22. $x^3 - 4x^2 + 7x - 28$
23. $x^3 + 4x^2 - 4x - 16$
24. $2x^3 - 3x^2 - 18x + 27$
25. $3x^3 - x^2 - 18x + 6$
26. $3x^3 + 2x^2 - 75x - 50$
27. $x^2 + 25x + 24$
28. $x^2 + 23x - 24$
29. $6y^3z + 7y^2z^2 + 2yz^3$
30. $2m^2 - m - 21$
31. $5x^2 - 10xy - 15y^2$
32. $5y^2 - 28y + 15$
33. $9z^2 - 12z + 4$
34. $9a^2 - 9ab - 4b^2$
35. $5b^3 - \frac{5}{27}$
36. $b^3 + b^2 - b - 1$
37. $8c^2 - 20cd + 8d^2$
38. $8m^2 + 12mn - 8n^2$
39. $4c^3 + 500d^3$
40. $(x - 2y)^2 - (x + 2y)^2$
41. $(a - b)^3 + b^3$
42. $6x^4 - 19x^2 - 20$
43. $6a^2 + 13ab + 6b^2$
44. $c^4 - 4c^2 + 4$
45. $3d^6 - 192c^6$
46. $12a^4 - 11a^2b^2 - 36b^4$
47. $x^4 + 2x^3 - x - 2$
48. $x^4 - 3x^3 + 8x - 24$
49. $x^5 - 4x^3 - 8x^2 + 32$
50. $4x^5 - 9x^3 + 4x^2 - 9$

SECTION 1.6. RATIONAL EXPRESSIONS

Any algebraic expression that is the quotient of two other algebraic expressions is called a **rational expression**. Some examples are

$$\frac{2}{3}, \quad \frac{x^2 - 2x - 3}{x^2 - 3x - 4}, \quad \frac{1 - \frac{1}{x}}{x - 1}$$

As they are algebraic, they can be evaluated as discussed in Sec. 1.4. However, with fractions there is the danger of dividing by zero. This is, of course, not allowed, because division by zero is undefined. So if you substitute a number into an expression and find you are dividing by zero, then the expression is **undefined** at that number.

EXAMPLE 1 Find the value of the third expression if $x = -2, -1, 0, 1, 3, 8.47$.

SOLUTION For $x = -2$,

$$\frac{1 - \frac{1}{x}}{x - 1} = \frac{1 - \frac{1}{-2}}{-2 - 1} = \frac{\frac{3}{2}}{-3} = -\frac{1}{2}$$

For $x = -1$,

$$\frac{1 - \frac{1}{x}}{x - 1} = \frac{1 - \frac{1}{-1}}{-1 - 1} = \frac{2}{-2} = -1$$

For $x = 0$, the numerator equals $1 - (1/0)$, which is undefined, so the whole expression is undefined. For $x = 1$, the denominator equals $1 - 1 = 0$, so again the expression is undefined. For $x = 3$,

$$\frac{1 - \frac{1}{x}}{x - 1} = \frac{1 - \frac{1}{3}}{3 - 1} = \frac{\frac{2}{3}}{2} = \frac{1}{3}$$

For $x = 8.47$,

$$\frac{1 - \frac{1}{x}}{x - 1} = \frac{1 - \frac{1}{8.47}}{8.47 - 1} \approx .11806375$$

using a calculator. Note that $.11806375 \approx 1/8.47$.

Example 1 illustrates two things. First, it is important to examine *all* denominators to determine where the expression is defined. Second, the preceding evaluations suggest that this expression can be simplified to $1/x$. Indeed (see Example 5 in this section) it does equal $1/x$ for $x \neq 0$ *and* for $x \neq 1$.

Such expressions can often be simplified. Since rational expressions are quotients containing symbols which represent real numbers, they can be manipulated the same way fractions can. We examine the various manipulations.

EXAMPLE 2 Simplify the following expressions and determine where they are defined:

(a) $\dfrac{2x^2 - 5x - 3}{x^2 - 9}$ (b) $\dfrac{x^2 + x - 2}{3 - x - 2x^2}$

SOLUTION We use the principle that $\dfrac{ac}{bc} = \dfrac{a}{b}$ for $c \neq 0$. For (a),

$$\frac{2x^2 - 5x - 3}{x^2 - 9} = \frac{(2x + 1)(x - 3)}{(x - 3)(x + 3)} = \frac{2x + 1}{x + 3}$$

By the factorization of the denominator, the original expression is defined for $x \neq \pm 3$. For (b),

$$\frac{x^2 + x - 2}{3 - x - 2x^2} = \frac{(x + 2)\overset{-1}{\cancel{(x - 1)}}}{(3 + 2x)\cancel{(1 - x)}} = -\frac{x + 2}{3 + 2x}$$

The original expression is defined for $x \neq 1, -3/2$.

Warning: It is very easy to overlook the fact that $1 - x = -(x - 1)$, and hence it is easy to miss seeing that the quotient of these two factors reduces to -1.

EXAMPLE 3 Perform the following operation, simplify, and determine where the expression is defined:

$$\frac{x^2 - x - 2}{6x - 8} \div \frac{x^2 - 4}{12x^2 - 16x}$$

SOLUTION We use the same principles as multiplying and dividing fractions of real numbers: $\frac{a}{b} \cdot \frac{c}{d} = \frac{ac}{bd}$, $\frac{a}{b} \div \frac{c}{d} = \frac{ad}{bc}$. However, just as with fractions of real numbers, we should cancel out all possible common factors before multiplying, so that the expressions are as simple as possible.

$$\frac{x^2 - x - 2}{6x - 8} \div \frac{x^2 - 4}{12x^2 - 16x} = \frac{(x - 2)(x + 1)}{2(3x - 4)} \div \frac{(x - 2)(x + 2)}{4x(3x - 4)}$$

$$= \frac{\cancel{(x - 2)}(x + 1)}{\cancel{2}\cancel{(3x - 4)}} \cdot \frac{\overset{2}{\cancel{4x}}\cancel{(3x - 4)}}{\cancel{(x - 2)}(x + 2)}$$

$$= \frac{x + 1}{1} \cdot \frac{2x}{x + 2} = \frac{2x(x + 1)}{x + 2}$$

In the long run, it is usually more useful to leave your answers in factored form. The original expression is defined for $x \neq 0, \pm 2, 4/3$, which can easily be seen by examining the second quotient.

EXAMPLE 4 Perform the following operations, simplify, and determine where the expression is defined:

$$\frac{x}{x + 1} + \frac{2}{3 - x} - \frac{x^2 + 3}{x^2 - 2x - 3}$$

SOLUTION To add or subtract fractions, we put each fraction over a least common denominator (l.c.d.) and use the principle $\frac{a}{d} \pm \frac{b}{d} = \frac{a \pm b}{d}$. Here, the l.c.d. is

$(x-3)(x+1)$. [Do not overlook the fact that $3 - x = -(x-3)$.]

$$\frac{x}{x+1} + \frac{2}{3-x} - \frac{x^2+3}{x^2-2x-3} = \frac{x}{x+1} \cdot \frac{x-3}{x-3} + \frac{2}{3-x} \cdot \frac{-(x+1)}{-(x+1)}$$
$$- \frac{x^2+3}{(x-3)(x+1)}$$
$$= \frac{x(x-3) - 2(x+1) - (x^2+3)}{(x-3)(x+1)}$$
$$= \frac{x^2 - 3x - 2x - 2 - x^2 - 3}{(x-3)(x+1)} = \frac{-5x-5}{(x-3)(x+1)}$$
$$= \frac{-5\cancel{(x+1)}}{(x-3)\cancel{(x+1)}} = \frac{-5}{x-3}$$

The original expression is defined for $x \neq 3, -1$.

EXAMPLE 5 Simplify and determine where the expression is defined:

(a) $\quad \dfrac{1 - \dfrac{1}{x}}{x-1} \quad$ (b) $\quad \dfrac{\dfrac{1-x}{x} - \dfrac{x}{1+x}}{\dfrac{1+x}{x} - \dfrac{x}{1-x}}$

SOLUTION These are called **complex fractions**. They are simplifed using the principle

$$\frac{\dfrac{a}{b}}{\dfrac{c}{d}} = \frac{a}{b} \div \frac{c}{d} = \frac{a}{b} \cdot \frac{d}{c}$$

For this principle to apply, the numerator and denominator must be fractions, so we usually first have to find common denominators and add. For (a),

$$\frac{1-\dfrac{1}{x}}{x-1} = \frac{\dfrac{x}{x} \cdot 1 - \dfrac{1}{x}}{x-1} = \frac{\dfrac{x-1}{x}}{x-1} = \frac{x-1}{x} \div \frac{x-1}{1} = \frac{\cancel{x-1}}{x} \cdot \frac{1}{\cancel{x-1}} = \frac{1}{x}$$

The original expression is defined for $x \neq 0, 1$. For (b),

$$\frac{\dfrac{1-x}{x} - \dfrac{x}{1+x}}{\dfrac{1+x}{x} - \dfrac{x}{1-x}} = \frac{\dfrac{1-x}{x} \cdot \dfrac{1+x}{1+x} - \dfrac{x}{1+x} \cdot \dfrac{x}{x}}{\dfrac{1+x}{x} \cdot \dfrac{1-x}{1-x} - \dfrac{x}{1-x} \cdot \dfrac{x}{x}} = \frac{\dfrac{1-x^2 - x^2}{x(1+x)}}{\dfrac{1-x^2 - x^2}{x(1-x)}}$$
$$= \frac{1-2x^2}{x(1+x)} \div \frac{1-2x^2}{x(1-x)} = \frac{1-2x^2}{x(1+x)} \cdot \frac{x(1-x)}{1-2x^2} = \frac{1-x}{1+x}$$

The original expression is defined for $x \neq 0, \pm 1, \pm(1/\sqrt{2})$. This can be seen by examining the fourth quotient (and remembering that you cannot divide by zero so that the expression is not defined when $1 - 2x^2 = 0$).

EXERCISES

In Exercises 1–4, evaluate the expression at $x = -2, -1, 0, 1, 2, .01,$ and 10^3, if possible, using your calculator only where necessary. Determine where the expression is defined.

1. $\dfrac{1-x}{1-\dfrac{1}{x}}$

2. $\dfrac{x^2 - 5x + 6}{x^2 - 4}$

3. $\dfrac{x+1}{x+2+\dfrac{1}{x}}$

4. $\dfrac{x^3 - 1}{x^2 - 1}$

In Exercises 5–38, perform the indicated operations and simplify.

5. $\dfrac{x^2 + 3x + 2}{x^2 + 6x + 8}$

6. $\dfrac{x+y}{x-y} \cdot \dfrac{x-y}{x}$

7. $\dfrac{x+1}{x} - \dfrac{2}{x-1}$

8. $\dfrac{3a + 12}{2a - 8} \div \dfrac{a+4}{a-4}$

9. $\dfrac{6x^2 + 5x - 4}{4x^2 - 4x + 1}$

10. $\dfrac{x^2 - y^2}{3x} \cdot \dfrac{x^2 + xy}{2x - 2y}$

11. $\dfrac{a}{a-b} - \dfrac{b}{a+b}$

12. $\dfrac{3s^2 + 14s - 5}{4s^2 - 9} \div \dfrac{12s - 4}{8s + 12}$

13. $\dfrac{3}{2x+1} - \dfrac{5}{3x-2}$

14. $\dfrac{t^2 - 2t - 35}{2t^2 - 3t} \cdot \dfrac{4t^3 - 9t}{7t - 49}$

15. $\dfrac{x^3 + 8}{x^2 - 4} \div \dfrac{x^2 - 2x + 4}{x^2 - 4x + 4}$

16. $\dfrac{6}{y^2 - y - 2} - \dfrac{5}{y^2 + y - 6}$

17. $\dfrac{2z^2 - 7z + 6}{4z^2 + 27z - 7} \cdot \dfrac{2z^2 - 3z}{6z^2 - 21z + 18}$

18. $\dfrac{s}{s-t} + \dfrac{t}{t-s}$

19. $\dfrac{1}{x+1} - \dfrac{x}{x-2} + \dfrac{x^2 + 2}{x^2 - x - 2}$

20. $\dfrac{a+15}{a^2 - 9} - \dfrac{a+12}{a^2 - a - 6}$

21. $[9 - (1/x^2)]/(3x+1) + 1/x^2$

22. $\dfrac{1-x}{1-\dfrac{1}{x}}$

23. $\dfrac{x^2 - 5x + 6}{x^2 - 4}$

24. $\dfrac{x+1}{x+2+\dfrac{1}{x}}$

25. $\dfrac{x^3 - 1}{x^2 - 1}$

26. $\dfrac{x - \dfrac{1}{x}}{1 - \dfrac{1}{x}}$

27. $\dfrac{\dfrac{x}{y} - \dfrac{y}{x}}{\dfrac{1}{y} + \dfrac{1}{x}}$

28. $\dfrac{1 - \dfrac{3}{z} + \dfrac{2}{z^2}}{\dfrac{1}{z} - \dfrac{2}{z^2}}$

29. $\dfrac{x + \dfrac{8}{x^2}}{1 + \dfrac{2}{x}}$

30. $\dfrac{x^2 + x + 1}{x^2 - \dfrac{1}{x}}$

31. $\dfrac{1 - \dfrac{2}{y+1}}{\dfrac{1}{y} - y}$

32. $\dfrac{\dfrac{z}{z-4} + \dfrac{1}{z-1}}{\dfrac{z}{z-1} + \dfrac{2}{z-3}}$

*33. $1 - \dfrac{1}{1 + \dfrac{1}{x}}$

34. $\dfrac{\dfrac{x}{1-x} + \dfrac{1+x}{x}}{\dfrac{1-x}{x} + \dfrac{x}{1+x}}$

35. $\dfrac{\dfrac{1}{x^2} + \dfrac{2}{xy} + \dfrac{1}{y^2}}{\dfrac{1}{x^2} - \dfrac{1}{y^2}}$

36. $x - \dfrac{x}{x - \dfrac{1}{x}}$

37. $\dfrac{1}{1 + x^{m-n}} + \dfrac{1}{1 + x^{n-m}}$ (*Hint*: Write x^{m-n} as $\dfrac{x^m}{x^n}$, etc.)

38. $\dfrac{1}{1 + x^{m-n} + x^{p-n}} + \dfrac{1}{1 + x^{n-m} + x^{p-m}} + \dfrac{1}{1 + x^{m-p} + x^{n-p}}$

SECTION 1.7. RATIONAL EXPONENTS AND RADICALS

The Definition of $x^{m/n}$

Let x be a real number. In Sec. 1.2, we discussed what is meant by x^n for any integer power n. Now we would like to define x to a rational power, i.e., to define $x^{m/n}$ where m, n are integers, $n \neq 0$. We want to do so in a way that is consistent with the definitions of Sec. 1.2 and also so that the laws of exponents (3) observed there still hold.

We first consider $x^{1/n}$ (which is sometimes written as $\sqrt[n]{x}$), n a positive integer. If $(x^r)^s = x^{rs}$ is to hold, we must have

$$(x^{1/n})^n = x^{(1/n)n} = x^1 = x$$

Hence, $x^{1/n}$ must be a number whose nth power is x. But are there any such real numbers? Are there more than one? We examine the situation by cases.

Case I. Suppose n is an even positive integer. We ask, are there real numbers y whose nth power is x? If $x = 0$, there is one and only one such y, namely 0. If x is negative, the answer is no. For example, if $n = 2$ and $x = -1$, there is no real number y with $y^2 = -1$, since the square of any positive or negative number is positive. Hence, $\sqrt[n]{x} = x^{1/n}$ cannot be defined for $x < 0$, n even. If x is positive, there are real numbers y with $y^n = x$, but there are two such numbers, one positive and one negative. For example, if $x = 9$, $n = 2$, then $(+3)^2 = 9$ and $(-3)^2 = 9$. If $x = 64$ and $n = 6$, then $(+2)^6 = 64$ and $(-2)^6 = 64$. As it would be awkward to have $\sqrt[n]{x} = x^{1/n}$ be two numbers, we pick the positive one (which is the one we want most often).

DEFINITION

If n is an even positive integer and $x > 0$, then $\sqrt[n]{x} = x^{1/n}$ is the (unique) positive real number y such that $y^n = x$. We call $x^{1/n}$ the **principal nth root** of x.

If $x = 0$, $\sqrt[n]{0} = 0^{1/n} = 0$.

The notation $\sqrt[n]{}$ is referred to as a **radical**. When $n = 2$ and radical notation is being used, the 2 is usually omitted: $\sqrt{x} = \sqrt[2]{x} \; (= x^{1/2})$.

Warning: In the case where n is even, you must remember that $\sqrt[n]{x} = x^{1/n}$ means only the *positive* root. If you want *both* real roots, you *must* write $\pm \sqrt[n]{x}$ or $\pm x^{1/n}$. For example, in solving $x^2 = 9$ or $x^8 = 256$, you must write $x = \pm \sqrt{9} (= \pm 3)$ or $x = \pm \sqrt[8]{256} \; (= \pm 2)$. Similarly, if you want only the negative root, you must write $-\sqrt[n]{x}$ or $-x^{1/n}$.

Case II. Suppose n is odd. Then for any real number x, there is one and only one real number y such that $y^n = x$. For example, -5 is the only real number such that $(-5)^3 = -125$, and 2 is the only real number such that $2^5 = 32$.

DEFINITION

If n is an odd positive integer and x is any real number, $\sqrt[n]{x} = x^{1/n}$ is that (unique) real number y such that $y^n = x$. This is called the **nth root** of x. It should be emphasized that $x^{1/n}$ and $\sqrt[n]{x}$ are different notations for the same thing. For algebraic manipulations, $x^{1/n}$ is more useful.

Now that we know what is meant by $x^{1/n}$, n a positive integer, we can easily define $x^{m/n}$, m any integer, since we want $(x^r)^s = x^{rs}$. [Note first that any rational number can be written as m/n, n positive; e.g., $-\frac{2}{3} = (-2)/3$.]

(7)   **DEFINITION**

If n is a positive integer and m is any integer, $x^{m/n} = (x^{1/n})^m$, provided $x^{1/n}$ is defined and m/n is reduced to lowest terms.

EXAMPLE 1   Simplify if possible: **(a)** $(-27)^{5/3}$; **(b)** $4^{-5/2}$; **(c)** $64^{4/6}$; **(d)** $(-64)^{4/6}$; **(e)** $(64^2)^{1/3}$; **(f)** $(-64)^{5/6}$.

SOLUTION
(a) $(-27)^{5/3} = [(-27)^{1/3}]^5 = (-3)^5 = -243$
(b) $4^{-5/2} = (4^{1/2})^{-5} = 2^{-5} = 1/2^5 = 1/32$
(c) $64^{4/6} = 64^{2/3} = (\sqrt[3]{64})^2 = 4^2 = 16$
(d) $(-64)^{4/6} = (-64)^{2/3} = (\sqrt[3]{-64})^2 = (-4)^2 = 16$

(e) $(64^2)^{1/3} = 4096^{1/3} = 16$

(f) $(-64)^{5/6} = (\sqrt[6]{-64})^5$, which is undefined, since $\sqrt[6]{-64}$ is undefined.

By examining parts (c) and (e), you can see that

$$(x^{1/n})^m = (x^m)^{1/n} \quad \text{whenever } x > 0$$

When $x < 0$, this is true when everything is defined.

Rational Exponents on a Calculator

The [y^x] button on your calculator works perfectly well for x a rational number. As before, you must remember that this button works only for $y > 0$. So if you wanted to use your calculator to compute $(-8)^{5/3} = -32$, you would have to know your algebra, compute $8^{5/3}$ on the calculator, and put in the signs yourself.

We first describe how to compute $y^{1/n}$, which is a special case. It can be computed using the [y^x] and [$1/x$] buttons. For example, to compute $2^{1/20} = \sqrt[20]{2}$, press

Alg.: [2][y^x][20][$1/x$][=]; RPN: [2][ENT][20][$1/x$][y^x]

Some calculators have a [$y^{1/x}$] button. This combines the [y^x] and [$1/x$] buttons, and its use saves one push.

The general case of $y^{m/n}$ is computed one of two ways:

A. Using the definition, $y^{m/n} = (y^{1/n})^m$. For example, to compute $9^{11/13} = (9^{1/13})^{11}$, press

Alg.: [9][y^x][13][$1/x$][=][y^x][11][=]
(the first [=] may not be necessary)

RPN: [9][ENT][13][$1/x$][y^x][11][y^x]

B. Computing m/n and raising y to this number.

On an RPN calculator, there is no problem. For example, to compute $9^{11/13}$, press [9][ENT][11][ENT][13][÷][y^x].

Warning: On an algebraic calculator there is a trap. To see it, suppose you wanted to compute $9^{3/2}$ (which, in fact, $= 27$). If you press

[9][y^x][3][÷][2][=]

you get the wrong answer of 364.5. Here is what happens: When [÷] is pressed, the calculator computes what is in the machine so far, namely, $9^3 = 729$. Then you get $(9^3) \div 2 = 729 \div 2 = 364.5$. (*Note:* The [$1/x$] button does not cause pending operations to complete. The same is usually true of the [x^2], [$\sqrt{\ }$], and any other button that operates on *one* number at a time.)

In the preceding computation, you want $9^{(3 \div 2)}$, and there are several ways of getting it:

1. *Hardest way:* If your calculator has parentheses, press
$$[9][y^x][(][3][\div][2][)][=]$$
If your calculator has no parentheses but has a memory, press
$$[3][\div][2][=][M+][9][y^x][MR][=]$$

2. *Easier way:* Compute 3/2 in your head. As $9^{3/2} = 9^{1.5}$, simply press
$$[9][y^x][1.5][=]$$
(Of course, if the problem had been $9^{11/13}$, you would have to do it as in method 1, or, which for this case is easier, as in method A.)

3. *Easiest way:* Do the whole problem in your head:
$$9^{3/2} = (9^{1/2})^3 = (\sqrt{9})^3 = 3^3 = 27$$

One of the important things a calculator does is to relieve you of the drudgery of arithmetic so that you can focus your attention on the important aspects of the problem. Unfortunately, though, it is tempting just to push buttons without really thinking about the calculations. The preceding example demonstrates how this often results in doing things in a more complicated way than is necessary. *You should think about two things as you do calculations*:

1. *Can they be simplified* (which consequently decreases the likelihood of mistakes)?
2. *Are you pushing the right buttons* to do the calculations? (For example, on an algebraic calculator, does $[9][y^x][3][\div][2][=]$ give $9^{3/2}$?)

While the calculator is a great computational aid, it cannot do the thinking for you.

EXAMPLE 2 Use your calculator to compute (a) $15^{7/11}$; (b) $3^{3/20}$; (c) $8^{1/15}$.

SOLUTION (a) Use the definition $15^{7/11} = (15^{1/11})^7$ as in method A or as in method 1. $15^{7/11} \approx 5.60301$. (b) As $3/20 = 15/100 = .15$, $3^{3/20} = 3^{.15}$, so you can do this as in method 2: $3^{.15} \approx 1.17915$. (c) Use the $[1/x]$ button. $8^{1/15} \approx 1.14870$.

Algebraic Properties of Rational Exponents

> **(8)**   **THEOREM**
>
> With the preceding definition of rational exponents, the laws of exponents [(3) in Sec. 1.2] hold if m and n are rational numbers provided x and y are positive.

This can be proved using the definitions and Theorem (4), but we shall not do it here. *Note:* $x > 0$ is necessary. For example, $[(-1)^{2/3}]^{1/2} \neq (-1)^{1/3}$.

The important consequence of this theorem is the following:

> If you have an expression involving exponents and radicals, the easiest way to handle it is to convert all radicals to fractional exponents and to use the laws of exponents to simplify it.

The following examples illustrate this. Since radicals are equivalent to exponents, the laws of exponents (appropriately modified) apply to them. For example,
$$\sqrt{a}\,\sqrt{b} = \sqrt{ab} \quad \text{because } a^{1/2}b^{1/2} = (ab)^{1/2}$$
In simpler expressions (such as (d) below), it is easier to work with radicals without changing to fractional exponents.

EXAMPLE 3 (1) Evaluate the following expressions. (2) Then simplify the expressions algebraically (using the laws of exponents) and evaluate the resulting expressions. Use your calculator only when necessary.

(a) $(\sqrt[3]{64})^{3/2}$; (b) $\sqrt{7\sqrt[5]{7^3}}$; (c) $(-2)^{2/3}(-2)^5$; (d) $\sqrt{12}\,\sqrt{75}$.

SOLUTION As RPN calculators should pose no difficulties, we give solutions only for algebraic calculators, where necessary.

(a) For (1), $(\sqrt[3]{64})^{3/2} = 4^{3/2} = (\sqrt{4})^3 = 2^3 = 8$. For (2), $(\sqrt[3]{64})^{3/2} = 64^{(1/3)(3/2)} = 64^{1/2} = 8$.

(b) First, rewrite it: $\sqrt{7\sqrt[5]{7^3}} = (7 \cdot 7^{3/5})^{1/2}$. For (1), use the calculator: [7][y^x][.6][=][×][7][=][$\sqrt{}$]. For (2), simplify: $(7 \cdot 7^{3/5})^{1/2} = (7^{1+3/5})^{1/2} = (7^{8/5})^{1/2} = 7^{4/5}$. Now using the calculator, press [7][y^x][.8][=]. The answer is 4.74328.

(c) Take care of the negative signs first: $(-2)^{2/3} = 2^{2/3}$; $(-2)^{-5} = -(2^{-5}) = -2^{-5}$. So $(-2)^{2/3}(-2)^{-5} = -(2^{2/3}2^{-5})$. For (1), use your calculator to evaluate what is inside the parentheses. You get .049606, so the answer is $-.049606$. For (2), $-(2^{2/3} \cdot 2^{-5}) = -2^{2/3-5} = -2^{-13/3} = -.049606$ (by computing $2^{-13/3}$ on the calculator and then putting in the sign).

(d) For (1), you must use your calculator, using the [$\sqrt{}$] button (or the [y^x] button with $x = 1/2 = .5$): [12][$\sqrt{}$][×][75][$\sqrt{}$][=]. For (2), $\sqrt{12}\,\sqrt{75} = \sqrt{12(75)} = \sqrt{900} = 30$. Alternatively, $\sqrt{12}\,\sqrt{75} = (2\sqrt{3})(5\sqrt{3}) = 10 \cdot 3 = 30$.

Algebraic Expressions

These same laws of exponents apply to algebraic expressions.

EXAMPLE 4 Simplify the following. Assume all letters represent positive real numbers.

(a) $x^{1/3}(x^4)^{1/6}$ (b) $\left(\dfrac{-27x^{-6}y^{3/2}}{z^{-4/5}w}\right)^{-1/3}$ (c) $\sqrt[7]{x\sqrt{x^5}}$

42 FUNDAMENTAL CONCEPTS OF ALGEBRA

SOLUTION (a) $x^{1/3}(x^4)^{1/6} = x^{1/3}x^{2/3} = x^{3/3} = x^1 = x$.

(b) $\left(\dfrac{-27x^{-6}y^{3/2}}{z^{-4/5}w}\right)^{-1/3} = \dfrac{(-27)^{-1/3}(x^{-6})^{-1/3}(y^{3/2})^{-1/3}}{(z^{-4/5})^{-1/3}w^{-1/3}}$

$= \dfrac{(-3)^{-1}x^2 y^{-1/2}}{z^{4/15}w^{-1/3}} = -\dfrac{x^2 w^{1/3}}{3y^{1/2}z^{4/15}}$

(c) $\sqrt[7]{x\sqrt{x^5}} = (x \cdot x^{5/2})^{1/7} = (x^{7/2})^{1/7} = x^{1/2} = \sqrt{x}$

Rationalizing the Denominator

Rationalizing the denominator is removing radicals (or fractional exponents) from the denominator. This sometimes makes an expression easier to handle.

EXAMPLE 5 Rationalize the denominator of $\sqrt{5/12}$.

SOLUTION The method is to find the smallest number with which to multiply the denominator, 12, to make it a perfect square. Since $12 = 4 \cdot 3 = 2^2 \cdot 3$, we see that the smallest number is 3. We then multiply both the numerator and denominator by 3 and then simplify.

$$\sqrt{\dfrac{5}{12}} = \sqrt{\dfrac{5}{12} \cdot \dfrac{3}{3}} = \dfrac{\sqrt{15}}{\sqrt{36}} = \dfrac{\sqrt{15}}{6} = \dfrac{1}{6}\sqrt{15}$$

EXAMPLE 6 Rationalize the denominator of $\sqrt[3]{3z/4x^5y}$.

SOLUTION We now want to find the smallest powers of the factors with which to make the denominator a perfect cube. Since $4x^5y = 2^2x^3x^2y$, we see that we want $2xy^2$:

$$\sqrt[3]{\dfrac{3z}{4x^5y}} = \sqrt[3]{\dfrac{3z}{4x^5y} \cdot \dfrac{2xy^2}{2xy^2}} = \dfrac{\sqrt[3]{6xy^2z}}{\sqrt[3]{8x^6y^3}} = \dfrac{\sqrt[3]{6xy^2z}}{2x^2y}$$

EXAMPLE 7 Assume all letters represent positive real numbers. Rationalize the denominator of

$$\dfrac{\sqrt{x} - \sqrt{y}}{\sqrt{x} + \sqrt{y}}$$

SOLUTION We use the principles that $(u + v)(u - v) = u^2 - v^2$ (where $u = \sqrt{x}$ and $v = \sqrt{y}$) and that $(\sqrt{x})^2 = x$ since x is positive. Then multiplying the numerator and denominator by $\sqrt{x} - \sqrt{y}$ will eliminate radicals from the denominator.

$$\frac{\sqrt{x} - \sqrt{y}}{\sqrt{x} + \sqrt{y}} = \frac{\sqrt{x} - \sqrt{y}}{\sqrt{x} + \sqrt{y}} \cdot \frac{\sqrt{x} - \sqrt{y}}{\sqrt{x} - \sqrt{y}} = \frac{(\sqrt{x})^2 - 2\sqrt{x}\sqrt{y} + (\sqrt{y})^2}{(\sqrt{x})^2 - (\sqrt{y})^2}$$

$$= \frac{x - 2\sqrt{xy} + y}{x - y}$$

EXERCISES

In Exercises 1–11, (1) evaluate and (2) then simplify (using the laws of exponents) and evaluate the resulting expression.

1. $(9^{1/2})^4$
2. $8^{1/3} 8^{-2}$
3. $25^{1/2} 4^{1/2}$
4. $(.04^{-1/2})^4$
5. $[(1/27)^{5/3}]^{-1/5}$
6. $(32^{1/3})^{3/5}$
7. $27^{1/4} 48^{1/4}$
8. $[(-64)^{-5/3}]^{-1/5}$
9. $\sqrt[3]{5\sqrt[8]{5}}$
10. $\sqrt[4]{\sqrt[3]{\sqrt{10}}}$
11. $\sqrt[4]{6^3 \sqrt[9]{6}}$

In Exercises 12–40, simplify (see p. 18 for the definition of *simplify*). Assume all variables represent positive real numbers.

12. $x^{1/2} x^{1/3}$
13. $y^{1/2} y^{-1/3}$
14. $\sqrt[3]{-8x^{-6}y^3}$
15. $(y^{-1/3})^{3/4}$
16. $\sqrt[3]{-27x^5 y^3}$
17. $\sqrt{6xy^3}\sqrt{24xy}$
18. $\sqrt[4]{.0016 x^{-4} y^{-8}}$
19. $\sqrt[4]{x^3}\sqrt[4]{x^5}$
20. $x^{-1/6}/x^{1/3}$
21. $(a^{-2/3} b^{1/2})^{-2}$
22. $\sqrt[3]{-4x^6 y^{1/2}}\sqrt[3]{2x^3 y^{-7/2}}$
23. $\sqrt{\sqrt[3]{x}}$
24. $\sqrt{\sqrt[4]{x^{16}}}$
25. $\sqrt[5]{x^{4/3}\sqrt{x^4}}$
26. $(\sqrt[4]{xy}\sqrt{x^7 y})^{1/3}$
27. $(\sqrt{x^4}\sqrt{x^5 y^2})^{-1/3}$
28. $z^{1/2}(z^{3/2} - z^{1/2} + z^{-1/2})$
29. $x^{1/5}(x^{1/5} - x^{-1/5})$
30. $(8x^3 y^{-6})^{-2/3}$
31. $\sqrt[3]{(27x^{3/4} y^{-6/7})^{-2}}$
32. $\left(\dfrac{8x^5 y^{-3}}{27x^{-4} y^3}\right)^{-1/3}$
33. $\left(\dfrac{4x^{-7/3} y^{-8}}{25 x^{5/3} y^{-6}}\right)^{-1/2}$
34. $\sqrt[3]{x^{1/2}}\sqrt{x^{-1/3}}$
35. $\sqrt[7]{a\sqrt{a^5}}$
36. $b^{1/2} c^{1/2}(b^{3/2} + b^{1/2})(c^{-1/2} + c^{-3/2})$
37. $(m^{1/2} + n^{1/2})(m^{1/2} - n^{1/2})$

38. $\sqrt{x\sqrt[3]{x\sqrt[4]{x}}}$

39. $\left(\dfrac{-y^{1/3}}{y^{-3/2}}\right)^{-3}$

40. $\dfrac{\sqrt{25x^2y^{-3}}}{(2x^{-2}y^{3/2})^3}$

In Exercises 41–56, rationalize the denominator. Assume all variables represent positive real numbers.

41. $\sqrt{\dfrac{2}{3}}$

42. $\sqrt{\dfrac{5}{18}}$

43. $\sqrt[3]{\dfrac{9}{16}}$

44. $\sqrt[3]{\dfrac{7}{144}}$

45. $\sqrt{\dfrac{1}{x}}$

46. $\sqrt[3]{\dfrac{1}{y}}$

47. $\sqrt[3]{\dfrac{x}{y^2}}$

48. $\sqrt[3]{\dfrac{2a}{3a^2b^3}}$

49. $\dfrac{1}{1+\sqrt{2}}$

50. $\dfrac{1}{3+\sqrt{2}}$

51. $\dfrac{1}{\sqrt{5}-3}$

52. $\dfrac{\sqrt{2}}{\sqrt{2}-\sqrt{3}}$

53. $\dfrac{\sqrt{3}-\sqrt{2}}{\sqrt{3}+\sqrt{2}}$

54. $\dfrac{\sqrt{x}-\sqrt{2}}{\sqrt{x}+\sqrt{2}}$

55. $\dfrac{1}{1+\sqrt[3]{x}}$

56. $\dfrac{1}{1-\sqrt{2}+\sqrt{3}}$

In Exercises 57–60, rationalize the numerator. Do this in a way similar to that done with denominators. Assume all variables represent positive real numbers.

57. $\dfrac{\sqrt{2}-\sqrt{3}}{\sqrt{2}+\sqrt{3}}$

58. $\dfrac{\sqrt{5}+\sqrt{7}}{\sqrt{5}-\sqrt{7}}$

59. $\dfrac{\sqrt{2+x}-\sqrt{2}}{x}$

60. $\dfrac{\sqrt{7+h}-\sqrt{7}}{3h}$

In Exercises 61–68, solve for x. Use your calculator only when necessary.

61. $x^2 = 4$

62. $x^{1/2} = 4$

63. $x^3 = -8$

64. $x^{1/3} = -8$

65. $x^{2/3} = 8$

66. $x^{20} = 2$

67. $x^{3/5} = 8$

68. $(1+x)^4 = 81$

REVIEW EXERCISES

1. State which property of the real numbers is being used.

   a. $3(-2) = (-2)3$

   b. $43(102) = 43(100 + 2) = 4300 + 86 = 4386$

2. Determine if the given number is rational or irrational. If the number is rational, express it as a ratio of two integers.

   a. 0

   b. $\sqrt{3} + 2$

   c. $-\dfrac{1}{5}\sqrt{16}$

   d. 3.14

3. Express the given number as an infinite decimal.

    a. $\dfrac{1}{4}$ **b.** $\dfrac{7}{6}$

    c. $\dfrac{3}{11}$ **d.** $\dfrac{6}{13}$

4. Which of the following numbers can be entered exactly on a calculator?

    a. $5/6$ **b.** $6/5$

    c. $5.21 \times 10^{53} + 4.87 \times 10^{53}$

    d. $(5.21 \times 10^{53})(4.87 \times 10^{53})$

    e. $\sqrt{.9}$ **f.** $\sqrt{.09}$

In Exercises 5–16, simplify.

5. $(4x^3 - 3x^2 + x - 5) + (x^3 - 5x^2 + 3x - 2)$
6. $(3a^4 - 2a^2 + 1) - a(3a^3 - 2a^2 + 2a - 1)$
7. $(b - 4)(b - 3) - (b + 4)(b + 3)$
8. $(2x + 3)(3x^2 - 2x + 1)$
9. $y(y + 2) + y - y(3y - 2)$
10. $(3z - 1)(2z - 3)(3z + 1)$
11. $(c - d)(c^3 + 3c^2d + 3cd^2 + d^3)$
12. $(2a + 5)^2$
13. $(3s^2 - 4s^{-2})^2$
14. $(.2c^2 + .3d^{-2})^2$
15. $(2x - y)^2(2x + y)^2$
16. $(a + b)^3$

In Exercises 17–26, factor.

17. $8xy - 16xz$
18. $3m^2y^3 - 12m^5y^2$
19. $6a^2 + 11ab - 35b^2$
20. $28c^2 - 4c - 9$
21. $9s^4 + 24s^2t^3 + 16t^6$
22. $24x^3 - 50x^2y + 24xy^2$
23. $x^3 - 3x^2 - 2x + 6$
24. $2y^3 + 5y^2 - 8y - 20$
25. $s^6 - t^6$
26. $256 - a^8$

In Exercises 27–52, simplify.

27. $(2a^3b)^2(3ab^2)^3$
28. $(12a^5b)/(4a^{-2}b^3)$
29. $(4a^{-2}b^3)^{-2}/(a^3b^{-3})$
30. $(144c^{-6}d^4)^{1/2}/(.09c^4d^{-6})^{-1/2}$
31. $(8m^{-3}n^{3/5})^{1/3}/(27mn^{4/5})^{-1}$
32. $(8x^2y^3)^0$
33. $\{[(s^{3/4}t^{1/2})^4]^{-3}\}^{-1}$
34. $[(x^{5/3}/y^{2/3})^{-4/3}]^{9/5}$
35. $(x^{-1} + y^{-1})^{-1}$
36. $(a^{-2} + b^{-2})^{-1}$
37. $\sqrt[3]{125a^6b^{-3}} \; \sqrt[3]{.001a^{-9}b^{12}}$
38. $\sqrt[4]{16c^8d^{-8}} \; / \; \sqrt[4]{10{,}000c^{-1}d^2}$
39. $\sqrt{7 \sqrt[5]{7^3}}$
40. $\sqrt[5]{9^6 \sqrt[4]{9}}$
41. $\sqrt[3]{x^2y^4} \; \sqrt[4]{x^3y^2}$
42. $\sqrt{\sqrt{u^3v^{-2}} \; \sqrt[3]{u^{3/2}v^{-6}}}$
43. $(a + b)^4(a + b)^{-3}$
44. $\sqrt{c + d} \, (c + d)^{3/2}$
45. $\dfrac{6x^2 - 5x - 6}{9x^2 + 18x + 8}$
46. $\dfrac{6c^2 - 5cd - 6d^2}{c^2 - 4d^2} \cdot \dfrac{c + 2d}{4c^2 - 6cd}$
47. $\dfrac{4x^2 - 16x + 15}{2x^2 - 9x + 9} \div \dfrac{4x^2 - 20x + 25}{x^2 - 9}$
48. $\dfrac{a^3 - b^3}{a^2 - b^2}$
49. $\dfrac{z - z^{-2}}{1 - z^{-2}}$
50. $\left(1 - \dfrac{1}{x^2}\right) \bigg/ \left(\dfrac{1}{x} - 1\right)$
51. $\dfrac{3}{5x - 4} - \dfrac{10}{10x - 8}$
52. $\dfrac{3}{x - 3} + \dfrac{2}{x^2 - 9} - \dfrac{4}{x + 3}$

In Exercises 53–56, rationalize the numerator or denominator.

53. $\sqrt{\dfrac{2}{11}}$

54. $\dfrac{3}{\sqrt{2} - \sqrt{3}}$

55. $\dfrac{8}{\sqrt{x} - 2}$

56. $\dfrac{\sqrt{3+h} - \sqrt{3}}{h}$

Equations and Inequalities

CHAPTER TWO

Early mathematics was not divorced from the concrete problems for which it was used. This is reflected in some of our modern terminology. For example, we refer to $x \cdot x$ as "x squared" because early mathematicians thought of this only in terms of the area of a square. Similarly $x \cdot x \cdot x$ is "x cubed" because they thought of this in terms of the volume of a cube. The strictly concrete approach gave geometric insight into algebraic relationships, but it also inhibited the development of mathematics. For example, since lengths, areas, etc., are positive or zero, it was considered "obvious" that all numbers are either positive or zero. Early Greek mathematicians would have said that the equation $x + 2 = 0$ has no solution. Even though their algebra was sophisticated enough to solve quadratic equations, they could only find positive (or zero) roots, and their "formulas" were divided into several cases. They could not give the single easy formula which we present in Sec. 2.3. The stigma of negative numbers was to last even to the seventeenth century when Descartes would refer to *positive* and *negative* roots of polynomials as "true" and "false" roots.

The essence of all the mathematics in this chapter has been known for over 2000 years. However, modern notation and abstract methods make the mathematics much easier to understand and use and at the same time allow it to be applicable to a much broader variety of problems.

SECTION 2.1. BASIC ALGEBRAIC EQUATIONS

An **algebraic equation** is a statement that two algebraic expressions are equal. Letters in these equations are variables or **unknowns**. For example, the following are algebraic equations in x:

(1) $(x + 3)^2 = x^2 + 6x + 9$ **(2)** $(\sqrt{x})^2 = x$ **(3)** $x^2 + 1 = 0$

(4) $x^2 - 4 = 0$ **(5)** $\dfrac{x}{x+1} = 3$

An equation in x is called an **identity** if equality holds for every value of x for which all terms in the equation are defined. For example, equations 1 and 2 are identities. Equation 1 is true for all values of x; equation 2 is true whenever the left-hand side is defined, i.e., when $x \geq 0$.

An equation which is not an identity is called a **conditional equation**. In this case, those values for which equality holds are called **solutions**. To **solve** an equation means to find all its solutions. In this chapter, we shall be solving only for solutions which are real numbers. Equations 3, 4, and 5 are conditional equations.

When an equation is solved, there are various ways of indicating the answer. One way is just to list the solutions. Another is to put the solutions in a set called the **solution set**. A **set** is just a collection of objects and is usually described by listing its members in braces, $\{-3, 0, 1\}$, or in the form $\{x | x$ satisfies a certain condition$\}$. This is read "the set of all x such that x satisfies a certain condition." This may be abbreviated by $\{x$ satisfies a condition$\}$. Generally, when there are just a very few solutions, set notation is minimally used. When there are many solutions, for example, in solving inequalities, set notation is used more often.

EXAMPLE 1 Describe the solutions to $x^2 + 1 = 0$.

SOLUTION Since $x^2 \geq 0$ for all real numbers, $x^2 + 1 \geq 1$, so $x^2 + 1 = 0$ has no solution. We write "no solution" or "the solution set is the empty set" or simply "the empty set." The symbol $\emptyset$ denotes the empty set.

EXAMPLE 2 Describe the solutions to $x^2 - 4 = 0$.

SOLUTION There are two solutions, 2 and -2. We write "$x = 2, -2$" or "$x = \pm 2$."

A **linear equation** (in x) is an equation in which x appears only to the first power and only in the numerator of any fractions. It is solved by "getting all the x's on one side and everything else on the other."

EXAMPLE 3 Solve $2x - 3(4 - x) = 8x - 3$.

SOLUTION

$$2x - 3(4 - x) = 8x - 3$$
$$2x - 12 + 3x = 8x - 3$$
$$5x - 8x = -3 + 12$$
$$-3x = 9, \quad x = -3$$

EXAMPLE 4 Solve $ax + b = cx + d$, $a \neq c$, for x.

SOLUTION

$$ax + b = cx + d, \quad a \neq c$$
$$ax - cx = d - b$$
$$(a - c)x = d - b$$
$$x = \frac{d - b}{a - c}$$

This exists since $a - c \neq 0$.

EXAMPLE 5 Solve $21.31 + 41.29x = 17.81x - 8.17$.

SOLUTION

$$21.31 + 41.29x = 17.81x - 8.17$$
$$41.29x - 17.81x = -8.17 - 21.31$$
$$(41.29 - 17.81)x = -8.17 - 21.31$$
$$x = \frac{-8.17 - 21.31}{41.29 - 17.81}$$

Estimating first, we obtain
$$x \approx \frac{-8-21}{41-18} \approx \frac{-30}{20} = -1.5$$
Using a calculator, $x \approx -1.25554$.

Remember, in problems with numbers for which you would use your calculator, wherever possible save your computations for one big calculation at the end (and estimate the calculation first if reasonable).

A **quadratic equation** (in x) is an equation which can be put in the form
$$ax^2 + bx + c = 0, \quad a \neq 0$$
This is called the **standard form** of a quadratic equation. To solve a quadratic equation, we first get it into standard form and then try to factor it. (Quadratic equations that cannot be easily factored will be treated in Sec. 2.3.) If we can get an equation into factored form on one side and zero on the other, then we use the principle that if a product of real numbers is zero, then one of the factors is zero. Hence it is crucial that a *zero* be on one side of the equation.

EXAMPLE 6 Solve $3x^2 - 3x + 1 = 5x + 4$.

SOLUTION
$$3x^2 - 3x + 1 = 5x + 4$$
$$3x^2 - 8x - 3 = 0$$
$$(3x + 1)(x - 3) = 0$$

Since the only way a product of two real numbers is zero is if one of the numbers is zero, we must have either

$3x + 1 = 0$, i.e., $x = -\frac{1}{3}$ or $x - 3 = 0$, i.e., $x = 3$

Thus the solutions are $x = -\frac{1}{3}, 3$.

EXAMPLE 7 Solve $(x + 1)(x + 4) = 4$.

SOLUTION $(x + 1)(x + 4) = 4$. We must have a zero on one side.
$$x^2 + 5x + 4 = 4$$
$$x^2 + 5x = 0$$
$$x(x + 5) = 0$$
$$x = 0 \text{ or } x + 5 = 0$$
$$x = 0, -5$$

Often we have to solve equations with nontrivial denominators. When these arise, it is usually easiest to multiply both sides of the equation by the least common denominator (l.c.d.) and cancel the denominators. The resulting equation can sometimes be put into linear or quadratic form.

EXAMPLE 8 Solve

$$\frac{x}{2} + \frac{2}{3} = \frac{x}{4} - \frac{5}{6}$$

SOLUTION

$$\frac{x}{2} + \frac{2}{3} = \frac{x}{4} - \frac{5}{6}$$

The l.c.d. = 12.

$$\cancel{12}^{6}\frac{x}{\cancel{2}} + \cancel{12}^{4}\frac{2}{\cancel{3}} = \cancel{12}^{3}\frac{x}{\cancel{4}} - \cancel{12}^{2}\frac{5}{\cancel{6}}$$

$$6(x) + 4(2) = 3(x) - 2(5)$$

$$6x + 8 = 3x - 10$$

$$3x = -18, \quad x = -6$$

EXAMPLE 9 Solve

$$\frac{x}{x+3} - \frac{1}{x-2} = \frac{-5}{x^2 + x - 6}$$

SOLUTION Multiply both sides of the equation by the l.c.d. $= (x+3)(x-2)$:

$$\cancel{(x+3)}(x-2)\frac{x}{\cancel{x+3}} - (x+3)\cancel{(x-2)}\frac{1}{\cancel{x-2}} = \cancel{(x+3)(x-2)}\frac{-5}{\cancel{x^2+x-6}}$$

At this point we must be extremely careful, because we have multiplied by a term with an x in it. What this might do is introduce new solutions, called **extraneous solutions**, into the equation which were not solutions to the original equation. This might happen whenever the term we are multiplying by is zero. Hence we must see where this term is zero and compare this with the answer(s).

Set $(x + 3)(x - 2) = 0$; we get $x = -3, 2$. Hence if either -3 or 2 is an answer we get at the end, we must check this back in the original equation.

Back to the original problem.

$$(x - 2)x - (x + 3) \cdot 1 = -5$$
$$x^2 - 2x - x - 3 = -5$$
$$x^2 - 3x + 2 = 0$$
$$(x - 2)(x - 1) = 0, \quad x = 1, 2$$

However, the value $x = 2$ is one of the two values we said we must check in the original equation. Since 2 gives a 0 in a denominator, it is not a solution. Thus the only answer is $x = 1$.

When working with formulas, it is sometimes necessary to solve for one letter in terms of the others. The same principles still apply.

EXAMPLE 10 Solve

$$\frac{1}{R} = \frac{1}{a} + \frac{1}{b}$$

for a. Assume all letters represent positive real numbers.

SOLUTION The l.c.d. $= abR$.

$$abR\frac{1}{R} = abR\frac{1}{a} + abR\frac{1}{b}$$

(Since $a, b, R > 0$, their product is never zero, so we have not introduced new solutions.)

$$ab = bR + aR$$
$$ab - aR = bR$$
$$a(b - R) = bR$$
$$a = \frac{bR}{b - R}$$

EXAMPLE 11 Solve

$$\frac{1}{32.76} = \frac{1}{x} + \frac{1}{18.91}$$

SOLUTION Since the numbers are "messy," we save all calculations for one big computation at the end to be done on a calculator (after estimating first). We follow the steps in Example 10, multiplying both sides of the equation with the l.c.d. $= 32.76(18.91)x$.

$$18.91x = 32.76(18.91) + 32.76x$$
$$18.91x - 32.76x = 32.76(18.91)$$
$$x = \frac{32.76(18.91)}{18.91 - 32.76}$$

Estimating, we obtain

$$x \approx \frac{30(20)}{20 - 30} = \frac{600}{-10} = -60$$

Using a calculator, $x \approx -42.5183$.

EXERCISES

In Exercises 1–42, solve the equations.

1. $4x + 9 = 0$ **2.** $3.15x - 2.17 = 0$

3. $2z^2 + 8z = 0$
4. $w^2 = w + 2$
5. $3u - 4 = 7u - 2$
6. $2u - (2 - u) = 8$
7. $\frac{1}{2} + 5a = 3a + \frac{1}{4}$
8. $3 - \frac{1}{3}b = 7 + b$
9. $\frac{1}{3}x^2 = -\frac{1}{3}x + 4$
10. $\frac{1}{2}x^2 - 4 = \frac{7}{3}x$
11. $3(y - 4) - (y - 1) = 4$
12. $5s^2 - 4 = 5s - 3 - s^2$
13. $2t^2 - 6 = 5t - 4t^2$
14. $(2p - 1)(p - 3) = (3p - 1)(p - 2)$
15. $x - 3(x - \frac{1}{3}) = (x - 1)^2 - x^2$
16. $(x - 2)^2 - 4 = x(x - 4)$
17. $\frac{y}{2} - \frac{1}{3} = \frac{y}{3} - \frac{1}{2}$
18. $\frac{1}{2} + \frac{2}{a} = \frac{1}{3} + \frac{3}{a}$
19. $\frac{1}{b} + \frac{1}{2} = \frac{11}{3b} + \frac{1}{3}$
20. $\frac{2x}{x - 3} = 1 + \frac{6}{x - 3}$
21. $\frac{6x}{3x - 1} - 5 = \frac{2}{3x - 1}$
22. $\frac{y^2}{y - 3} = \frac{9}{y - 3}$
23. $\frac{y^2 - y}{y - 5} = \frac{25 - y}{y - 5}$
24. $\frac{1}{2s} - \frac{2}{5s} = \frac{1}{10s} - 1$
25. $\frac{1}{2s} - \frac{2}{3s} = \frac{1}{2} - \frac{1}{6s}$
26. $(2x - 1)^2 = (4x + 1)(x - 2)$
27. $(2x - 3)(2 - 3x) = 6 - 6x(x + 1)$
28. $(x + 1)^2 = (3x - 1)(x - 1)$
29. $\frac{1}{3} + \frac{2}{6u + 3} = \frac{3}{2u + 1}$
30. $\frac{1 + w}{1 - w} - \frac{1 - w}{1 + w} = \frac{1}{1 - w^2}$
31. $1 - \frac{3}{s} = \frac{40}{s^2}$
32. $2 - \frac{3}{s^2} = \frac{5}{s}$
33. $\frac{t^2 + 2}{2t^2 + t - 1} = \frac{3}{2t - 1} - \frac{1}{t + 1}$
34. $\frac{4t^2 + 5}{3t^2 + 5t - 2} = \frac{4}{3t - 1} - \frac{3}{t + 2}$
35. $4.12x^2 - .81x = 0$
36. $.0014x^2 + 11.9x = 0$
37. $\frac{231.2}{x} - 3.1 = \frac{41.7}{2x}$
38. $\frac{11.2}{x} - \frac{1}{4} = \frac{15.7}{x}$
39. $\frac{1}{.2131} = \frac{1}{x} + \frac{1}{.7218}$
40. $\frac{2}{476.1} + \frac{1}{x} = \frac{3}{919.7}$
41. $18.34x - \frac{14.21}{3} = \frac{33.41}{3} - 46.82x$
42. $\frac{423.1}{51.71} + \frac{53.19x}{21.72} = x$

In Exercises 43–54, solve the formula for the variable given.

43. $F = \frac{9}{5}C + 32$ for C
44. $V = \frac{1}{3}\pi r^2 h$ for h
45. $S = a + (n - 1)d$ for n
46. $\frac{1}{F} = \frac{1}{a} + \frac{1}{b}$ for F
47. $(m + 1)^2 x^2 + (m + 1)x = 2$ for x
48. $\frac{a}{z} + \frac{1}{b} = \frac{c}{z}$ for z
49. $P = 2l + 2w$ for w
50. $S = a + (n - 1)d$ for d
51. $F = mv_1 - mv_2$ for m
52. $S = \frac{a - rl}{1 - r}$ for r

53. $k^4y^2 = 3k^2y + 4$ for y

54. $\dfrac{m}{a-1} = \dfrac{n}{a+1}$ for a.

55. Two equations are called **equivalent** if they have the same solutions. When solving equations, we attempt at each step to replace the previous equation with one which is equivalent to it. When we multiply or divide by terms which might be zero (i.e., if they have variables in them), then the resulting equation may not be equivalent. For each of the following sequences of equations, find the equation not equivalent to the preceding equation. Then solve the initial equation.

(a)
$$x^2 - 2x - 3 = x^2 - 9$$
$$(x-3)(x+1) = (x-3)(x+3)$$
$$x + 1 = x + 3$$
$$1 = 3$$

(b)
$$x^2 - x - 2 = 2x^2 - 3x - 2$$
$$(x-2)(x+1) = (x-2)(2x+1)$$
$$x + 1 = 2x + 1$$
$$0 = x$$

SECTION 2.2. APPLICATIONS

In our complex society, many problems involving numbers arise naturally. To solve these problems, the tools of mathematics are used. It is seldom, however, that actual problems are phrased in terms of equations to be solved. Real problems are usually given orally or written in words (and hence are **word problems**.) To be solved, such problems must be translated into the mathematical language of equations and solved using mathematical tools, and the results must be interpreted back into the terms of the original problem.

There are two types of difficulties in this process: translating the problem from words to mathematics (and the answer back again) and solving the equations which arise. The first can be the more difficult.

GENERAL GUIDELINES FOR SOLVING APPLIED PROBLEMS

1. Skim through the problem once, getting a rough idea of what is going on but *mainly trying to determine what the problem is looking for.* Usually, the last sentence tells you what you are trying to find.
2. Choose letters to represent the unknown quantities. Write down *exactly* what the letters represent. Failure to do this can cause confusion.
3. Draw a picture, if appropriate, labeling the unknown items with the chosen variables, or make a table. This step cannot be overemphasized.
4. Determine what equations are appropriate by translating the relationships which are stated in words into mathematical symbols. Write these equations down.
5. Solve the equations formulated in step 4.
6. Check your mathematical answers, and interpret them back in terms of the original question.

The first type of problem we shall consider involves percent. In the language of percent, the word "is" translates as "equals," the word "of" translates as "multiplication," and the word "what" usually expresses the item we are seeking.

EXAMPLE 1   17% of what is 15.64?

SOLUTION   Translate: $17\% \cdot x = 15.64$

Solve: $.17x = 15.64, \quad x = \dfrac{15.64}{.17} = 92$

Check: $17\%(92) = .17(92) = 15.64.$

The next example expands on this technique.

EXAMPLE 2   Due to inflation, the price of a textbook increased 5%. The new price is $15.12. What was the original price?

SOLUTION   From the last sentence, we see the unknown is the original price. Hence, let x be the original price. We must read the first sentence and think:

The original price plus 5% of the original price is the new price.

Translate: $x + 5\% \cdot x = 15.12$

Solve:
$$x + .05x = 15.12$$
$$1.05x = 15.12$$
$$x = \dfrac{15.12}{1.05} = 14.40$$

The next example involves simple interest (compound interest is discussed in Chapter 6). Suppose a sum of money P (the principal) is invested at a simple interest rate of r per year. Then the interest I earned after t years is

> Interest = Principal × rate × time
> or
> $I = Prt$

EXAMPLE 3   Suppose a woman invested a sum of money at 6% simple interest. If at the end of 2 years she had $2576, how much did she start with?

SOLUTION   From the last sentence, we see that the unknown should be $P =$ her original principal. From the simple interest formula, the interest earned for the 2 years is $I = P(.06)(2) = .12P$. We must read the problem and think:

Principal plus interest is total.

Translate: $P + .12P = 2576$

56   EQUATIONS AND INEQUALITIES

Solve: $\qquad 1.12P = 2576, \qquad P = \dfrac{2576}{1.12} = 2300$

So she started with $2300.

As noted earlier, sometimes translating a problem from words to algebra can be a difficult task. A technique that sometimes makes the translation process easier is to introduce a second variable to help set up the algebraic equations and then to eliminate the extra variable as soon as the equations are found. The second variable is not really necessary and the problems can be done with just one variable. Two variables are used because it makes the translation process a little easier.*

EXAMPLE 4 The sum of two numbers is 19, and their product is 78. What are the numbers?

SOLUTION *Two variables.* From the last sentence, we see that we should let x be one number and y be the other number. From the first sentence,

the sum is 19;   the product is 78.

Translate: $\qquad x + y = 19, \qquad xy = 78.$

We solve the first equation for y and substitute it into the second equation:

$$y = 19 - x, \qquad x(19 - x) = 78$$

Solve:
$$19x - x^2 = 78$$
$$0 = x^2 - 19x + 78$$
$$0 = (x - 13)(x - 6), \qquad x = 6, 13$$

Solving for $y = 19 - x$, if $x = 6$, $y = 13$; if $x = 13$, $y = 6$. Thus the two numbers are 6 and 13.

One variable. From the last sentence, we see that we should let x be one of the two numbers. From the first sentence, the sum of two numbers is 19, so we conclude the other number is $19 - x$. From the phrase "their product is 78," we think:

One number times the other number is 78.

Translate: $\qquad x \cdot (19 - x) = 78.$

Solve: $\qquad 19x - x^2 = 78, \qquad 0 = x^2 - 19x + 78$

$$0 = (x - 6)(x - 13), \qquad x = 6, 13$$

As in the first solutions, if one number is 6, the other is $19 - 6 = 13$ and conversely, so the two numbers are 6 and 13.

*Because two variables are used, you may prefer to study Sec. 6.1 first.

Hereafter where there is a choice between one or two variables, we shall give the solution in terms of two variables. You are encouraged to try to do the same problem with one variable.

A few of the basic geometric facts you should know are the areas and circumferences of circles, triangles, squares, and rectangles.

	Rectangle	Square	Triangle	Circle
Area	lw	s^2	$\frac{1}{2}bh$	πr^2
Perimeter	$2l + 2w$	$4s$	$a + b + c$	$2\pi r$

EXAMPLE 5 A city builds a neighborhood park and puts in a circular pool which has a radius of 12 meters. The workers now want to build a concrete walk of uniform width around the pool. If they have enough concrete left to pour a total area of 52π square meters of concrete, how wide should they make the walk?

SOLUTION

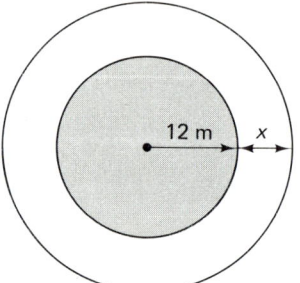

From the last sentence, we let x be the width of the walk. We draw a picture, labeling it with the information given.

From the last sentence, the equation is

$$\text{The area of the strip} = 52\pi.$$

By the geometry of the situation,

$$\text{Area of the strip} = [\text{area of big circle}] - [\text{area of small circle}]$$

We know the area of a circle is πr^2, where r is the radius. The radius of the small circle is 12, while that of the larger circle is $12 + x$. Hence the equation is

$$\pi(12 + x)^2 - \pi 12^2 = 52\pi$$

Solve (divide by π first):

$$(12 + x)^2 - 12^2 = 52$$
$$144 + 24x + x^2 - 144 = 52$$
$$x^2 + 24x - 52 = 0$$
$$(x + 26)(x - 2) = 0, \qquad x = 2, -26$$

58 EQUATIONS AND INEQUALITIES

Since -26 meters is not a solution to the original problem, we see that the workers must make the walk 2 meters wide.

Many problems are about things moving at constant speeds (airplanes, sound, boats, etc.). If something is moving at a constant rate R, then the distance D traveled in a time T is

> **Rate times Time equals Distance**
>
> or
>
> $RT = D$

Of course, the units must agree; e.g., if R is in miles per hour, T must be in hours (not minutes) and D in miles.

When several things are going on in a motion problem, there are two things which help to organize the information so that you can write down the equations. The first is to *draw a picture* and label it with the appropriate information. The second is to *make a table,* as is done in the following example.

EXAMPLE 6 Suppose a motorboat goes 15 miles downstream at its top speed and then turns around and returns the 15 miles upstream at its top speed. The trip upstream took twice as long as the trip downstream. If the boat's top speed is 10 miles per hour in still water, what is the rate of the stream?

SOLUTION We let r be the rate of the stream and draw a "picture":

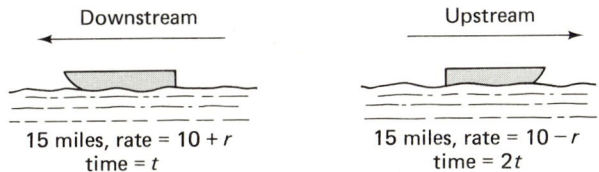

The rate of the boat downstream is $10 + r$, and its rate upstream is $10 - r$. Since we do not know how long the trip took either way, we introduce a second variable t, representing the time it took to go downstream. From the second sentence, the trip upstream took $2t$.

From the information we now have, we fill in the following table:

	R	T	D
Downstream	$10 + r$	t	15
Upstream	$10 - r$	$2t$	15

Using rate $\times$ time = distance, we obtain two equations:

$$(10 + r)t = 15 \quad \text{and} \quad (10 - r)2t = 15$$

Since we want *r* (and not *t*), we solve the first equation for *t*,

$$t = \frac{15}{10 + r}$$

substitute that in the second,

$$(10 - r)2\frac{15}{10 + r} = 15$$

divide by 15 (always simplify whenever possible),

$$(10 - r)2\frac{1}{10 + r} = 1$$

multiply by $10 + r$,

$$(10 - r)2 = 10 + r$$

and now solve for *r*:

$$20 - 2r = 10 + r$$
$$10 = 3r$$
$$r = \frac{10}{3} = 3\frac{1}{3}$$

Thus the stream is flowing at $3\frac{1}{3}$ miles per hour.

Warning: Although using tables is very helpful when you are first learning to do "rate times time equals distance" problems, it can be cumbersome and slow you down as you become better. After you begin to get the idea of how to set up these problems, you are strongly encouraged just to draw the pictures and to try to set up the equations directly from the pictures.

Another common type of problem is where the same job can be done at differing rates by different things (different people, machines, etc.) and you want to know how quickly it can be done if they work together. The easiest way to solve this type of problem is as follows:

> Set up an equation stating that the sum of the work done by each separately in one unit of time is equal to the work they do together in one unit of time.

We use the relationship that if a whole job can be done in time *t*, then $(1/t)$th of the job can be done in one unit of time. For example, if a woman can do a job in 4 hours, then she can do one-fourth of it in 1 hour.

EXAMPLE 7 Suppose Jane and John are construction engineers. Their boss, who has a job to be done, knows Jane can do it in 3 hours and John can do it in 4 hours. How quickly will it get done if they are assigned to do it together?

Warning: If you first take a guess, you might guess somewhere in between, like $3\frac{1}{2}$ hours. Sometimes intuition can be misleading. With the two of them working, it must be done quicker than the faster of the two; the answer must be less than 3 hours.

SOLUTION Let t be the time (in hours) it takes them to do it together. Then together they can do $1/t$ of it in 1 hour. In 1 hour, Jane can do one-third of it and John can do one-fourth of it. So together they can do $\frac{1}{3} + \frac{1}{4}$ of the job in 1 hour. Therefore, the equation is

$$\frac{1}{3} + \frac{1}{4} = \frac{1}{t}$$

Solve:

$$\frac{1}{3}(12t) + \frac{1}{4}(12t) = \frac{1}{t}(12t)$$

$$4t + 3t = 12,$$

$$7t = 12, \quad t = \tfrac{12}{7} = 1\tfrac{5}{7} \approx 1.71429 \text{ hr}$$

Thus together they can do it in just under $1\frac{3}{4}$ hours. (*Note*: Since we multiplied by $12t$, we would have had to check an answer of zero.)

EXERCISES

1. 22 is what percent of 48?
2. What percent of 54 is 10?
3. What percent of 1500 is 2?
4. .01 is what percent of 20?
5. The sum of two numbers is 24. If one number is three times the other, what are the numbers?
6. The sum of two numbers is 20, and their product is 96. Find the numbers.
7. The sum of Kim's and Kevin's ages is 18. In 3 years, Kim will be twice as old as Kevin. What are their ages now?
8. Four years ago, Bill was twice as old as Ruth. Now he is 4 years older than Ruth. What are their ages?
9. A man gets a 5% salary raise. His new salary is $14,910 per year. What was his original salary?
10. A woman gets a 6% salary raise, which is $1098 per year. What was her salary before the raise?
11. All dresses in a certain store are on sale at 20% off. One particular dress now costs $34.36. What was its original price?
12. Because of an imperfection, a new coat was reduced 15%. It now costs $42.33. What was it originally?
13. After a 3% increase, the population of a city is 156,560. What was its original population?
14. A woman has invested $5000 at 7% annual simple interest. What additional amount should she invest at 9% to realize an average rate of $7\frac{1}{4}$% interest?
15. A man bought stock in two companies for a total investment of $44,000. When he sold them, he made 12% on the first company and lost 10% on the second company for a total profit of $2200. How much did he invest in each company?

16. A woman has an amount invested at $6\frac{1}{2}\%$ annual simple interest and $3000 more than the first amount invested at 8%. If together she earns $1545 in a year, how much is invested at each rate?

17. A college student has loans from two different accounts totaling $9000. Five years after the student graduates, the first account starts charging $3\frac{1}{2}\%$ and the second $4\frac{1}{2}\%$ simple annual interest. If nothing is paid back by then, the interest per year will be $330. How much is each loan?

18. A student has test scores of 75, 83, and 78. What must be the score on the next test so that the average will be 80?

19. A student has test scores of 75, 83, and 78. The final test counts half the total grade. What must be the minimum (integer) score on the final so that the average will be 80?

20. In Exercise 19, if the final counted only one-third of the total, what must be the minimum (integer) score on the final then?

21. In a triangle ABC, the length of side AB is twice the length of AC and 3 less than the length of BC. If the total perimeter is 33, what is the length of each side?

22. The length of a rectangle is 15 centimeters more than the width. What are the dimensions if the perimeter is 398 centimeters?

23. If the radius of a circle were increased by 2 inches, the area would be increased by 32π square inches. What is the radius?

24. The length of a rectangle is three times its width. If each were increased by 3 centimeters, the area would be increased by 105 square centimeters. What are the dimensions?

25. The outside dimensions of a framed picture are 12 by 16 inches. If the frame is of uniform width and the area of the actual picture is half of the total area, how wide is the frame?

26. A triangle has base 8 feet and height 4 feet. How much do you have to increase each dimension equally so that the area is tripled?

27. A car is driven 2 miles at 40 miles per hour and then 3 miles at 50 miles per hour. What is its average speed?

28. In Exercise 27, if it was 2 miles at each speed, what is the average speed then? (*Note*: It is *not* 45 miles per hour, since it takes longer to go the 2 miles at 40 miles per hour.)

29. A boat goes 62 kilometers downstream in the same time it takes to go 38 kilometers upstream. If the speed of the stream is 4 kilometers per hour, what is the speed of the boat in still water?

30. The cruising speed of an airplane is 420 miles per hour. Find the speed of the wind if against the wind the plane can cover 10/11 of the distance that it can cover with the wind.

31. The speed of a boat in still water is 8 miles per hour. It goes 30 miles upstream and then comes the 30 miles back in a total of 8 hours. What is the speed of the stream?

32. A boat goes downstream in 1 hour and returns in $1\frac{1}{2}$ hours. If the stream is flowing at 3 miles per hour, what is the speed of the boat in still water?

33. A motorist drove 144 miles. If he had gone 6 miles per hour faster, the trip would have taken 20 fewer minutes. What was his speed?

34. A man rides his bike 5 miles to work at a steady rate. If he were to pedal 2 miles per hour faster, he would make the trip in 5 fewer minutes. What is his speed?

35. A jogger starts a course at the steady rate of 8 kilometers per hour. Five minutes later a second jogger starts the same course at 10 kilometers per hour. How long will it take the second jogger to catch the first?

36. Two cars, 155 miles apart, start at the same time heading toward each other. How long will it be before they pass if one is going at 50 and the other at 55 miles per hour?

37. It takes 2 hours for John to paint a wall and $2\frac{1}{2}$ hours for Jim. How long will it take for the two of them working together?

38. One pipe can fill a pool in 25 minutes, but a second pipe can fill it in 20 minutes. How long will it take if both pipes are used?

39. A boy's older sister can mow their family's yard 30 minutes quicker than he can. If working together using two lawnmowers they can do the job in 36 minutes, how long does it take for each to do it separately?

40. Two secretaries working together can type a certain large report in 2.4 hours. If either secretary had to type the report all alone, it would take the first secretary 2 hours longer than the second secretary. How long would it take each secretary?

41. One secretary can type a certain large report in 6 hours. A second secretary can type that report in 8 hours. How long will it take if they work together?

42. Julie can paint an apartment in 10 hours and Jackie can paint it in 12 hours. How long would it take them working together to paint the apartment?

43. Working together, Alice and Dick can paint their apartment in $7\frac{1}{2}$ hours. If either had to paint alone, it would take Dick 8 hours longer to paint the apartment than Alice. How long would it take either person working alone?

44. A squirrel and a half eats a nut and a half in a day and a half. How many nuts do six squirrels eat in 6 days?

45. A woman drives home at 50 miles per hour and is 2 minutes early for dinner. If she drives home at 45 miles per hour, she is 2 minutes late. How far does she drive?

SECTION 2.3. QUADRATIC EQUATIONS AND APPLICATIONS

A **quadratic equation** is an equation that can be written in the form

$$ax^2 + bx + c = 0$$

where $a, b,$ and c are real numbers, $a \neq 0$. This is called the standard form of a quadratic equation. We discussed solving these equations in Sec. 2.1 by factoring. There are several ways of solving quadratic equations, but if it factors readily, that way is the *easiest* way. Thus

> When solving quadratic equations, always try factoring first.

We now develop a method for solving *any* quadratic equation. We first consider the equation $x^2 = d, d \geq 0$. This equation can be rewritten as

$$x^2 - d = 0, \quad d \geq 0$$

and then factored:

$$(x - \sqrt{d})(x + \sqrt{d}) = 0$$

Setting each factor equal to zero and solving, we obtain the solutions $x = \sqrt{d}$ and $x = -\sqrt{d}$. This is frequently abbreviated as $x = \pm\sqrt{d}$. Thus we have shown the following:

> **The solutions to $x^2 = d$, $d \geq 0$, are $x = \pm\sqrt{d}$.**

This way of solving $x^2 = d$ (by just writing $x = \pm\sqrt{d}$ as the next step) is more easily thought of as "taking the square root of both sides." However, do not forget to write the $\pm$.

If $d = 0$, then technically the equation $x^2 = 0$ has two solutions; they are both the same number, 0. In this case, the number 0 is called a **double root** or a **root of multiplicity two**. In general, the equation

$$x^2 = d \text{ has } \begin{cases} \text{two unequal real roots} & \text{if } d > 0 \\ \text{a double root} & \text{if } d = 0 \\ \text{no real solutions} & \text{if } d < 0 \end{cases}$$

We shall now use our knowledge of how to solve quadratic equations of the special form $x^2 = d$, $d \geq 0$, to derive a general formula which will allow us to solve any quadratic equation which has real solutions. We shall proceed by trying to get a square involving the variable on the left-hand side of an equation and a constant on the right. This technique is called **completing the square**.

Suppose that we have a general quadratic equation in the form

$$ax^2 + bx + c = 0, \quad a \neq 0$$

Divide by a, $x^2 + (b/a)x + c/a = 0$, and subtract c/a from both sides:

$$x^2 + \frac{b}{a}x = -\frac{c}{a}$$

We now want to add a constant term to both sides of the equation so that the left-hand side of the equation is a perfect square, i.e., of the form $(x + d)^2$ (in doing so, we shall have "completed the square"). Since

$$(x + d)^2 = x^2 + 2dx + d^2$$

we must have $2d = b/a$, or $d = b/2a$. Thus the term we have to add is $d^2 = b^2/4a^2$. We obtain

$$x^2 + \frac{b}{a}x + \frac{b^2}{4a^2} = \frac{b^2}{4a^2} - \frac{c}{a}$$

Rewriting the left-hand side as a square and putting the right-hand side over a common denominator, we obtain

$$\left(x + \frac{b}{2a}\right)^2 = \frac{b^2 - 4ac}{4a^2}$$

Then (by the preceding special case),

$$x + \frac{b}{2a} = \pm\sqrt{\frac{b^2 - 4ac}{4a^2}}$$

or
$$x + \frac{b}{2a} = \pm \frac{\sqrt{b^2 - 4ac}}{2a}$$

or
$$x = -\frac{b}{2a} \pm \frac{\sqrt{b^2 - 4ac}}{2a}$$

or
$$x = \frac{-b \pm \sqrt{b^2 - 4ac}}{2a}$$

The last formula gives us all the real roots of any quadratic equation which is written in standard form. The formula, called the **quadratic formula**, is very important.

---

(1)   **THE QUADRATIC FORMULA**

The roots of the equation $ax^2 + bx + c = 0$, $a \neq 0$, are given by
$$x = \frac{-b \pm \sqrt{b^2 - 4ac}}{2a}$$

The technique of completing the square may be used directly to solve quadratic equations, but it is easier to solve them by factoring or by using the quadratic formula. However, we shall use completing the square later when we are graphing circles and quadratic functions.

EXAMPLE 1   Use the quadratic formula to solve $2x^2 + x = 3$.

SOLUTION   We rewrite the equation in standard form, $ax^2 + bx + c = 0$:
$$2x^2 + x - 3 = 0$$
We see $a = 2$, $b = 1$, $c = -3$. Write the formula and substitute:
$$x = \frac{-b \pm \sqrt{b^2 - 4ac}}{2a}$$
$$= \frac{-1 \pm \sqrt{1^2 - 4(2)(-3)}}{2(2)}$$
$$= \frac{-1 \pm \sqrt{25}}{4} = \frac{-1 \pm 5}{4}$$

Hence,
$$x = \frac{-1 + 5}{4} = \frac{4}{4} = 1 \quad \text{or} \quad x = \frac{-1 - 5}{4} = \frac{-6}{4} = -\frac{3}{2}$$

Of course, we could have solved this by factoring much more easily:
$$2x^2 + x - 3 = 0, \quad (2x + 3)(x - 1) = 0, \quad x = 1, -\frac{3}{2}.$$

This illustrates that you should *always try to solve a quadratic equation by factoring first*. If you can factor readily, that way is much easier than using the quadratic formula.

EXAMPLE 2 Use the quadratic formula to solve $19.1x^2 - 231.2x + 5.012 = 0$.

SOLUTION This equation is already in standard form. We see that $a = 19.1$, $b = -231.2$, $c = 5.012$. Then

$$x = \frac{-(-231.2) \pm \sqrt{(-231.2)^2 - 4(19.1)(5.012)}}{2(19.1)}$$

$$= \frac{231.2 \pm \sqrt{231.2^2 - 4(19.1)(5.012)}}{2(19.1)}$$

Estimate first:

$$x \approx \frac{200 \pm \sqrt{200^2 - 4(20)(5)}}{2(20)} = \frac{200 \pm \sqrt{40,000 - 400}}{40}$$

Thus,

$$x \approx \frac{200 + \sqrt{40,000}}{40} = \frac{200 + 200}{40} = 10$$

$$x \approx \frac{200 - \sqrt{39600}}{40} \approx \frac{200 - 198}{40} = \frac{2}{40} \approx .05$$

Note that you have to be a little careful with estimations when the answers are close to zero. Possibly $x \approx 0$ would have been a good enough estimation for the second answer.

To compute the answer on your calculator, compute $\sqrt{b^2 - 4ac}$ first (by $-4ac$ and then $+b^2$ and finally taking the square root) and then store until needed. For example, on an algebraic calculator, press

$$[5.012][\times][19.1][\times][-4][+][231.2][x^2][=][\sqrt{\ }][M+].$$

The memory now contains $\sqrt{b^2 - 4ac}$. Finish by pressing:

[231.2][+][MR][=][÷][2][÷][19.1][=]   (first answer)
[231.2][−][MR][=][÷][2][÷][19.1][=]   (second answer)

The two answers are $x \approx 12.0830, .0217172$.

EXAMPLE 3 Use the quadratic formula to solve $s = -16t^2 + v_0 t$ for t.

SOLUTION We rewrite it in standard form, $at^2 + bt + c = 0$:

$$16t^2 - v_0 t + s = 0$$

66 EQUATIONS AND INEQUALITIES

Then $a = 16$, $b = -v_0$, $c = s$, so

$$t = \frac{-b \pm \sqrt{b^2 - 4ac}}{2a}$$

$$= \frac{-(-v_0) \pm \sqrt{(-v_0)^2 - 4(16)(s)}}{2(16)}$$

$$= \frac{v_0 \pm \sqrt{v_0^2 - 64s}}{32}.$$

This is all we can do until we know more about v_0 and s.

EXAMPLE 4 Use the quadratic formula to solve $x^2 + x + 1 = 0$.

SOLUTION It is already in standard form. We see that $a = b = c = 1$, so

$$x = \frac{-1 \pm \sqrt{1^2 - 4(1)(1)}}{2} = \frac{-1 \pm \sqrt{-3}}{2}$$

Since $\sqrt{-3}$ is not a real number, this equation has no real solutions.

The last example illustrates the fact that if $ax^2 + bx + c = 0$ has real solutions, then $b^2 - 4ac$ must be ≥ 0. Just like the case $x^2 = d$, there are three possibilities for the nature of the solutions to $ax^2 + bx + c = 0$. The number $b^2 - 4ac$ is called the **discriminant**, because it discriminates among the possibilities.

(2)

If $b^2 - 4ac > 0$, the equation has two distinct roots.
If $b^2 - 4ac = 0$, the equation has a double root, $-b/2a$.
If $b^2 - 4ac < 0$, the equation has no real solutions.

EXAMPLE 5 Determine the nature of the solutions but do not solve:
(a) $3x^2 + 2\sqrt{3}\, x - 1 = 0$; (b) $3x^2 + 2\sqrt{3}\, x + 2 = 0$; (c) $3x^2 + 2\sqrt{3}\, x + 1 = 0$.

SOLUTION (a) Here $a = 3$, $b = 2\sqrt{3}$, $c = -1$, so $b^2 - 4ac = (2\sqrt{3})^2 - 4(3)(-1) = 12 + 12 > 0$. There are two real solutions.
(b) Here $a = 3$, $b = 2\sqrt{3}$, $c = 2$, so $b^2 - 4ac = (2\sqrt{3})^2 - 4(3)(2) = 12 - 24 < 0$. There are no real solutions.
(c) Here $a = 3$, $b = 2\sqrt{3}$, $c = 1$, so $b^2 - 4ac = (2\sqrt{3})^2 - 4(3)(2) = 12 - 12 = 0$. There is a double root. [Actually, $3x^2 + 2\sqrt{3}\, x + 1 = (\sqrt{3}\, x + 1)^2$, so the double root is $-1/\sqrt{3} = -b/2a$.]

EXAMPLE 6 For what value of k does the equation $2x^2 + kx + 1 = 0$ have a double root?

SOLUTION We need $b^2 - 4ac = 0$. Here $a = 2$, $b = k$, $c = 1$. Thus, $k^2 - 4(2)(1) = 0$, $k^2 = 8$, $k = \pm \sqrt{8} = \pm 2\sqrt{2}$.

Given two numbers m and n, you can readily obtain a quadratic equation with m and n as roots. You form the equation $(x - m)(x - n) = 0$ and then multiply the two factors together.

EXAMPLE 7 Find a quadratic equation with **(a)** $1, -2/3$ as roots; **(b)** $-\frac{7}{2}$ as a double root.

SOLUTION **(a)** The equation $(x - 1)[x - (-\frac{2}{3})] = 0$ satisfies the condition, so the answer is $(x - 1)(x + \frac{2}{3}) = 0$ or $x^2 - \frac{1}{3}x - \frac{2}{3} = 0$. An equivalent equation with integer coefficients is $3x^2 - x - 2 = 0$. **(b)** The equation $[x - (-\frac{7}{2})][x - (-\frac{7}{2})] = 0$ satisfies the condition, so $(x + \frac{7}{2})^2 = 0$ or $4x^2 + 28x + 49 = 0$ is the answer.

Applications

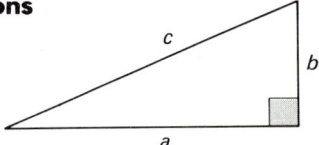

One of the ways in which quadratic equations arise is when the relationship between quantities is expressed in terms of right triangles. Then by the Pythagorean theorem, if a and b are the lengths of the legs and c is the length of the hypotenuse, then $a^2 + b^2 = c^2$.

EXAMPLE 8 A ladder 13 feet long leans against a (vertical) wall. The bottom of the ladder is 5 feet from the wall. If the bottom of the ladder is pulled 1 foot farther from the wall, how much does the top of the ladder slide down the wall?

SOLUTION

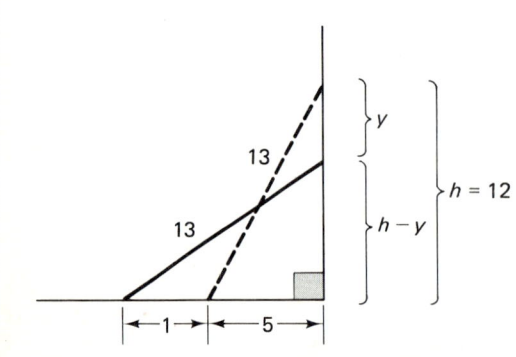

We let y be the distance that the ladder drops, and we draw a picture, as shown. We are not given the original height h where the top of the ladder leans against the wall. The original placement of the ladder, the ground, and the wall form a right triangle.

By the Pythagorean theorem,
$$h^2 + 5^2 = 13^2, \qquad h = \pm\sqrt{13^2 - 5^2} = \pm 12$$
So $h = 12$. Since the new position of the ladder, the wall, and the ground form another right triangle,
$$6^2 + (12 - y)^2 = 13^2, \qquad 36 + 144 - 24y + y^2 = 169, \qquad y^2 - 24y + 11 = 0$$
Thus,
$$y = \frac{-(-24) \pm \sqrt{(-24)^2 - 4(1)(11)}}{2(1)}$$
$$= \frac{24 \pm \sqrt{24^2 - 44}}{2}$$
Estimating, we see that
$$y \approx \frac{24 + \sqrt{24^2}}{2} = 24, \qquad y \approx \frac{24 - 23}{2} \approx \frac{1}{2}$$
(The 23 for the $\sqrt{24^2 - 44}$ is just a very rough guess, an integer slightly smaller than $\sqrt{24^2}$.) The first answer is clearly too big, so the second answer is the one we want. Using a calculator, we obtain $y \approx .467437$ foot.

When an object is dropped or thrown upwards or downwards (near the surface of the earth), the object's height s after t seconds is given by
$$s = -4.9t^2 + v_0 t + s_0 \qquad \text{if } s \text{ is measured in meters}$$
$$s = -16t^2 + v_0 t + s_0 \qquad \text{if } s \text{ is measured in feet}$$
Here v_0 is the initial velocity (measured in meters per second or feet per second as appropriate), and s_0 is its initial height above the surface of the earth (measured in meters or feet). Of course, v_0 is positive if the object is thrown upwards and negative if thrown downwards, and the formula makes sense only until the object hits the ground (or something else).

EXAMPLE 9 Suppose you are standing on a cliff 200 feet high. If you have a rock in your hand, how long will it take to reach the ground if **(a)** you drop it; **(b)** you throw it downwards at 40 feet per second?

SOLUTION In both cases, the initial height s_0 is 200, and you wish to find the time $t > 0$ when the height $s = 0$. **(a)** Since you drop it, the initial velocity v_0 is 0. The equation $s = -16t^2 + v_0 t + s_0$ becomes $0 = -16t^2 + 200$. Solving,
$$16t^2 = 200, \qquad 2t^2 = 25, \qquad t^2 = \frac{25}{2}, \qquad t = \pm\sqrt{\frac{25}{2}} = \pm\frac{5}{\sqrt{2}}$$
Disregarding the negative answer and estimating, we obtain $t \approx 3$ seconds. Using a calculator, we obtain $t \approx 3.53553$ seconds. **(b)** Since you throw it

downwards at 40 feet per second, $v_0 = -40$. From $s = -16t^2 + v_0 t + s_0$,
$$0 = -16t^2 - 40t + 200$$
$$0 = 2t^2 + 5t - 25,$$
$$0 = (2t - 5)(t + 5), \quad t = -5, 5/2$$

Disregarding the negative answer, we obtain $t = 5/2$ seconds.

EXERCISES

In Exercises 1–4, first solve by the quadratic formula and then by factoring.

1. $2x^2 + 3x - 2 = 0$
2. $2x^2 + 3x = 0$
3. $2x^2 - 18 = 0$
4. $4x^2 - 10x - 6 = 0$

In Exercises 5–10, find a quadratic equation whose solutions are the given numbers.

5. $-\frac{9}{2}, 10$
6. $-\frac{2}{3}, \frac{2}{3}$
7. 18.1, double root
8. $\frac{5}{3}, \frac{7}{3}$
9. $\sqrt{2}, -2\sqrt{2}$
10. $2 + \sqrt{3}, 2 - \sqrt{3}$

In Exercises 11–22, solve by the easiest possible method. Approximate any calculator computations first. If you use the quadratic formula and get rational roots, work backward to see how to factor.

11. $2x^2 - x - 3 = 0$
12. $2x^2 - x - 6 = 0$
13. $3x^2 - 27 = 0$
14. $2(x^2 - 20) = 10 - x^2$
15. $4y - 1 = 2/y$
16. $\dfrac{1}{3} = \dfrac{2y - 1}{y^2}$
17. $15.1z^2 = 23.2 - 4.1z$
18. $.1z^2 = .01z + .001$
19. $(2x - 3)^2 = 15 - 17x$
20. $3x^2 - (x - 1)(x + 1) = 4$
21. $2 = \dfrac{31}{x} - \dfrac{92}{x^2}$
22. $\dfrac{5 - 2x^2}{x} = \dfrac{43}{2}$

In Exercises 23–28, solve for the indicated variable. All letters denote positive real numbers.

23. $V = \frac{1}{3}\pi r^2 h$ for r
24. $F = \dfrac{kM_1 M_2}{d^2}$ for d
25. $A = \pi r^2 + 2\pi r h$ for r
26. $Ls^2 + Rs = c$ for s
27. $x^2 - 3x + kx + 1 = kx^2$ for x
28. $A = 2\pi r(r + h)$ for r

In Exercises 29–32, determine the nature of the solutions, but do not solve.

29. $2x^2 - 20x + 51 = 0$
30. $3x^2 - 4x + \sqrt{3} = 0$
31. $5x^2 + 7x + \sqrt{2} = 0$
32. $4x^2 + 4\sqrt{3}\,x + 3 = 0$

33. One root of the equation $kx^2 + 2x + 6 = 0$ is 1. Find k and the other root.
34. One root of the equation $kx^2 + 3x - k = 0$ is -2. Find k and the other root.
35. Prove that each solution of $ax^2 + bx + c = 0$ is a reciprocal of a solution of $cx^2 + bx + a = 0$, where $a \neq 0$, $c \neq 0$.
36. Prove that each solution of $ax^2 + bx + c = 0$ is the negative of a solution of $ax^2 - bx + c = 0$, where $a \neq 0$, $c \neq 0$.

37. A 10-foot-long ladder leans against the wall. The bottom of the ladder is 6 feet from the wall. If the bottom is pulled out 1 foot, how far down the wall does the ladder slide?

38. A 10-foot-long ladder leans against the wall; its top is 8 feet up the wall.
   a. How far do you have to pull the bottom out to get the top to slide down 1 foot?
   b. How far do you have to push the bottom in to get the top to slide up 1 foot?

39. A 5-meter-long ladder leans against the wall; its bottom is 3 meters from the wall. How much would the lower end have to be pulled away from the wall so that the top of the ladder would slide down the same amount?

40. A 5-meter extension ladder leans against the wall; the bottom is 3 meters from the wall. If the bottom stays at the same place, how much should the ladder be extended so that the top would lean against the wall 1 meter higher?

41. The area of a triangle is 14 square inches; its height is 3 inches less than its base. Find the base and height.

42. The area of a rectangle is 28 square centimeters; its length is twice its width. Find the dimensions.

43. A target with a black circular center and a white ring of uniform width around it is to be made. If the radius of the center is to be 3 inches, how wide should the ring be so that the area of the ring is the same as the area of the center?

44. A target is to be made similar to the one in Exercise 43. If the width of the strip is to be 3 inches, what should the radius of the center be so that the two areas are the same?

45. A baseball diamond is a square 90 feet on a side. How far must a catcher throw from home plate to second base?

46. The diagonal of a square is 2 centimeters longer than a side. How long is a side?

47. Trains A and B leave the same city at the same time traveling at right angles to each other. Train B travels 10 kilometers per hour faster than train A. After 2 hours, they are 100 kilometers apart. Find the speed of each train.

48. Two airplanes, flying at right angles, leave the same city at the same time. If the second airplane is twice as fast as the first and after 3 hours they are 750 miles apart, find the speed of each airplane.

49. Two cyclists are at a crossroads. The first leaves heading south. A half an hour later, the second leaves heading west 7 miles per hour faster than the first. One and a half hours after that, they are 30 miles apart. Find the speed of each cyclist.

50. Two trains leave the same city at the same time traveling at right angles to each other at the same speed. After $\frac{1}{2}$ hour the first train stops; 42 minutes later the second train stops. How fast were they traveling if they are now 39 miles apart?

51. A box with an open top is to be made by taking a rectangular piece of tin 8 by 10 inches and cutting a square of the same size out of each corner and folding up the sides. If the area of the base is to be 24 square inches, what should the length of the sides of the squares be?

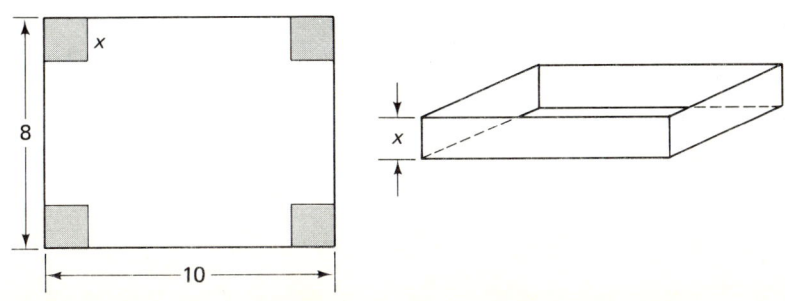

52. A box with an open top is to be made from a square piece of tin by cutting 4-centimeter squares from each corner. If the box is to hold 576 cubic centimeters, find the size of the piece of tin to be used.

53. The Gateway Arch in St. Louis is 195 meters high. How long will it take a baseball to reach the ground if, from the top of the arch, it is
   a. Dropped?
   b. Thrown downward at 10 meters per second?
   c. Thrown upward at 10 meters per second?

54. a. A mortar, on a cliff, 192 feet high, fires a shell with an upwards velocity of 160 feet per second. How long will it be before the shell hits the ground?
   b. Another mortar on the ground fires a shell with an upwards velocity of 120 feet per second in such a way that the shell goes up to its highest point and then hits the cliff on the way back down. How long does it take to hit the cliff?

55. A ball is thrown upward at 112 feet per second.
   a. How high is it after 2 seconds?
   b. How long does it take for the ball to reach (1) 96 feet on its way up? (2) 96 feet on its way down? (3) the ground again?
   c. For what heights s can you solve $s = -16t^2 + 112t$ for real numbers t? What does this say about the highest point the ball will reach?

56. It takes John an hour longer to mow the yard than Jim. Together, with two lawnmowers, they do the job in 1 hour and 12 minutes. How fast can they each do it separately?

The final exercises indicate one type of numerical difficulty that can arise when using calculators or computers and one strategy for dealing with that problem.

57. By rationalizing the numerator, show that

$$\frac{-b + \sqrt{b^2 - 4ac}}{2a} = \frac{2c}{-b - \sqrt{b^2 - 4ac}} \quad \text{and} \quad \frac{-b - \sqrt{b^2 - 4ac}}{2a} = \frac{2c}{-b + \sqrt{b^2 - 4ac}}$$

In Exercises 58–61 solve to six significant figures. You may need to use the alternate forms of the quadratic formula given in Exercise 57.

58. $x^2 + 10^{10}x + 1 = 0$ **59.** $x^2 - 10^{10}x + 1 = 0$
60. $10^{-5}x^2 - x - 10^{-8} = 0$ **61.** $10^{-5}x^2 + x - 10^{-8} = 0$

SECTION 2.4. EQUATIONS WITH RADICALS AND EQUATIONS QUADRATIC IN FORM

Equations with radicals or rational exponents can sometimes be relatively easily solved by factoring or raising both sides of the equation to an appropriate power. Care must be taken, however, so that roots are not eliminated and to check all answers back in the original equation to see if any extraneous solutions were introduced during the solving process. (Recall, extraneous solutions were discussed in Sec. 2.1.)

EXAMPLE 1 Solve $x = x^{1/2}$.

SOLUTION Square both sides, getting $x^2 = x$. Solving this, we obtain $x^2 - x = 0$, $x(x - 1) = 0$, $x = 0, 1$. Both answers check.

Note: A common error is to divide both sides of the original equation by $x^{1/2}$, getting $x^{1/2} = 1$ or $x = 1$. This loses an answer, because the equation is divided by a quantity which could be zero.

EXAMPLE 2 Solve $\sqrt{x - 1} + 7 = x$.

SOLUTION We first isolate the radical

$$\sqrt{x - 1} = x - 7$$

Next, square both sides [remembering that $(x - 7)^2$ is *not* $x^2 - 49$ or $x^2 + 49$, which are very common errors in this context]:

$$(\sqrt{x - 1})^2 = (x - 7)^2$$
$$x - 1 = x^2 - 14x + 49$$
$$0 = x^2 - 15x + 50$$
$$0 = (x - 5)(x - 10), \quad x = 5, 10$$

Checking $x = 5$ in the *original* equation, we obtain $\sqrt{5 - 1} = 5 - 7$, $\sqrt{4} = -2$, $2 = -2$. This is false, so 5 is an extraneous root and is not an answer. Checking $x = 10$, we obtain $\sqrt{10 - 1} = 10 - 7$, $3 = 3$. So $x = 10$ is the only answer.

Warning: It is necessary to isolate the radical first. Otherwise, squaring both sides would give $(\sqrt{x - 1} + 7)^2 = x - 1 + 14\sqrt{x - 1} + 49$ on the left, and this does not eliminate the radical.

EXAMPLE 3 Solve $D = \sqrt{(a - 1)^2 + b^2}$ for a.

SOLUTION Square both sides, $D^2 = (a - 1)^2 + b^2$, then solve:

$$D^2 - b^2 = (a - 1)^2, \quad \pm\sqrt{D^2 - b^2} = a - 1$$
$$a = 1 \pm \sqrt{D^2 - b^2}$$

These check.

In the previous examples, we squared both sides because that is the *opposite* or *inverse* of taking square roots; i.e., doing that eliminated the radical. The same principle applies to other rational powers.

EXAMPLE 4 Solve $(4 - x^2)^{3/5} = -8$.

SOLUTION We raise both sides to the power 5/3 to eliminate the 3/5:
$$\left[(4 - x^2)^{3/5}\right]^{5/3} = (-8)^{5/3}$$
$$4 - x^2 = -32$$
$$36 = x^2, \quad x = \pm 6$$

Both answers check.

For some equations involving radicals, it is necessary to raise both sides to a power more than once.

EXAMPLE 5 Solve $\sqrt{x + 4} - \sqrt{3x + 1} + 1 = 0$.

SOLUTION We first isolate one of the radicals and then square:
$$(\sqrt{x + 4})^2 = (\sqrt{3x + 1} - 1)^2$$
$$x + 4 = 3x + 1 - 2\sqrt{3x + 1} + 1$$

Now we isolate (a multiple of) the radical that is left and then simplify, square, and solve:
$$2\sqrt{3x + 1} = 2x - 2$$
$$(\sqrt{3x + 1})^2 = (x - 1)^2$$
$$3x + 1 = x^2 - 2x + 1$$
$$0 = x^2 - 5x, \quad 0 = x(x - 5), \quad x = 0, 5$$

Checking, we find that 5 works but that 0 is extraneous.

Substitution is one of the powerful tools of mathematics. Its use is to transform an expression or equation into a form more easily handled. Its use in this section is to transform an equation so that it can be put in standard quadratic form, $au^2 + bu + c = 0, a \neq 0$, and then solved by techniques already discussed.

EXAMPLE 6 Solve $x^4 + x^2 - 6 = 0$.

SOLUTION Let $u = x^2$. Then $u^2 = (x^2)^2 = x^4$. Substituting in, we get
$$u^2 + u - 6 = 0, \quad (u + 3)(u - 2) = 0, \quad u = -3 \quad \text{or} \quad u = 2$$

Substituting back $u = x^2$, we obtain
$$x^2 = -3 \quad \text{or} \quad x^2 = 2$$

Since $x^2 \geq 0$ if x is any real number, $x^2 = -3$ has no real solutions. Solving $x^2 = 2$, we obtain $x = \pm \sqrt{2} \approx \pm 1.41421$.

Note: You are encouraged as much as possible to do substitution in your head. For example, in Example 6, you may just think "$u = x^2$" and write

$$(x^2)^2 + x^2 - 6 = 0$$
$$(x^2 + 3)(x^2 - 2) = 0, \qquad x^2 = -3, \qquad x^2 = 2$$

$x^2 = -3$ has no real solution; $x = \pm \sqrt{2}$. Although many of the problems in this section can be done this way, do not hesitate to write out the whole substitution process whenever you might need it.

EXAMPLE 7 Solve $x^{1/2} = 3x^{1/4} - 2$.

SOLUTION Sometimes it may not be obvious at first what to let u represent. The key to remember is that we want to let u be one thing so that u^2 can be the other. In this case, since

$$(x^{1/2})^2 = x^1 \quad \text{and} \quad (x^{1/4})^2 = x^{1/2}$$

we see that we should let $u = x^{1/4}$ so that $u^2 = x^{1/2}$. Substituting, we obtain

$$u^2 = 3u - 2, \qquad u^2 - 3u + 2 = 0$$
$$(u - 2)(u - 1) = 0, \qquad u = 1, 2$$

Substituting back $u = x^{1/4}$, and solving, we obtain

$$x^{1/4} = 2 \qquad \text{or} \qquad x^{1/4} = 1$$
$$x = 16 \qquad \text{or} \qquad x = 1$$

Checking,

$$16^{1/2} = 3(16^{1/4}) - 2 \qquad\qquad 1^{1/2} = 3(1^{1/2}) - 2$$
$$4 = 3(2) - 2 \qquad\qquad\qquad 1 = 3(1) - 2$$
$$4 = 4 \qquad\qquad\qquad\qquad 1 = 1$$

Both check.

Sometimes equations need to be manipulated before we see they are quadratic in form.

EXAMPLE 8 Solve $x^2 - x^{-2} = 4$.

SOLUTION First, rewrite this as $x^2 - 1/x^2 = 4$. Next, multiply by the l.c.d. $= x^2$; later we shall have to check any number where the l.c.d. $= 0$, in this case only $x = 0$:

$$x^2(x^2) - x^2\left(\frac{1}{x^2}\right) = x^2(4)$$
$$x^4 - 1 = 4x^2$$
$$x^4 - 4x^2 - 1 = 0$$

Let $u = x^2$ so that $u^2 = x^4$. Substituting, we obtain

$$u^2 - 4u - 1 = 0$$

Section 2.4. Equations With Radicals and Equations Quadratic in Form

This does not (easily) factor, so we use the quadratic formula:

$$u = \frac{-(-4) \pm \sqrt{(-4)^2 - 4(1)(-1)}}{2(1)}$$

$$= \frac{4 \pm \sqrt{20}}{2} = 2 \pm \sqrt{5}$$

Substituting back $u = x^2$, we obtain

$$x^2 = 2 + \sqrt{5} \quad \text{or} \quad x^2 = 2 - \sqrt{5}$$

Since $4 < 5$ implies $2 < \sqrt{5}$, we have that $2 - \sqrt{5}$ is negative. Thus, $x^2 = 2 - \sqrt{5}$ has no real solution. However, solving $x^2 = 2 + \sqrt{5}$ yields

$$x = \pm\sqrt{2 + \sqrt{5}} \approx \pm 2.05817$$

Both check.

An Application

Different things obey different laws of motion. For example, near sea level, sound travels in air at the constant speed of about 1100 feet per second in all directions. So after a given time t, sound has traveled $s = 1100t$ feet. On the other hand, the distance in feet an object has risen or fallen after time t is given by the formula $s = -16t^2 + v_0 t + s_0$, as discussed in Sec. 2.3. Solving some problems which combine these laws requires the techniques presented in this section.

EXAMPLE 9 A rock is dropped into a well. Three seconds later, the sound of the splash is heard at the top of the well. How deep is the well? (Assume the data are correct to two significant figures.)

SOLUTION

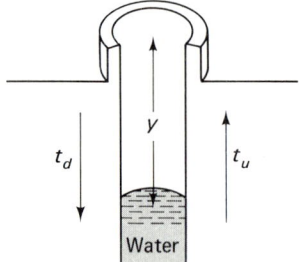

Let y be the depth (in feet) of the well. ($y > 0$.) We must first find an equation expressing the total time, 3 seconds, it took for the rock to fall to the water and the sound to reach the top. Let t_d be time it took for the rock to fall down. Let t_u be the time it took for the sound to get up.

Then

$$t_d + t_u = 3$$

From the above discussion about sound,

$$y = 1100 t_u \quad \text{or} \quad t_u = \frac{y}{1100}$$

In the formula $s = -16t^2 + v_0 t + s_0$ for the rock, we have $v_0 = 0$, since the rock is dropped, and $s_0 = 0$, by taking the altitude to be zero at the top of the well. When $t = t_d$, the rock is at the bottom of the well so $s = -y$ (the negative sign for so many feet *below* the top of the well). Therefore,

$$-y = -16t_d^2, \qquad t_d^2 = \frac{y}{16}, \qquad t_d = \frac{\sqrt{y}}{4}$$

Since $t_d > 0$, we take the positive square root. The equation $t_d + t_u = 3$ now becomes

$$\frac{\sqrt{y}}{4} + \frac{y}{1100} = 3$$

Substituting $u = \sqrt{y}$, $u^2 = y$, we obtain

$$\frac{u}{4} + \frac{u^2}{1100} = 3$$

Multiply by 1100 and solve:

$$275u + u^2 = 3300, \qquad u^2 + 275u - 3300 = 0$$

Using the quadratic formula to solve for u, we obtain

$$u = \frac{-275 \pm \sqrt{275^2 - 4(1)(-3300)}}{2(1)}$$

Since $u = \sqrt{y}$ is positive, $u = \left[-275 + \sqrt{275^2 + 4(3300)} \right]/2 \approx 11.517616$. Solving for $y = u^2$, we obtain

$$y \approx 11.517616^2 \approx 132.655 \text{ ft}$$

Thus the well is about 130 feet deep (to two significant figures).

EXERCISES

In Exercises 1–56, solve.

1. $x^{3/2} = x^{1/2}$
2. $y^{3/2} = 8y$
3. $\dfrac{1}{\sqrt{z}} = \dfrac{1}{z}$
4. $w^{5/2} = w^{1/2}$
5. $\sqrt{5x+1} = 4$
6. $\sqrt[3]{4t-1} = -3$
7. $\sqrt[9]{5a+3} = -2$
8. $(4b-1)^{2/3} = 9$
9. $(7-3m^3)^{5/3} = 32$
10. $(n^2-1)^{2/3} = 4$
11. $\sqrt{a+1} + 4 = 0$
12. $\sqrt{2b-3} + 5 = 4$
13. $\sqrt{2x-1} + 2 = x$
14. $\sqrt{3x-8} - x = -6$
15. $z - \sqrt{10 + z/2} = -5$
16. $\sqrt{7+3z} - 3 = z$
17. $3\sqrt{6+5x} - 6 = 4x - 2\sqrt{6+5x}$
18. $\sqrt{3x+2} + 9x = 5\sqrt{3x+2} - 2$
19. $\sqrt{4p+1} - \sqrt{p+2} = 1$
20. $2\sqrt{q+5} + \sqrt{q} = 8$

21. $\sqrt{1-5w} + \sqrt{1-w} = 6$
22. $\sqrt{3v+7} + 2 = \sqrt{8v+25}$
23. $\sqrt{7-2a} + \sqrt{5+a} = \sqrt{24-a}$
24. $2\sqrt{b} - \sqrt{b-5} = \sqrt{2b-2}$
25. $\dfrac{x + \sqrt{x+1}}{x - \sqrt{x+1}} = \dfrac{11}{5}$
26. $\dfrac{2x + 3\sqrt{2x+1}}{3x - 2\sqrt{2x+1}} = \dfrac{17}{6}$
27. $\sqrt{\sqrt{y+21} - \sqrt{y-3}} = 2$
28. $\sqrt{2\sqrt{y+1}} = \sqrt{3y-5}$
29. $x^4 - 13x^2 + 36 = 0$
30. $x^4 - 7x^2 + 12 = 0$
31. $4x^4 - 3x^2 = 1$
32. $36x^4 = 5x^2 + 1$
33. $x - 10\sqrt{x} + 9 = 0$
34. $x - 8\sqrt{x} = 9$
35. $3x^{2/3} = 8x^{1/3} + 3$
36. $t^{-1/5} - t^{1/5} = \tfrac{3}{2}$
37. $5y^{1/6} = 2y^{1/3} + 2$
38. $\sqrt{s} = \sqrt[4]{s} + 2$
39. $7x + \sqrt{6x} - 2 = 0$
40. $y - \sqrt{y} - 1 = 0$
41. $t^{1/3} + 1 = 6t^{-1/3}$
42. $x + 2 = \tfrac{9}{2}\sqrt{x}$
43. $2\left(\dfrac{x}{1+x}\right)^2 - 7\left(\dfrac{x}{1+x}\right) + 3 = 0$
44. $\left(\dfrac{y}{y-1}\right)^2 + \dfrac{2y}{y-1} = 15$
45. $\dfrac{2}{(x-1)^2} + \dfrac{3}{x-1} = 2$
46. $\dfrac{x}{x-1} = 7\sqrt{\dfrac{x}{x-1}} + 8$
47. $2^{2x} - 9(2^x) + 8 = 0$
48. $100^x - 11(10^x) + 10 = 0$
49. $9(9^x) + 3 = 28(3^x)$
50. $4(4^x) + 1 = 5(2^x)$
51. $\sqrt{2z} - \sqrt[4]{4z} - \sqrt{2} = 0$
52. $\sqrt[3]{w} - \sqrt[6]{8w} - 2 = 0$
53. $12x^2 - 18x^{-2} = 19$
54. $25x^2 = 2x^{-2} - 49$
55. $3x^2 - 3 = x^{-2}$
56. $x^2 + x^{-2} = 4$

In Exercises 57–64, solve for the indicated variable. Assume all letters represent positive real numbers.

57. $T = 2\pi\sqrt{\dfrac{m}{g}}$ for m
58. $A = \sqrt{1 + \dfrac{a^2}{b^2}}$ for a
59. $x^{2/3} + 4^{2/3} = a^{2/3}$ for x
60. $D = \sqrt{(x-a)^2 + (x+a)^2}$ for x
61. $T = 2\pi\sqrt{\dfrac{m}{g}}$ for g
62. $A = \sqrt{1 + \dfrac{a^2}{b^2}}$ for b
63. $y = (\sqrt{a} - \sqrt{x})^4$ for x
64. $\pi R\sqrt{1 + (1-a)^2} = T$ for a

65. A stone is dropped into a well. Two seconds later a splash is heard at the top. How deep is the well?

66. A stone is dropped off a cliff. In 4 seconds, the sound of the stone hitting the ground is heard at the top of the cliff. How high is the cliff?

67. A stone is dropped into a crevasse in a glacier. Six and a half seconds later, the sound of the stone hitting the bottom is heard at the top. How deep is the crevasse?

68. A racing car, starting from rest, accelerated so that its distance from the starting line was $20t^2$ feet, where t is in seconds. After 3 seconds, a bang was heard back at the starting line, because a tire blew out on the car. How far was the car from the starting line when the blowout occurred?

69. A racing car, starting from rest, accelerated so that its distance from the starting line was $25t^2$ feet, where t is in seconds. After 4.5 seconds, a bang was heard back at the starting line, because a tire blew out on the car. How far was the car from the starting line when the blowout occurred?

SECTION 2.5. INEQUALITIES AND SETS

You are probably familiar with a **number line**. This is a line where each point is associated with a unique real number (called the **coordinate** of the point), and *vice versa*. The line is (usually) drawn horizontal and the association made as follows: First, choose any point on the line, label it 0 (zero), and call it the **origin**. Then choose a distance to be **one unit**. The points one unit, two units, three units, etc., to the right of the origin are labeled 1, 2, 3, etc., respectively. Similarly, the points one unit, two units, three units, etc., to the left of the origin are labeled $-1, -2, -3$, etc., respectively. Other real numbers "fit in" exactly where you would expect. That is, the point x units to the right of 0 is labeled x, while the point x units to the left of 0 is labeled $-x$. We tend to identify the points with their coordinates and say "3 is to the right of 2" as an abbreviation for "the point whose coordinate is 3 is to the right of the point whose coordinate is 2."

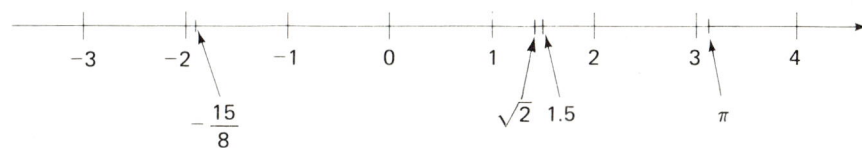

A real number is positive, zero, or negative, and these three categories are mutually exclusive. The **positive** numbers correspond to points to the right of 0, while the **negative** numbers correspond to points to the left of 0.

> **(3)**   **DEFINITION**
>
> If a and b are real numbers, then either of the inequalities $a > b$ (read "a is greater than b") or $b < a$ (read "b is less than a") means $a - b$ is positive.

> **(4)**   **DEFINITION**
>
> The notation $a \geq b$ (read "a is greater than or equal to b") means either $a > b$ or $a = b$, and similarly for $b \leq a$.

For example, we may write

$$6 > 2, \quad -5 < -1, \quad -3 \leq 2, \quad 5 \geq 5, \quad \sqrt{3} \leq \pi$$

Geometrically, $a > b$ (or $b < a$) means the point (whose coordinate is) a lies to the right of the point (whose coordinate is) b. If $a > b$ and $b > c$, then a lies to the right of b, and b lies to the right of c. It follows that a lies to the right of c, or $a > c$. Thus we have a geometric demonstration of one of the fundamental properties of inequalities called **transitivity**:

> **(5)**
>
> If $a > b$ and $b > c$, then $a > c$.

For an algebraic proof of (5), see Exercise 64.

Another fundamental property of inequalities is

> **(6)**
>
> If $a > b$, then $a + c > b + c$.

This can be seen geometrically, since c translates (shifts) both a and b the same distance (to the right if c is positive and to the left if c is negative). Thus their relative positions are the same both before and after translation. For example,

$$5 > 1 \quad \text{so} \quad 5 + 3 > 1 + 3 \quad \text{and} \quad 5 - 2 > 1 - 2$$

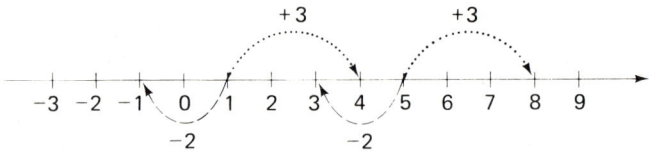

For an algebraic proof of (6), see Exercise 65. The following is a corollary to (5) and (6):

> **(7)**
>
> If $a > b$ and $c > d$, then $a + c > b + d$.

See Exercise 66 for a proof.

The next two fundamental properties of inequalities are the following:

> **(8)**
>
> i. If $a > b$ and $c > 0$, then $ac > bc$.
> ii. If $a > b$ and $c < 0$, then $ac < bc$.

The two follow from the fact that the product of two positive numbers is positive. We prove the second and leave the first as an exercise. See Exercise 67.

If $a > b$ and $c < 0$, then $a - b$ and $0 - c = -c$ are positive, so $(a - b)(-c) = bc - ac$ is positive. Hence, $ac < bc$ by (3).

Note: What (8ii) says is that when an inequality is multiplied (or divided) by a negative number, the inequality is turned around or "reversed." We refer to this property as **reversing the inequality**.

The final fundamental property is the following:

---

(9)

Suppose a and b are either both positive or both negative.

If $a > b$, then $\dfrac{1}{a} < \dfrac{1}{b}$

Thus whenever a and b both have the same sign, taking reciprocals also reverses the inequality. See Exercise 68 for a proof. Of course, if $a > b$ and b is negative and a is positive, then $1/b$ is still negative and $1/a$ is still positive, so $1/a > 1/b$.

We now consider inequalities between algebraic expressions, such as

$$3x - 2 < 7x + 4 \quad \text{or} \quad 5.1x - 4.2 \leq 2.9x - 13.7$$

Just as with equalities, when x is replaced by a specific number, that number is called a **solution** if the resulting inequality is actually true. To **solve** an inequality means to find all solutions. This is done very similarly to solving equations, except that the fundamental properties (5)–(9) are used. In particular, you must be careful to remember that multiplying (or dividing) both sides of an inequality by a negative number **reverses** the inequality.

EXAMPLE 1 Solve $2.9x - 13.7 \geq 5.1x - 4.2$.

SOLUTION For this first example, we shall include all the steps to illustrate the properties being used, but thereafter we shall combine obvious steps.

$$2.9x - 13.7 \geq 5.1x - 4.2$$
$$(2.9x + 13.7) + (-13.7 - 5.1x) \geq (5.1x - 4.2) + (-13.7 - 5.1x) \quad \text{[by (6)]}$$
$$-2.2x \geq -17.9$$
$$\left(-\frac{1}{2.2}\right)(-2.2x) \leq \left(-\frac{1}{2.2}\right)(-17.9) \quad \text{[by (8ii)]}$$
$$x \leq \frac{17.9}{2.2} = \frac{179}{22}$$

Estimating first,

$$x \lessapprox \frac{180}{20} = 9$$

Using a calculator, we obtain $x \lessapprox 8.13636$.

The set of all solutions, the **solution set**, of an inequality is often a very large set of numbers. It is useful and instructive to have geometric interpretations for these sets. By the **graph** of a set of real numbers we mean the collection of all points on the real line whose coordinates are in the set. To **sketch a graph**, we darken the appropriate portion of the line. It is usually important to indicate if an endpoint is or is not included. To include a point, we put a square bracket; to exclude a point, we put a parenthesis.

EXAMPLE 2 $\{x \mid x > -3\}$:

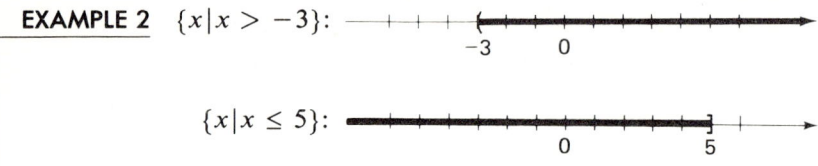

$\{x \mid x \leq 5\}$:

These two sets in Example 2 are **infinite intervals**, and there is a standard and convenient notation for such sets.

(10)   NOTATION

$$(-\infty, a) = \{x \mid x < a\}, \quad (a, \infty) = \{x \mid x > a\}$$
$$(-\infty, a] = \{x \mid x \leq a\}, \quad [a, \infty) = \{x \mid x \geq a\}$$

The symbols ∞ and $-\infty$ are read "infinity" and "minus infinity." It is important to remember that these symbols are *notational devices* and do *not* represent any real numbers.

Very often inequalities are part of compound statements.

EXAMPLE 3   Sketch the graph of the set of all numbers which satisfy $-3 < x$ and $x \leq 5$.

SOLUTION   For a number x to satisfy this statement, it must be greater than -3 and less than or equal to 5 at the same time. Thus we get the set of all numbers between -3 and 5 including 5 but excluding -3. We sketch:

The set in Example 3 is a **finite interval** and is denoted by $(-3, 5]$. It is standard practice to abbreviate "$-3 < x$ and $x \leq 5$" by "$-3 < x \leq 5$." There are four types of finite intervals with corresponding notation:

(11)   NOTATION

$$(a, b) = \{x \mid a < x < b\}, \quad [a, b] = \{x \mid a \leq x \leq b\}$$
$$(a, b] = \{x \mid a < x \leq b\}, \quad [a, b) = \{x \mid a \leq x < b\}$$

If $a < b$, they are sketched as follows:

(a, b): ⊢————⊣ $[a, b]$: ⊢————⊣
     a   b   a   b

$(a, b]$: ⊢————⊣ $[a, b)$: ⊢————⊣
     a   b   a   b

The interval (a, b) is called the **open interval** from a to b; $[a, b]$ is called the **closed interval** from a to b. The other two intervals are called **half-open intervals.**

Note that all four intervals are the empty set if $a > b$.

Another kind of compound statement is of the form

$$x < a \quad \text{or} \quad b < x.$$

To satisfy this statement, a real number x must make one or the other (or both) of the inequalities true.

EXAMPLE 4 The set $\{x \mid x < 2 \text{ or } 5 < x\}$ is made up of two infinite intervals, $(-\infty, 2)$ and $(5, \infty)$, as indicated in the figure which follows. Note that in this figure (and subsequent figures) we omit unnecessary hash marks, but we keep the number lines aligned.

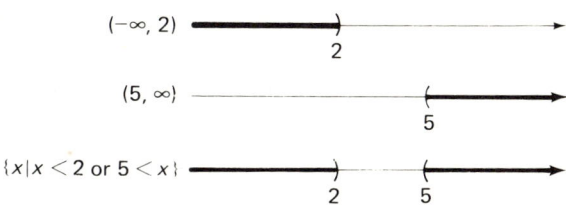

For sets of the type in Example 4, we often use the union symbol ∪ and write

$$(-\infty, 2) \cup (5, \infty)$$

to denote the set of all real numbers which are in $(-\infty, 2)$ *or* which are in $(5, \infty)$. In general,

---

(12)   **DEFINITION**

If A and B are sets,

$$A \cup B = \{x \mid x \text{ is in } A \text{ or } x \text{ is in } B\}$$

The notation $A \cup B$ is read "A union B."

Warning: It is important to remember that the inequality $a < x < b$ means $a < x$ and $x < b$. If you want the two infinite intervals as in Example 4, you must write "or." If you carelessly write "$5 < x < 2$," this implies "$5 < 2$," which is false. Thus, $\{x \mid 5 < x < 2\} = \emptyset$ and *not* $(-\infty, 2) \cup (5, \infty)$.

We now turn to solving compound inequalities.

EXAMPLE 5 Solve and graph
$$2 \le \frac{3x - 1}{2} < 4$$

SOLUTION First rewrite the one statement into two inequalities and then solve:
$$2 \le \frac{3x - 1}{2} < 4$$

$2 \le \frac{3x-1}{2}$	and	$\frac{3x-1}{2} < 4$
$4 \le 3x - 1$	and	$3x - 1 < 8$
$5 \le 3x$	and	$3x < 9$
$5/3 \le x$	and	$x < 3$

Thus a solution must be greater than or equal to $5/3$ and less than 3 *at the same time*, so the solution set is $\{x | 5/3 \le x < 3\} = [5/3, 3)$:

This problem has a shorter method of solution by combining the two parts into one:
$$2 \le \frac{3x - 1}{2} < 4$$
$$4 \le 3x - 1 < 8$$
$$5 \le 3x < 9$$
$$5/3 \le x < 3$$

The next problem *cannot* be done by this shorter method; it must be broken into two inequalities which are solved separately.

EXAMPLE 6 Solve and graph
(a) $2x - 1 \le 4x - 3 < 3 + 6x$; (b) $2x - 1 \le 4x - 3$ or $4x - 3 < 3 + 6x$.

SOLUTION

(a) $2x - 1 \le 4x - 3$	and	$4x - 3 < 3 + 6x$
$2 \le 2x$	and	$-6 < 2x$
$1 \le x$	and	$-3 < x$
(b) $2x - 1 \le 4x - 3$	or	$4x - 3 < 3 + 6x$
$2 \le 2x$	or	$-6 < 2x$
$1 \le x$	or	$-3 < x$

For (a), a solution must be greater than or equal to 1 *and* greater than -3 *at the same time*, so the solution set is $\{x \mid x \geq 1\} = [1, \infty)$. See the diagram.

For (b), a solution must satisfy one or the other (or both) of the inequalities, so the solution set is $\{x \mid -3 < x\} = (-3, \infty)$. See the diagram.

$1 \leq x$:

$-3 < x$:

(a): $1 \leq x$ and $-3 < x$:

(b): $1 \leq x$ or $-3 < x$:

Unions and Intersections

We have already defined **union**, which corresponds to "or"; we now define **intersection**, which corresponds to "and."

---

(14)

DEFINITION

If A and B are sets,

$A \cap B = \{x \mid x \text{ is in } A \text{ and } x \text{ is in } B\}$

The notation $A \cap B$ is read "A intersect B."

EXAMPLE 7 Find **(a)** $(2, \infty) \cap (-\infty, 5)$ and **(b)** $(-\infty, 2) \cap (5, \infty)$.

SOLUTION

(a) $(2, \infty) \cap (-\infty, 5)$
$= \{x \mid 2 < x \text{ and } x < 5\} = (2, 5)$:

$(2, \infty)$

$(-\infty, 5)$

$(2, \infty) \cap (-\infty, 5)$
$= (2, 5)$

These are the points in both $(-\infty, 5)$ and $(2, \infty)$

(b) $(-\infty, 2) \cap (5, \infty)$
$= \{x \mid x < 2 \text{ and } 5 < x\} = \emptyset$:

$(-\infty, 2)$

$(5, \infty)$

$(-\infty, 2) \cap (5, \infty) = \emptyset$

There are no points in both $(-\infty, 2)$ and $(5, \infty)$

EXAMPLE 8 Describe the solutions of Example 6 in terms of intersections or unions.

SOLUTION The solutions in Example 6 were as follows:
For (a), $\{x|1 \leq x \text{ and } -3 < x\} = [1, \infty) \cap (-3, \infty) = [1, \infty)$.
For (b), $\{x|1 \leq x \text{ or } -3 < x\} = [1, \infty) \cup (-3, \infty) = (-3, \infty)$.

EXAMPLE 9 Suppose $A = \{-1, 1, 3\}$, $B = \{0, 1, 2, 3\}$, $C = \{0, 2, 4\}$. Find $A \cap B$, $A \cap C$, $A \cup B$, $A \cup C$, $B \cap (A \cup C)$, and $(B \cap A) \cup (B \cap C)$.

SOLUTION $A \cap B = \{1, 3\}$, $A \cap C = \emptyset$, $A \cup B = \{-1, 0, 1, 2, 3\}$, $A \cup C = \{-1, 0, 1, 2, 3, 4\}$, $B \cap (A \cup C) = \{0, 1, 2, 3\} \cap \{-1, 0, 1, 2, 3, 4\} = \{0, 1, 2, 3\} = B$, and $(B \cap A) \cup (B \cap C) = \{1, 3\} \cup \{0, 2\} = \{0, 1, 2, 3\} = B$.

EXERCISES

1. It is true that $-4 < -2$, which can be seen in the following graph:

What inequality is obtained if

   a. 3 is added to both sides?   **b.** 3 is subtracted from both sides?
   c. Both sides are multiplied by 2?   **d.** Both sides are divided by -2?
   e. Both sides are multiplied by 0?

In Exercises 2–12, verify the inequality with as little computation as possible and *without* using calculators. Where appropriate, use $3.14 < \pi < 3.15 = 315/100 = 63/20$. State which of the properties (5)–(9) you used.

2. $8123 - 6713 < 8713 - 6713$
3. $11{,}864 - 6919 < 5000$
4. $213/54 < 4$
5. $6 \cdot 8 \cdot 9 \cdot 11 < 50 \cdot 100$
6. $96 \cdot 97 \cdot 98 < 1{,}000{,}000$
7. $100\sqrt{2} > 140$
8. $100\sqrt{3} > 170$
9. $1/\pi > 1/4$
10. $6.28 < 2\pi < 6.30$
11. $\pi^2 < 10$
12. $\pi^6 < 1000$

In Exercises 13–15, express the inequality in interval notation, and sketch the graph of the interval.

13. a. $-2 < x < 3$ **b.** $-2 < x \leq 3$ **c.** $-2 \leq x < 3$ **d.** $-2 \leq x \leq 3$
14. a. $x < 5$ **b.** $x \leq 5$ **c.** $x > 5$ **d.** $x \geq 5$
15. a. $x < -3$ and $-7 < x$ **b.** $x < -3$ or $-7 < x$
    c. $x < -7$ and $-3 < x$ **d.** $x < -7$ or $-3 < x$

In Exercises 16–18, express the intervals as an inequality in the variable x.

16. a. $(-4, 7)$ **b.** $(3, 9]$ **c.** $[-5, -2)$ **d.** $[-1, 1]$
17. a. $(-\infty, 5)$ **b.** $(-\infty, -2]$ **c.** $[-3, \infty)$ **d.** $(0, \infty)$
18. a. $(-\infty, 2) \cup (4, \infty)$ **b.** $(-\infty, -2) \cup [3, \infty)$ **c.** $(-\infty, -1] \cup (0, \infty)$ **d.** $\emptyset$

In Exercises 19–47, solve the inequality, expressing the solution in terms of intervals, and graph the result.

19. $-3x < 6$
20. $4 - 3x \geq 7$
21. $5 + x < 3x$
22. $2.148 - 3.719x > 5.417 + 2.918x$
23. $341.2x + 517.5 \geq 918.3 - 491.2x$
24. $15.3y - 7.1 < 18.2y + 23.8$
25. $22.1 - 18.9y \leq .2 + 13.1y$
26. $7 - \frac{1}{2}s \leq 3 + \frac{1}{4}s$
27. $\frac{1}{6}s + \frac{1}{3} \geq \frac{1}{4}s - \frac{1}{2}$
28. $(x - 1)(x + 3) < (x - 2)^2$
29. $(x - 3)^2 \geq x + x(x + 4)$
30. $-3 \leq 2x + 1 \leq 3$
31. $-4 \leq \frac{2x - 1}{3} \leq 4$
32. $0 < 3 - \frac{2}{3}x \leq \frac{1}{3}$
33. $-\frac{1}{3} \leq \frac{1}{2}x - 6 < 0$
34. $-3 \geq 2x + 1$ or $2x + 1 \geq 3$
35. $-4 \geq \frac{2x - 1}{3}$ or $\frac{2x - 1}{3} \geq 4$
36. $0 > 3 - \frac{2}{3}x$ or $3 - \frac{2}{3}x \geq \frac{1}{3}$
37. $-\frac{1}{3} \geq \frac{1}{2}x - 6$ or $\frac{1}{2}x - 6 > 0$
38. $3x - 1 < 2x - 2 < 4x + 3$
39. $0 \geq 3x + 5 \geq -1 - x$
40. $4 - 3x \geq 9 + 2x > 4$
41. $(x - 1)^2 \leq 1 - x(4 - x) < (4 + x)^2 - 3x$
42. $3x - 1 < 2x - 2$ or $2x - 2 < 4x + 3$
43. $0 \geq 3x + 5$ or $3x + 5 \geq -1 - x$
44. $2x - 3 < 4x - 5 < 6x - 9$
45. $\frac{1}{2}x - 1 \geq \frac{1}{4}x + 1 \geq \frac{1}{6}x - \frac{1}{3}$
46. $2x - 3 < 4x - 5$ or $4x - 5 < 6x - 9$
47. $\frac{1}{2}x - 1 \geq \frac{1}{4}x + 1$ or $\frac{1}{4}x + 1 \geq \frac{1}{6}x - \frac{1}{3}$

48. Find the values for k for which the equation $3x^2 - 2x - 3k = 0$ has no real roots.
49. Find the values for k for which the equation $kx^2 + 2x + 9 = 0$ has two distinct real roots.
50. To get an A in a course a student's average must be at least 90 (and at most 100). If a particular student's grades so far are 84, 88, and 93, what scores must that student get on the last test to earn an A if all tests count equally?
51. In Exercise 50, if the last test counts one-third of the total grade, what must be the score then?
52. In Exercise 50, if the scores were 94, 78, and 83, what must be the score then? Is it possible? What would be the answer if the last test counted one-third of the total grade?

In working with problems with gas mileage, use (mi/gal)(no. of gal) = (no. of mi).

53. A small dairy has two trucks to deliver milk on a 350-mile route. Truck A gets 10 miles per gallon, but truck B gets only 6 miles per gallon. For economic reasons, they want to use at most 40 gallons of gas each day. Of course they would want to use truck A for the whole route, but because of the time involved, they can use it for at most 250 miles. What are the lengths of routes they can give truck A and still remain within their restrictions? Is it possible?
54. In Exercise 53, if they fix up truck B so that it gets 8 miles per gallon, what are the answers now?
55. The relationship between the Fahrenheit and Celsius temperature scales is $F = \frac{9}{5}C + 32$.
   a. If $60 \leq F \leq 80$, find the corresponding range for C.
   b. If $-40 \leq C \leq 0$, find the corresponding range for F.
56. Ohm's law for electricity states that $RI = E$.
   a. If $I = 30$ and $10 \leq R \leq 20$, find the corresponding range for E.
   b. If $E = 100$ and $10 \leq R \leq 20$, find the corresponding range for I.

57. Boyle's law for a certain gas is $pv/t = 3.12$.

   a. If $v = 20$ and $15 \leq p \leq 20$, what is the corresponding range for t?
   b. If $t = 50$ and $15 \leq p \leq 20$, what is the corresponding range for v?

58. If R_1 and R_2 are two resistances in parallel in an electric circuit, then their total resistance R is

$$\frac{1}{R} = \frac{1}{R_1} + \frac{1}{R_2}$$

   a. If $R_1 = 20$ and $5 \leq R_2 \leq 10$, what is the corresponding range for R?
   b. If $R_1 = 20$ and $5 \leq R \leq 10$, what is the corresponding range for R_2?

59. If $a = b$, what are (a, b), $(a, b]$, $[a, b)$, and $[a, b]$?

60. If $A = \{1, 2, 3, 4\}$, $B = \{2, 4, 6\}$, and $C = \{1, 3, 5\}$, find $A \cup B$, $A \cap B$, $B \cup C$, $B \cap C$, $(A \cup B) \cap C$, and $(A \cap C) \cup (B \cap C)$.

61. If $A = \{-7, -5, -1\}$, $B = \{-3, -1, 1, 3\}$, $C = \{-2, -1, 0, 1\}$, and $D = \{0, 1, 2, 3\}$, find $A \cap B$, $A \cap D$, $A \cup C$, $(B \cup C) \cap D$, and $(B \cap D) \cup (C \cap D)$.

62. If $A = (-\infty, -3)$, $B = (-5, \infty)$, $C = (-\infty, 3]$, and $D = [2, \infty)$, find $A \cap B$, $A \cap D$, $A \cup C$, $B \cup C$, and $(A \cap B) \cup (C \cap D)$.

63. If $A = [1, \infty)$, $B = (3, \infty)$, and $C = (-\infty, 2)$, find $A \cup B$, $A \cap B$, $B \cap C$, $B \cup C$, $A \cap (B \cup C)$, and $(A \cap B) \cup (A \cap C)$.

64. Give an algebraic proof of (5): If $a > b$ and $b > c$, then $a > c$. [*Hint*: Apply Definition (3) to $a > b$ and $b > c$ and then add.]

65. Give an algebraic proof of (6): If $a > b$, then $a + c > b + c$. [*Hint*: Apply the definition (3) to $a > b$ and then add $c + (-c)$.]

66. Give an algebraic proof of (7): If $a > b$ and $c > d$, then $a + c > b + d$. [*Hint*: Use (6), adding c to $a > b$ and b to $c > d$.]

67. Give an algebraic proof of (8i): If $a > b$ and $c > 0$, then $ac > bc$. [*Hint*: See the proof of (8ii).]

68. Give an algebraic proof of (9): Suppose a and b are either both positive or both negative. If $a > b$, then $1/a < 1/b$. [*Hint*: First argue that $1/ab > 0$; then multiply $a > b$ by $1/ab$, using (8i).]

SECTION 2.6.   EXPRESSIONS, EQUATIONS, AND INEQUALITIES WITH ABSOLUTE VALUE

From Sec. 2.5, any two real numbers are the coordinates of two unique points on the real line. When we refer to the distance between two numbers, we really mean the distance between those two corresponding points. Distance is always a nonnegative real number.

(14) **DEFINITION**

If a is a real number, then $|a|$, called the **absolute value** of a, is the distance between a and 0.

For example, $|0| = 0$, $|2| = 2$, and $|-2| = 2$, as can be seen in the following figure:

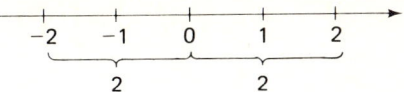

Absolute value signs can be used to express the distance between any two numbers. To begin with, the distance between two numbers is the larger number minus the smaller number. However, if absolute value signs are used, then we can write the numbers in either order. For example, the distance between -1 and 3 is either

$$|3 - (-1)| = |4| = 4 \quad \text{or} \quad |-1 - 3| = |-4| = 4$$

(15)

If a and b are any two real numbers, then the distance between them is
$$|b - a| = |a - b|$$

Many equations and inequalities involving absolute values can be solved using this geometric aspect of absolute value.

EXAMPLE 1 Solve and graph $|x - 3| = 2$.

SOLUTION We want all numbers whose distance from 3 is 2. These are $x = 1, 5$:

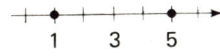

EXAMPLE 2 Solve and graph $|x - 3| \leq 2$.

SOLUTION We want all numbers whose distance from 3 is less than or equal to 2. These are the numbers between 1 and 5, including the endpoints. Thus, $|x - 3| \leq 2$ if and only if $1 \leq x \leq 5$, so the solution set is $[1, 5]$:

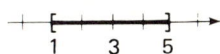

EXAMPLE 3 Solve and graph $|x + 4| > 5$.

SOLUTION To use distances, this must be in the form $|x - a|$. We rewrite this as
$$|x - (-4)| > 5$$

The distance from x to -4 is greater than 5 if and only if $x < -9$ or $x > 1$. Therefore the solution set is $\{x \mid x < -9 \text{ or } x > 1\} = (-\infty, -9) \cup (1, \infty)$:

The principles used in the preceding examples can be summarized:

> **(16)**
>
> If $c > 0$,
> i. $|u| = c$ is equivalent to $u = c$ or $u = -c$.
> ii. $|u| < c$ is equivalent to $-c < u < c$.
> iii. $|u| > c$ is equivalent to $u < -c$ or $c < u$.

There are similar statements for $\leq$ and $\geq$.

The easiest way of solving more complicated equations and inequalities involving absolute values can be a straight algebraic interpretation using (16).

EXAMPLE 4 Solve and graph $|4x - 1| \leq \frac{1}{2}$.

SOLUTION By a statement similar to (16ii),

$$-\tfrac{1}{2} \leq 4x - 1 \leq \tfrac{1}{2}$$

Adding 1, we obtain

$$\tfrac{1}{2} \leq 4x \leq \tfrac{3}{2}$$

and dividing by 4, we obtain $\tfrac{1}{8} \leq x \leq \tfrac{3}{8}$. The solution set is $\left[\tfrac{1}{8}, \tfrac{3}{8}\right]$:

EXAMPLE 5 Solve and graph

$$\left|\frac{2x + 3}{4}\right| > 3$$

SOLUTION By (16iii),

$$\frac{2x + 3}{4} < -3 \quad \text{or} \quad \frac{2x + 3}{4} > 3$$

Solving, we obtain

$$\begin{aligned} 2x + 3 &< -12 & &\text{or} & 2x + 3 &> 12 \\ 2x &< -15 & &\text{or} & 2x &> 9 \\ x &< -15/2 & &\text{or} & x &> 9/2 \end{aligned}$$

Thus the solution set is $\{x \mid x < -15/2 \text{ or } x > 9/2\} = (-\infty, -15/2) \cup (9/2, \infty)$:

We are now ready for a straight algebraic definition of absolute value. In general, if a is greater than or equal to 0, $|a| = a$. If a is less than 0, we change its sign to find $|a|$. The easiest way algebraically to indicate this is to put a negative sign in front of the number; for example,

$$|-2| = -(-2) = 2, \qquad |\sqrt{3} - 3| = -(\sqrt{3} - 3) = 3 - \sqrt{3}$$

Summarizing, we have

(17)
$$|a| = \begin{cases} a & \text{if } a \geq 0 \\ -a & \text{if } a < 0 \end{cases}$$

EXAMPLE 6 Simplify **(a)** $|14.2|$; **(b)** $|-3.5|$.

SOLUTION **(a)** Since $14.2 \geq 0$, $|14.2| = 14.2$. **(b)** Since $-3.5 < 0$, $|-3.5| = -(-3.5) = 3.5$.

EXAMPLE 7 Simplify $|\pi^3 - 3^\pi|$.

SOLUTION Using a calculator, we obtain $3^\pi \approx 31.5443$, $\pi^3 \approx 31.0063$. Hence, $\pi^3 - 3^\pi < 0$, so $|\pi^3 - 3^\pi| = -(\pi^3 - 3^\pi) = 3^\pi - \pi^3$.

One of the uses of absolute value is to simplify radicals containing variables which may be negative. For example,

$$\sqrt{2^2} = \sqrt{4} = 2 \qquad \text{but} \qquad \sqrt{(-2)^2} = \sqrt{4} = 2 = |-2|$$

Thus if a is allowed to be a negative number, then $\sqrt{a^2} \neq a$. But the following *is* true:

(18)

If a is any real number, $\sqrt{a^2} = |a|$.

For similar reasons, the corresponding statement is true for any even root.

(19)

If a is any real number and n is an *even* integer, $\sqrt[n]{a^n} = (a^n)^{1/n} = |a|$.

Of course, **if for some reason you know that** $a \geq 0$, then $|a| = a$, so $\sqrt{a^2} = a$, and, in general for n even, $\sqrt[n]{a^n} = a$.

EXAMPLE 8 Simplify $\sqrt{(x-3)^2}$.

SOLUTION $\sqrt{(x-3)^2} = |x-3|$. Until more is known about x, this is all that can be written.

EXAMPLE 9 Simplify $\sqrt{(-3)^2 x^2 y^4}$.

SOLUTION $\sqrt{(-3)^2 x^2 y^4} = \sqrt{(-3)^2} \sqrt{x^2} \sqrt{y^4} = |-3||x||y^2| = 3|x|y^2$.

EXAMPLE 10 Simplify $\sqrt[6]{a^6(b+2)^{12}}$.

SOLUTION $\sqrt[6]{a^6(b+2)^{12}} = \sqrt[6]{a^6} \sqrt[6]{(b+2)^{12}} = |a||(b+2)^2| = |a|(b+2)^2$.

EXERCISES

In Exercises 1–12, simplify. Variables may be negative.

1. $|\sqrt{7} - 7|$
2. $|(-8)^{5/3}|$
3. $|(\sqrt{2})^2 - 2^{\sqrt{2}}|$
4. $|(\sqrt{3})^3 - 3^{\sqrt{3}}|$
5. $\sqrt{(-5)^2}$
6. $\sqrt{(-2x)^2}$
7. $\sqrt[4]{(-3)^4(-4)^2}$
8. $\sqrt[6]{(-4)^6 a^{18}}$
9. $\sqrt{(-3)^2 x^2 (z-1)^2}$
10. $\sqrt[4]{(-2)^4 c^4 e^6}$
11. $\sqrt[4]{\dfrac{(-4)^2 p^{12}}{9^4 r^{16}}}$
12. $\sqrt[6]{\dfrac{(-3)^6 (x-1)^{12}}{4^3 (x+4)^{18}}}$

In Exercises 13–22, solve by interpreting absolute value in terms of distance. Graph the solution.

13. $|x| = 3$
14. $|y| \leq 4$
15. $|z| \geq \frac{1}{2}$
16. $|x - 1| = 2$
17. $|y + 3| < 3/2$
18. $|z - 2| > 2$
19. $|x + 4| = 0$
20. $|x - \frac{1}{2}| \leq -2$
21. $|x - 5| \geq -4$
22. $|x + 7| \geq 0$

In Exercises 23–34, solve. Graph the solution.

23. $|3x| = 1$
24. $|4x| = 8$
25. $|\frac{1}{2} x| < 3$
26. $|\frac{1}{3} x| \geq 4$
27. $|2x - 1| \geq 3/2$
28. $\left|\dfrac{14 - x}{2}\right| \leq 23$
29. $\left|\dfrac{15x - 1.1}{32}\right| = .41$
30. $\left|\dfrac{1.7x + 3.8}{21.1}\right| \leq 8.14$

92   EQUATIONS AND INEQUALITIES

31. $|2x - \tfrac{1}{5}| < 8/5$

32. $\left|\dfrac{3x-4}{10}\right| \geq \dfrac{1}{2}$

33. $\left|\dfrac{14x+1}{5}\right| \leq 3$

34. $|\tfrac{4}{5}x - \tfrac{1}{2}| > \tfrac{1}{10}$

35. For what values of x and y is it true that $|x| = |y|$?

36. For what values of x and y is $|x| \neq |y|$?

37. a. Find a value for x and value for y so that $|x + y| \neq |x| + |y|$.
  b. Find all values of x and y for which $|x + y| \neq |x| + |y|$.
  c. If $|x + y| \neq |x| + |y|$, is there a relationship between $|x + y|$ and $|x| + |y|$?

38. a. Find a value for x which is negative and a value for y for which $|x + y| = |x| + |y|$.
  b. Find all values of x and y for which $|x + y| = |x| + |y|$.
  c. Using part b and Exercise 37c, what is the relationship between $|x + y|$ and $|x| + |y|$?

39. Compute $|-a|$ when (a) $a < 0$ and (b) $a > 0$.

SECTION 2.7. QUADRATIC AND OTHER INEQUALITIES

In this section, we devise a method which helps us solve quadratic and other inequalities. The technique used is explained in Example 1. The point to keep in mind is that we have to determine where a product is positive (or negative) and not just where the factors are.

EXAMPLE 1 Solve and graph $x^2 > x + 2$, writing the answer as an interval or union of intervals.

SOLUTION First rewrite to get 0 on one side and *then* factor:

$$x^2 > x + 2$$
$$x^2 - x - 2 > 0$$
$$(x - 2)(x + 1) > 0$$

We now have to determine where the product $(x - 2)(x + 1)$ is positive. Let us first examine the factors separately. See Fig. 1.

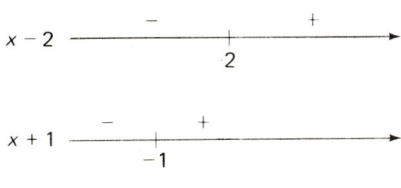

FIG. 1

We see that $x - 2$ is 0 at 2, positive to the right of 2, and negative to the left of 2 and that $x + 1$ is 0 at -1, positive to the right of -1, and negative to the left of -1. Consequently, when we consider the product $(x - 2)(x + 1)$, we see that

For $x = -1$ or 2, the product is 0.
For $x > 2$, both factors are positive, so the product is positive.

For $-1 < x < 2$, one factor is negative, and the other is positive, so the product is negative.

For $x < -1$, both factors are negative, so the product is positive.

Therefore, the signs of the product over each interval are as shown in Fig. 2.

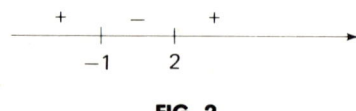

FIG. 2

From Fig. 2, we can sketch where the product is positive; see Fig. 3.

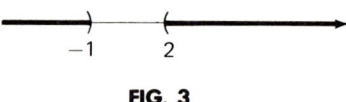

FIG. 3

This is the graph of the solution, and from this graph we easily determine the answer:

$$(-\infty, -1) \cup (2, \infty)$$

Once you understand the above reasoning, the process can be shortened into one diagram which combines Figs. 1, 2, and 3: see Fig. 4. Although this compact diagram requires only one number line to be drawn before you graph the solution, all the preceding reasoning is required to complete the display. The vertical lines of dashes are in the display only to help us keep the signs aligned.

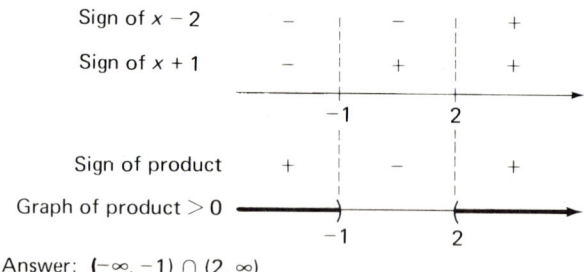

FIG. 4

EXAMPLE 2 Solve and graph $2x^2 + x \leq 3$, writing the answer as an interval or union of intervals.

SOLUTION We rewrite the inequality and factor, as we did in Example 1, and then display the compact diagram (Fig. 5):

$$2x^2 + x \leq 3$$
$$2x^2 + x - 3 \leq 0$$
$$(2x + 3)(x - 1) \leq 0$$

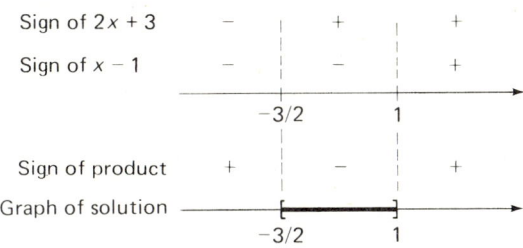

FIG. 5

The solution set is $[-\frac{3}{2}, 1]$. In this case, the endpoints are included because the inequality is $\leq$; in Example 1 there was no equal sign.

The technique used to solve the preceding inequalities applies to quotients, but care must be taken about dividing by zero.

EXAMPLE 3 Solve and graph

$$\frac{x - 1}{3x - 8} \leq 2$$

Warning: You *cannot* solve this with a first step of multiplying through by $3x - 8$ to obtain $x - 1 \leq 2(3x - 8)$, because the quantity $3x - 8$ is negative for some values of x (actually $x < 8/3$), and the inequality should be reversed for these values. In fact, the solution to $x - 1 \leq 2(3x - 8)$ is $x \geq 3$, and you should compare this with the actual solution that follows. The correct procedure is to obtain zero on one side and a factored form on the other side, just as before.

SOLUTION

$$\frac{x - 1}{3x - 8} \leq 2$$
$$\frac{x - 1}{3x - 8} - 2 \leq 0$$
$$\frac{x - 1}{3x - 8} - 2\frac{3x - 8}{3x - 8} \leq 0$$
$$\frac{x - 1 - 6x + 16}{3x - 8} \leq 0$$
$$\frac{15 - 5x}{3x - 8} \leq 0$$

Section 2.7 Quadratic and Other Inequalities

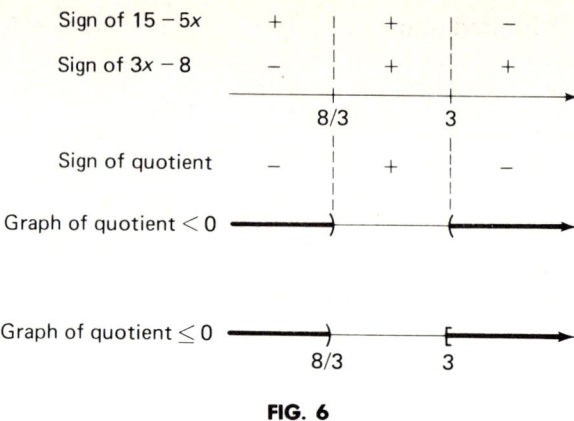

FIG. 6

Answer: $(-\infty, 8/3) \cup [3, \infty)$.

Note: With quotients and inequalities of ≤ 0 or ≥ 0, it is better to be on the safe side and sketch the solution to < 0 or > 0 first, before the answer. Otherwise it is very easy to include endpoints you should not. In the case of Example 3, endpoint 3 is included because it gives a zero in the numerator, but endpoint 8/3 is not included, because it gives a zero in the denominator.

The same techniques apply to more complicated expressions. When a quantity occurs to a power, it is usually best to determine the sign of that whole expression. You just have to remember the following:

$$(\text{Positive})^{\text{any power}} \text{ is positive.}$$

$$(\text{Negative})^{\text{even power}} \text{ is positive.}$$

$$(\text{Negative})^{\text{odd power}} \text{ is negative.}$$

For example,

Sign of $(x - 2)^2$ $\quad\xrightarrow{\;+\;\;\;\;+\;}$ $\quad$ Sign of $(x + 3)^3$ $\quad\xrightarrow{\;-\;\;\;\;+\;}$
$\qquad\qquad\qquad\quad\;\;2$ $\qquad\qquad\qquad\qquad\qquad\qquad\;\;-3$

Also, be careful when graphing factors such as $3 - x$:

Sign of $3 - x$: $\quad\xrightarrow{\;+\;\;\;\;-\;}$
$\qquad\qquad\qquad\;\;\;3$

96   EQUATIONS AND INEQUALITIES

EXAMPLE 4 Solve and graph $(3 - x)(x + 1)(x - 2)^2(x + 3)^3 \leq 0$.

SOLUTION

	−3	−1	2	3	
Sign of $(3 - x)$	+	+	+	+	−
Sign of $(x + 1)$	−	−	+	+	+
Sign of $(x - 2)^2$	+	+	+	+	+
Sign of $(x + 3)^3$	−	+	+	+	+
Sign of product	+	−	+	+	−

Graph of product ≤ 0

Note that all four numbers $-3, -1, 2,$ and 3 make the expression zero, so they are all solutions. The solution set is $[-3, -1] \cup \{2\} \cup [3, \infty)$.

EXERCISES

In Exercises 1–42, solve and graph.

1. $x^2 - x - 2 > 0$
2. $x^2 + 2x - 3 \leq 0$
3. $x^2 - 6x + 5 < 0$
4. $x^2 + 7x + 12 \geq 0$
5. $6x^2 \geq x + 1$
6. $12x^2 + 5x > 2$
7. $12x^2 \geq 7x + 12$
8. $x^2 + 1 < 13x/6$
9. $x^2 \geq 9$
10. $x^2 < 4$
11. $x^2 - 1 \leq 0$
12. $x^2 - 12 > 0$
13. $\dfrac{x + 4}{2x + 1} \geq 1$
14. $\dfrac{x - 1}{x - 2} \leq 3$
15. $\dfrac{3x - 1}{x - 2} < 1$
16. $\dfrac{2x + 3}{3x + 4} > 2$
17. $x^2 > x$
18. $x^3 \geq 4x$
19. $x^4 - x^3 \leq 0$
20. $x^4 < x^2$
21. $\dfrac{1}{x} \leq \dfrac{1}{x^2}$
22. $\dfrac{1}{x^2} \leq \dfrac{1}{x^3}$
23. $(x + 1)(x - 2)(x - 3) > 0$
24. $(x + 4)(x - 1)(x - 4) \leq 0$
25. $x^3 - 4x < 0$
26. $x^3 + 3x^2 \geq 4x$
27. $\dfrac{1}{x - 2} \leq \dfrac{3}{x + 1}$
28. $\dfrac{4}{x - 1} \geq \dfrac{3}{x + 2}$

Section 2.7 Quadratic and Other Inequalities

29. $\dfrac{x^2 - 9}{4x^2 - 25} \le 0$

30. $\dfrac{3}{x^2 - 16} \le 0$

31. $x + \dfrac{4}{x} \le 4$

32. $x + \dfrac{2}{x} \ge 1$

33. $\dfrac{(x - 1)}{(x + 1)(x + 2)} < 0$

34. $\dfrac{(x + 2)(x + 3)(x + 4)}{(x + 1)} \le 0$

35. $\dfrac{(x + 2)^2(1 - x)}{(x + 3)(x - 2)^3} \ge 0$

36. $\dfrac{x^3(x - 2)^2}{(x + 3)(x - 4)} \le 0$

37. If a ball is thrown straight up from ground level at 56 feet per second, its height (in feet) after t seconds is $y = -16t^2 + 56t$. During what time interval will the ball be at least 24 feet above the ground?

38. A projectile is fired from ground level with a vertical velocity of 320 feet per second. After t seconds its height is $y = -16t^2 + 320t$. During what time interval will it be at least 1200 feet above the ground?

39. For what values of k will the equation $x^2 + kx + 1 = 0$ have
   a. No real roots?   b. Two distinct real roots?

40. For what values of k will the equation $x^2 + kx + k = 0$ have
   a. No real roots?   b. Real roots?

REVIEW EXERCISES

In Exercises 1–15, solve the equations.

1. $\dfrac{x^2}{2} = 3x - 4$

2. $\dfrac{1}{x} + \dfrac{3}{2} = \dfrac{5}{2} - \dfrac{3}{x}$

3. $643.2x + 885.3 = 54.7 - 91.32x$

4. $s = a + (n - 1)d$, for n

5. $x^2 + 3x + 1 = 0$

6. $3.2x^2 - 6.8x - 4.59 = 0$

7. $A = \pi r^2 + 2\pi rh + \tfrac{2}{3}\pi r^2$, for r

8. $\sqrt{3x + 10} - 4 = x$

9. $(5x^2 - 1)^{3/2} = 8$

10. $\sqrt{x - 2} + \sqrt{x + 2} = 4$

11. $\sqrt{2x + 1} - \sqrt{x + 1} = 2$

12. $x^4 + 13x^2 + 36 = 0$

13. $x^4 - 26x^2 + 25 = 0$

14. $3\left(\dfrac{x}{x + 1}\right)^2 - 5\left(\dfrac{x}{x + 1}\right) + 2 = 0$

15. $w^{1/3} - 6w^{-1/3} = -1$

In Exercises 16–19, simplify. Variables may be negative.

16. $|(\sqrt{5})^5 - 5^{\sqrt{5}}|$

17. $|\sqrt{6} - 3|$

18. $\sqrt{16x^2}$

19. $\left(\dfrac{(-9)^8 x^{12} y^{16}}{(-6)^4}\right)^{1/4}$

In Exercises 20–23, determine the nature of the solutions without solving.

20. $x^2 + x + 1 = 0$

21. $4x^2 - 12x + 9 = 0$

22. $3.1x^2 + 1.2x - 4.1 = 0$

23. $3.1x^2 + 1.2x + 4.1 = 0$

In Exercises 24–26, verify the inequality using the fundamental properties (5)–(9) of inequalities.

24. $47 \cdot 48 < 2500$

25. $-\dfrac{256}{26} > -10$

26. $\dfrac{1}{8.63} > \dfrac{1}{10}$

In Exercises 27–37, solve the inequality, expressing the solutions in terms of intervals, and graph.

27. $-5x \geq 15$

28. $5x - 1 \geq 6x + 3 > 4x - 9$

29. $2x + 5 < 5x - 4$ or $3 - x < 7 - 2x$

30. $-1 \leq 3x + 8 \leq 7$

31. $|x - 1| < 3$

32. $|3 - 2x| > 5$

33. $x^2 - 5x - 6 > 0$

34. $\dfrac{(x + 1)(x - 2)(x^2 - 3x)}{x + 3} \leq 0$

35. $\dfrac{3}{x - 1} \geq \dfrac{1}{x + 2}$

36. $\dfrac{(x - 3)^2(2x - 3)^3}{2 - x} < 0$

37. $|x^2 - 4| \geq 1$

38. A plane leaves an airport and flies at 250 miles per hour. An hour and a half later a second plane leaves the same airport and flies on the same course at 400 miles per hour. How long does it take the second plane to catch the first?

39. A plane leaves an airport and flies east. One hour later, a second plane leaves the same airport and flies south at a speed 50 kilometers per hour greater than the first. Two hours later they are 850 kilometers apart. Find the speed of each plane.

40. A 15.4-foot guy wire runs from the top of a pole to the ground at a point 3.6 feet from the base of the pole. It is decided that in order to keep the pole perpendicular to the ground, it would be better to attach the wire to the middle of the pole instead of the top. How far from the pole will the wire be secured to the ground?

41. A worker has just received a 9.5% raise. If the new salary is $18,450.75, what was the old salary, and how much was the raise?

42. An investor has bought stock in two companies for a total of $24,000. When sold, the profits were 5.6% on the first company, 16.4% on the second, and 12.125% on the total. How much was invested in each company?

43. A student has test scores of 67, 95, 78, and 90. If the final test counts as one-third of the total grade, what must be the score on the final so that the average will be 85?

44. An elevator is stuck partway down a mine shaft. If a stone is dropped from the top of the shaft, 5 seconds later the sound of the stone hitting the elevator can be heard at the top. How far down is the elevator?

45. The gas tank on your car holds 21 gallons. If you get between 16 and 28 miles per gallon, depending on conditions, write an inequality stating how far you can go on a tank of gas.

Relations, Functions, and Graphs

CHAPTER THREE

The Cartesian coordinate system is named after its inventor, René Descartes (1596–1650), who is usually credited with being the founder of analytic geometry. The mathematical thrust of this gifted Frenchman's work was to use algebra to free geometry from the total reliance on diagrams (which he felt could be encumbering and misleading) but at the same time to give geometric interpretations to algebraic operations.

Without question, the most fundamental concept in modern mathematics is that of a function. It has been central to analysis since a 1748 treatise on infinite analysis by the Swiss mathematician Leonhard Euler (1707–1783), the most prolific mathematician in history.

A rudimentary concept of a function, *variable forms*, is in the works of Aristotle. This concept was used by scholars for the next 15 centuries to discuss, for example, temperatures that vary over distance and velocities that vary over time. Around 1361, Oresme had the brilliant idea to draw a picture of the way in which things vary. Hence he was essentially the first person to graph a function, though of course he did not have Descartes' work to aid him.

The German mathematician Gottfried Leibniz (1646–1716), co-founder of calculus, was the first to use the word *function* somewhat as we do today, though definitions were vague. They were still vague at the time of Euler, who described functions as any analytic expressions made up of variable quantities, constant quantities, and numbers (closer to our definition of a relation).

In 1837, Lejeune Dirichlet (1805–1859) made the definition more precise by describing a function as a rule which for every value of x determined a unique value of y. This definition is very close to one we use today.

The ordered pair definition of a function was formulated much later and did not make its way into calculus and lower-level textbooks until the 1950s.

SECTION 3.1. THE COORDINATE SYSTEM

In Sec. 2.1, we discussed how coordinates are associated with points on the real line. Coordinates can also be introduced into the plane using ordered pairs of real numbers. By an **ordered pair of real numbers**, denoted by (a, b), we mean two real numbers a and b of which one, a, is first and the other, b, is second. By equality of ordered pairs we mean the following:

(1)
$$(a, b) = (c, d) \text{ if and only if } a = c \text{ and } b = d.$$

In particular, $(a, b) \neq (b, a)$ if $a \neq b$.

Note: The notation (a, b) is used to denote both ordered pairs and open intervals, but these uses occur in such different situations that no confusion should arise.

A **rectangular** or **Cartesian coordinate system** is introduced into the plane using two perpendicular lines, usually one horizontal and the other vertical. Coordinates are associated with each line so that the origin O on each line is their point of intersection, and usually increasing numbers go to the right on the horizontal line and up on the vertical line. The two lines are called **coordinate axes** and the common point O is called the **origin** of the plane. Usually the horizontal axis is called the **x-axis** and the vertical axis, the **y-axis**, and they are labeled x and y, respectively. The coordinate axes divide the plane into four regions, called the **first, second, third,** and **fourth quadrants**, and may be labeled $Q_1, Q_2, Q_3,$ and Q_4, respectively, as shown in Fig. 1.

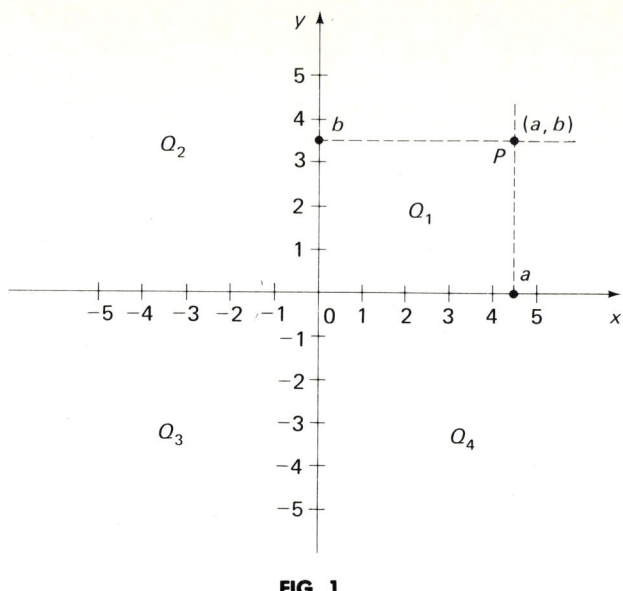

FIG. 1

Let P be a point in the plane. Suppose the vertical and horizontal lines through P intersect the x-axis at a and the y-axis at b, respectively. Then associate P with the ordered pair (a, b). In this way, each point P in the plane is associated with a unique ordered pair of real numbers (a, b) and conversely. The number a is called the **x-coordinate**, or **abscissa**, of P, and b is called the **y-coordinate**, or **ordinate**, of P. Together, a and b are the **coordinates** of P. We sometimes identify points and their coordinates and write "$P = (a, b)$" when we really mean "P is the point whose coordinates are a and b." Fig. 2 shows the coordinates of several points.

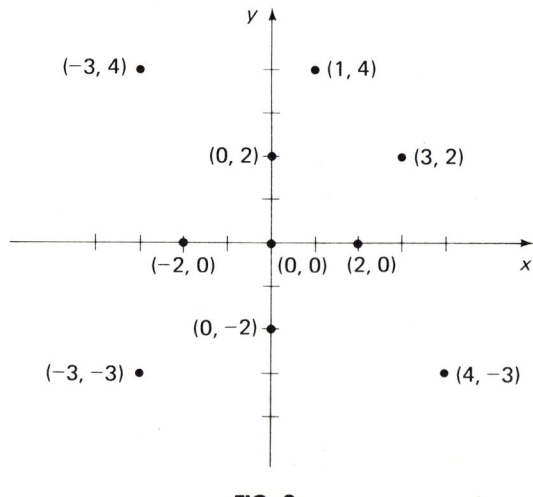

FIG. 2

102 RELATIONS, FUNCTIONS, AND GRAPHS

We now derive the formula for finding the distance between two points in the plane, $P = (x_1, y_1)$ and $Q = (x_2, y_2)$. We shall use the following notation:

---

(2)

The distance between the points P and Q is denoted by $d(P, Q)$.

We first consider the special case where P and Q are on the same horizontal line. Then their y-coordinates are the same, so $y_2 = y_1$. The distance between $P = (x_1, y_1)$ and $Q = (x_2, y_1)$ is the same as the distance between $A = (x_1, 0)$ and $B = (x_2, 0)$, because the line segments PQ and AB are opposite sides of a rectangle. See Fig. 3(a). But by Sec. 2.7, $d(A, B) = |x_1 - x_2| = |x_2 - x_1|$. Hence,

---

(3)

If $P = (x_1, y)$ and $Q = (x_2, y)$ are on the same line parallel to the x-axis, then the distance between them is

$$d(P, Q) = |x_1 - x_2| = |x_2 - x_1|$$

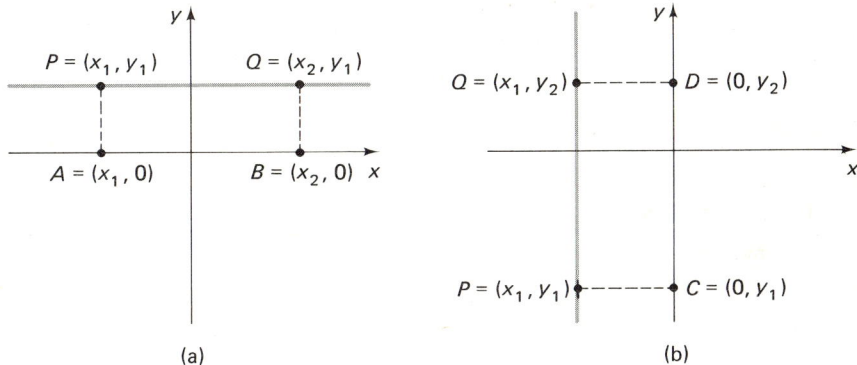

FIG. 3

Similarly,

---

(4)

If $P = (x, y_1)$ and $Q = (x, y_2)$ are on the same line parallel to the y-axis, then the distance between them is

$$d(P, Q) = |y_1 - y_2| = |y_2 - y_1|$$

We now turn to the general case.

Let R be the point (x_2, y_1), and consider the right triangle PQR. See Fig. 4 for two possibilities.

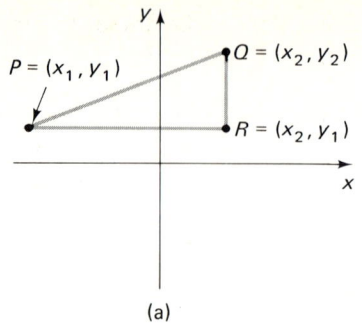
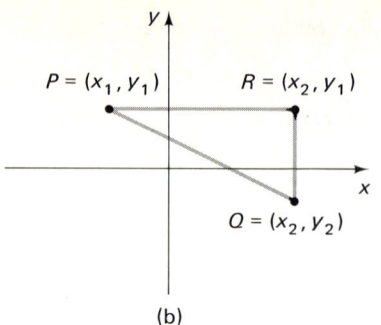

FIG. 4

Now by the previous cases, $d(P,R) = |x_2 - x_1|$ and $d(R,Q) = |y_2 - y_1|$, so by the Pythagorean theorem,

$$d(P,Q)^2 = |x_2 - x_1|^2 + |y_2 - y_1|^2$$

Since $|u|^2 = u^2$ for any real number u,

$$d(P,Q)^2 = (x_2 - x_1)^2 + (y_2 - y_1)^2$$

or

---

(5)   **DISTANCE FORMULA**

If $P = (x_1, y_1)$ and $Q = (x_2, y_2)$, then

$$d(P,Q) = \sqrt{(x_2 - x_1)^2 + (y_2 - y_1)^2}$$

Of course, $(x_1 - x_2)^2 = (x_2 - x_1)^2$ and $(y_1 - y_2)^2 = (y_2 - y_1)^2$, so $d(P,Q) = d(Q,P)$.

EXAMPLE 1 Find the distance d between $(3, -2)$ and $(-1, -7)$.

SOLUTION Let $P = (-1, -7)$, $Q = (3, -2)$, but it does not make any difference in the answer which of the two points we denote as P and which as Q. By (5),

$$d = \sqrt{[3 - (-1)]^2 + [-2 - (-7)]^2} = \sqrt{4^2 + 5^2} = \sqrt{41} \approx 6.40312$$

EXAMPLE 2 Find the distance d between $(a, 2)$ and $(2, a)$.

SOLUTION By (5),

$$d = \sqrt{(a - 2)^2 + (2 - a)^2} = \sqrt{2(a - 2)^2} = |a - 2|\sqrt{2}$$

Note that $(a - 2)\sqrt{2}$ would not be correct, since possibly $a < 2$. Then $(a - 2)\sqrt{2}$ would be negative, but distance is always greater than or equal to zero.

EXAMPLE 3 Use the distance formula to determine if the three points $P = (-5, 3)$, $Q = (1, -1)$, and $R = (4, -3)$ are collinear. See Fig. 5.

SOLUTION

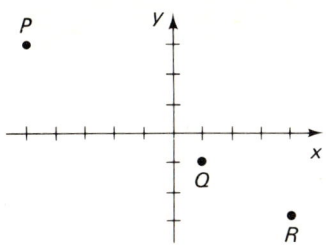

FIG. 5

The triangle inequality states for any three points P, Q, and R, $d(P, R) \leq d(P, Q) + d(Q, R)$, and equality holds if and only if the three points are on a straight line (with Q between P and R), i.e., if and only if P, Q, and R are collinear. We compute the three distances:

$$d(P, R) = \sqrt{(-5-4)^2 + [3-(-3)]^2} = \sqrt{9^2 + 6^2} = \sqrt{117} = 3\sqrt{13}$$

$$d(P, Q) = \sqrt{(-5-1)^2 + [3-(-1)]^2} = \sqrt{6^2 + 4^2} = \sqrt{52} = 2\sqrt{13}$$

$$d(Q, R) = \sqrt{(1-4)^2 + [-1-(-3)]^2} = \sqrt{3^2 + 2^2} = \sqrt{13}$$

Since $d(P, R) = d(P, Q) + d(Q, R)$, they are collinear.

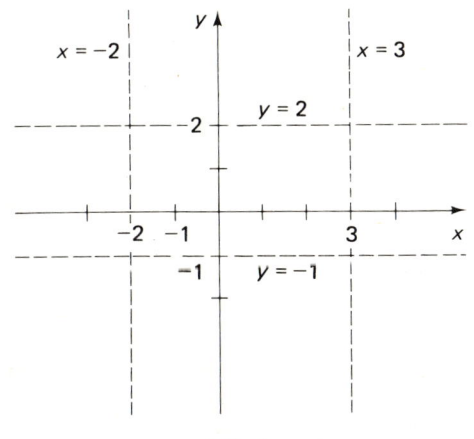

FIG. 6

Before doing the next example, we must describe vertical and horizontal straight lines. See Fig. 6. A vertical line consists of all points with the same x-coordinate. If the x-coordinate is 3, we say "the line $x = 3$." The y-axis is the line $x = 0$. Similarly, a horizontal line consists of all points with the same y-coordinate, such as the line $y = -1$. The x-axis is the line $y = 0$.

EXAMPLE 4 Find the point on the line $x = 2$ equidistant from the points $(1, 2)$ and $(5, -1)$. See Fig. 7.

SOLUTION Let $P = (1, 2)$ and $Q = (5, -1)$. Then we wish to find the point R on the line $x = 2$ with $d(R, P) = d(R, Q)$. Since R lies on the line $x = 2$, it must have coordinates $(2, y)$.

Using the distance formula, we obtain

$$\sqrt{(2-1)^2 + (y-2)^2} = \sqrt{(2-5)^2 + [y-(-1)]^2}.$$

Section 3.1 The Coordinate System

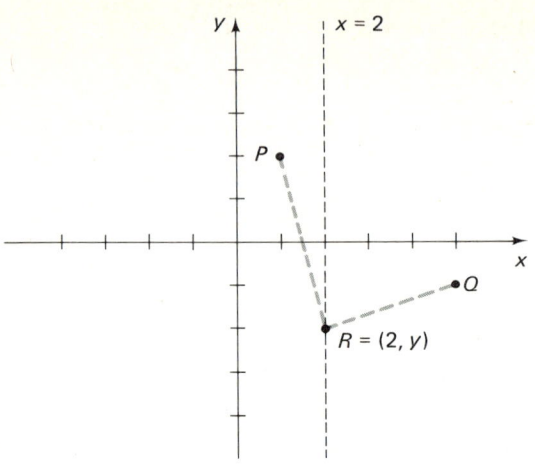

FIG. 7

Squaring both sides and solving,

$$1^2 + (y - 2)^2 = 3^2 + (y + 1)^2$$
$$1 + y^2 - 4y + 4 = 9 + y^2 + 2y + 1$$
$$-5 = 6y, \quad y = -\tfrac{5}{6}$$

The point is $(2, -\tfrac{5}{6})$.

Suppose a and b are any two real numbers such that $a < b$. See Fig. 8(a). If m is the midpoint of the line segment between a and b, then

$$m - a = b - m$$

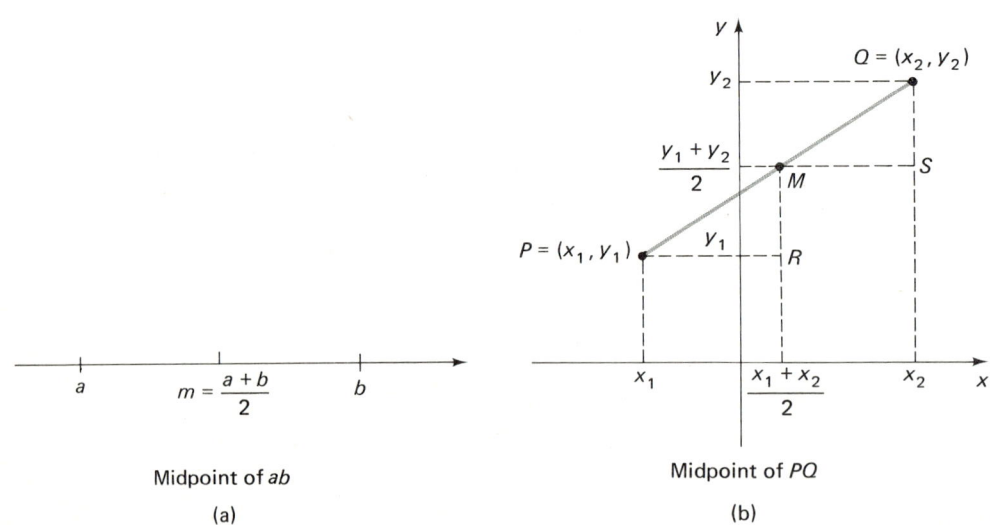

Midpoint of ab

(a)

Midpoint of PQ

(b)

FIG. 8

Solving for m, we get

$$2m = a + b \quad \text{or} \quad m = \frac{a + b}{2}$$

Now let us find the coordinates of the midpoint of the line segment between any two points $P = (x_1, y_1)$ and $Q = (x_2, y_2)$ in the plane. In Fig. 8(b), we draw this segment and then drop perpendiculars from P and Q to the x- and y-axes. On the x-axis, the midpoint of the segment between x_1 and x_2 is $(x_1 + x_2)/2$, while on the y-axis, the midpoint of the segment between y_1 and y_2 is $(y_1 + y_2)/2$. We claim that the midpoint of the line segment between P and Q is
$$M = \left(\frac{x_1 + x_2}{2}, \frac{y_1 + y_2}{2}\right).$$

This can be seen two ways. You can use the distance formula to check that $d(P, M) = d(M, Q)$ and that P, M, and Q are collinear, or you can plot the points $R = \left(\frac{x_1 + x_2}{2}, y_1\right)$ and $S = \left(x_2, \frac{y_1 + y_2}{2}\right)$ and observe that triangles PRM and MSQ are congruent. Hence,

---

(6) **MIDPOINT FORMULA**

The midpoint of the segment between (x_1, y_1) and (x_2, y_2) is
$$M = \left(\frac{x_1 + x_2}{2}, \frac{y_1 + y_2}{2}\right)$$

EXAMPLE 5 Find the midpoint of the segment between $(-3, 1)$ and $(2, -5)$. Sketch.

SOLUTION

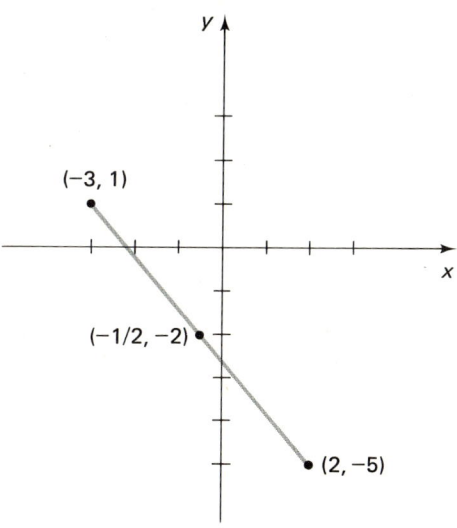

By (6), the midpoint is
$$\left(\frac{-3 + 2}{2}, \frac{1 + (-5)}{2}\right) = \left(-\frac{1}{2}, -2\right)$$

EXERCISES

In Exercises 1 and 2, plot the points on a rectangular coordinate system.

1. $A = (5, -3)$, $B = (-5, -3)$, $C = (3, 2)$, $D = (-3, 2)$, $E = (-2, 0)$, $F = (0, -2)$
2. $A = (4, 5)$, $B = (-4, 5)$, $C = (3, -2)$, $D = (-3, -2)$, $E = (3, 0)$, $F = (0, 3)$

In Exercises 3–6, choose suitable scales on the axes, and plot and label the points.

3. $A = (300, 1)$, $B = (350, 2)$, $C = (400, -1)$
4. $A = (.1, -.002)$, $B = (.2, -.004)$, $C = (-.1, -.005)$
5. $A = (-.01, 25)$, $B = (-.04, 35)$, $C = (-.05, -15)$
6. $A = (-1, -5000)$, $B = (0, 2000)$, $C = (2, 3000)$

In Exercises 7–14, plot the points and then find (a) the distance between the given points and (b) the coordinates of the midpoint of the line segment between them.

7. $(5, 2)$, $(-7, -3)$
8. $(-4, 3)$, $(2, 2)$
9. $(7, -2)$, $(7, 5)$
10. $(-3, 7)$, $(0, -5)$
11. $(a + b, a - b)$, $(b - a, a + b)$
12. $(c^{1/2}, d^{1/2})$, $(-c^{1/2}, -d^{1/2})$
13. $(s^{3/2}, \frac{3}{2})$, $(s^{-3/2}, -\frac{1}{2})$
14. $(t, |t|)$, $(-|t|, t)$

In Exercises 15–17, plot the three points, and use the distance formula to determine if they are collinear.

15. $(5, -4)$, $(0, 0)$, $(-6, 5)$
16. $(3, 2)$, $(0, 3)$, $(-\frac{1}{2}, 3\frac{1}{3})$
17. $(-2, -3)$, $(2, -\frac{1}{3})$, $(1, -1)$

18. The points $(-2, -1)$, $(3, 1)$, $(0, 4)$ are vertices of a triangle. Show that the triangle is isosceles and find its area.
19. The points $(1, 2)$ and $(1, 6)$ are vertices of a square. Find the other two vertices. (*Note*: There are three possible pairs.)
20. Answer Exercise 19 if the two vertices are (n, p) and (n, q), $q > p$.
21. Find all points on the y-axis that are a distance of 5 from the point $(3, 5)$.
22. Find all points on the x-axis that are a distance of 13 from the point $(3, 5)$.
23. Find all points on the line $x = -2$ that are a distance of 13 from the point $(3, 5)$.
24. Find all points on the x-axis equidistant from the points $(1, -2)$ and $(3, 1)$.
25. Find all points on the y-axis equidistant from the points $(-2, -1)$ and $(-1, 1)$.
26. Find all points on the line $x = 4$ equidistant from $(-1, -1)$ and $(1, 3)$.

27. Let $O = (0,0)$, $A = (a,0)$, and $B = (0,b)$, $a, b > 0$, be the vertices of a right triangle.
   a. Find the midpoint M of AB.
   b. Prove that M is equidistant from the three vertices.

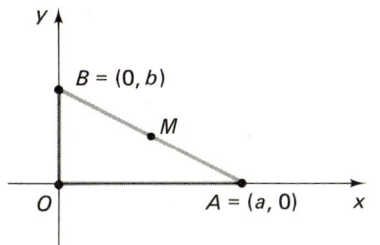
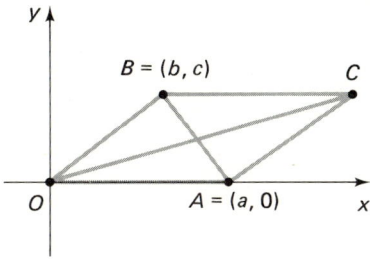

28. Let $O = (0,0)$, $A = (a,0)$, $B = (b,c)$, and C be vertices of a parallelogram, where $a, c > 0$.
   a. Determine the coordinates of C (from the fact $OABC$ is a parallelogram).
   b. Show that the midpoints of AB and OC are the same and hence that the diagonals of a parallelogram bisect each other.
29. Referring to Fig. 8(b), use the distance formula to show that
   a. $d(P,M) = d(M,Q)$
   b. $d(P,Q) = d(P,M) + d(M,Q)$

SECTION 3.2. RELATIONS, GRAPHS, AND SYMMETRY

Subsets of the plane, i.e., sets of ordered pairs of real numbers, arise in many different contexts, and it becomes convenient to have a name for such sets.

(7) **DEFINITION**

A **relation** is a set of ordered pairs of real numbers.

Usually relations are given in terms of conditions that the ordered pairs must satisfy. For example,

$$\{(x,y) | y \leq x\}, \quad \{(x,y) | y = x^2 + 1\}, \quad \{(x,y) | |x| < 1 \text{ or } |y| < 2\}$$

In most cases, the set notation is omitted, and we just write "the relation $y \leq x$" or "the relation $y = x^2 + 1$," etc. However, sometimes we just have to list the members of the set, for example $\{(1,2), (3,5), (-2,6), (1,-4)\}$.

To understand a particular relation, it is helpful to sketch its graph.

(8) **DEFINITION**

The **graph of a relation** is the set of points in the plane whose coordinates are in the relation.

EXAMPLE 1 Sketch the graph of the following relations: **(a)** $y = x$; **(b)** $y \leq x$; **(c)** $y > x$.

SOLUTION
  (a) By plotting a few points, it is easy to see the graph of $y = x$ is just the diagonal line through the first and third quadrants. See Fig. 9(a).
  (b) A point (x,y) is in the graph of $y \leq x$ if and only if its second coordinate is less than or equal to its first coordinate, so that it lies on or below the line $y = x$. See Fig. 9(b).
  (c) Similar to (b), this is all points above, but not on, the line $y = x$. See Fig. 9(c).

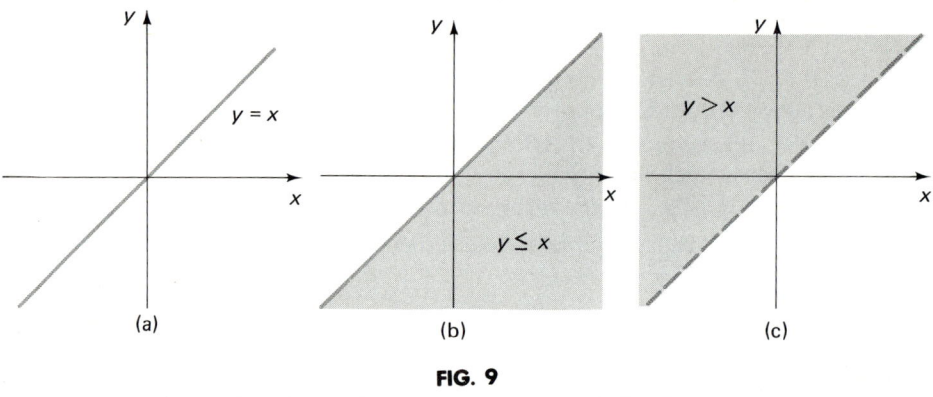

FIG. 9

Note that the boundary of a region is indicated by a solid or dotted line, depending on whether it is included or not. See Fig. 9.

EXAMPLE 2 Sketch the graph of $|x + 2| \leq 1$.

SOLUTION The solution to $|x + 2| \leq 1$ is $-3 \leq x \leq -1$, so a point (x,y) is on this graph if and only if its first coordinate is in the interval $[-3, -1]$. See Fig. 10(a).

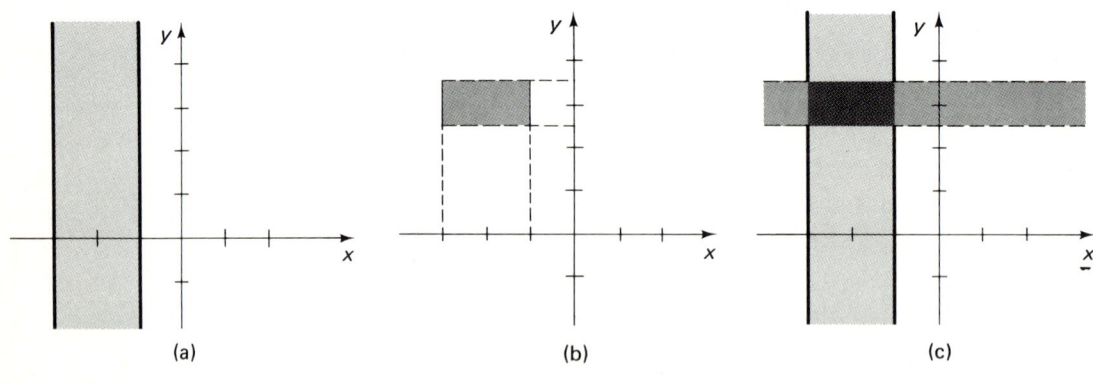

FIG. 10

EXAMPLE 3 Sketch the graph of $|x + 2| \leq 1$ and $|y - 3| < \frac{1}{2}$.

SOLUTION For (x,y) to satisfy $|x + 2| \leq 1$ and $|y - 3| < \frac{1}{2}$, we have x in $[-3, -1]$ and y in $(2.5, 3.5)$. The graph is a box with part of its boundary. See Fig. 10(b).

EXAMPLE 4 Sketch the graph of $|x + 2| \leq 1$ or $|y - 3| < \frac{1}{2}$.

SOLUTION This is similar to Example 4 except that "and" is replaced by "or." The graph is two strips forming a cross. See Fig. 10(c).

Note that Fig. 10(b) and (c) can be described as the intersection and union, respectively, of the two strips.

Since graphs often give us a better understanding of relations, any concept that facilitates graphing is useful. One such concept is symmetry.

EXAMPLE 5 Sketch the graphs of **(a)** $y = x^2$ and **(b)** $y^2 = x$.

SOLUTION

(a)

x	$y = x^2$
-2	4
-1	1
$-\frac{1}{2}$	$\frac{1}{4}$
0	0
$\frac{1}{2}$	$\frac{1}{4}$
1	1
2	4

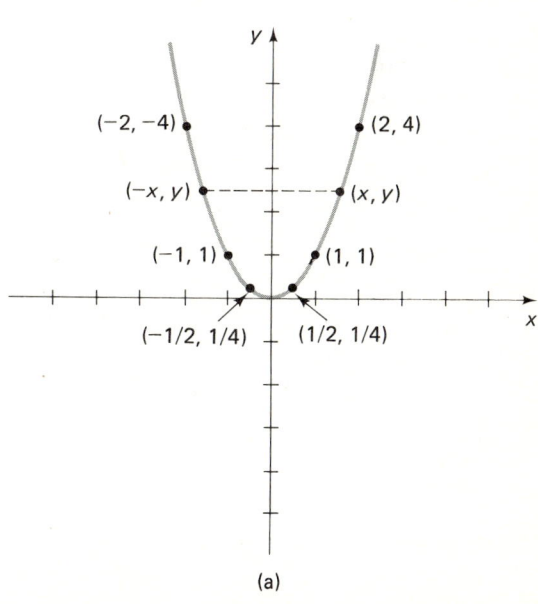

(a)

FIG. 11

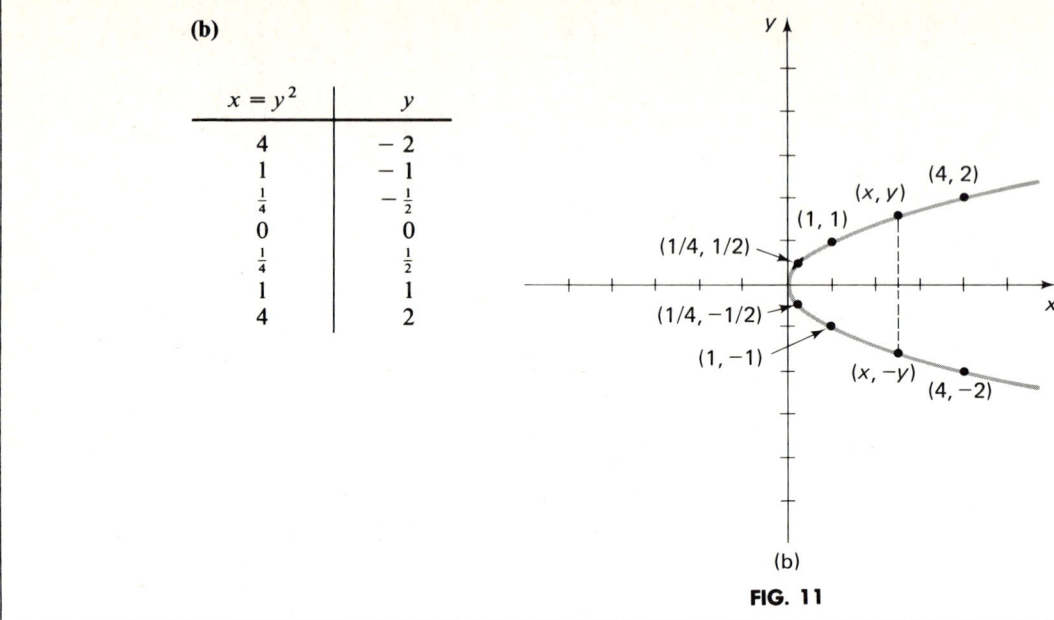

FIG. 11

If the plane in Fig. 11(a) is folded along the y-axis, then the left half of the graph of $y = x^2$ coincides with the right half. When this happens, we say the graph is **symmetric with respect to** (or **about**) **the y-axis**. Thus the graph of $y = x^2$ is symmetric about the y-axis. Further, we can see in Fig. 11(a) that the point $(-x, y)$ is on the graph whenever the point (x, y) is on the graph.

Similarly, if the plane in Fig. 11(b) is folded along the x-axis, then the upper half of the graph of $y^2 = x$ coincides with the lower half. When this happens, we say the graph is **symmetric with respect to** (or **about**) **the x-axis**. Thus the graph of $y^2 = x$ is symmetric about the x-axis. We can see in Fig. 11(b) that the point $(x, -y)$ is on the graph whenever the point (x, y) is on the graph.

These examples illustrate the general situation.

---

(9)

When a relation is given by an equation,
1. The graph of the relation is symmetric with respect to the x-axis if and only if replacing y with $-y$ produces an equivalent equation.
2. The graph of the relation is symmetric with respect to the y-axis if and only if replacing x with $-x$ produces an equivalent equation.

EXAMPLE 6 Determine if the relation $y = x^4 - x^2$ is symmetric about either the x- or y-axis.

SOLUTION If we replace x by $-x$, we get $y = (-x)^4 - (-x)^2$, which is equivalent to $y = x^4 - x^2$. If we replace y by $-y$, we get $-y = x^4 - x^2$ or $y = x^2 - x^4$, which is not equivalent. Thus, $y = x^4 - x^2$ is symmetric about the y-axis but not about the x-axis.

EXAMPLE 7 Determine if the relation $|y| = x$ is symmetric about either axis.

SOLUTION $|y| = x$. If we replace x by $-x$, we get $|y| = -x$, which is not equivalent. If we replace y by $-y$, we get $|-y| = x$, which is equivalent to $|y| = x$. Thus, $|y| = x$ is symmetric about the x-axis but not about the y-axis.

Another type of symmetry is **symmetry with respect to** (or **about**) **the origin**. In this case, whenever the point (x,y) is on the graph, the point $(-x, -y)$ is also on it.

EXAMPLE 8 Graph $y = x^3$.

SOLUTION

x	$y = x^3$
-2	-8
-1	-1
$-\frac{1}{2}$	$-\frac{1}{8}$
0	0
$\frac{1}{2}$	$\frac{1}{8}$
1	1
2	8

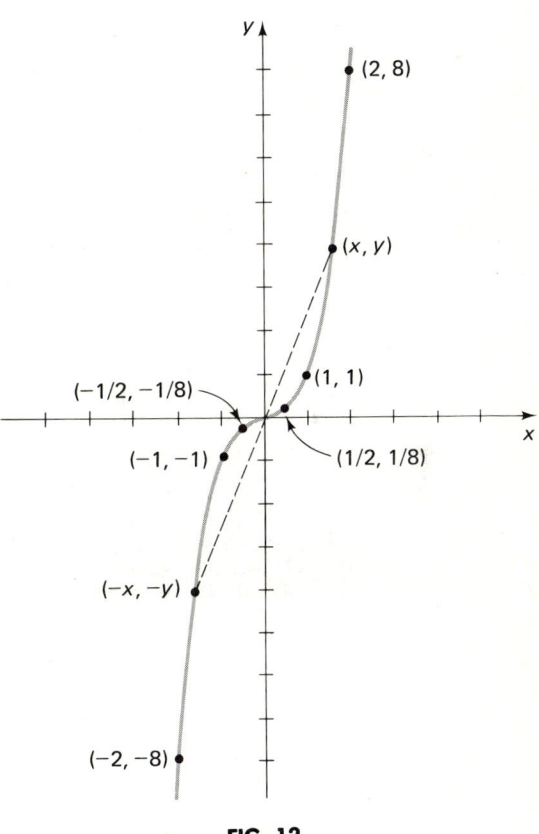

FIG. 12

You can see that the graph (Fig. 12) in Example 8 is symmetric with respect to the origin. Note that substituting $-x$ for x or $-y$ for y, but not both, gives an equation which is not equivalent to $y = x^3$. However, substituting both gives $-y = (-x)^3$, which is equivalent to $y = x^3$. This illustrates the general situation.

> **(10)**
>
> When a relation is given by an equation, the graph of the relation is symmetric with respect to the origin if and only if replacing x with $-x$ and simultaneously y with $-y$ produces an equivalent equation.

Note: If a relation is symmetric with respect to any two of the x-axis, y-axis, and origin, it is also symmetric with respect to the third.

We now discuss how to use the knowledge that the graph is symmetric to aid in graphing the relation. We shall need the concept of a reflection.

> **(11)**   **DEFINITION**
>
> To **reflect** a graph **through the y-axis** means to replace each point (x, y) of the graph with the point $(-x, y)$.
> To **reflect** a graph **through the x-axis** means to replace each point (x, y) of the graph with the point $(x, -y)$.
> To **reflect** a graph **through the origin** means to replace each point (x, y) of the graph with the point $(-x, -y)$.

See Fig. 13 for examples of reflections.

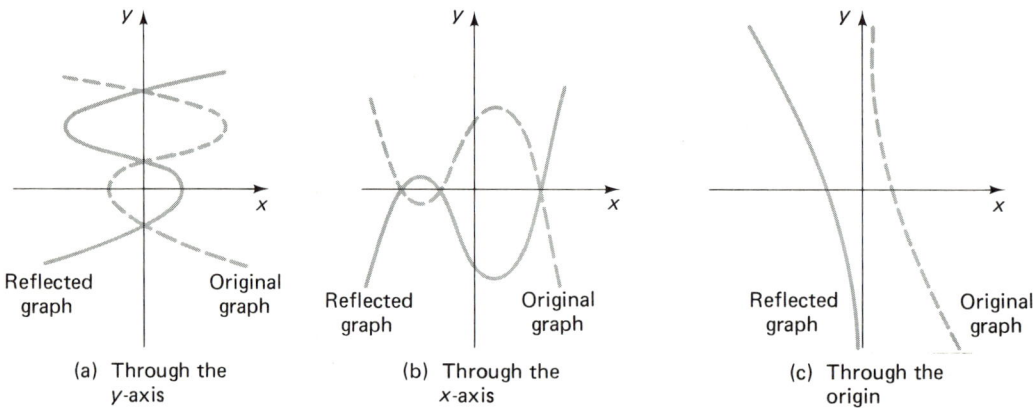

(a) Through the y-axis   (b) Through the x-axis   (c) Through the origin

Suppose that a relation is symmetric with respect to the x-axis. We first graph all points (x, y) in the relation with $y \geq 0$ and then reflect this part of the graph through the x-axis to get all points (x, y) in the relation with $y \leq 0$. (The whole graph is then these two parts put together.) If a relation is symmetric with respect to the y-axis, first graph all points (x, y) with $x \geq 0$, and then reflect this through the y-axis to get all the points (x, y) with $x \leq 0$. If the graph is symmetric with respect to the origin, first graph all points (x, y) with $x \geq 0$ and then reflect this through the origin to get all the points (x, y) with $x \leq 0$. We shall give some examples.

EXAMPLE 9 Figure 14 shows portions of graphs of relations in the first quadrant. Find more of the graphs if the relation is symmetric about
(a) the x-axis, (b) the y-axis, (c) the origin.

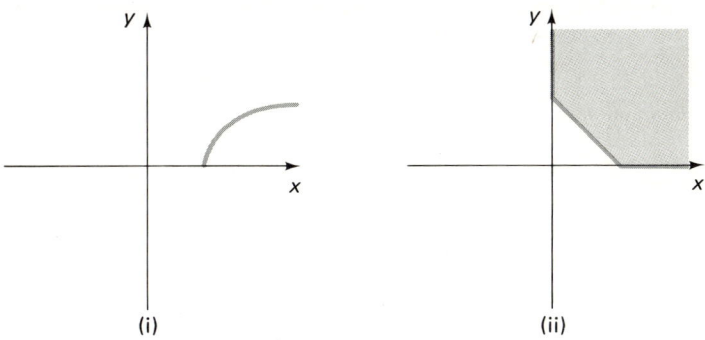

FIG. 14

SOLUTION See Fig. 15.

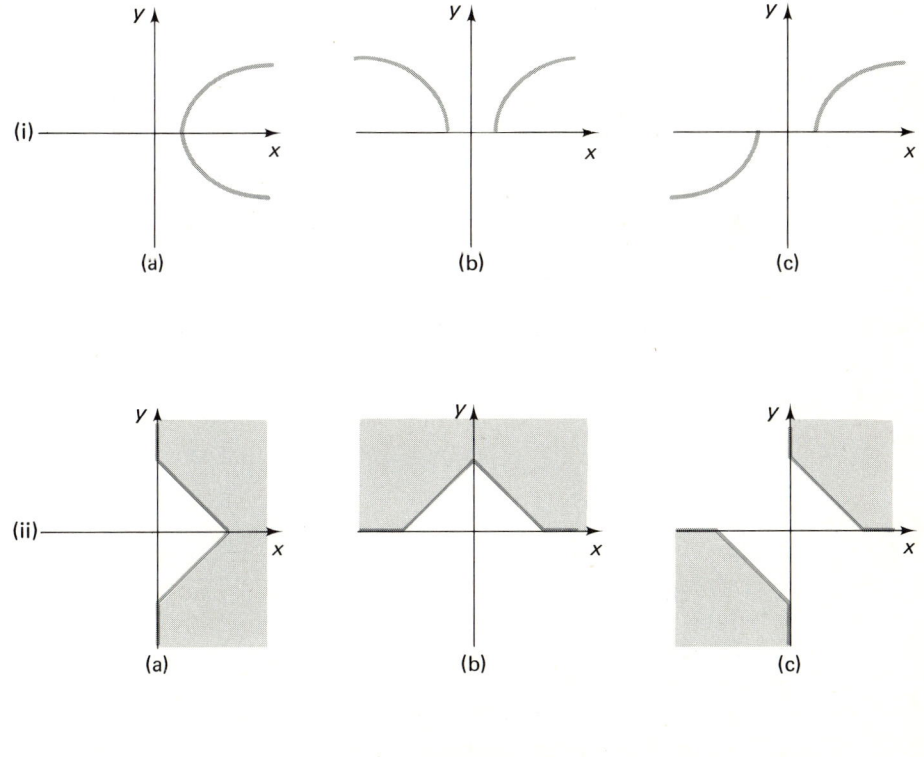

FIG. 15

EXAMPLE 10 Discuss the symmetry of $y = 1/x$. Graph the relation for $x \geq 0$, and then use symmetry to find the complete graph.

SOLUTION First substituting $-x$ for x, then $-y$ for y, and then both, we see that the relation is symmetric about the origin but not about either axis. See Fig. 16 for the graphs.

x	$y = 1/x$
0	Undefined
10	$\frac{1}{10}$
5	$\frac{1}{5}$
1	1
$\frac{1}{5}$	5
$\frac{1}{10}$	10

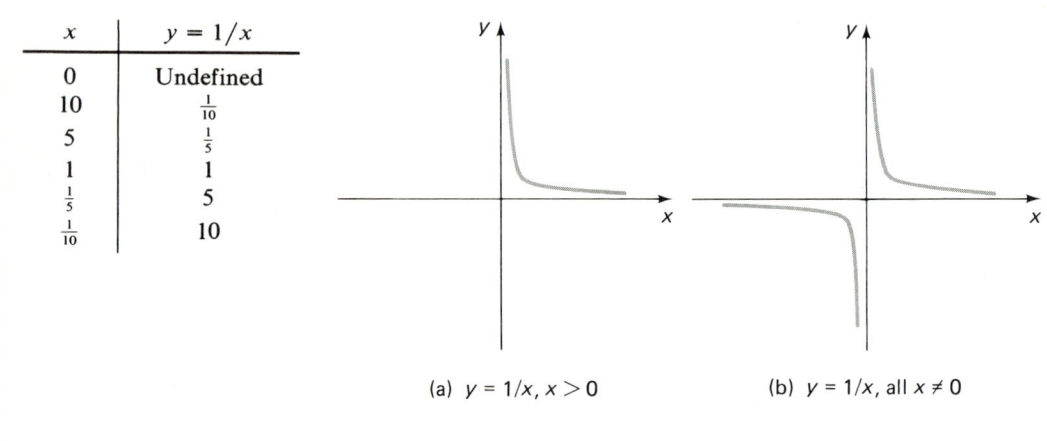

(a) $y = 1/x$, $x > 0$

(b) $y = 1/x$, all $x \neq 0$

FIG. 16

Figure 16 illustrates a phenomenon we have not observed before, namely, that the graph approaches fixed lines. As x gets very large positively or very large negatively, the graph approaches the x-axis. When this happens, we call the x-axis a **horizontal asymptote**. As x gets close to zero, the graph approaches the y-axis. When this happens, we call the y-axis a **vertical asymptote**. We shall see in the following sections that asymptotes are a valuable aid in graphing.

One very important relation is a circle. A **circle** is the set of all points in the plane at a given distance, called the **radius**, from a fixed point, called the **center**. (See Fig. 17.)

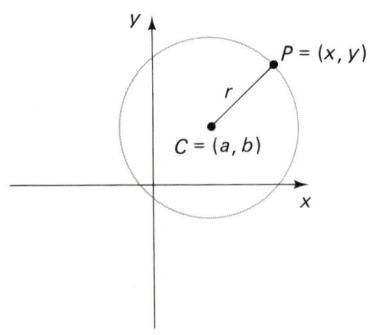

If $C = (a, b)$ is the center and r is the radius of a circle, then a point $P = (x, y)$ is on the circle if and only if

$$d(P, C) = r$$

By the distance formula,

$$\sqrt{(x - a)^2 + (y - b)^2} = r$$

FIG. 17

By squaring both sides, we get the following very useful form:

---

(12)

The standard form of the equation for a circle of radius r and center (a, b) is

$$(x - a)^2 + (y - b)^2 = r^2$$

An important special case is when $a = 0 = b$:

---

(13)

The equation of the circle of radius r centered at the origin is

$$x^2 + y^2 = r^2$$

A circle centered at the origin is symmetric with respect to both axes and the origin.

EXAMPLE 11 Describe and graph the relation $2(x + 3)^2 + 2(y - 1)^2 = 4.7$.

SOLUTION First, rewrite this equation in standard form,

$$(x + 3)^2 + (y - 1)^2 = 2.35 \quad \text{or} \quad [x - (-3)]^2 + (y - 1)^2 = (\sqrt{2.35})^2$$

This is the equation of a circle of radius $\sqrt{2.35} \approx 1.53297$ and center $(-3, 1)$. See Fig. 18(a).

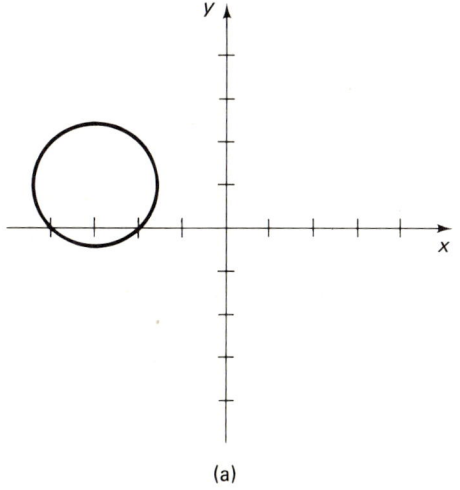
(a)

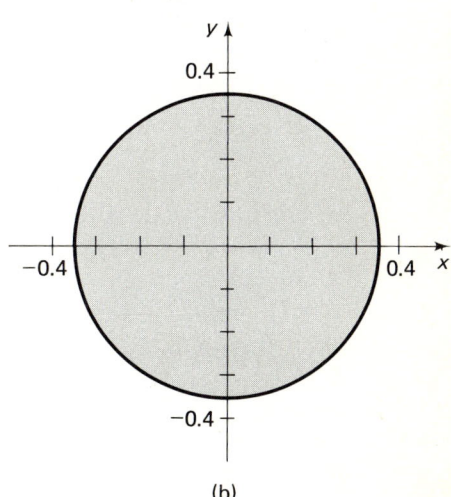
(b)

FIG. 18

EXAMPLE 12 Describe and graph the relation $x^2 + y^2 \leq .11$.

SOLUTION First, rewrite this as $x^2 + y^2 \leq (\sqrt{.11})^2$ and let $r = \sqrt{x^2 + y^2}$. Then r is the distance from the point (x,y) to the origin, and (x,y) satisfies this relation if and only if $0 \leq r \leq \sqrt{.11}$. Therefore, the graph is the set of all points whose distance from the origin is less than or equal to $\sqrt{.11}$, i.e., all points on or inside the circle centered at the origin of radius $\sqrt{.11} \approx .331662$. See Figure 18(b).

EXAMPLE 13 Find the equation of the circle with center $C = (2, -3)$ and through the point $P = (4, 1)$.

SOLUTION The equation for the circle is $(x - a)^2 + (y - b)^2 = r^2$. Since $C = (2, -3)$, it follows that $a = 2$ and $b = -3$. Hence we only need to find r. Since P is on the circle, the radius r of the circle is the distance between C and P. Thus

$$r = d(C,P) = \sqrt{(2-4)^2 + (-3-1)^2} = \sqrt{2^2 + 4^2} = \sqrt{20}$$

Therefore, the equation is $(x - 2)^2 + (y + 3)^2 = 20$.

Sometimes an equation for a circle is not given in standard form. Since we need to know its center and radius in order to graph the circle, we must rewrite the equation in standard form. The algebraic technique used to do this is called *completing the square* (and was used to derive the quadratic formula in Sec. 2.3). The process is illustrated in Example 14.

EXAMPLE 14 Put the equation $2x^2 + 8x + 2y^2 - 3y - 1 = 0$ into standard form. Determine the center and radius of the circle with this equation.

SOLUTION We first divide by the coefficient of x^2 and y^2 (making their coefficients 1),

$$x^2 + 4x + y^2 - \tfrac{3}{2}y - \tfrac{1}{2} = 0$$

and then bring the constant to the other side:

$$x^2 + 4x + y^2 - \tfrac{3}{2}y = \tfrac{1}{2}$$

We now want to add constants to both sides of the equation, so that the left-hand side is the sum of two perfect squares; i.e., the equation is of the form

$$(x - a)^2 + (y - b)^2 = \text{a constant}$$

(In doing so, we shall have "completed the square," twice.) Since

$$(x - a)^2 = x^2 - 2ax + a^2, \qquad (y - b)^2 = y^2 - 2by + b^2$$

we must have

$$-2a = 4, \qquad -2b = -\tfrac{3}{2} \qquad \text{or} \qquad a = -2, \qquad b = \tfrac{3}{4}$$

Thus the terms we have to add are

$$a^2 = (-2)^2 = 4, \qquad b^2 = \left(\tfrac{3}{4}\right)^2 = \tfrac{9}{16}$$

Therefore, we get

$$x^2 + 4x + 4 + y^2 - \tfrac{3}{2}y + \tfrac{9}{16} = \tfrac{1}{2} + 4 + \tfrac{9}{16}$$

Factoring and combining, we obtain

$$(x + 2)^2 + \left(y - \tfrac{3}{4}\right)^2 = \frac{8 + 64 + 9}{16} = \frac{81}{16}$$

or

$$[x - (-2)]^2 + \left(y - \tfrac{3}{4}\right)^2 = \left(\tfrac{9}{4}\right)^2$$

Therefore this is a circle with center $(-2, \tfrac{3}{4})$ and radius $\tfrac{9}{4}$.

EXERCISES

In Exercises 1–44, sketch the graphs of the relation. Use symmetry where possible. Determine the center and radius when the graph is a circle.

1. $y = -x$
2. $y < -x$
3. $y \geq -x$
4. $y = 3$
5. $y = -4$
6. $x = -7$
7. $x = 1$
8. $xy > 0$
9. $xy \leq 0$
10. $(x - 3)(y + 4) = 0$
11. $(x + 2)(y - 3) = 0$
12. $|x - 2| < 3$
13. $|x + 1| \geq 4$
14. $|y + 2| > 1$
15. $|y - 7| < 2$
16. $|x - 3| < 1$ and $|y + 1| < 2$
17. $|x + 4| \geq 4$ and $|y - 1| \leq 2$
18. $|x - 2| \leq 3$ or $|y + 4| < 1$
19. $|x - 7| \geq 1$ or $|y - 2| \geq 2$
20. $y = x^2 - 1$
21. $x = y^2 + 1$
22. $y = |x + 1|$
23. $y \leq |x| + 1$
24. $x = |y| - 1$
25. $y = -x^3$
26. $y \geq x^3 + 1$
27. $y = -1/x$
28. $y = 1/x^2$
29. $y = x + |x|$
30. $y = |x| - x$
31. $|x| + |y| = 0$
32. $|y| = |x|$
33. $x^2 + y^2 = 8$
34. $(x - 2)^2 + (y - 4)^2 = 1$
35. $(x + 2)^2 + (y + 1)^2 \geq 1$
36. $(x - 1)^2 + (y - 3)^2 < 4$
37. $x^2 + y^2 - 2x + 4y + 1 = 0$
38. $2x^2 + 2y^2 + 2x - 4y - 1 = 0$
39. $3x^2 + 3y^2 - 12x + 18y \geq 0$
40. $x^2 + y^2 + x + 6y - \tfrac{1}{4} < 0$
41. $3x^2 + 3y^2 - 6x - 10y < 0$
42. $2x^2 + 2y^2 + 5x + 6y - 4 = 0$
43. $2x^2 + 2y^2 + 3x + 5y + 4 > 0$
44. $-4x^2 - 4y^2 + 3x + 7y \geq 0$

In Exercises 45–64 determine if the relation given by the equation is symmetric about (a) the x-axis, (b) the y-axis or (c) the origin.

45. $y = x^2 - 2$
46. $3x^2 + y = 1$
47. $x + y = 2$
48. $3x + 3y = 7$

49. $y^2 + x = 4$
51. $y^4 = x^3$
53. $x = -y$
55. $3x^2 + 4y^2 = 2$
57. $x = \frac{5}{4}$
59. $x = y^5$
61. $x + 2y = 3$
63. $y = |x|$

50. $y^3 = x^2$
52. $4x + 4y = 0$
54. $|x| + |y| = 1$
56. $4xy = 7$
58. $y = |2x|$
60. $3x - 2y = 1$
62. $5x + 5y = 6$
64. $|y| = |x|$

In Exercises 65–70, part of the graph of a relation is given. Complete the graph if it is symmetric about **(a)** the *x*-axis, **(b)** the *y*-axis, **(c)** the origin, or **(d)** both the *x*- and *y*-axes.

65.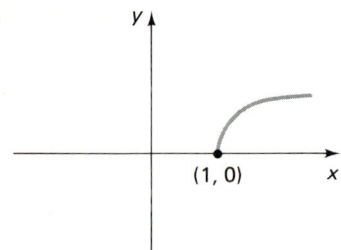

66.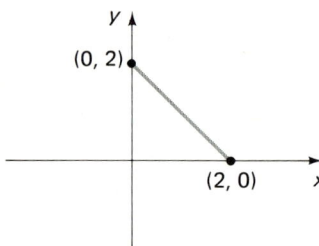

67.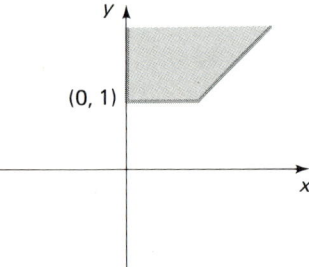

68.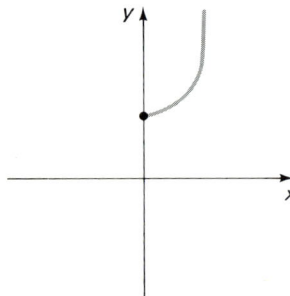

69.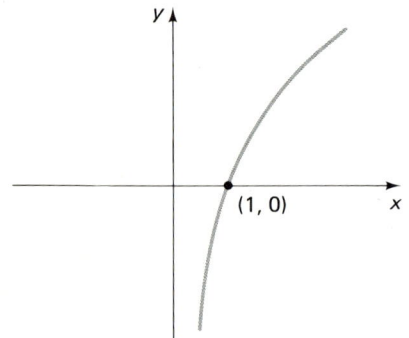

70.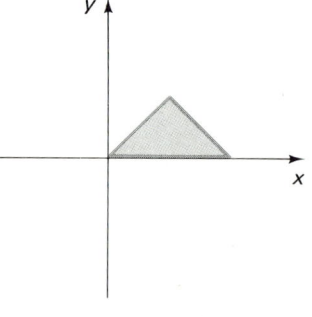

71. Find the equation of the circle centered at $(-1, -3)$ and through the point $(3, 0)$.
72. Find the equation of the circle centered at $(1, -2)$ and tangent to the line $x = 4$.

73. Is the point $(3, -4)$ inside, on, or outside the circle centered at the origin and passing through the point $(-4, 5)$?

74. Is the point $(3, -2)$ inside, on, or outside the circle centered at $(0, 2)$ and tangent to the line $y = 6$?

SECTION 3.3.   FUNCTIONS

The concept of a function is one of the most central concepts in modern mathematics. It is also used frequently in everyday life. For instance, the statements "each student in the course will be assigned a grade at the end of the course" and "each item in the store has a price marked on it" describe functions. If we analyze these statements, we shall find the essential ingredients of a function.

For the first statement, there is a set of students, a set of possible grades, and a rule which assigns to each member of the first set (to each student) a unique member of the second set (a grade). For the second statement, there is a set of items for sale, a set of possible prices, and a rule which assigns to each member of the first set (to each item) a unique member of the second set (a price).

---

(14)   **DEFINITION**

A **function** f from a set D to a set R is a rule which assigns to each element of D a unique element of R. The set D is called the **domain** of the function, while the set R is called the **range** of the function.

Besides f, we also use the letters F, g, G, etc., to denote functions.

If a is an element of D, then the unique element in R which f assigns to a is called the **value of f at a** or the **image of a under f** and is denoted by $f(a)$. We will understand that the range R is the set of all values of the function.* We sometimes represent functions pictorially as in the accompanying figure.

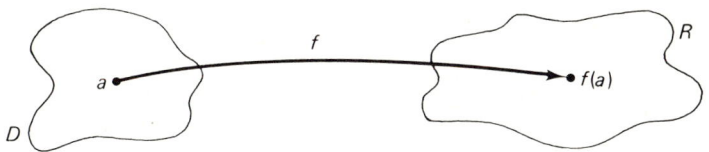

We often think of x as representing an arbitrary element of D and y representing the corresponding value of f at x. We then call x the **independent variable**, because a value for x can be selected arbitrarily from D, and call y the **dependent variable**, because the value of y depends on the value of x selected.

---

*Some texts allow the range to be larger than the set of all values of the function.

We also write
$$y = f(x)$$
which is read "y is a function of x" or "y is f of x."

The rule of a function gives the value of the function at each element of the domain. Often the rule is a formula, but it can be other things such as a list of ordered pairs, a table, or a set of instructions.

You can think of some of the buttons on your calculator as providing rules for different functions. One that you already know about is [√]. If you press [2][√], you obtain [1.4142135]; your calculator is giving you the (approximate) value of the square root function at the number 2. In this way, the calculator illustrates what has been one traditional description of a function: A function is like a black box into which you can put any number from the domain and out of which comes the corresponding value in the range. See Fig. 19.

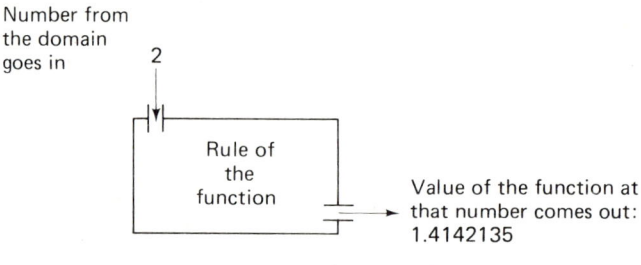

FIG. 19

If the rule of a function f is a formula giving y in terms of x, say $y = x^2 + 2x$, then the statements "the function $y = x^2 + 2x$," "$y = f(x)$ where $f(x) = x^2 + 2x$," or just "$f(x) = x^2 + 2x$" all indicate the same thing. Given any of these, to find the value of f at a number, you substitute that number for x wherever x occurs in the formula (and then simplify).

EXAMPLE 1 If f is a function with domain the set of all real numbers and with rule $f(x) = x^2 + 2x$, find the following values of f:
(a) $f(3)$; (b) $f(-11.21)$; (c) $f(1/a)$, $a \neq 0$; (d) $f(-1 + h)$; (e) $f(f(3))$.

SOLUTION (a) $f(3) = 3^2 + 2(3) = 9 + 6 = 15$.
(b) $f(-11.21) = (-11.21)^2 + 2(-11.21) = 125.6641 - 22.42 = 103.2441$.
(c) $f\left(\dfrac{1}{a}\right) = \left(\dfrac{1}{a}\right)^2 + 2\dfrac{1}{a} = \dfrac{1}{a^2} + \dfrac{2}{a} = \dfrac{1 + 2a}{a^2}$.
(d) $f(-1 + h) = (-1 + h)^2 + 2(-1 + h) = 1 - 2h + h^2 - 2 + 2h = h^2 - 1$.
(e) By part (a), $f(3) = 15$, so $f(f(3)) = f(15) = 15^2 + 2(15) = 255$.

There is a standard notation indicating a function together with its domain and range.

> **(15) NOTATION**
>
> We write "$f: D \to R$" for "f is a function with domain D and range R" or, equivalently, "f is a function from D to R."

For example, a statement like

"Let $f: \mathbb{R} \to [1, \infty)$ by $f(x) = x^2 + 1$"

gives the domain ($\mathbb{R}$), range ($[1, \infty)$), and rule ($f(x) = x^2 + 1$). However, since you can compute the range from the domain and the rule, the range is often not given. In fact, most of the time the domain is also not given. All that is written is something like

"Let $f(x) = x^2 + 1$"   or   "Let $y = x^2 + 1$"

When this is written, the domain is understood to be the largest possible set of real numbers for which the rule applies (i.e., for which the rule gives a real number).

EXAMPLE 2 Find the largest possible set of real numbers which can be the domain of $h(x) = \sqrt{1 - 1/x}$.

SOLUTION

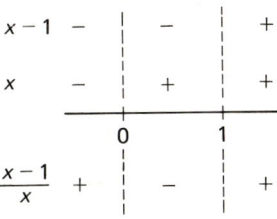

Since $\sqrt{u}$ is defined only when $u \geq 0$, $g(x)$ is defined whenever $1 - 1/x \geq 0$. We rewrite this inequality as $(x-1)/x \geq 0$ and solve it as in Sec. 2.7. The domain is $\{x \mid x < 0 \text{ or } x \geq 1\} = (-\infty, 0) \cup [1, \infty)$.

To understand a function better, it is often helpful to draw its graph. The graph of a function f is the set of all points in the plane whose coordinates satisfy the equation $y = f(x)$. One way to sketch the graph of a function is to make a table of several values of the function, plot these points on a graph, and sketch in the curve between these points. (Be sure to pick points which give the true nature of the graph.) See Fig. 20 where this is done for the function $f(x) = \sqrt{x}$.

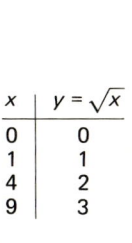
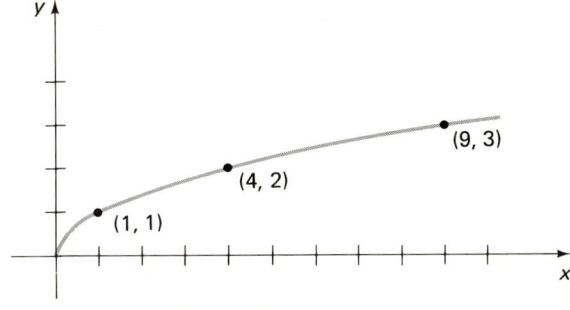

FIG. 20

Section 3.3 Functions

It has probably occurred to you that this is what you did when graphing some relations. Indeed, the equation $y = f(x)$ determines a relation; this relation is the set of all ordered pairs (x,y) where $y = f(x)$. In this way, a function is a special kind of relation.

> **(16)** **(ALTERNATE) DEFINITION**
>
> A **function** is a set of ordered pairs such that no two different pairs have the same first element.

With this definition, the domain D is the set of first elements from the ordered pairs, and the range R is the set of second elements. The phrase "no two different pairs have the same first element" corresponds to "assigns to each element of D a unique element of R" in the original definition.

Since every function is a relation, it is natural to ask how you recognize whether a particular relation is a function. It is easy to do this from its graph. If there are two or more points of the graph of a relation on the same vertical line, then the coordinates of these points have the same first coordinate but different second coordinates. These ordered pairs are in the relation, so the relation is not a function. If there is no vertical line which contains two points of the graph of the relation, the relation is a function. This is the **vertical line test**.

EXAMPLE 3 From the graphs in Fig. 21, determine which of the relations are functions.

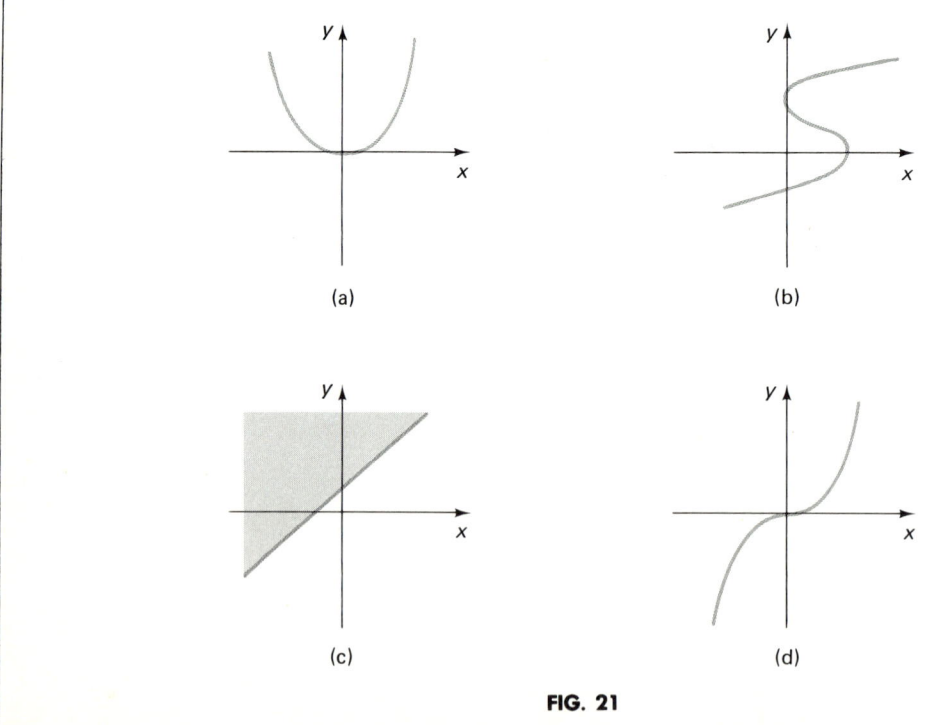

FIG. 21

SOLUTION We see in (a) and (d) that there is no vertical line containing two points of these graphs. So (a) and (d) "pass" the vertical line test and hence are functions. In (b) and (c), however, there are vertical lines which contain two (or more) points of the graph. Hence (b) and (c) "fail" the vertical line test and are not functions.

It is important to recognize how graphs of functions can be used to estimate values.

EXAMPLE 4

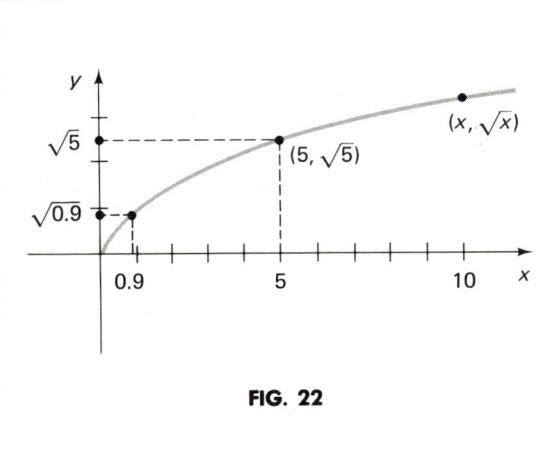

FIG. 22

Consider the graph of $y = \sqrt{x}$ in Fig. 22. Suppose we start at $x = 5$ on the x-axis and go straight up to a point on the graph. See Fig. 22. Any point on the graph has coordinates $(x, \sqrt{x})$, so this point has coordinates $(5, \sqrt{5})$. Therefore, if we go straight over to the y-axis (see Fig. 22), the number we reach on the y-axis must be $\sqrt{5}$. This will allow us to estimate the value of $f(5)$. In this case we can see that $\sqrt{5} \approx 2.2$. When we do this, we say we **estimate** the value **graphically**. As another example, we can see from Fig. 22 that $\sqrt{.9}$ is slightly larger than .9.

Such estimates can be used to check yourself when you use a calculator.

When functions are used in applied situations, often the relationships are described in words, and the formula or rule for the function has to be found.

EXAMPLE 5 Find the rule for the function described, and give the domain: A piece of wire is 60 inches long. Part of the wire is cut off and bent into a rectangle whose length is twice its width. Find the area of the rectangle as a function of the width.

SOLUTION Let w be the width of the rectangle. Then the length is $2w$, and the area $A = l \cdot w = 2w \cdot w$ or $A = 2w^2$. Since $w + 2w + w + 2w \leq 60$, $6w \leq 60$ or $w \leq 10$. Thus the domain is $0 \leq w \leq 10$, and the rule is $A = 2w^2$.

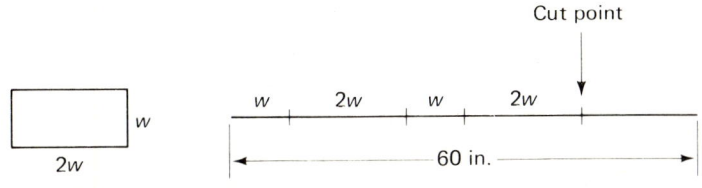

EXAMPLE 6 Find the rule for the function described, and give the domain: A group wishes to charter a bus, which holds at most 48 people, to go to a ball game. The bus company requires at least 30 people to go. It charges $2.50 per person if up to 40 people go. If more than 40 people go, it charges each person $2.50 less 10 cents times the number more than 40 who go. Find the total cost as a function of the number of people who go.

SOLUTION Let x be the number of people who go. Then $30 \leq x \leq 48$, and x is an integer. The formula is

$$\text{Total cost} = (\text{cost per person})(\text{number of people})$$

If between 30 and 40 people go, the cost per person is 2.50, so the total cost is $y = 2.50x$. If between 41 and 48 people go, the cost per person is $2.50 - .10(x - 40) = 6.50 - .10x$. Then the total cost is $y = (6.50 - .10x)x = 6.50x - .10x^2$. So altogether the rule is

$$y = \begin{cases} 2.50x, & 30 \leq x \leq 40 \\ 6.50x - .10x^2, & 41 \leq x \leq 48 \end{cases}$$

where x is an integer. The domain is $\{30, 31, \ldots, 48\}$.

EXERCISES

1. If $f(x) = 3x^2 - 2x + 1$, find the values $f(-3)$, $f(-1)$, $f(0)$, and $f(2)$.

2. If $g(x) = x^3 + 1$, find the values $g(-2)$, $g(-1)$, $g(0)$, and $g(1)$.

3. If $h(x) = \sqrt{x + 1}$, find the values $h(8)$, $h(3)$, $h(0)$, and $h(-1)$.

4. If $k(x) = x/(x + 1)$, find $k(-2)$, $k(0)$, $k(3)$, and $k(9)$.

5. If $F(x) = 7x + 3$, find
   a. $F(0)$
   b. $F(-3)$
   c. $F(1/a)$
   d. $F(a + 3)$
   e. $\dfrac{F(2 + h) - F(2)}{h}$

6. If $G(x) = 3x^2 - 2$, find
   a. $G(-1)$
   b. $G(2)$
   c. $G(2a)$
   d. $G(a - 1)$
   e. $\dfrac{G(-1 + h) - G(-1)}{h}$

7. If $H(x) = 3|x| + 2x$, find
   a. $H(-2)$
   b. $H(2)$
   c. $H(\frac{1}{2})$
   d. $H(a^2)$
   e. $H(H(-2))$

8. If $K(x) = \sqrt{2x + 3}$, find
   a. $K(3)$
   b. $K(11)$
   c. $K(-1)$
   d. $K(1/a^2)$
   e. $K(K(11))$

9. If the graph of f is the sketch shown,

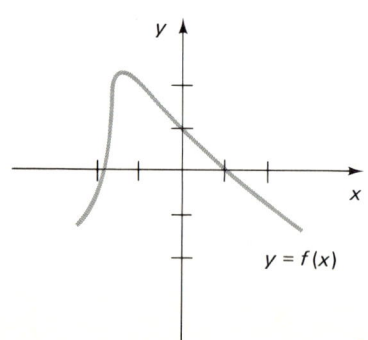

$y = f(x)$

10. If the graph of g is the sketch shown,

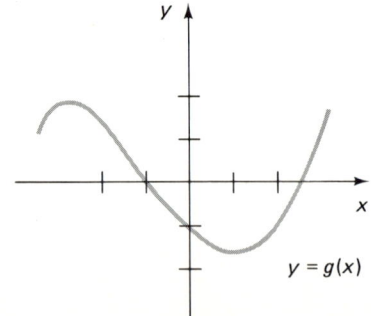

$y = g(x)$

graphically estimate the given values.
  a. $f(-2)$ b. $f(-1)$
  c. $f(0)$ d. $f(1)$
  e. $f(2)$

11. The graph of $y = \sqrt{x}$ is shown. Use the graph to estimate graphically the given values.
  a. $\sqrt{2}$ b. $\sqrt{4.5}$
  c. $\sqrt{.8}$ d. $\sqrt{\pi}$
  e. $\sqrt{\sqrt{2}}$

graphically estimate the given values.
  a. $g(-2)$ b. $g(-1)$
  c. $g(0)$ d. $g(1)$
  e. $g(2)$

12. Use the graph of $y = \sqrt[3]{x}$ as shown to estimate graphically the given values.
  a. $\sqrt[3]{2}$ b. $\sqrt[3]{6}$
  c. $\sqrt[3]{.5}$ d. $\sqrt[3]{-4}$
  e. $\sqrt[3]{-5}$

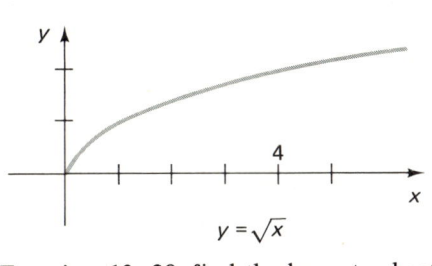
$y = \sqrt{x}$

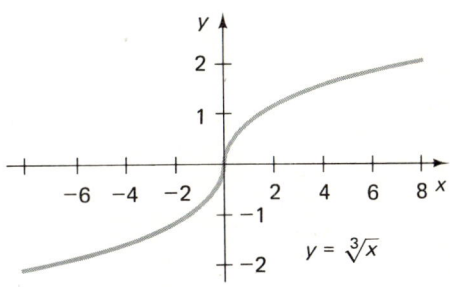
$y = \sqrt[3]{x}$

In Exercises 13–28, find the largest subset of $\mathbb{R}$ that can serve as the domain.

13. $f(x) = \sqrt{x - 4}$

14. $g(x) = \sqrt{x + 3}$

15. $h(x) = \sqrt{-x}$

16. $k(x) = \sqrt{\frac{3}{2} - x}$

17. $F(x) = \dfrac{1}{x^2 - 9}$

18. $G(x) = \dfrac{2x - 1}{3x^2 - 12}$

19. $H(x) = \dfrac{x^2 + 1}{(x + 1)(x - 2)(x + 3)}$

20. $K(x) = \sqrt{x^2 - 2x - 3}$

21. $f(x) = \sqrt{4x^2 - 9}$

22. $g(x) = \sqrt{\dfrac{x}{x^2 - 1}}$

23. $h(x) = \sqrt{\dfrac{x - 1}{x^2 - 4x - 12}}$

24. $k(x) = \dfrac{\sqrt{x}}{2x^2 + 5x - 12}$

25. $F(x) = \dfrac{\sqrt{1 - x}}{4x^2 - 9}$

26. $G(x) = \dfrac{x^3 - 1}{x^2 - 1}$

27. $H(x) = \dfrac{x^6 - 1}{x^4 - 1}$

28. $K(x) = \dfrac{\sqrt{x + 3}}{(x^2 - 1)(x^2 - 4)}$

29. Which of the following are graphs of functions?

a. 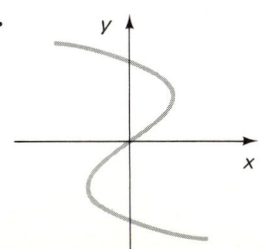   b. 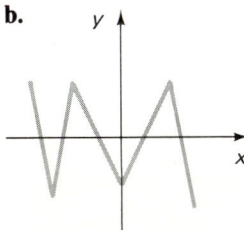   c. 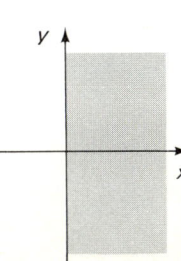   d.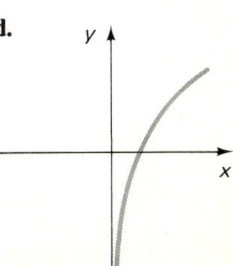

30. Which of the following are graphs of functions?

a. 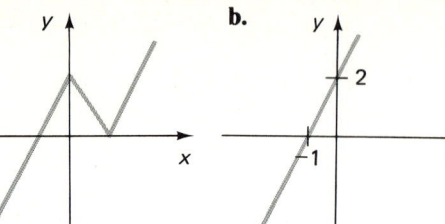 b. c. 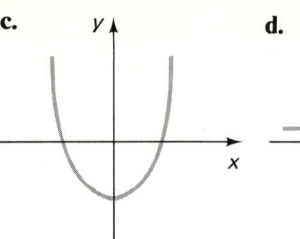 d.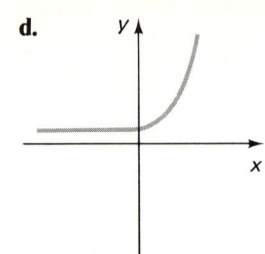

31. Reflect each of the graphs in Exercise 29 through the *x*-axis, and determine which of these reflected graphs is a function.

32. Reflect each of the graphs in Exercise 30 through the *y*-axis, and determine which of these reflected graphs is a function.

33. Find a formula that expresses the radius of a circle as a function of its circumference.

34. Find a formula that expresses the area of a circle as a function of its circumference.

35. Find a formula that expresses the area of a cube as a function of its volume.

36. Find a formula that expresses the volume of a sphere as a function of its surface area.

37. An open box is to be made from a rectangular piece of tin 12 inches long and 10 inches wide by cutting pieces x inches square from each corner and bending up the sides.

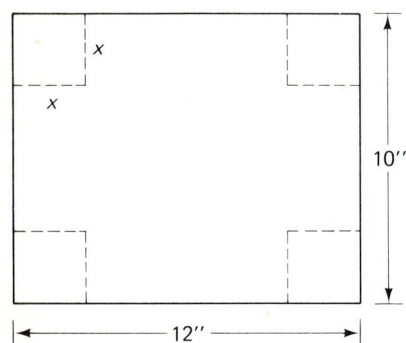

a. Find a formula that expresses the volume V of the box as a function of x.

b. Find the domain of the function.

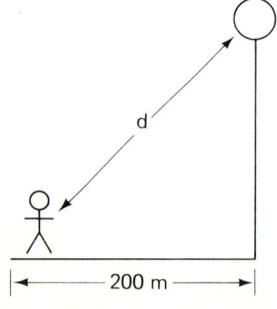

38. A balloon is released from the ground 200 meters from an observer and rises steadily at the rate of 3 meters per second.

a. Find a formula which expresses the distance d between the observer and the balloon as a function of the time t after the balloon was released.

b. What is the value of d twenty seconds after the balloon was released?

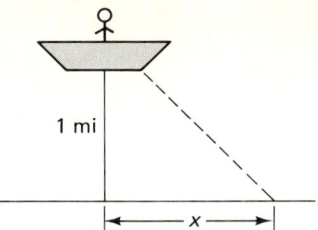

39. A man is in a rowboat 1 mile from a straight shore. He starts out rowing toward a point x miles up the shore. Suppose he rows at a steady 3 miles per hour.

a. Find a formula that expresses the time T that it takes him to reach the shore as a function of x.

b. How long does it take him if x is 2 miles?

40. A rectangular region of 6000 square meters is to be fenced. On three of the sides the fencing used will cost $2 per meter, and on the fourth side the fencing will cost $5 per meter.

    a. If x represents the length of the fourth side, find a formula which expresses the total cost C as a function of x.

    b. Find the total cost when $x = 30$, 60, or 100.

41. Two ships leave a port on the same morning, one at 8 a.m. heading south at 15 miles per hour and the other at 10 a.m. heading west at 20 miles per hour. Find formulas expressing the distance d between them as a function of the time t, in hours, after 8 a.m. Include the domain of each formula.

42. A manufacturer sells a certain item at a price of $30 each if fewer than 20 items are ordered. If 20 or more items are ordered (up to 300), the price per article is $31 less 5 cents times the total number ordered. Find formulas which express the total cost as a function of the number n of items ordered. Include the domain of each formula.

43. A travel agency offers a tour. It charges $20 per person if fewer than 25 people go. If 25 people or more, up to a maximum of 110, take the tour, they charge each person $22.40 less 10 cents times the number of people who go. Find formulas which express the total charge C as a function of the number n who go. Include the domain of each formula.

44. A piece of wire, 90 centimeters long, is cut once, and each piece is bent into a square. Find a formula giving the total area A of the two squares as a function of the distance x that the cut is made from one end.

SECTION 3.4. GRAPHING TECHNIQUES

One of the things that helps us understand a particular function is its graph. Fortunately, there are several techniques that can make finding a graph easier. Symmetry, which we studied in Sec. 3.2, is one such technique. The techniques that we shall study in this section will allow us to obtain the graph we want from one we already know by applying what are called translations, expansions, contractions, and reflections to the known graph. When we introduce each such technique, we shall first look at a simple example, plotting several points so that you will have some understanding as to how and why the technique works.

However, the purpose of these techniques is to enable you to sketch the graph without plotting points. Thus when you want to sketch the graph of a function which is related to a graph you already know, it is much easier if you

> Apply the techniques instead of plotting points.

The first technique we describe is called a **translation**. This is shifting a graph without changing its shape or orientation. If the shift is up or down, it is a **vertical translation**; if the shift is to the right or left, it is a **horizontal translation**. We shall illustrate such translations with the graph of $y = x^2$; this is a nice and important example to keep in mind.

We first consider the graphs of $y = x^2 + 3$ and $y = x^2 - 2$. We do this by making a table of values in such a way that we can easily see how we get the values of $y = x^2 + 3$ or $y = x^2 - 2$ from the values of $y = x^2$. See Table 1 and Fig. 23.

TABLE 1

x	x^2	$x^2 + 3 = y$	$x^2 - 2 = y$
-2	4	$4 + 3 = 7$	$4 - 2 = 2$
-1	1	$1 + 3 = 4$	$1 - 2 = -1$
0	0	$0 + 3 = 3$	$0 - 2 = -2$
1	1	$1 + 3 = 4$	$1 - 2 = -1$
2	4	$4 + 3 = 7$	$4 - 2 = 2$

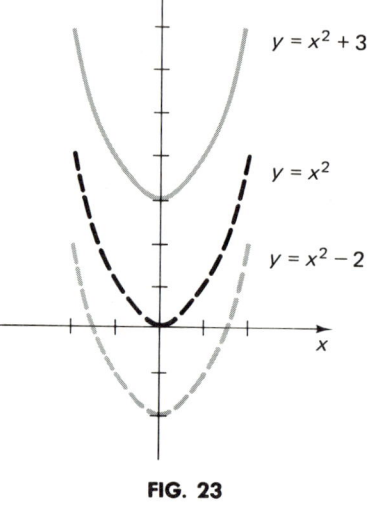

FIG. 23

We see the following:

1. The values of $y = x^2 + 3$ are 3 more than the corresponding values of $y = x^2$. Hence the graph of $y = x^2 + 3$ is just the graph of $y = x^2$ translated upward 3 units.
2. The values of $y = x^2 - 2$ are 2 less than the corresponding values of $y = x^2$. Hence the graph of $y = x^2 - 2$ is just the graph of $y = x^2$ translated downward 2 units.

These examples illustrate the general principle.

(17) If $c > 0$, then the graph of
$$\left\{ \begin{array}{l} y = f(x) + c \\ y = f(x) - c \end{array} \right\} \text{ is the graph of } y = f(x) \text{ translated } c \text{ units } \left\{ \begin{array}{l} \text{upward} \\ \text{downward} \end{array} \right\}.$$

As we consider other examples, remember that we apply the principles—we do *not* plot points.

EXAMPLE 1 Graph **(a)** $y = x^3 + 2$ and **(b)** $y = x^3 - \frac{1}{2}$.

SOLUTION We first graph $y = x^3$ in Fig. 24(a). Then in Fig. 24(b), we translate this up 2 for part (a) and down 1/2 for part (b).

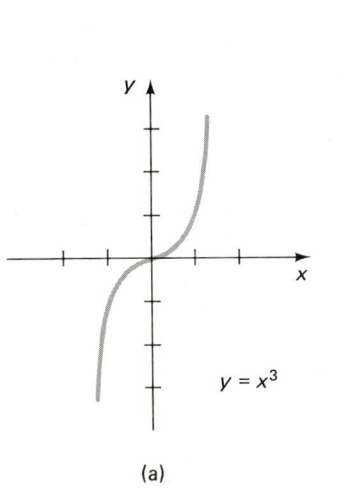

FIG. 24

EXAMPLE 2 If the graph of $y = F(x)$ is as given in Fig. 25(a), graph **(a)** $y = F(x) + 3$ and **(b)** $y = F(x) - 1$.

SOLUTION The graphs in Fig. 25(b) are obtained by translating the graph in Fig. 25(a) up 3 for part (a) and down 1 for part (b).

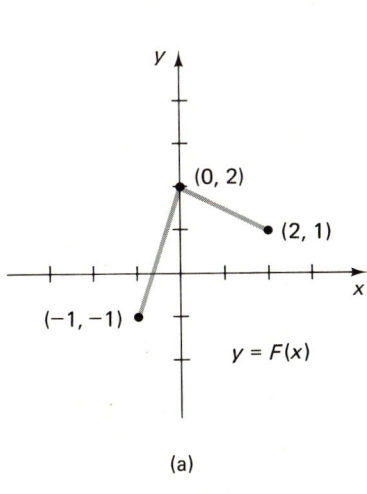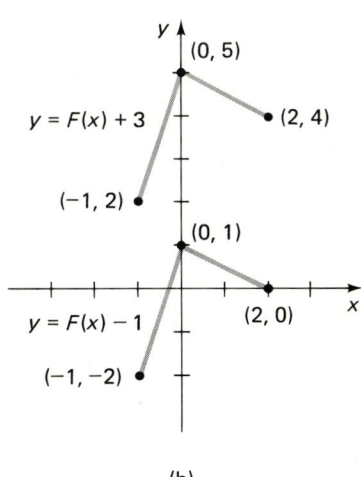

FIG. 25

Section 3.4 Graphing Techniques

We now turn to horizontal translations. As an example, we examine the graphs of $y = (x + 3)^2$ and $y = (x - 2)^2$. See Table 2 and Fig. 26.

TABLE 2

x	x^2	$x + 3$	$(x + 3)^2$	$x - 2$	$(x - 2)^2$
−5	25	−2	4	−7	49
−4	16	−1	1	−6	36
−3	9	0	0	−5	25
−2	4	1	1	−4	16
−1	1	2	4	−3	9
0	0	3	9	−2	4
1	1	4	16	−1	1
2	4	5	25	0	0
3	9	6	36	1	1
4	16	7	49	2	4

We see the following:

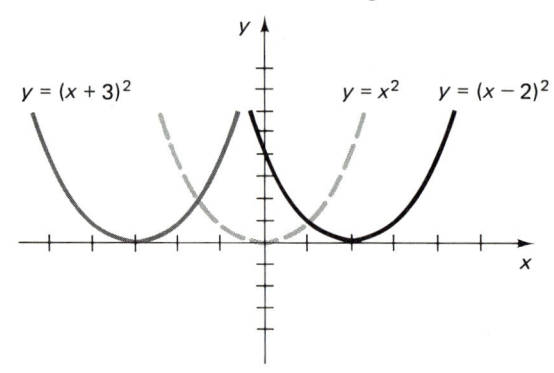

FIG. 26

1. The zero of $x + 3$ occurs at $x = -3$, so that the zero value for $y = (x + 3)^2$ occurs at $x = -3$ instead of at $x = 0$ for $y = x^2$. In the same way, all y values have been moved 3 units to the left. Thus the graph of $y = (x + 3)^2$ is just the graph of $y = x^2$ translated 3 units to the left.

2. The zero of $x - 2$ occurs at $x = +2$, so that the zero value for $y = (x - 2)^2$ occurs at $x = +2$ instead of at $x = 0$ for $y = x^2$. In the same way, all y values have been moved 2 units to the right. Thus the graph of $y = (x - 2)^2$ is just the graph of $y = x^2$ translated 2 units to the right.

These examples illustrate the general principles.

(18)

If $c > 0$, then the graph of

$\left\{\begin{array}{l} y = f(x + c) \\ y = f(x - c) \end{array}\right\}$ is the graph of $y = f(x)$ translated c units $\left\{\begin{array}{l} \text{to the left} \\ \text{to the right} \end{array}\right\}$.

Warning: If $c > 0$, the graph of $y = f(x) + c$ is obtained by translating the graph of $y = f(x)$ upward, in the *positive* direction, whereas the graph of

$y = f(x + c)$ is obtained by translating the graph of $y = f(x)$ to the left, in the *negative* direction, and similarly for $y = f(x) - c$ and $y = f(x - c)$. Many students at first feel that horizontal translations are backwards. The way to keep horizontal translations straight is to keep the preceding simple examples in mind and to *determine which x-value gives you zero* in the quantity $(x \pm c)$.

EXAMPLE 3 Graph **(a)** $y = (x + 2)^3$ and **(b)** $y = (x - \frac{1}{2})^3$.

SOLUTION

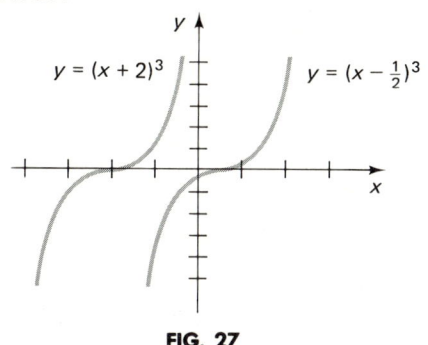

The graph of $y = x^3$ is given in Fig. 24(a). We translate this 2 units to the left for part (a) (since the zero of $x + 2$ is -2) and $\frac{1}{2}$ unit to the right for part (b) (since the zero of $x - \frac{1}{2}$ is $+\frac{1}{2}$). See Fig. 27.

FIG. 27

EXAMPLE 4

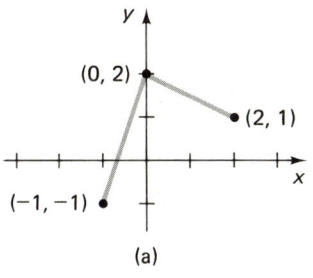

(a)

If the graph of $y = F(x)$ is as given in Fig. 28(a), graph **(a)** $y = F(x + 4)$ and **(b)** $y = F(x - 1)$.

SOLUTION

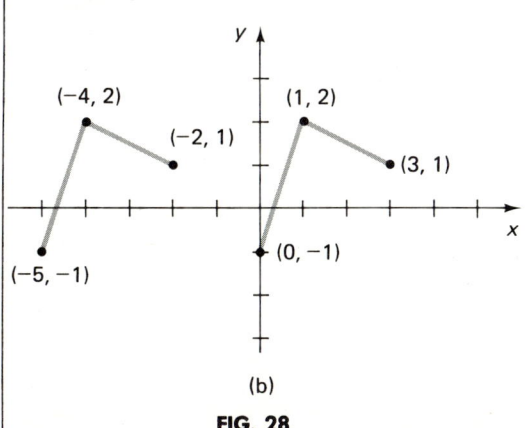

Translate the graph 4 units to the left for part (a) and 1 unit to the right for part (b). See Fig. 28(b).

(b)

FIG. 28

Graphs can also be altered by expansions, contractions, and reflections. We illustrate this with the graph of $f(x) = x^2 - 4$. First consider what happens when $y = f(x)$ is replaced by $y = cf(x)$. See Table 3 and Fig. 29.

TABLE 3
$f(x) = x^2 - 4$

x	f(x)	2f(x)	½f(x)	−f(x)
−3	5	10	$\frac{5}{2}$	−5
−2	0	0	0	0
−1	−3	−6	$-\frac{3}{2}$	3
0	−4	−8	−2	4
1	−3	−6	$-\frac{3}{2}$	3
2	0	0	0	0
3	5	10	$\frac{5}{2}$	−5

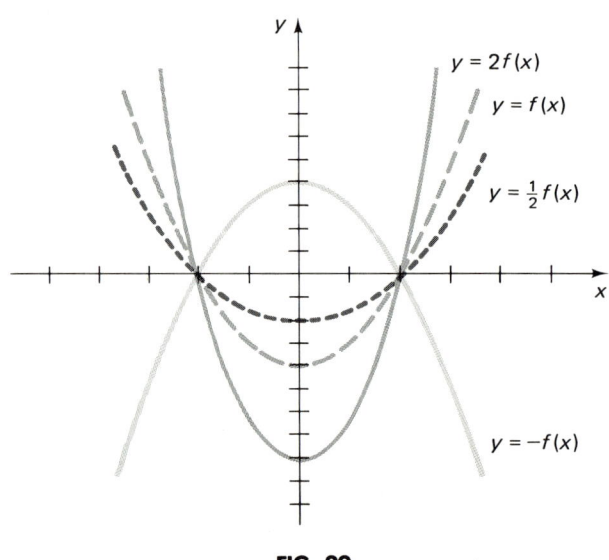

FIG. 29

We see the following:

1. The values of $y = 2f(x)$ are the corresponding values of $y = f(x)$ multiplied by 2. Hence the graph of $y = 2f(x)$ is just the graph of $y = f(x)$ stretched out or expanded vertically but crossing the x-axis in the same place.

2. The values of $y = \frac{1}{2}f(x)$ are the corresponding values of $y = f(x)$ multiplied by $\frac{1}{2}$. Hence the graph of $y = \frac{1}{2}f(x)$ is just the graph of $y = f(x)$ squeezed or contracted vertically but crossing the x-axis in the same place.

3. The values of $y = -f(x)$ are the negatives of the values of $y = f(x)$. Hence the graph of $y = -f(x)$ is just the graph of $y = f(x)$ reflected through the x-axis.

These examples illustrate the general principles.

(19)

If $c \neq 0$, the graph of $y = cf(x)$ is the graph of $y = f(x)$ except that all the y-coordinates have been multiplied by c. Hence the graph has been

$$\left\{ \begin{array}{l} \text{expanded or stretched} \\ \text{contracted or squeezed} \end{array} \right\} \text{vertically if } \left\{ \begin{array}{l} |c| > 1 \\ |c| < 1 \end{array} \right\}.$$

If $c < 0$, the graph is also reflected through the x-axis.

The places where the graph crosses the x-axis remain unchanged.

We now consider what happens when $y = f(x)$ is replaced by $y = f(cx)$. See Table 4 and Fig. 30.

TABLE 4
$f(x) = x^2 - 4$

x	$f(x) = x^2 - 4$	$2x$	$f(2x) = (2x)^2 - 4$	$\frac{1}{2}x$	$f(\frac{1}{2}x) = (\frac{1}{2}x)^2 - 4$	$-x$	$f(-x) = (-x)^2 - 4$
-4	12	-8	60	-2	0	4	12
-2	0	-4	12	-1	-3	2	0
-1	-3	-2	0	$-\frac{1}{2}$	$-\frac{15}{4}$	1	-3
0	-4	0	-4	0	-4	0	-4
1	-3	2	0	$\frac{1}{2}$	$-\frac{15}{4}$	-1	-3
2	0	4	12	1	-3	-2	0
4	12	8	60	2	0	-4	12

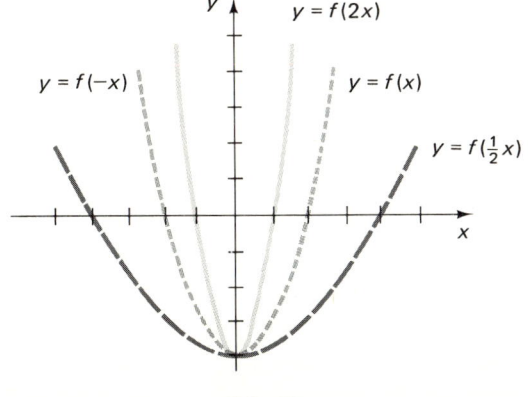

FIG. 30

We see the following:

1. If you take the graph of $y = f(x)$ and divide each x-coordinate by 2, you obtain the graph of $y = f(2x)$. Hence the graph of $y = f(2x)$ is just the graph of $y = f(x)$ squeezed or contracted horizontally but crossing the y-axis in the same place.

2. If you take the graph of $y = f(x)$ and multiply each x-coordinate by 2, you obtain the graph of $y = f(\frac{1}{2}x)$. Hence the graph of $y = f(\frac{1}{2}x)$ is just the graph of $y = f(x)$ stretched or expanded horizontally but crossing the y-axis in the same place.
3. If you take the graph of $y = f(x)$ and multiply each x-coordinate by -1, you obtain the graph of $y = f(-x)$. Hence you obtain the graph of $y = f(x)$ reflected through the y-axis.

These examples illustrate the general principles.

---

(20)

If $c \neq 0$, the graph of $y = f(cx)$ is the graph of $y = f(x)$ with all the x-coordinates divided by c. Hence the graph has been

$\left\{ \begin{array}{l} \text{contracted or squeezed} \\ \text{expanded or stretched} \end{array} \right\}$ horizontally if $\left\{ \begin{array}{l} |c| > 1 \\ |c| < 1 \end{array} \right\}$.

If $c < 0$, the graph has also been reflected through the y-axis.

The place where it crosses the y-axis remains unchanged.

Warning: If $c > 1$, observe that the graph of $y = cf(x)$ is the graph of $y = f(x)$ *expanded* vertically, while the graph of $y = f(cx)$ is the graph of $y = f(x)$ *contracted* horizontally. If $0 < c < 1$, the graph of $y = cf(x)$ is the graph of $y = f(x)$ *contracted* vertically, while the graph of $y = f(cx)$ is the graph of $y = f(x)$ *expanded* horizontally.

Example 5 illustrates horizontal and vertical expansions, contractions, and reflections.

EXAMPLE 5 If the graph of $y = F(x)$ is as given in Fig. 31(a), graph **(a)** $y = 2F(x)$, **(b)** $y = \frac{1}{2}F(x)$, **(c)** $y = -F(x)$, **(d)** $y = F(-x)$, **(e)** $y = F(2x)$, and **(f)** $y = F(\frac{1}{2}x)$.

SOLUTION For part (a), we multiply all the y-coordinates by 2; for part (b), we multiply them all by $1/2$. These graphs are given in Fig. 31(b). Note that the graph of $y = F(x)$ crosses the x-axis at $-1/2$, so that $F(-1/2) = 0$. Since $2 \cdot 0 = 0$ and $\frac{1}{2} \cdot 0 = 0$, the graphs of $y = 2F(x)$ and $y = \frac{1}{2}F(x)$ still cross the x-axis in the same place.

The graph of $y = -F(x)$ is the graph of $y = F(x)$ reflected across the x-axis; the graph of $y = F(-x)$ is the graph of $y = F(x)$ reflected across the y-axis. These are given in Fig. 31(c).

The graph of $y = F(2x)$ is the graph of $y = F(x)$ contracted by 2 horizontally (i.e., all the x-coordinates are divided by 2); the graph of $y = F(\frac{1}{2}x)$ is the graph of $y = F(x)$ expanded by 2 horizontally (i.e., all the x-coordinates are multiplied by 2). These are given in Fig. 31(d).

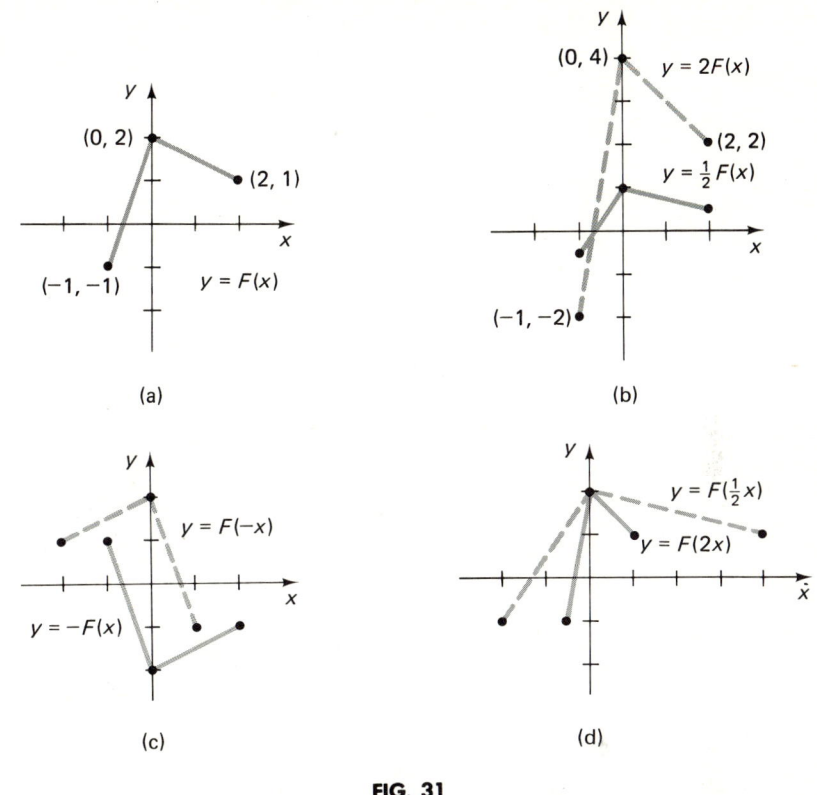

FIG. 31

We have now discussed horizontal and vertical translations, expansions, contractions, and reflections, but so far our examples have involved only one of these operations at a time. However, many problems involve combinations of these operations. Some of these combinations pose no difficulties since the operations involved can be performed in any order without affecting the outcome.

EXAMPLE 6 Graph

$$y = \frac{1}{x - 2} - 3$$

SOLUTION The graph of $y = 1/x$ is given in Fig. 32(a). We can now approach the problem in either of two ways:

1. First, translate the graph to the right 2, obtaining the graph of $y = 1/(x - 2)$. Then translate the new graph down 3, obtaining the graph of $y = 1/(x - 2) - 3$. See Fig. 32(b).

2. First, translate the graph down 3, obtaining the graph of $y = 1/x - 3$. Then translate this new graph to the right 2, obtaining the graph of $y = 1/(x - 2) - 3$. See Fig. 32(c).

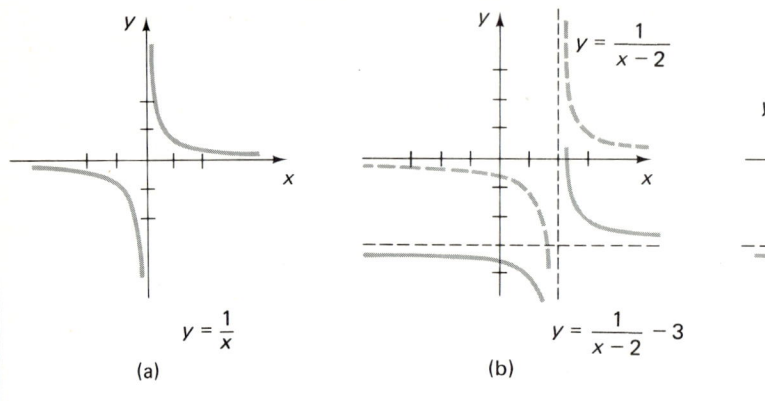
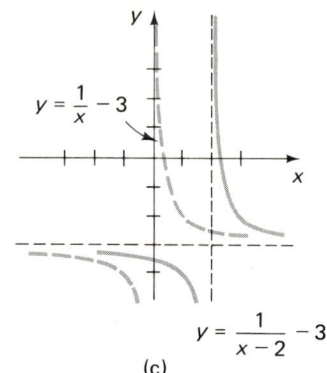

FIG. 32

You can see that the order does not affect the results in this example. Note that the asymptotes are translated also.

There are combinations which pose difficulties because the order of the operations involved does affect the outcome.

EXAMPLE 7 Graph $y = |2x - 1|$.

SOLUTION The graph of $y = |x|$ is given in Fig. 33(a). There are two ways to use this graph to obtain the graph of $y = |2x - 1|$:

1. First, replace x by $x - 1$ (which gives a translation of 1 to the right), obtaining $y = |x - 1|$. Then in this, replace x by $2x$ (which gives a horizontal contraction of 2), obtaining $y = |2x - 1|$. See Fig. 33(b).

2. First, replace x by $2x$ (which gives a horizontal contraction of 2), obtaining $y = |2x|$. Then in this, replace x by $x - \frac{1}{2}$ (which gives a translation of 1/2 to the right), obtaining $y = |2(x - \frac{1}{2})| = |2x - 1|$. See Fig. 33(c).

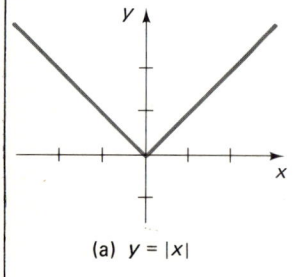
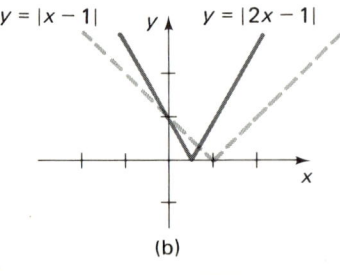

FIG. 33

138 RELATIONS, FUNCTIONS, AND GRAPHS

Note that the order affects how much we translate and that the contraction is easier if we contract first (when we can use the symmetry). Consequently, contracting (or expanding) first is recommended. When we did this in Example 7, the amount of contraction was obvious, but the amount of translation was not. The amount of translation would be more obvious if we first factored out the coefficient of x in $y = |2x - 1|$ to obtain $y = |2(x - \frac{1}{2})|$. Such a procedure should be followed in general.

EXAMPLE 8 Each of the following functions are of the form $y = af(bx + c) + d$. Rewrite them in the form $y = af(b(x + \frac{c}{b})) + d$.
(a) $y = -2(3x - 5)^2 + 4$; (b) $y = 3/(2x + 4) + 2$; (c) $y = 3\sqrt{2 - x}$.

SOLUTION (a) $y = -2[3(x - \frac{5}{3})]^2 + 4$. (b) $y = 3/2(x + 2) + 2$.
(c) $y = 3\sqrt{-(x - 2)}$.

We are now ready to give the general procedure for performing transformations in combination.

---

(21)

The preferred order in which to perform transformations in combination is as follows:

1. First, rewrite $y = af(bx + c) + d$ as $y = af\left(b\left(x + \frac{c}{b}\right)\right) + d$.

2. Perform any expansions, reflections, or contractions as dictated by a and b (in any order).

3. Perform any translations as dictated by c/b and d (in any order).

EXAMPLE 9 Graph $y = 3\sqrt{2 - x}$.

SOLUTION First, rewrite this as $y = 3\sqrt{-(x - 2)}$, as in Example 8(c).
Next, start with the graph of $y = \sqrt{x}$. Reflect it in the y-axis, obtaining the graph of $y = \sqrt{-x}$. Then expand that graph vertically (by multiplying all the y-coordinates by 3), obtaining the graph of $y = 3\sqrt{-x}$.
Finally, translate this graph to the right 2 units, obtaining the graph of $y = 3\sqrt{-(x - 2)}$. See Fig. 34.

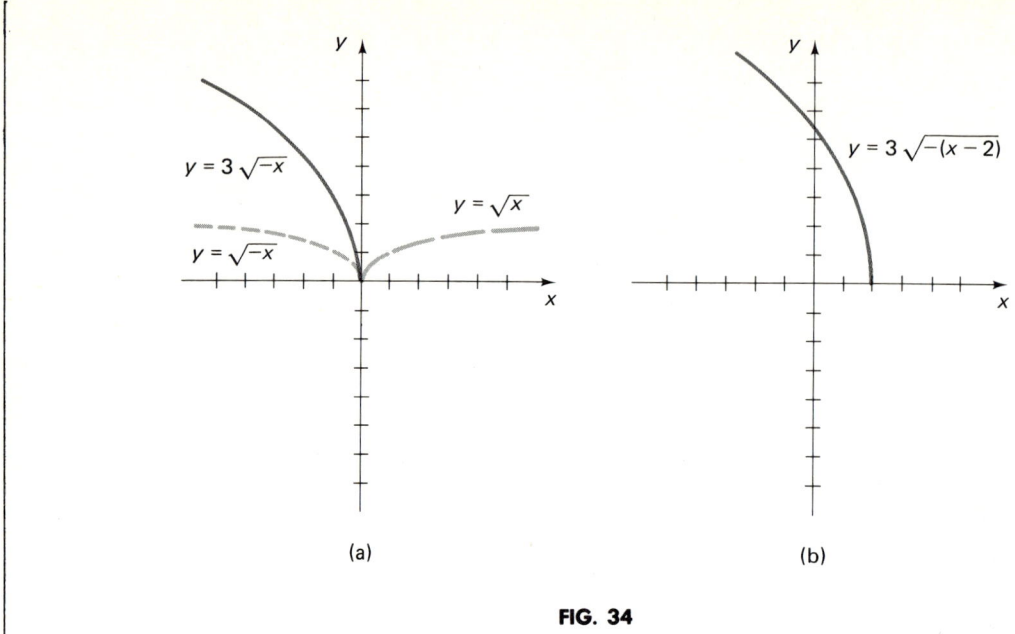

(a)   (b)

FIG. 34

EXAMPLE 10 Graph $y = F(\frac{1}{2}x - 1) + 1$ if the graph of $y = F(x)$ is as given in Fig. 35.

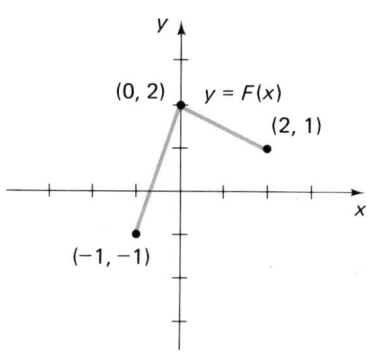

FIG. 35

SOLUTION We first rewrite this as $y = F(\frac{1}{2}(x - 2)) + 1$. Next, we expand the graph of $y = F(x)$ horizontally by multiplying all the x-coordinates by 2, obtaining the graph of $y = F(\frac{1}{2}x)$. Finally, we translate this graph to the right 2 and up 1.

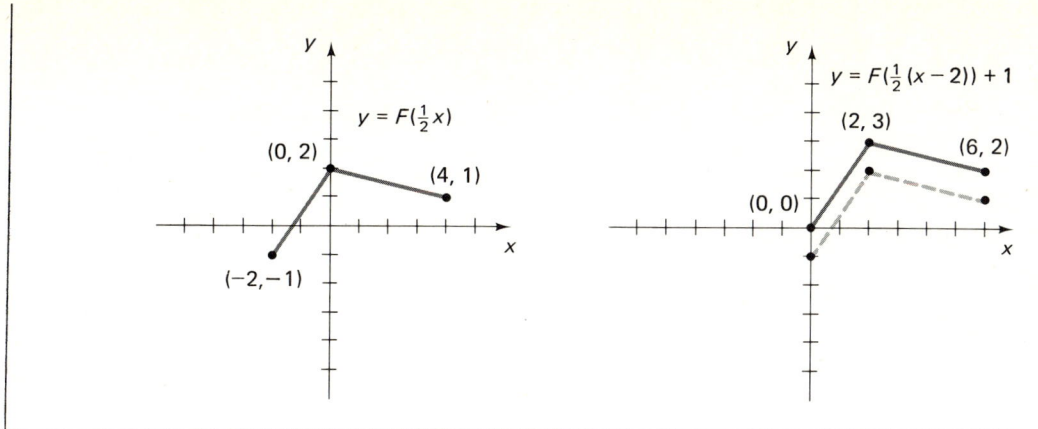

EXERCISES

In Exercises 1–10, rewrite each function in the form $y = af\left(b\left(x + \frac{c}{b}\right)\right) + d$.

1. $y = (2x - 3)^2$
2. $y = |4x - 2|$
3. $y = \dfrac{1}{2x + 4}$
4. $y = \dfrac{5}{(3x - 2)^2}$
5. $y = -3(2 - \frac{1}{2}x)^3$
6. $y = \frac{1}{2}\sqrt{3 - 2x} - 4$
7. $y = 5|1 - 3x| - 1$
8. $y = \dfrac{2}{(3 - \frac{1}{2}x)^3} + 2$
9. $y = -\sqrt{2x} - 4$
10. $y = 2(x - 2)^3 + 3$

In Exercises 11–40, graph.

11. $y = (x + 1)^2$
12. $y = -2x^2$
13. $y = 2(x - 1)^2$
14. $y = -x^2 + 4$
15. $y = 2(-x + 1)^2$
16. $y = (\frac{1}{2}x - \frac{1}{2})^2 - \frac{1}{2}$
17. $y = \dfrac{1}{x^2}$
18. $y = \dfrac{1}{(x - 2)^2}$
19. $y = -\dfrac{1}{x^2} + 3$
20. $y = \dfrac{-1}{(x + 1)^2}$
21. $y = \dfrac{1}{(x - 1)^2} - 2$
22. $y = \dfrac{-1}{(2x + 1)^2}$
23. $y = \sqrt{x}$
24. $y = \sqrt{x - 2}$
25. $y = \sqrt{-x}$
26. $y = -\sqrt{-x} + 1$
27. $y = 3\sqrt{x + 4}$
28. $y = \sqrt{-2x + 2} - 1$
29. $y = |x - 4|$
30. $y = |x + 3| - 1$
31. $y = 2|x - 1|$
32. $y = |-2x + 1|$
33. $y = -|3x| + 2$
34. $y = -2|x + 1| - 1$

35. $y = x^3$
37. $y = (-x + 2)^3$
39. $y = (-x + 1)^3 + 2$

36. $y = -(x + 2)^3$
38. $y = -2x^3 - 4$
40. $y = -\frac{1}{2}(x - 1)^3 - 1$

In Exercises 41–46, assume the graph of $y = G(x)$ is as given in Fig. 36, and graph.

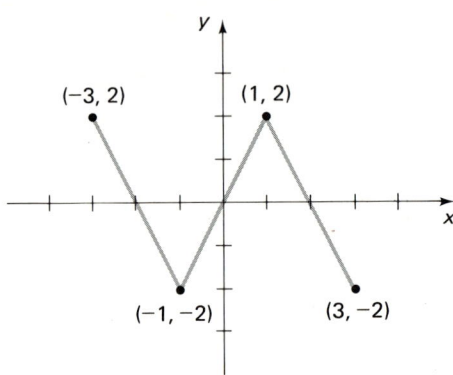

FIG. 36

41. $y = G(x) + 1$
43. $y = 2G(-x)$
45. $y = -G(\frac{1}{2}x + 2)$

42. $y = G(x + 1)$
44. $y = G(2x - 3)$
46. $y = -2G(x - 2) - 1$

In Exercises 47–54, assume the graph of $y = H(x)$ is as given in Fig. 37, and graph.

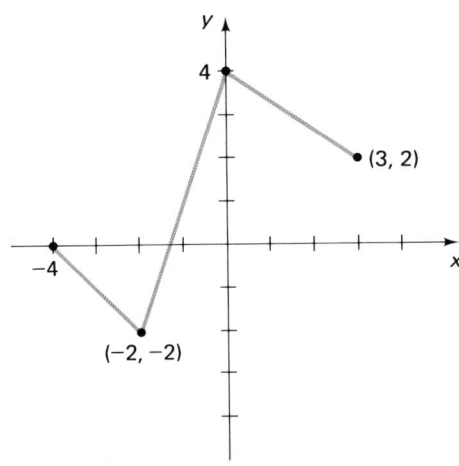

FIG. 37

47. $y = \frac{1}{2}H(x)$
49. $y = -H(-x)$
51. $y = 2H(x - 1)$
53. $y = -H(2 - x) + 2$

48. $y = -H(x) + 1$
50. $y = H(3x) - 1$
52. $y = 3H(-\frac{1}{2}x)$
54. $y = -H(\frac{1}{3}x + 1)$

SECTION 3.5. LINEAR FUNCTIONS AND LINES

In this section we consider a very important class of functions called *linear functions*. Not only do linear functions arise naturally in many places but also they are used to approximate more complicated functions.

---

(22)   **DEFINITION**

A **linear function** is a function whose rule is of the form

$$f(x) = mx + b$$

where m and b are real numbers, $m \neq 0$.

The word *linear* is derived from the word *line*. We shall see later that the graph of a linear function is a straight line. Before discussing linear functions further, we must discuss straight lines in the plane and their equations.

One fundamental property of a line is its slope.

---

(23)   **DEFINITION**

If l is a line which is not vertical and if $P = (x_1, y_1)$ and $Q = (x_2, y_2)$ are two distinct points on the line, then the **slope** of the line l, usually denoted by m, is given by

$$m = \frac{y_2 - y_1}{x_2 - x_1} = \frac{\text{change in } y}{\text{change in } x}$$

The slope of a vertical line is not defined.

There are three important remarks to make concerning slopes. First, since $(y_2 - y_1)/(x_2 - x_1) = (y_1 - y_2)/(x_1 - x_2)$, it is immaterial which point is labeled P and which Q.

EXAMPLE 1   Find the slope m of the straight line through $(-1, 2)$ and $(3, -1)$. See Fig. 38.

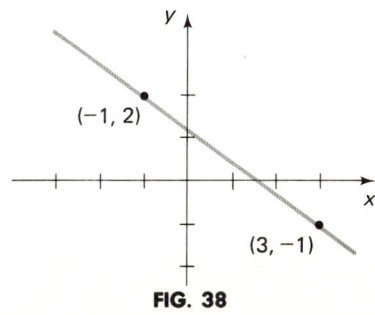

FIG. 38

SOLUTION We can compute either

$$m = \frac{2-(-1)}{-1-3} = \frac{3}{-4} = -\frac{3}{4} \quad \text{or} \quad m = \frac{-1-2}{3-(-1)} = \frac{-3}{4} = -\frac{3}{4}$$

Warning: Just be careful that if you write the *y*-coordinate of one point first in the numerator, you also write the *x*-coordinate of that same point first in the denominator. If you mix the order, you will get the wrong sign.

Before stating the second remark, we shall discuss an example.

EXAMPLE 2 Sketch the line through the given points and find the slope of each line:
(a) $(-1, 2), (4, 1)$; (b) $(-1, 2), (1, 6)$; (c) $(-1, 2), (3, 2)$; (d) $(-1, 2), (-1, -1)$.

SOLUTION

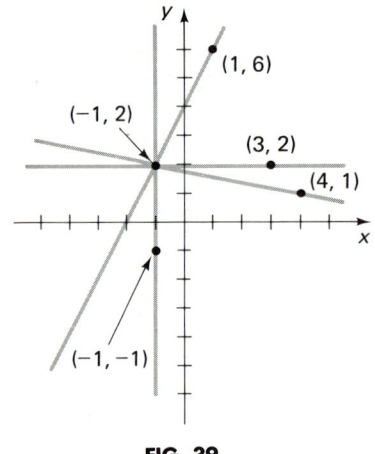

FIG. 39

We first sketch the lines; see Fig. 39. We now compute the slopes.

(a) $\quad m = \dfrac{y_2 - y_1}{x_2 - x_1} = \dfrac{1-2}{4-(-1)}$

$\qquad = \dfrac{-1}{5} = -\dfrac{1}{5}$

Note that the slope is negative and that the line goes downward as it goes to the right. Also, $|m| < 1$, and the drop is gradual.

(b) $\quad m = \dfrac{y_2 - y_1}{x_2 - x_1} = \dfrac{6-2}{1-(-1)} = \dfrac{4}{2} = 2$

Note that the slope is positive and that the line goes upward as it goes to the right.

(c) $\quad m = \dfrac{y_2 - y_1}{x_2 - x_1} = \dfrac{2-2}{3-(-1)} = \dfrac{0}{4} = 0$

Note that the slope is 0 and that the line is horizontal.

(d) $\quad m = \dfrac{y_2 - y_1}{x_2 - x_1} = \dfrac{-1-2}{-1-(-1)} = \dfrac{-3}{0}, \quad$ undefined

Note that the slope is undefined and that the line is vertical.

Our second important remark is that the slope always gives us the preceding kind of geometric information.

(24)

Suppose a line has slope m.

1. If m is $\begin{Bmatrix} \text{positive} \\ \text{negative} \end{Bmatrix}$, then the line goes $\begin{Bmatrix} \text{upward} \\ \text{downward} \end{Bmatrix}$ as it goes to the right.
2. If $m = 0$, the line is horizontal.
3. If m is undefined, the line is vertical.

The third (and most) important remark to make concerning the slope of a (nonvertical) line l is that it does not make any difference which two (different) points you pick to compute the slope. If you pick $R = (x_3, y_3)$ and $S = (x_4, y_4)$ instead of $P = (x_1, y_1)$ and $Q = (x_2, y_2)$, then (assuming $x_2 > x_1$ and $x_4 > x_3$), let $A = (x_2, y_1)$ and $B = (x_4, y_3)$. See Fig. 40. Since l is a straight line, the triangles PAQ and RBS are similar. Thus the ratios of their corresponding sides are equal, so

$$\frac{y_2 - y_1}{x_2 - x_1} = \frac{y_4 - y_3}{x_4 - x_3}$$

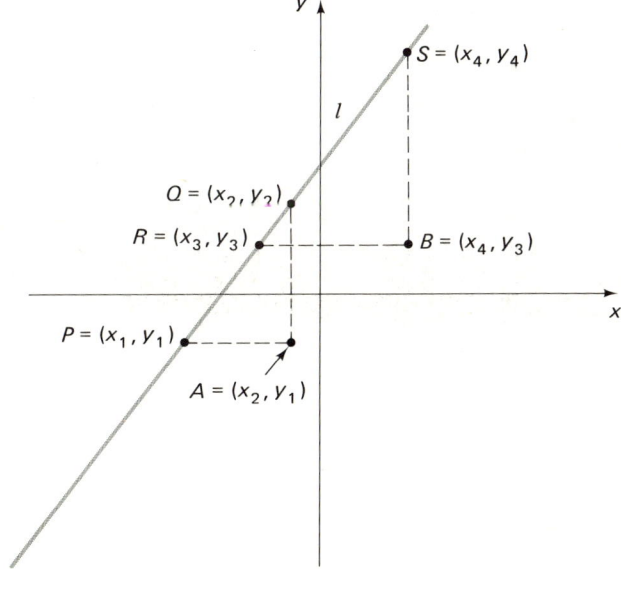

FIG. 40

Suppose now that we know the slope of a line and a point on it. Then using the fact that slope = (change in y)/(change in x), we can sketch the line.

EXAMPLE 3 Find two other points on the line through $(-1, -2)$, and sketch if the slope is **(a)** $3/2$ or **(b)** -3.

SOLUTION **(a)** Since the slope is $3/2$, for every change of two units in the x direction, there must be a change of three units in the y direction. Since the slope is positive, either both changes must be increases or both changes must be decreases. Using Fig. 41(a), we increase x by 2 and y by 3 to get $(1, 1)$ or decrease x by 2 and y by 3 to get $(-3, -5)$.

(b) Since (change in y)/(change in x) $= -3 = (-3)/1 = 3/(-1)$, we see that as x increases 1, y decreases 3 or that as x decreases 1, y increases 3. Hence two other points on the line are $(-2, 1)$ and $(0, -5)$. See Fig. 41(b).

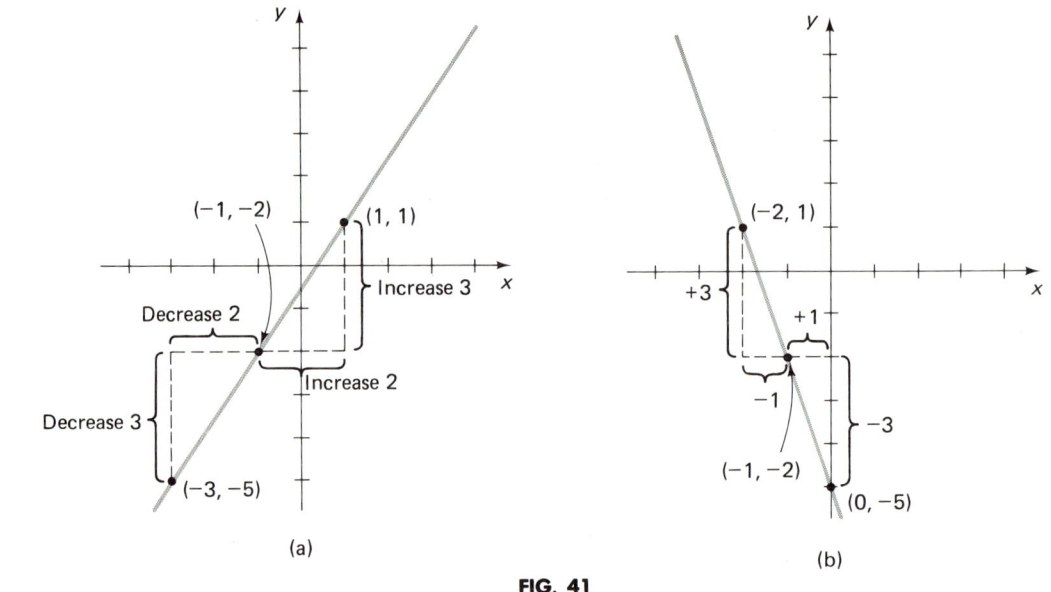

FIG. 41

We now consider this situation in general, which will lead us to one of the general equations for a straight line. Suppose we know the slope m of a (nonvertical) straight line and a point $P = (x_1, y_1)$ on the line. (This completely determines a unique straight line.) Then another point $Q = (x, y)$ is on this line if and only if

$$\frac{y - y_1}{x - x_1} = m$$

Multiplying by $x - x_1$, we get the following:

---

(25) **THE POINT-SLOPE FORM FOR THE EQUATION OF A LINE**

An equation for the line through the point $P = (x_1, y_1)$ and with slope m is

$$y - y_1 = m(x - x_1)$$

EXAMPLE 4 Find an equation of the line through $(-\frac{1}{3}, 3)$ and $(1, -5)$.

SOLUTION We first use (23) to find the slope:
$$m = \frac{y_2 - y_1}{x_2 - x_1} = \frac{-5 - 3}{1 - (-\frac{1}{3})} = \frac{-8}{\frac{4}{3}} = -6$$

We now use (25), $y - y_1 = m(x - x_1)$, picking either of the given points for (x_1, y_1), say $(1, -5)$. We obtain
$$y - (-5) = -6(x - 1), \qquad y + 5 = -6x + 6, \qquad y - 6x - 1 = 0$$

Besides slope, another fundamental property of a straight line is its y-intercept.

---

(26)   **DEFINITION**

The **y-intercept** of a nonvertical straight line *l* is the *y*-coordinate of the point where the line crosses the *y*-axis.

The y-intercept is usually denoted by the letter b. The **x-intercept** of a line is defined similarly.

Suppose you know the slope m of a line and its y-intercept b. Then the line contains the point $(0, b)$, so by (25), its equation is
$$y - b = m(x - 0)$$

Solving for y, we get the most useful form for an equation of a straight line.

---

(27)   **THE SLOPE-INTERCEPT FORM FOR THE EQUATION OF A LINE**

An equation for the line having slope *m* and *y*-intercept *b* is
$$y = mx + b.$$

There are several other forms for equations of lines which are useful in special circumstances. One of these is given in Exercise 55. However, in general it is best to have as few formulas as possible to memorize and to understand how to use these few well.

EXAMPLE 5 Put the equation $2x + 3y + 3 = 0$ into slope-intercept form, and then graph the line. Find the x-intercept of this line.

SOLUTION We want the equation in the form $y = mx + b$, so we solve the given equation for y:
$$2x + 3y + 3 = 0, \qquad 3y = -2x - 3, \qquad y = -\tfrac{2}{3}x - 1$$

By comparing $y = -\frac{2}{3}x - 1$ with $y = mx + b$, we see that the y-intercept is $b = -1$ and that the slope is $m = -\frac{2}{3}$. See Fig. 42.

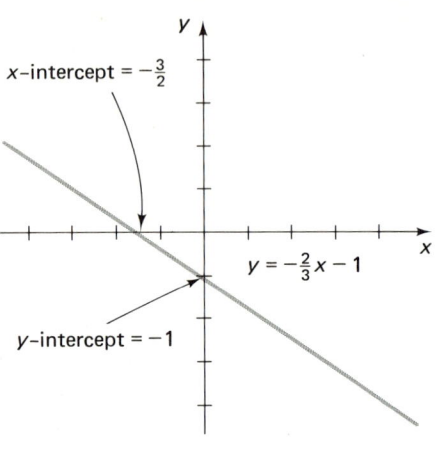

FIG. 42

To find the x-intercept, we want to find the point on the line whose y-coordinate is 0. Thus we set $y = 0$ in the original equation and solve for x:

$$2x + 3y + 3 = 0, \quad 2x + 0 + 3 = 0, \quad x = -\tfrac{3}{2}$$

---

(28)   **DEFINITION**

Two lines are **parallel** if both have the same slope or if both are vertical.

EXAMPLE 6   Find the equation of the line through the point $(1, 2)$ and parallel to
$$2x - 4y + 5 = 0.$$

SOLUTION   We first need to find the slope of the given line, so we put it in slope-intercept form:

$$2x - 4y + 5 = 0, \quad -4y = -2x - 5, \quad y = \tfrac{1}{2}x + \tfrac{5}{4}$$

We now want the equation of the line through $(1, 2)$ and with slope $1/2$. We use (25):

$$y - y_1 = m(x - x_1), \quad y - 2 = \tfrac{1}{2}(x - 1), \quad y = \tfrac{1}{2}x + \tfrac{3}{2}$$

We now return to linear functions. Since the rule of a linear function is of the form $f(x) = mx + b$, the graph is the set of all points (x, y) such that $y = mx + b$. Hence,

> **(29)** The graph of a linear function is a straight line.

Therefore, all we now know about finding equations of straight lines can be applied to finding the rule of a linear function.

EXAMPLE 7 If a projectile is shot straight up, its velocity v is a linear function of time t. Suppose a shell is shot straight up with an initial velocity of 920 feet per second and that after 5 seconds it is going up at a rate of 760 feet per second. **(a)** Find the equation which expresses v as a linear function of time t. **(b)** How long does it take for the shell to reach its highest point?

SOLUTION For part (a): When you are given that v is a linear function of t, the *first* thing to write is
$$v = mt + b$$
Next, translate the given conditions to the following: When $t = 0$, $v = 920$, and when $t = 5$, $v = 760$. Thus we want the equation of the line $v = mt + b$ through $(0, 920)$ and $(5, 760)$. With these conditions, we can just read off the y-intercept, or really the v-intercept, as $b = 920$. Next, compute the slope:
$$m = \frac{760 - 920}{5 - 0} = \frac{-160}{5} = -32$$
Thus, $v = -32t + 920$.

To answer part (b), when the projectile reaches its highest point, its velocity is momentarily zero. Thus we solve
$$0 = -32t + 920, \quad 32t = 920, \quad t = \frac{920}{32} = 28.75 \text{ sec}$$

EXERCISES

In Exercises 1–6, **(a)** find the slope of the line through the given points, then **(b)** find the equation of the line through the given points.

1. $(2, 1), (5, -2)$
2. $(-2, 1), (3, 2)$
3. $(-3, -5), (3, -4)$
4. $(-4, 5), (2, 5)$
5. $(2, 5), (2, -5)$
6. $(4, 1), (1, 4)$

In Exercises 7–9, use slopes to determine if the given points lie on a straight line.

7. $(-4, 1), (2, -2), (4, -3)$
8. $(-3, -1), (-1, 2), (\frac{1}{2}, \frac{17}{4})$
9. $(-4, -10), (-1, 30), (4, 90)$

In Exercises 10–15, find two other points on the line through the given point and with the given slope.

Section 3.5 Linear Functions and Lines

10. $(-1, -2)$, slope $= 3$
11. $(4, 3)$, slope $= -3/4$
12. $(-4, 1)$, slope $= -1/2$
13. $(6, -2)$, slope $= 2/3$
14. $(-1, 5)$, slope $= -1/2$
15. $(2, -1)$, slope $= -1/3$

In Exercises 16–28, find an equation of the line satisfying the given conditions. *Note:* Two lines are **perpendicular** if either one line is horizontal and the other is vertical or else $m_1 = -(1/m_2)$, where m_1 and m_2 are the slopes of the two lines.

16. Through $(4, -1)$, slope $= 2/3$
17. Through $(-2, 5)$, slope $= -2$
18. Through $(-1, 4)$, $(2, 7)$
19. Through $(5, 1)$, $(-1, 6)$
20. Slope $= -3$, y-intercept 4
21. Slope $= 2/3$, x-intercept -2
22. Through $(-2, 3)$ and parallel to the **(a)** x-axis, **(b)** y-axis
23. Through $(4, -2)$ and perpendicular to the **(a)** x-axis, **(b)** y-axis
24. Through $(-3, -2)$ and perpendicular to the line $y = 3x - 4$
25. Through $(1, -5)$ and perpendicular to the line $2x - 3y - 1 = 0$
26. Coinciding with the **(a)** x-axis, **(b)** y-axis
27. Through $(5, -1)$ and parallel to the line with equation $y = \frac{2}{3}x - 1$
28. Through $(-\frac{1}{3}, 2)$ and parallel to the line through $(0, -2)$ and $(2, 0)$

In Exercises 29–36, put the given equation into slope-intercept form, and find the slope and y-intercept. Sketch the graph.

29. $3x + 4y + 6 = 0$
30. $3y - 2x - 5 = 0$
31. $4x + 3y = 0$
32. $2y - 3x = 0$
33. $2y = 5$
34. $y = 0$
35. $42.18x + 15.1y = 16.24$
36. $.412x - .913y - 11.91 = 0$

In Exercises 37–40, sketch the graph of the given linear function, labeling the y- and x-intercepts.

37. $f(x) = 3x - 6$
38. $g(x) = -4x + 7$
39. $h(x) = -\frac{2}{3}x + 4$
40. $k(x) = -\frac{2}{5}x$

41. Sketch the graph of $f(x) = -\frac{1}{2}x + b$ if

    a. $b = 0$ **b.** $b = -2$
    c. $b = 3$ **d.** $b = -5$

42. Sketch the graph of $f(x) = mx + 2$ if

    a. $m = 0$ **b.** $m = 2$
    c. $m = 1/2$ **d.** $m = -3$

43. If the graph of a linear function g contains the points $(3, 1)$ and $(-1, 3)$, find $g(x)$.
44. If the graph of the linear function k contains the points $(0, -3)$ and $(2, -2)$, find $k(x)$.
45. If the graph of the linear function F is parallel to $3x + 2y + 1 = 0$ and $F(0) = 4$, find $F(x)$.
46. If the graph of the linear function G is parallel to $2x = 1 - 5y$ and $G(1) = 1$, find $G(x)$.
47. Fahrenheit temperature F is a linear function of Celsius, or centigrade, temperature C. When C is 0, F is 32, and when C is 100, F is 212. Express F as a linear function of C.
48. From Exercise 47, express C as a linear function of F, using C is 0 when F is 32 and C is 100 when F is 212.
49. The velocity v of a baseball thrown straight up is a linear function of the time t after it is thrown. If the ball is thrown up at 90 feet per second and after 2 seconds it is still going up at 26 feet per second, then

    a. Express v as a linear function of t.

b. Determine how long it takes to reach its highest point.

c. What is its velocity after 4 seconds?

50. If a projectile is fired straight up with an initial velocity of 60.2 meters per second and after 8 seconds it is falling *down* at the rate of 18.2 meters per second,

a. Express its velocity v as a linear function of the time t after it was fired.

b. Determine how long it took to reach its highest point.

c. What is its velocity after 10 seconds?

51. Assume that the temperature T is a linear function of height h above the surface of the earth. Suppose the temperature at the surface is 20 °C and at 1200 meters it is 15 °C.

a. Express T as a linear function of h.

b. What is the temperature at 2000 meters?

52. Suppose the quantity of heat Q (measured in calories) that it takes to change ice at 0 °C to water at T °C is a linear function of T for $0 \leq T \leq 100$. For 1 gram of ice, $Q = 100$ when $T = 20$, and $Q = 120$ when $T = 40$.

a. Express Q as a linear function of T.

b. How many calories does it take just to melt the ice?

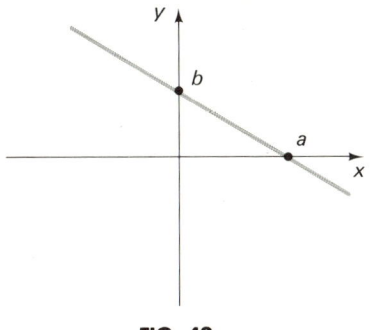

FIG. 43

53. Suppose a line l has nonzero x- and y-intercepts a and b, respectively. See Fig. 43. Prove that an equation for l is

$$\frac{x}{a} + \frac{y}{b} = 1$$

This is called the **intercept** form for an equation of a line. (*Hint*: Start with $y = mx + b$, and determine m in terms of a and b).

54. Find both the intercept form and the slope-intercept form for the equation of the line through

a. $(0, 1), (-2, 0)$ b. $(0, \frac{1}{2}), (3, 0)$ c. $(0, -2), (\frac{1}{2}, 0)$

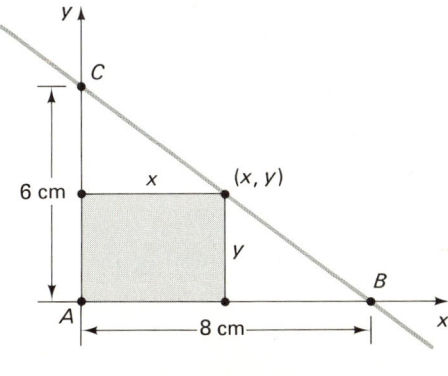

FIG. 44

55. A rectangle with an area of 9 square centimeters is inscribed in triangle ABC as shown in Fig. 44. Find the possible dimensions of the rectangle. (*Hint*: Use the equation of the line through B and C.)

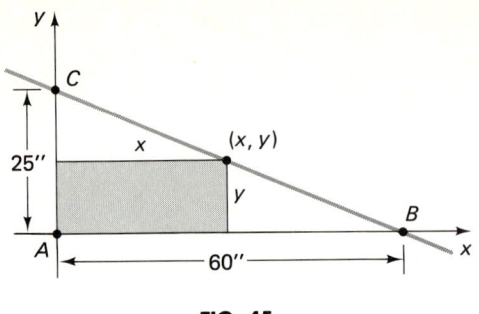

FIG. 45

56. A rectangle with an area of 375 square inches is inscribed in triangle ABC as shown in Fig. 45. What are its dimensions?

SECTION 3.6. VARIATION

When functions are described in the natural and social sciences, often the word *function* is not used but is replaced by terminology describing relationships among variables. In this section we introduce three particular examples of this kind of terminology.

---

(30)   **DEFINITION**

The phrase **u varies directly as v**, or **u is directly proportional to v**, means

$$u = kv$$ for some fixed real number k.

In this definition, the letters u and v may represent single quantities such as "time, t" or "distance, d," or they may represent more complicated expressions such as "distance squared, d^2" or "the square root of the length, $\sqrt{l}$," etc.

The number k is often called the **constant of variation** or the **constant of proportionality**.

This definition should be thought of as a means of translating a relationship from words into symbols. The following are examples of the translation process.

EXAMPLE 1

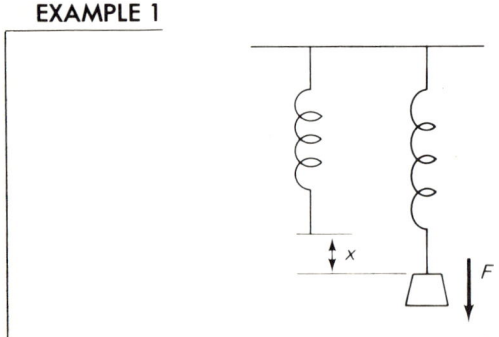

FIG. 46

Hooke's law. If a spring, which is hanging naturally, is stretched, then "the force F required to stretch the spring is directly proportional to the distance stretched x" within certain limits. This translates as

$$F = kx$$

See Fig. 46. The constant k depends on the nature and composition of the spring.

152 RELATIONS, FUNCTIONS, AND GRAPHS

EXAMPLE 2 If a simple pendulum is swinging back and forth, then "the time required for one complete oscillation (this is called the **period** P of the pendulum) is directly proportional to the square root of its length l." This translates as

$$P = k\sqrt{l}$$

This k depends on the nature of the pendulum.

---

(31)   DEFINITION

The phrase *u* **varies inversely as** *v*, or *u* **is inversely proportional to** *v*, means

$$u = \frac{k}{v} \quad \text{for some fixed real number } k.$$

Again, u and v may represent single quantities or more complicated expressions. The number k is still called the **constant of variation** or the **constant of proportionality**.

EXAMPLE 3 If the temperature remains constant, then "the pressure P of a fixed amount of gas varies inversely as the volume V." This translates as

$$P = \frac{k}{V}$$

EXAMPLE 4 "The force F of attraction due to gravity of two objects on each other is inversely proportional to the square of the distance D between them." This translates as

$$F = \frac{k}{D^2}$$

In many applied problems, the constant of variation can be determined by examining the experimental facts. The resulting equation can then be used to predict outcomes.

EXAMPLE 5 The electrical resistance R of a wire of a fixed length is inversely proportional to the square of the diameter D of the wire. Suppose 1 meter of a certain type of wire has a resistance of 210 ohms when the diameter is 5 millimeters. **(a)** Express this statement as a formula, and determine the constant of proportionality. **(b)** How much resistance would 1 meter of this same kind of wire 6 millimeters in diameter have? **(c)** If we wanted 1 meter of this kind of wire to have a resistance of exactly 250 ohms, what should be its diameter?

SOLUTION
    (a) Translating, we get

$$R = \frac{k}{D^2}$$

The information given is

$$\text{When } D = 5, \quad R = 210$$

Substituting in and solving for k, we obtain

$$210 = \frac{k}{5^2}, \quad k = 210(25) = 5250$$

Thus the formula is $R = 5250/D^2$.

(b) The question translates as

$$\text{If } D = 6, \quad \text{what is } R?$$

Substituting in and solving, we obtain

$$R = \frac{5250}{6^2} \approx 145.833$$

(c) The question translates as

$$\text{If } R = 250, \quad \text{what is } D?$$

Substituting in and solving, we obtain

$$250 = \frac{5250}{D^2}, \quad D^2 = \frac{5250}{250} = 21, \quad D = \sqrt{21} \approx 4.58258 \text{ mm}$$

(*Note*: We neglect the negative sign, as the diameter is always greater than or equal to zero.)

When an equation relating variables is found, it is sometimes useful to translate the relationship into words.

EXAMPLE 6 If x varies directly as the fourth power of y and y varies inversely as the square root of z, what is the relationship between x and z?

SOLUTION Translating these statements yields

$$x = ky^4 \quad \text{and} \quad y = \frac{K}{\sqrt{z}}$$

(*Note*: We must use different letters for the different constants of proportionality.) Since we want one statement relating x and z, we substitute in for y:

$$x = k\left(\frac{K}{\sqrt{z}}\right)^4 = \frac{kK^4}{z^2}$$

Since k and K are constants, kK^4 is a constant, so this says "x varies inversely as z squared."

When there are several independent variables in a variation problem, this is expressed by using the word **jointly**.

As an illustration we define two such situations. There are many others, but the language is analogous.

> **(32)** **DEFINITION**
>
> The phrase *u* **varies jointly as *v* and *w*** means
>
> $$u = kvw \quad \text{for some fixed real number } k.$$
>
> The phrase *u* **varies jointly as *v* and the inverse of *w*** means
>
> $$u = k\frac{v}{w} \quad \text{for some fixed real number } k.$$

Example 7 is typical of joint variation situations.

EXAMPLE 7 The electrical resistance R of wire (of uniform material) varies jointly as the length L and the inverse of the square of the diameter D. Suppose a certain type of wire 1 meter long and 5 millimeters in diameter has a resistance of 210 ohms. (a) Express this statement as a formula, and determine the constant of proportionality. (b) What would be the resistance of a piece of wire (of the same material) 45 centimeters long and 4 millimeters in diameter?

SOLUTION

(a) Translating the statement, we get

$$R = \frac{kL}{D^2}$$

The information given is

$$R = 210 \quad \text{when} \quad L = 1 \quad \text{and} \quad D = 5$$

Substituting in and solving for k, we obtain

$$210 = \frac{k(1)}{5^2}, \quad k = 5^2(210) = 5250$$

Thus, the formula is

$$R = \frac{5250L}{D^2}$$

(b) The question translates as follows: What is R when $L = .45$ (45 centimeters = .45 meter) and $D = 4$. We substitute in and solve for R:

$$R = \frac{5250(.45)}{4^2} = 147.656$$

EXERCISES

In Exercises 1–12, express each statement as a formula, and determine the constant of variation from the given conditions.

1. y varies directly as x. If $x = 15$, then $y = 35$.
2. a is directly proportional to b. If $a = 42$, then $b = .7$.

3. m is inversely proportional to n. If $m = .12$, then $n = .9$.
4. u is inversely proportional to v. If $u = 810$, then $v = .32$.
5. w varies directly as z cubed. If $w = 36$, then $z = 3$.
6. s varies inversely as t cubed. If $s = 4$, then $t = 1.2$.
7. c varies inversely as the cube root of d. If $c = 18$, then $d = 8$.
8. p varies directly as the square root of q. If $p = 276$, then $q = 144$.
9. u varies jointly as v and w cubed. When $v = 5$ and $w = 2.2$, $u = 26.62$.
10. a varies jointly with x and y squared and the inverse of d to the $3/2$ power. When $b = 15$, $c = 32.1$, and $d = 16$, $a = 4$.
11. A is directly proportional to the sum of B and C. When $B = 4$ and $C = 5$, $A = 6$.
12. x is inversely proportional to the sum of y squared and z squared. When $y = 7$ and $z = 24$, $x = .2$.

In Exercises 13–16, $y = f(x)$, and the point $(9, 4)$ is on the graph. Find the formula for $f(x)$.

13. y is directly proportional to x.
14. y is inversely proportional to x.
15. y varies directly as $x^{3/2}$.
16. y varies inversely as x^2.
17. If y varies directly as x^2 and x varies directly as z^3, what is the relationship between y and z?
18. If y varies inversely as x^2 and x varies inversely as the square root of z, what is the relationship between y and z?
19. If y varies inversely as x, what is the relationship between y^{-1} and x?
20. If y varies jointly as x and z, x varies directly as t, and z varies inversely as t, what is the relationship between y and t?

In Exercises 21–30, first find the constant of variation and the formula involved. Then answer the question.

21. Hooke's law states that the force F required to stretch a spring x units beyond its natural length is directly proportional to x. Suppose a weight of 6 pounds (weight is the force due to gravity) stretches a spring 2 inches.
    a. How far will a weight of 8 pounds stretch the spring?
    b. How much weight will stretch the spring exactly $1\frac{1}{2}$ inches?
22. The force F of attraction of two oppositely charged bodies is inversely proportional to the square of the distance between them. Suppose two oppositely charged bodies which are 12 centimeters apart are attracted with a force of 42 dynes.
    a. How far apart should they be so that they are attracted with a force of 32 dynes?
    b. What would be the force of attraction if they were 20 centimeters apart?
23. The kinetic energy K of an object varies directly as the square of its velocity. A particular object traveling at 80 feet per second has a kinetic energy of 240 foot-pounds.
    a. What would be its kinetic energy if it were traveling at 100 feet per second?
    b. How fast would it be going if it had a kinetic energy of 50 foot-pounds?
24. The current I in a simple circuit varies inversely as the resistance R. A particular circuit has a current of 12 amperes when the resistance is 20 ohms.
    a. What resistance would give a current of 10 amps?
    b. What would the current be if the resistance were lowered to 10 ohms?

25. The volume V of wood in a tree varies jointly with the height h and the square of the girth g. (Girth is the distance around the tree.) A particular tree of height 100 feet and girth 8 feet is estimated to have a volume of 8000 cubic feet of wood. What would be the increase in wood if the tree were to grow another 10 feet in height and 1 foot in girth?

26. The volume V of a gas varies jointly with the temperature T and inversely as the pressure P. The volume of a certain gas is 20 when the temperature is 270 and the pressure is 15. What would be the volume if the temperature and pressure were each increased by 10?

27. The electrical resistance of a wire varies jointly with the length and the inverse of the square of the diameter. A particular wire 50 feet long with a diameter of .02 inch has a resistance of 25 ohms. How much resistance would there be if the length and diameter were halved?

28. The safe load S (the amount it can support without breaking) a beam with a rectangular cross section can support varies jointly with its width w, the square of its height h, and the inverse of its length L. A particular beam of height 8 inches, width 3 inches, and length 20 feet can support 30 tons. If each dimension were doubled, how much could it support then?

29. The distance d that you can see to the horizon varies directly with the square root of your height above sea level. Suppose a person 20 meters above sea level can see 32.5 kilometers.
  a. How far can the person see at 25 meters?
  b. How high does the person have to be to see 50 kilometers?

30. The number of oscillations per minute of a simple pendulum varies inversely as the square root of its length. Suppose a particular pendulum 20 centimeters long makes eight oscillations per minute. How many oscillations per minute would the pendulum make if it were cut to 15 centimeters?

SECTION 3.7. COMPOSITE AND INVERSE FUNCTIONS

In this section we first discuss an important method of combining two given functions to form a third function. Suppose that X, Y, and Z are sets of real numbers, and suppose that f is a function from X to Y and that g is a function from Y to Z. Then we can naturally form a function from X to Z as follows (see Fig. 47):

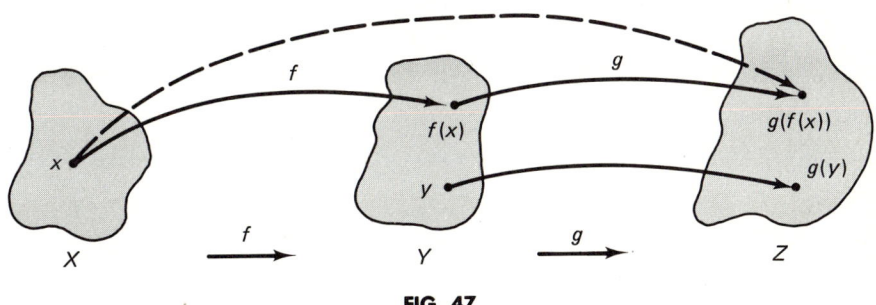

FIG. 47

For every number x in X, f associates a number $f(x)$ in Y.

For every number y in Y, g associates a number $g(y)$ in Z; in particular $f(x)$ is in Y, so g associates with $f(x)$ the number $g(f(x))$ in Z. By associating with each x in X the number $g(f(x))$ in Z, we obtain what is called the **composite function** of f and g. This function is denoted by $g \circ f$.

---

(33)   **DEFINITION**

If f is a function from X to Y and g is a function from Y to Z, then the **composite function** $g \circ f$ is the function from X to Z defined by

$$(g \circ f)(x) = g(f(x)) \quad \text{for every } x \text{ in } X.$$

EXAMPLE 1   If $f(x) = x^2$ and $g(x) = x + 1$, find $(g \circ f)(x)$ and $(f \circ g)(x)$.

SOLUTION   We apply the definitions and then simplify:

$$(g \circ f)(x) = g(f(x)) \quad \text{(def. of } g \circ f)$$
$$= g(x^2) \quad \text{(def. of } f)$$
$$= x^2 + 1 \quad \text{(def. of } g)$$
$$(f \circ g)(x) = f(g(x)) \quad \text{(def. of } f \circ g)$$
$$= f(x + 1) \quad \text{(def. of } g)$$
$$= (x + 1)^2 \quad \text{(def. of } f)$$

Note that $(g \circ f)(x)$ and $(f \circ g)(x)$ are different. This is often the case.

When the domains and ranges are not the set of all real numbers, we may have to be careful in stating the domains of the composites. In general, the domain of $g \circ f$ is the domain of f less any points x for which $f(x)$ is not in the domain of g. These points to be deleted usually can be determined from the formula for $(g \circ f)(x)$.

EXAMPLE 2   If $f(x) = \sqrt{x - 1}$, $x \geq 1$, and $g(x) = 4x^2 + 1$, find $(g \circ f)(x)$ and $(f \circ g)(x)$.

SOLUTION

$$(g \circ f)(x) = g(f(x)) \quad \text{(def. of } g \circ f)$$
$$= g(\sqrt{x - 1}) \quad \text{(def. of } f)$$
$$= 4(\sqrt{x - 1})^2 + 1 \quad \text{(def. of } g)$$
$$= 4(x - 1) + 1 = 4x - 3$$

Although the rule $4x - 3$ is defined for all x, $f(x) = \sqrt{x - 1}$ is defined only for $x \geq 1$, so $g(f(x))$ is defined only for $x \geq 1$. Thus,

$$(g \circ f)(x) = 4x - 3, \quad x \geq 1$$

For the other composite,

$$(f \circ g)(x) = f(g(x)) \qquad \text{(def. of } f \circ g\text{)}$$
$$= f(4x^2 + 1) \qquad \text{(def. of } g\text{)}$$
$$= \sqrt{(4x^2 + 1) - 1} \qquad \text{(def. of } f\text{)}$$
$$= \sqrt{4x^2} = 2|x|$$

The domain of g is the set of all real numbers. Since $2|x|$ is defined for all these, the domain of $f \circ g$ is the set of all real numbers.

A calculator illustrates composites nicely. The composite of functions on a calculator "means" using one button (or buttons) and then another, and of course the order is important.

EXAMPLE 3 Suppose $f(x) = x + 3$ and $g(x) = \sqrt{x}$. Then to use a calculator to compute $(g \circ f)(7) = g(f(7))$, press

$$\underbrace{}_{f(7)} \qquad \underbrace{}_{g(f(7))}$$

Alg.:   [7]   [+][3][=]   [√]

RPN: [7]   [ENT][3][+]   [√]

getting $\sqrt{7 + 3} \approx 3.16228$. To use the calculator to compute $(f \circ g)(7) = f(g(7))$, press

$$\underbrace{}_{g(7)} \qquad \underbrace{}_{f(g(7))}$$

Alg.:   [7]   [√]   [+][3][=]

RPN: [7]   [√]   [ENT][3][+]

getting $\sqrt{7} + 3 \approx 5.64575$.

We now turn to the very important topic of inverse functions. We begin with an example. Let $f(x) = x^3$ and $g(x) = \sqrt[3]{x}$. Since

$$g(f(x)) = \sqrt[3]{x^3} = x \quad \text{and} \quad f(g(x)) = (\sqrt[3]{x})^3 = x$$

we have

$$(g \circ f)(x) = x, \quad \text{all } x \quad \text{and} \quad (f \circ g)(x) = x, \quad \text{all } x$$

Thus applying the composite in either direction gets us back to where we started. When this happens, we say g is the **inverse function** of f, because g does the inverse of (i.e., reverses) what f does.

(34) **DEFINITION**

Suppose f and g are functions such that

$(g \circ f)(x) = g(f(x)) = x$ for all x in the domain of f, and

$(f \circ g)(x) = f(g(x)) = x$ for all x in the domain of g

Then we say g is the **inverse function** of f, and we write $g = f^{-1}$. (f^{-1} is read "f inverse.")

Thus suppose $f: U \to V$ and $f(u) = v$. If g is f^{-1}, then $g: V \to U$ and $g(v) = u$. See Fig. 48.

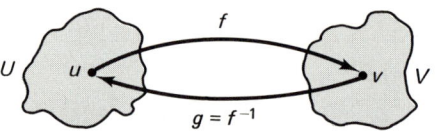

FIG. 48

Note the following:

(35)

The domain of f is the range of f^{-1} and vice versa.

It will be useful to rewrite the equations in (34) using f^{-1}.

(36)

If f has an inverse function f^{-1}, then

$f^{-1}(f(x)) = x$ for all x in the domain of f, and

$f(f^{-1}(x)) = x$ for all x in the domain of f^{-1}

Sometimes an inverse function can be found easily using (36).

EXAMPLE 4 By the preceding discussion, if $f(x) = x^3$, then $f^{-1}(x) = x^{1/3}$, since

$$f^{-1}(f(x)) = f^{-1}(x^3) = (x^3)^{1/3} = x$$

and

$$f(f^{-1}(x)) = f(x^{1/3}) = (x^{1/3})^3 = x$$

160 RELATIONS, FUNCTIONS, AND GRAPHS

EXAMPLE 5 If $g(x) = x + 2$, then $g^{-1}(x) = x - 2$ since
$$g^{-1}(g(x)) = g^{-1}(x + 2) = (x + 2) - 2 = x$$
and
$$g(g^{-1}(x)) = g(x - 2) = (x - 2) + 2 = x$$

Warning: $f^{-1}(x)$ is not $1/f(x)$, as can be seen in Examples 4 and 5. If we want $1/f(x)$, we write $[f(x)]^{-1}$.

It is very interesting and instructive to see how your calculator handles certain inverse functions. Your calculator has special keys that have not been explained:

$$[\sin], \quad [\cos], \quad [\tan], \quad [\log], \quad [\ln]$$

We shall use them here *only* to illustrate (36). (Of course, labeling varies on different calculators; for example [ln] might be [ln x], [LN], etc.) These five keys represent five functions which also have inverse functions (when the domain is appropriately restricted). To compute the inverse functions, your calculator has a key labeled

$$[\text{INV}] \quad \text{or} \quad [\text{ARC}] \quad \text{or} \quad [F]$$

(and there are a few other variations). If $g(x) = \sin x$, to compute $g^{-1}(.5)$, press

$$[.5][\text{INV}][\sin]$$

The other inverse functions work similarly (with a few exceptions for [log] and [ln]). We can now use a calculator to illustrate (36). To avoid difficulties that we do not need to consider for this illustration, we shall always start with numbers between 0 and 1.

EXAMPLE 6 If $h(x) = \cos x$, use your calculator to compute **(a)** $h^{-1}(h(.2))$ and **(b)** $h(h^{-1}(.3))$.

SOLUTION Of course, we should end up with what we started with. For (a), press

$$\overbrace{[.2] \quad [\cos]}^{h(.2)} \quad \overbrace{[\text{INV}][\cos]}^{h^{-1}(h(.2))}$$

getting .2. For (b), press

$$\overbrace{[.3] \quad [\text{INV}][\cos]}^{h^{-1}(.3)} \quad \overbrace{[\cos]}^{h(h^{-1}(.3))}$$

getting .3.

We now have three important questions to answer concerning inverse functions. Suppose we start with a function f.

1. When does f have an inverse function f^{-1}?

2. How do you find the rule for f^{-1} from the rule for f?
3. How do you find the graph for f^{-1} from the graph for f?

To give comprehensive and integrated answers to these three questions, we turn to the ordered pair definition of a function (16). First, consider again the function $f(x) = x^3$ and its inverse $f^{-1}(x) = \sqrt[3]{x}$. Since $f(2) = 2^3 = 8$, $(2, 8)$ is an ordered pair in f. However, because $2^3 = 8$, it follows that $\sqrt[3]{8} = 2$, so that $f^{-1}(8) = 2$ and $(8, 2)$ is in f^{-1}. This illustrates the general situation.

(37)

f^{-1} is the set of all ordered pairs (b, a) such that (a, b) is an ordered pair in f.

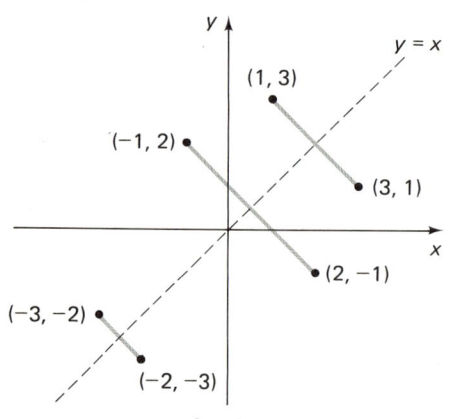

FIG. 49

This immediately tells us about the graphs. By (37), the graph of the inverse is the set of all points in the plane with coordinates (b, a), where (a, b) is in the original function. When $a \neq b$, the relationship between (a, b) and (b, a) is that they are *symmetric with respect to the line $y = x$*. See Fig. 49. When we start with a graph and replace each point of that graph with the point symmetric to it with respect to the line $y = x$, we say we have *reflected the graph through the line $y = x$*. Hence:

(38)

The graph of an inverse function is just the graph of the original function reflected through the line $y = x$.

When we reflect the graph of a function through the line $y = x$, the resulting graph may or may not pass the vertical line test. Hence it may or may not be the graph of a function. A function f has an inverse function f^{-1} if reflecting the graph of f through the line $y = x$ results in a graph of a function (and this will then be the graph of f^{-1}). There is a geometric test to determine this. The test corresponds to doing the vertical line test after reflecting the graph through the line $y = x$.

(39)   THE HORIZONTAL LINE TEST

1. If there are two or more points of the graph of a given function on the same horizontal line, the function does not have an inverse function.
2. If there is no horizontal line which contains two or more points of the graph of the function, it has an inverse function.

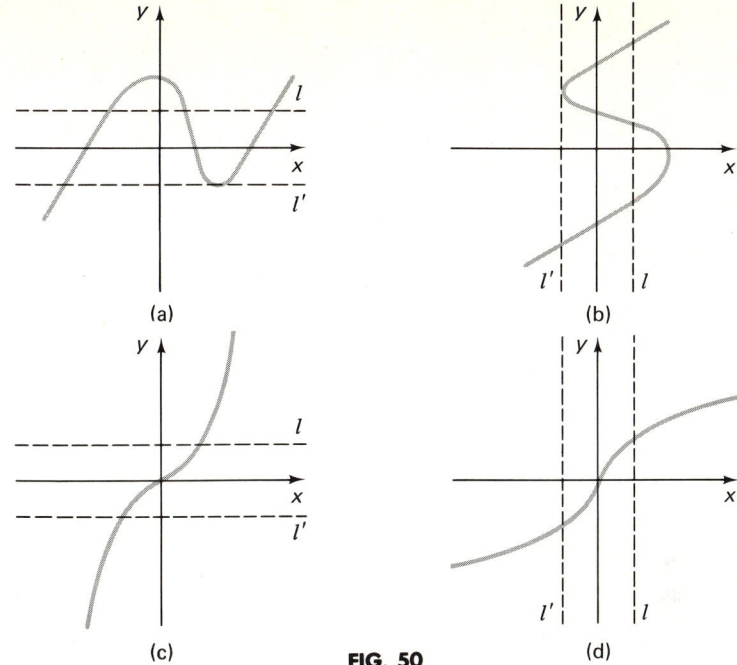

FIG. 50

EXAMPLE 7 If we examine Fig. 50, we see that the function graphed in Fig. 50(a) "fails" the horizontal line test; i.e., there is a horizontal line (either l or l') which intersects the graph in two or more places. So the reflected graph is not the graph of a function, as can be seen in Fig. 50(b). On the other hand, the function graphed in Fig. 50(c) "passes" the horizontal line test, so reflecting its graph through the line $y = x$ yields the graph of a function, as can be seen in Fig. 50(d).

There are two important and common types of functions which have inverse functions. A function f is called **strictly increasing** if whenever a and b are in the domain of f and $a < b$, then $f(a) < f(b)$; it is **strictly decreasing** if $f(a) > f(b)$ whenever $a < b$ and a and b are in the domain of f. See Fig. 51. Clearly, such functions pass the horizontal line test.

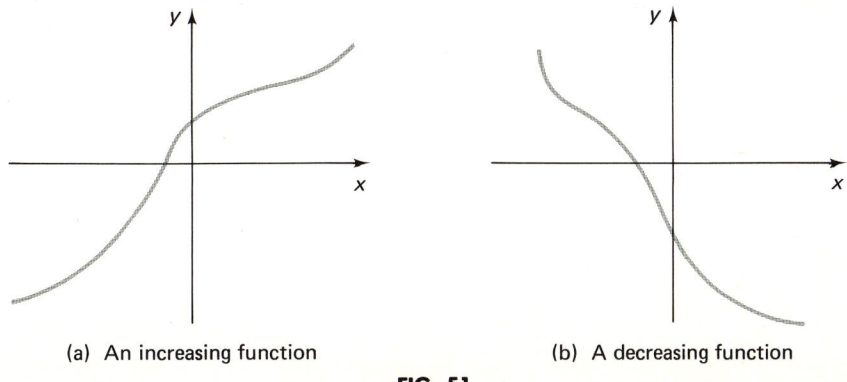

(a) An increasing function   (b) A decreasing function

FIG. 51

You should note that there are many functions which have inverse functions but which are neither strictly increasing nor strictly decreasing. See Example 10.

We are now ready to see how to obtain the rule for the inverse from the original function. Suppose that we start with a function f with rule $y = f(x)$ and that f has an inverse function f^{-1}. Then f is the set of all ordered pairs whose coordinates satisfy the equation $y = f(x)$. The ordered pairs of f^{-1} are obtained by interchanging all the x- and y-coordinates of f. Thus if we start with the equation $y = f(x)$ and interchange x and y, the new equation obtained must determine f^{-1}. Hence if we solve this new equation for y, we get an equation which both determines f^{-1} and gives a formula for y in terms of x. Such a formula is exactly what we mean by a rule for f^{-1}, so the last equation is of the form $y = f^{-1}(x)$.

EXAMPLE 8 If $f(x) = -2x + 4$, find $f^{-1}(x)$. Graph $y = f(x)$ and $y = f^{-1}(x)$.

SOLUTION Start with $y = f(x)$:

$$y = -2x + 4$$

Switch x and y (this equation determines the inverse function):

$$x = -2y + 4$$

Solve for y:

$$2y = -x + 4$$
$$y = -\tfrac{1}{2}x + 2$$

The last equation is of the form $y = f^{-1}(x)$, so we can just read off $f^{-1}(x)$:

$$f^{-1}(x) = -\tfrac{1}{2}x + 2$$

The graphs are in Fig. 52.

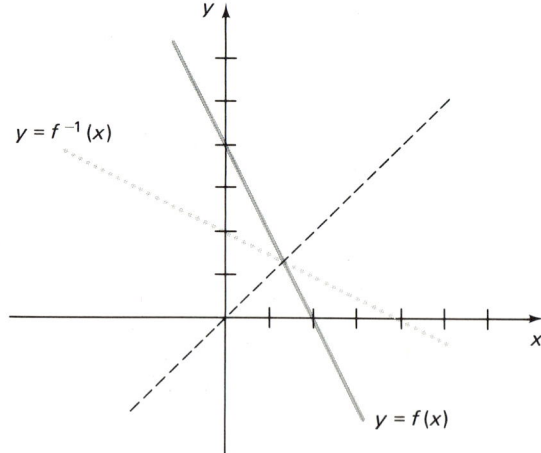

FIG. 52

164 RELATIONS, FUNCTIONS, AND GRAPHS

EXAMPLE 9 If $k(x) = x^3$, find $k^{-1}(x)$. Graph.

SOLUTION

Set $y = k(x)$:

Switch x and y: $\qquad y = x^3$

Solve for y: $\qquad x = y^3$

Read off $k^{-1}(x)$: $\qquad y = x^{1/3}$

The graphs are in Fig. 53. $k^{-1}(x) = x^{1/3}$

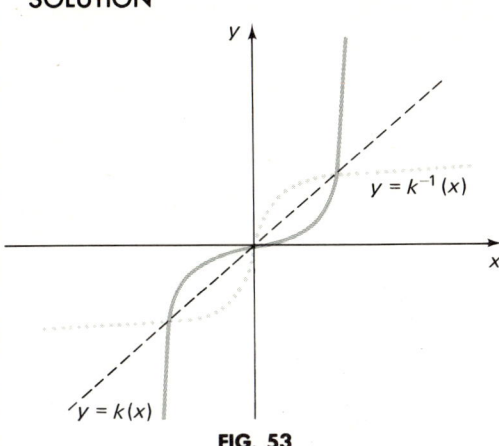

FIG. 53

EXAMPLE 10 If $h(x) = 1/(x - 1)$, find $h^{-1}(x)$. Graph.

SOLUTION Start with $y = h(x)$:

$$y = \frac{1}{x - 1}$$

Switch x and y:

$$x = \frac{1}{y - 1}$$

Solve for y:

$$y - 1 = \frac{1}{x}, \qquad y = \frac{1}{x} + 1$$

Read off $h^{-1}(x)$:

$$h^{-1}(x) = \frac{1}{x} + 1$$

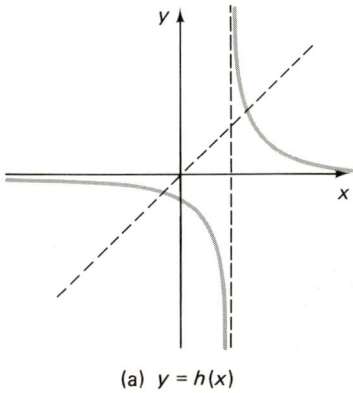

(a) $y = h(x)$

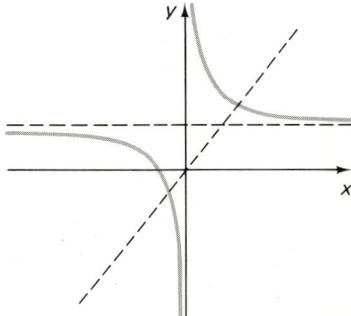

(b) $y = h^{-1}(x)$

FIG. 54

The graphs are in Fig. 54. Note that you can see from the graphs that
$$h: \{\text{nos.} \neq 0\} \to \{\text{nos.} \neq 1\}$$
and
$$h^{-1}: \{\text{nos.} \neq 1\} \to \{\text{nos.} \neq 0\}$$

This illustrates (35): The domain of a function is the range of its inverse and vice versa.

Sometimes a given function does not have an inverse, but you can obtain a function with the same rule which does have an inverse by restricting the domain.

EXAMPLE 11 Determine which of the functions have inverse functions:
(a) $f(x) = x^2$, all x; (b) $g(x) = x^2$, $x \geq 0$; (c) $h(x) = x^2$, $x \leq 0$.

SOLUTION These functions are graphed in Fig. 55. It is easy to see that f fails the horizontal line test but that g and h pass it. Thus g and h have inverse functions. They are $g^{-1}(x) = +\sqrt{x}$ and $h^{-1}(x) = -\sqrt{x}$. Note that you can read off the domains and ranges from the graphs, again illustrating that (35) holds:

$$g: [0, \infty) \to [0, \infty) \quad \text{and} \quad g^{-1}: [0, \infty) \to [0, \infty)$$
$$h: (-\infty, 0] \to [0, \infty) \quad \text{and} \quad h^{-1}: [0, \infty) \to (-\infty, 0]$$

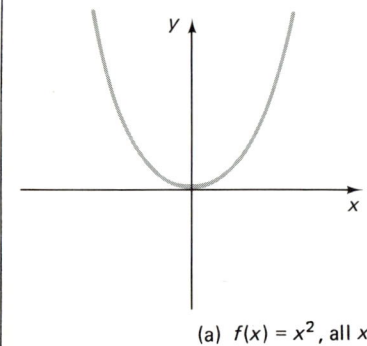

(a) $f(x) = x^2$, all x

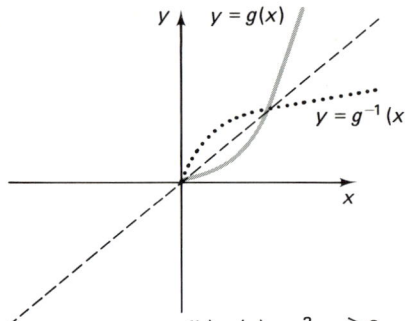

(b) $g(x) = x^2$, $x \geq 0$
$g^{-1}(x) = \sqrt{x}$, $x \geq 0$

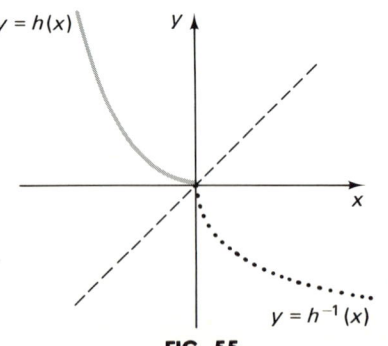

(c) $h(x) = x^2$, $x \leq 0$
$h^{-1}(x) = -\sqrt{x}$, $x \leq 0$

FIG. 55

EXERCISES

In Exercises 1–12, find $(f \circ g)(x)$ and $(g \circ f)(x)$.

1. $f(x) = 2x - 3$, $g(x) = 4x + 1$
2. $f(x) = -x + 1$, $g(x) = 2x - 1$
3. $f(x) = 3x^2 + 1$, $g(x) = 2x$
4. $f(x) = 1 - x^2$, $g(x) = 4x - 1$
5. $f(x) = x^2 - 1$, $g(x) = \sqrt{x + 2}$, $x \geq -2$
6. $f(x) = x^4 + 4$, $g(x) = \sqrt{x - 4}$, $x \geq 4$
7. $f(x) = x^3 - 1$, $g(x) = 2\sqrt[3]{x}$
8. $f(x) = 8x^3$, $g(x) = \sqrt[3]{x} + 1$
9. $f(x) = \dfrac{1}{x - 3}$, $x \neq 3$, $g(x) = \dfrac{1}{x}$, $x \neq 0$
10. $f(x) = \dfrac{x}{x - 4}$, $x \neq 4$, $g(x) = x + 4$
11. $f(x) = 1/x^2$, $x \neq 0$, $g(x) = x^2$
12. $f(x) = x - 4$, $g(x) = x + 5$

In Exercises 13–20, for the given function F, find functions f and g such that $F(x) = f(g(x))$. For example, if $F(x) = \sqrt{x + 1}$, then $f(x) = \sqrt{x}$ and $g(x) = x + 1$. Be careful of the order, for if $F(x) = \sqrt{x} + 1$, then $f(x) = x + 1$ and $g(x) = \sqrt{x}$.

13. $F(x) = (x + 1)^2$
14. $F(x) = 5x - 3$
15. $F(x) = x^2 + 1$
16. $F(x) = 5(x - 3)$
17. $F(x) = \sqrt[3]{1/x}$
18. $F(x) = 4^{x+3}$
19. $F(x) = 1/\sqrt[3]{x}$
20. $F(x) = 4^x + 3$

In Exercises 21–28, assume that $f(x) = x + 1$, $g(x) = x^2$, $h(x) = -x$, and $k(x) = \sqrt{x}$, and use your calculator, where necessary, to compute the answers.

21. $(f \circ g)(.5)$ and $(g \circ f)(.5)$
22. $(g \circ h)(.4)$ and $(h \circ g)(.4)$
23. $(f \circ h)(.8)$ and $(h \circ f)(.8)$
24. $(f \circ k)(.1)$ and $(k \circ f)(.1)$
25. $(g \circ k)(.7)$ and $(k \circ g)(.7)$
26. $(g \circ h)(.9)$ and $(h \circ g)(.9)$
27. $(f \circ h \circ g)(.4)$ and $(h \circ f \circ g)(.4)$
28. $(k \circ g \circ f)(.3)$ and $(g \circ k \circ f)(.3)$

In Exercises 29–38, use the horizontal line test to determine if the function whose graph is given has an inverse function.

29.

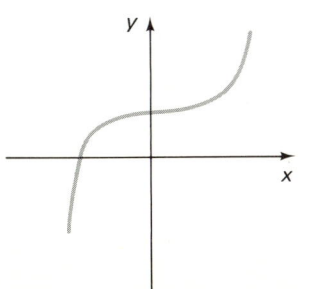

30.

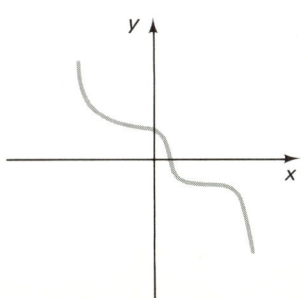

31.

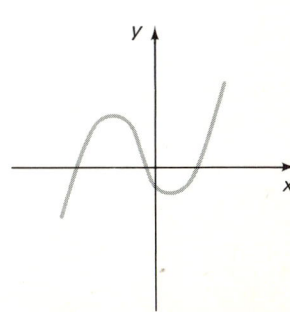

32.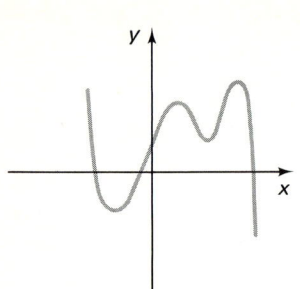

33.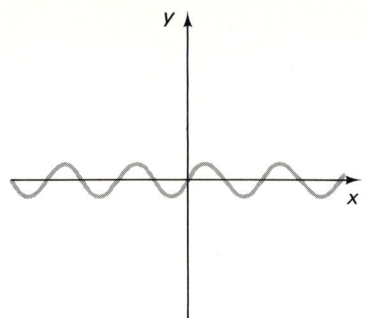

34.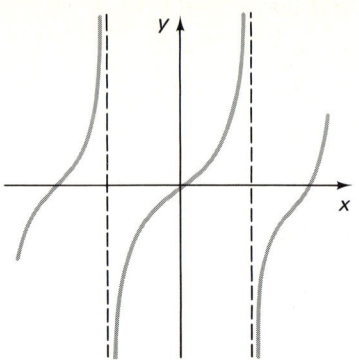

35.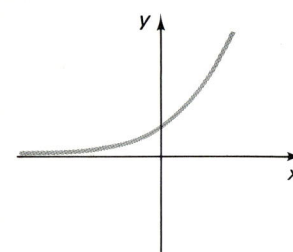

36.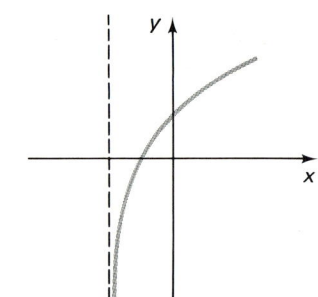

37.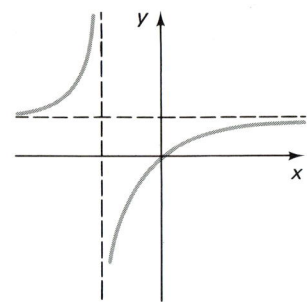

38.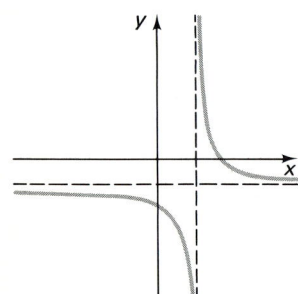

In Exercises 39–52, find the rule for the inverse and graph the function and its inverse.

39. $f(x) = x - 1$
40. $g(x) = x + 2$
41. $h(x) = -\frac{1}{2}x$
42. $k(x) = -3x$
43. $F(x) = 2x - 3$
44. $G(x) = \frac{1}{4}x + \frac{1}{2}$
45. $H(x) = -3x + \frac{1}{3}$
46. $K(x) = -2x - 4$
47. $f(x) = \dfrac{1}{x - 2}, x \neq 2$
48. $g(x) = \dfrac{1}{x + 3}, x \neq -3$
49. $h(x) = (x + 1)^2, x \geq -1$
50. $k(x) = \sqrt{x + 2}, x \geq -2$
51. $F(x) = (x - 2)^2 - 1, x \leq 2$
52. $G(x) = \sqrt{2 - x} + 1, x \leq 2$

In Exercises 53–60, for the given function f, find the rule for f^{-1}, and show that $f(f^{-1}(x)) = x$ and that $f^{-1}(f(x)) = x$ whenever this is defined.

53. $f(x) = x + 1$
54. $f(x) = 5x$
55. $f(x) = 2x + 3$
56. $f(x) = \frac{1}{3}x - \frac{5}{6}$
57. $f(x) = \frac{x}{x-1}, x \neq 1$
58. $f(x) = \frac{x+1}{x}, x \neq 0$
59. $f(x) = (x - 1)^2, x \geq 1$
60. $f(x) = (2x + 3)^2, x \leq -\frac{3}{2}$

In Exercises 61–66, use your calculator, if necessary, to show that $f^{-1}(f(a)) = a$ and that $f(f^{-1}(a)) = a$ for the given function f and number a.

61. $f(x) = x^2, a = .11$
62. $f(x) = \sqrt{x}, a = \pi$
63. $f(x) = \sin x, a = .92$
64. $f(x) = \cos x, a = .37$
65. $f(x) = \tan x, a = .18$
66. $f(x) = \ln x, a = .56$

In Exercises 67–72, from the given function and its graph, determine its domain and range. Then find the graph, domain, and range for its inverse.

67. $f(x) = x^2 - 1, x \geq 0$

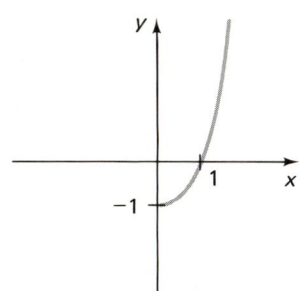

68. $g(x) = 1 - x^2, x \leq 0$

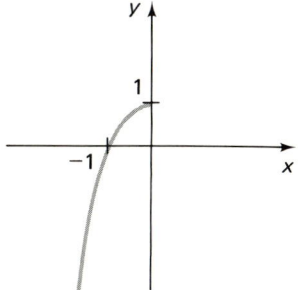

69. $h(x) = \frac{1}{x} - 2, x > 0$

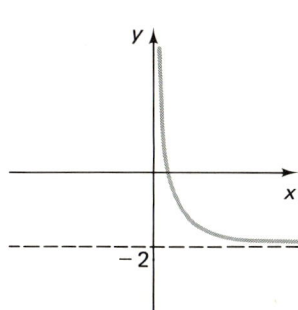

70. $k(x) = \frac{1}{x+1}, x < -1$

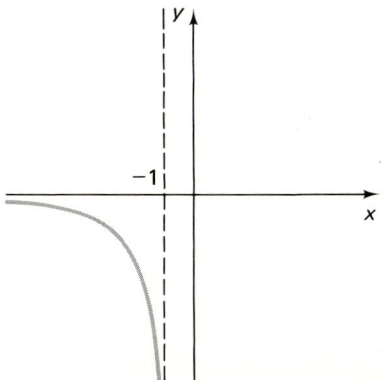

71. $F(x) = \dfrac{6-x}{3+x}$, $x \neq -3$ 72. $G(x) = \dfrac{x+2}{2x-1}$, $x \neq \dfrac{1}{2}$

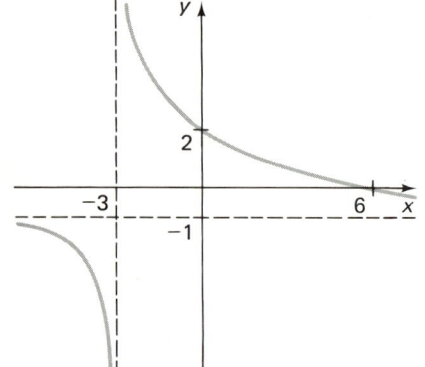
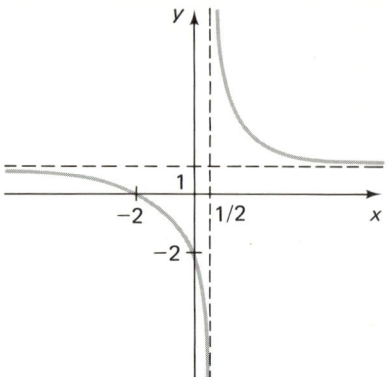

73. Let $f(x) = ax + b$ and $g(x) = cx + d$, $a \neq 0$, $c \neq 0$.

  a. Assume $a = b = 1$. If $(f \circ g)(x) = (g \circ f)(x)$, all x, what can you say about c and d?
  b. Assume $a = 1$, $b = 2$. If $(f \circ g)(x) = (g \circ f)(x)$, all x, what can you say about c and d?
  c. If $a = c$ and $f(g(x)) = g(f(x))$, all x, what can you say about b and d?

REVIEW EXERCISES

1. Plot the points $(-4, -3)$, $(1, -1)$, and $(-3, 9)$, and prove they are the vertices of a right triangle. What is the area of that triangle?
2. Let $P = (-4, 7)$ and $Q = (3, -2)$. Find **(a)** the midpoint M of the segment PQ and **(b)** a point R such that P is the midpoint of RQ.
3. Describe the set of all points in the plane such that
   a. $xy = 0$ **b.** $xy \leq 0$
4. Find the equation of the circle with center $C = (-3, -5)$ and passing through $(1, -2)$.
5. Find an equation of the circle with center $(3, -4)$ and tangent to the line $x = -4$.
6. Put the equation $6x + 5y + 30 = 0$ into slope-intercept form.
7. Find the equation of the line that passes through the points $(-1, 3)$ and $(3, 1)$.
8. Find the equation of the line through $(-2, 0)$ and parallel to $3x - 6y + 2 = 0$.

In Exercises 9–28, sketch the graph.

9. $3x + 2y + 7 = 0$
10. $x = 4y - 3$
11. $x + 7 = 0$
12. $3y - 5 = 0$
13. $y = \sqrt{3-x}$
14. $x - y^2 = 4$
15. $y - x^2 = 4$
16. $y^2 = 9 - x^2$
17. $(x-3)^2 + (y+4)^2 = 4$
18. $y = \dfrac{1}{x-2} + 3$
19. $y = -2|x+3|$
20. $y = (2x+6)^2 - 1$
21. $|x-2| \leq 1$
22. $|x+1| < 2$ and $|y-3| < 1$
23. $x^2 + y^2 + 2x - 4y \leq 0$
24. $x^2 + y^2 + 8x \geq 0$

25. $xy < 0$
27. $y + 4 \geq x^2$
26. $2y + 3x + 6 \geq 0$
28. $xy \leq 1$

29. If $f(x) = x/(x-2)$, find
  a. $f(1)$
  b. $f(-2)$
  c. $f(0)$
  d. $f(a^2)$
  e. $f(a)^2$
  f. $f(1+h)$
  g. $f\left(\dfrac{1}{a}\right)$
  h. $\dfrac{1}{f(a)}$

30. Find the domain of f if
  a. $f(x) = \sqrt{4-3x}$
  b. $f(x) = \dfrac{\sqrt{x+5}}{x^2-9}$

In Exercises 31 and 32, find $(f \circ g)(x)$ and $(g \circ f)(x)$.

31. $f(x) = x^2 - 1$, $g(x) = 2x - 3$
32. $f(x) = \dfrac{1}{x^2}$, $g(x) = \sqrt{x^2+1}$

In Exercises 33–35, find $f^{-1}(x)$, and sketch the graphs of f and f^{-1} on the same graph.

33. $f(x) = 3 - 2x$
34. $f(x) = \dfrac{1}{x-1}$
35. $f(x) = 8 - 2x^2$, $x \geq 0$

In Exercises 36–39, determine if the relation is symmetric about either axis, or the origin.

36. $3y + x = 4$
37. $y^4 + x^4 + 1 = 0$
38. $xy^2 = 8$
39. $x^2 + y^2 = 9$

40. If part of the graph of a relation is as shown, complete the graph if it is symmetric to (a) the x-axis, (b) the y-axis, or (c) the origin.

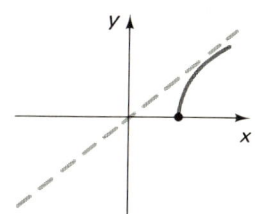

41. Find all points on the line $y = -2$ which are equidistant from the points $(3, -3)$ and $(5, 2)$.
42. Find the equation of the circle centered at $(3, -2)$ and through the point $(-1, 3)$.
43. Find a formula that expresses the circumference of a circle as a function of its area.
44. If the height of a right circular cylinder is twice the radius, express the volume of that cylinder as a function of its height.
45. The volume V of a gas varies jointly with the temperature T and the inverse of the pressure P. If the volume of a certain gas is 30 when the temperature is 200 and the pressure is 40, what would the temperature be if the volume were 35 and the pressure 45?

Graphs of Polynomial and Rational Functions and Conic Sections

CHAPTER FOUR

Without the benefit of Descartes and his coordinate system, the early Greeks had only two ways of describing curves: by combinations of uniform motions and by intersections of familiar geometric surfaces. It was in this later context that Menaechmus (ca. 350 B.C.), while trying to solve certain geometric constructability problems, discovered the parabola, ellipse, and hyperbola as sections of a cone (see Fig. 16). These curves have remained important since then, being used today to describe the shapes of gears and cams, paths of heavenly bodies and atomic particles, the curvature of mirrors and lenses, and many other things.

Once they had the Cartesian coordinate system, mathematicians could graph functions and hence better understand the relationships the functions represented. For example, maximizing or minimizing quantities is a very important mathematical application.

When graphing complicated functions, it is often helpful to graph asymptotes. The word *asymptote* was coined by Thomas Hobbes (1588–1679), using various Latin stems meaning roughly "to fall together, but not touch." However, asymptotic behavior was recognized much earlier; for example Menaechmus knew that hyperbolas have asymptotes. In describing asymptotes, the term *infinity* is often used. John Wallis (1616–1703) invented the symbol ∞ for infinity, though in a context of taking infinite sums.

SECTION 4.1. QUADRATIC FUNCTIONS

> **(1)**   **DEFINITION**
>
> A function f is a **quadratic function** if
>
> $$f(x) = ax^2 + bx + c, \qquad a \neq 0$$

This form is called the **standard form** for a quadratic function. Before discussing the graph of a quadratic function in general, we first consider two special cases, $f(x) = x^2$ and $g(x) = -x^2$. We have discussed these before, and we have seen that their graphs are as shown in Fig. 1.

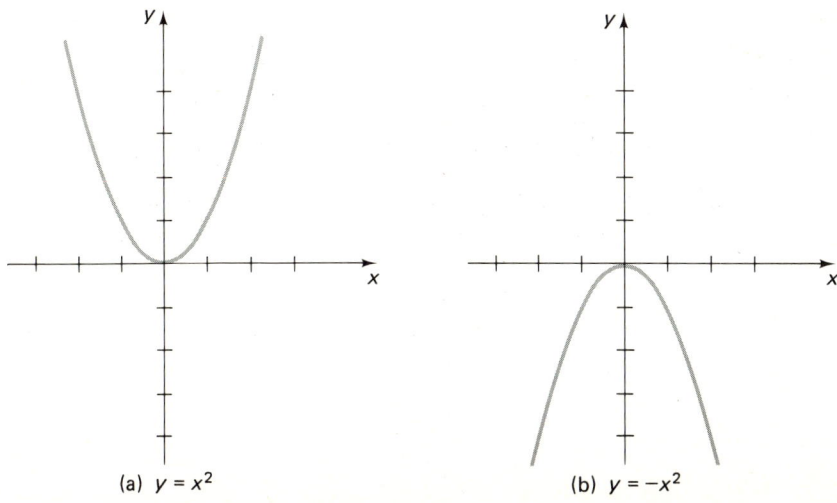

(a) $y = x^2$ (b) $y = -x^2$

FIG. 1

Each graph is called a **parabola**, and in both cases the origin $(0, 0)$ is called the **vertex**. For what is coming later, it is important to note the following.

The graph of $y = x^2$ opens upward, the vertex is a minimum (lowest) point on its graph, and the value $f(0) = 0^2 = 0$ is the minimum (smallest) value of all the values of $f(x) = x^2$. Similarly for $y = -x^2$, the graph opens downward, the vertex is a maximum (highest) point on its graph, and the value $g(0) = -0^2 = 0$ is the maximum (largest) value of all the values of $g(x) = -x^2$.

For the general case, if $f(x)$ is in the form

$$f(x) = a(x - h)^2 + k, \quad a \neq 0$$

then the techniques of translations and expansions discussed in Sec. 3.4 can be applied to the preceding special cases.

EXAMPLE 1 Graph $f(x) = 3(x - 2)^2 - 1$, label its vertex, determine if it opens upward or downward, and determine the maximum or minimum value.

SOLUTION Since $a = 3$ is positive, we start with the graph of $y = x^2$, so the resulting graph will open upward. We expand the graph of $y = x^2$ by 3 vertically to get the graph of $y = 3x^2$ (see Fig. 2), then translate this 2 to the right to get the graph of $y = 3(x - 2)^2$, and finally translate this down 1 to get the graph of $y = 3(x - 2)^2 - 1$. See Fig. 3.

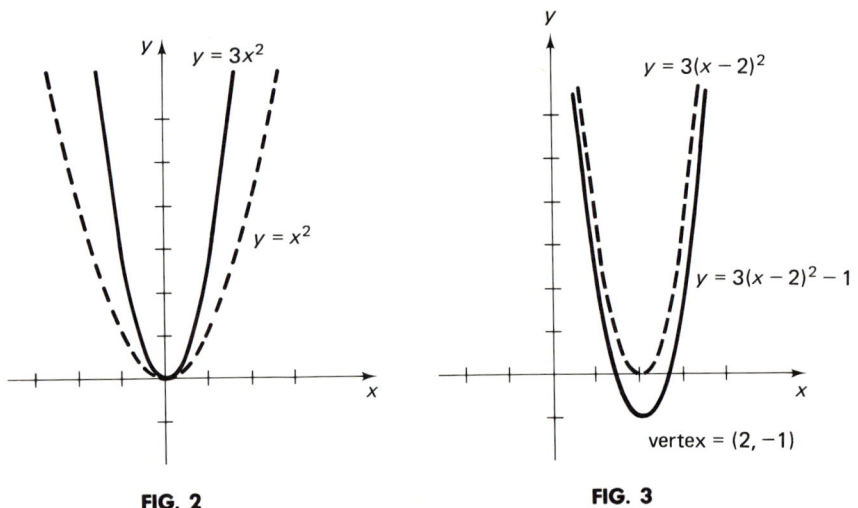

FIG. 2   FIG. 3

By the graph in Fig. 3, it is easy to see that the value $-1 = f(2)$ is the minimum value of the function $f(x) = 3(x - 2)^2 - 1$.

EXAMPLE 2 If in Example 1 the function had been $f(x) = -3(x - 2)^2 - 1$, we would have proceeded similarly, starting with $y = -x^2$. Thus the graph would open downward, the vertex would still be $(2, -1)$, and the value $f(2) = -1$ would be the maximum value of the function.

We have illustrated the following:

---

(2)   **THEOREM**

If f is a quadratic function with
$$f(x) = a(x-h)^2 + k, \quad a \neq 0$$
then the graph of $y = f(x)$ is a parabola with vertex at (h, k). If $\begin{Bmatrix} a > 0 \\ a < 0 \end{Bmatrix}$, then the graph opens $\begin{Bmatrix} \text{upward} \\ \text{downward} \end{Bmatrix}$ and the value $k = f(h)$ is a $\begin{Bmatrix} \text{minimum} \\ \text{maximum} \end{Bmatrix}$ value.

Quadratic functions are not usually given in the form used in Theorem (2). However,

---

(3)

A quadratic function in the form
$$f(x) = ax^2 + bx + c, \quad a \neq 0$$
can be transformed to the form
$$f(x) = a(x-h)^2 + k, \quad a \neq 0$$
by the method of completing the square.

Note: The a's in these two forms are the same number. We have used *completing the square* before in Secs. 2.3 and 3.2. We give several examples.

EXAMPLE 3   Write $g(x) = \frac{1}{2}x^2 - 2x - 1$ in the form $g(x) = a(x - h)^2 + k$ and graph.

SOLUTION   We begin by regrouping,

$$g(x) = \left(\tfrac{1}{2}x^2 - 2x \quad\quad\right) - 1$$

and then factoring out by $1/2$ to make the coefficient of x^2 equal to 1:

$$g(x) = \tfrac{1}{2}(x^2 - 4x \quad\quad) - 1$$

Warning: It is very easy to make an error in this factoring. Always multiply the factored form back out in your head and check with the preceding equation.
    Since $(x + h)^2 = x^2 + 2hx + h^2$, we add inside the parentheses the coefficient of x divided by 2 and then squared. That is, we add $(-4/2)^2 = (-2)^2 = 4$. Thus

$$g(x) = \tfrac{1}{2}(x^2 - 4x + 4) - 1 - (\quad)$$

Section 4.1   Quadratic Functions   175

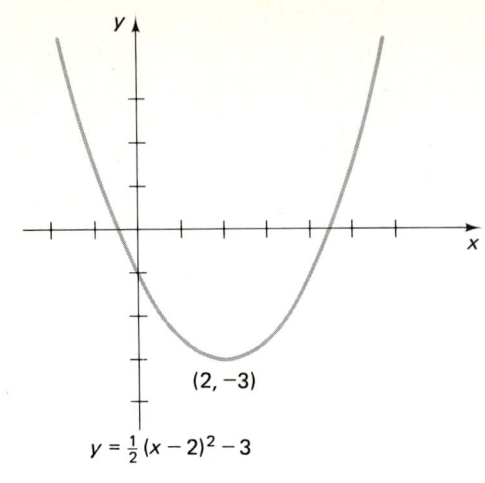

FIG. 4

To keep the equation the same, we must subtract exactly what we have added. This is a 4 inside the parentheses, but what is inside the parentheses is multiplied by 1/2. Thus we have really added $\frac{1}{2}(4)$ to the equation, so we must subtract this. Therefore

$$g(x) = \tfrac{1}{2}(x^2 - 4x + 4) - 1 - \tfrac{1}{2}(4)$$

or

$$g(x) = \tfrac{1}{2}(x - 2)^2 - 3$$

EXAMPLE 4 Write $h(x) = -2x^2 + 3x - 2$ in the form $h(x) = a(x - h)^2 + k$.

SOLUTION First regroup terms,

$$h(x) = (-2x^2 - 3x \quad) - 2$$

and then factor out the coefficient of x^2:

$$h(x) = -2\left(x^2 + \tfrac{3}{2}x \quad \right) - 2$$

To complete the square, we divide the coefficient of x by 2, getting $\tfrac{3}{4}$; square this, getting $\tfrac{9}{16}$; and add this result inside the parentheses:

$$h(x) = -2\left(x^2 + \tfrac{3}{2}x + \tfrac{9}{16}\right) - 2 - (\quad)$$

To keep equality, we must subtract what we have added, namely, $(-2)(\tfrac{9}{16})$. Thus

$$h(x) = -2\left(x^2 + \tfrac{3}{2}x + \tfrac{9}{16}\right) - 2 - (-2)\left(\tfrac{9}{16}\right)$$

or

$$h(x) = -2\left(x + \tfrac{3}{4}\right)^2 - \tfrac{7}{8}$$

One thing that is helpful in graphing functions is knowing the **intercepts**, i.e., where the graph crosses the two axes. To find the y-intercept, you set x equal to zero and solve for y. To find the x-intercept, you set y equal to zero and solve for x. Usually the standard form $y = ax^2 + bx + c$ is the easiest form to work with when finding intercepts.

EXAMPLE 5 Find the intercepts of $y = \tfrac{1}{2}x^2 - 2x - 1$. (See Example 3.)

SOLUTION First set $x = 0$, getting $y = -1$ as the y-intercept. Setting $y = 0$ and solving, we use the quadratic formula since the equation does not readily factor, obtaining $x = 2 \pm \sqrt{6}$. Thus $x = 2 + \sqrt{6} \approx 4.44949$ and $x = 2 - \sqrt{6} \approx -.449490$ are the x-intercepts. See Fig. 4.

We have seen that the technique of completing the square to find the vertex is a very useful aid to graphing. This technique is also useful in applications, particularly in maximizing or minimizing problems in which quadratic functions arise. Techniques are developed in higher mathematics to handle more complicated functions.

EXAMPLE 6

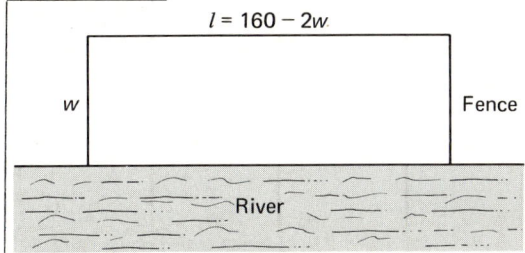

FIG. 5

A farmer wants to enclose a rectangular region next to a river. He has 160 meters of fencing to use and will not fence the side next to the river. See Fig. 5. **(a)** What are the dimensions of the largest region he can enclose? **(b)** What is the area of this region?

SOLUTION When solving maximizing or minimizing problems, there are two questions to ask.

> The first question to ask:
>
> What are we trying to maximize or minimize?
>
> The second question to ask:
>
> What is the quantity that is changing and affecting the item that we are maximizing or minimizing?
>
> You then express the answer to the first question as a function of the answer to the second.

In this case, we want to maximize the area in terms of its dimensions, so we write

$$A = lw$$

By the condition in the problem, $l + 2w = 160$ or $l = 160 - 2w$. Thus we want the maximum value of

$$A = (160 - 2w)w \quad \text{or} \quad A = -2w^2 + 160w$$

You should note that this function A only makes sense as area when $0 \leq w \leq 80$ (since $0 \leq 2w \leq 160$), so the domain of A should be restricted to this interval.

Section 4.1 Quadratic Functions

We now complete the square to write the function A in the form $A = a(w - h)^2 + k$:

$$A = -2w^2 + 160w$$
$$A = -2(w^2 - 80w)$$
$$A = -2(w^2 - 80w + 1600) - (-2)(1600)$$
$$A = -2(w - 40)^2 + 3200$$

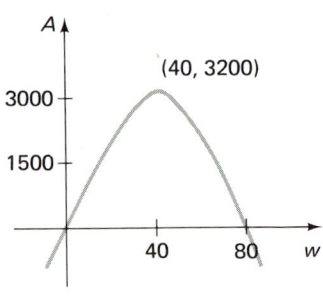

FIG. 6

This function is graphed in Fig. 6. We see from the graph that the maximum area that it is possible to surround is 3200 square meters and that this occurs when the width $w = 40$ meters and the length $l = 160 - 2w = 160 - 2(40) = 80$ meters.

EXERCISES

1. Given $f(x) = ax^2 + 1$, sketch the graph of f if
   a. $a = 3$ b. $a = 1/3$ c. $a = -1$ d. $a = -4$
2. Given $f(x) = 2x^2 + k$, sketch the graph of f if
   a. $k = 3$ b. $k = 1/3$ c. $k = -1$ d. $k = -4$
3. Given $f(x) = 2(x - h)^2$, sketch the graph of f if
   a. $h = 3$ b. $h = 1/3$ c. $h = -1$ d. $h = -4$

In Exercises 4–12, determine if the graph opens upwards or downwards, determine the maximum or minimum value, graph the function, and label its vertex.

4. $f(x) = 3x^2 - 2$
5. $g(x) = \frac{1}{2}x^2 + 5$
6. $h(x) = -\frac{1}{4}x^2 + 7$
7. $k(x) = -5x^2 - 8$
8. $F(x) = 5(x - 1)^2 + 2$
9. $G(x) = -2(x - 3)^2 - 4$
10. $H(x) = \frac{1}{3}(x + 1)^2 - 1$
11. $K(x) = -\frac{1}{2}(x + 3)^2 + 1$
12. $M(x) = 3(x - 2)^2 - 1$

In Exercises 13–20, find the x- and y-intercepts, if any.

13. $f(x) = x^2 - 5x + 4$
14. $g(x) = -x^2 - 7x + 8$
15. $h(x) = -3x^2 - 11x + 4$
16. $k(x) = 6x^2 + 13x + 6$
17. $F(x) = 4x^2 + 12x + 9$
18. $G(x) = -25x^2 + 20x - 4$
19. $K(x) = \dfrac{2x - 1}{x + 2} - \dfrac{x + 1}{x - 2}$
20. $L(x) = \dfrac{3x - 2}{x + 4} + \dfrac{x + 1}{x + 2}$

In Exercises 21–24, complete the square if necessary, and graph for several values of k. Determine the values of k for which the graph has two, one, or no x-intercepts.

21. $f(x) = (x - 1)^2 + k$
22. $g(x) = -(x + 1)^2 + k$
23. $h(x) = -x^2 + 4x + k$
24. $k(x) = x^2 - 6x + k$

In Exercises 25–34, write the function in the form $f(x) = a(x - h)^2 + k$, and graph, labeling the vertex.

25. $f(x) = x^2 - 6x + 1$
26. $g(x) = -x^2 - 2x + 1$
27. $h(x) = -2x^2 + 5x - 2$
28. $k(x) = 3x^2 - 9x - 4$
29. $m(x) = -x^2 - x + 1$
30. $n(x) = 2x^2 - x - 1$
31. $F(x) = \frac{1}{2}x^2 + 2x + 3$
32. $G(x) = -\frac{1}{3}x^2 + 3x - 7$
33. $H(x) = -\frac{2}{3}x^2 + \frac{1}{5}x - 8$
34. $K(x) = \frac{3}{4}x^2 + \frac{1}{2}x - \frac{1}{2}$

35. The sum of the base and height of a triangle is 30. Find the dimensions which will give maximum area, and determine that maximum area.

36. The perimeter of a rectangle is 20 centimeters. Find the dimensions which will give the maximum area, and determine that maximum area.

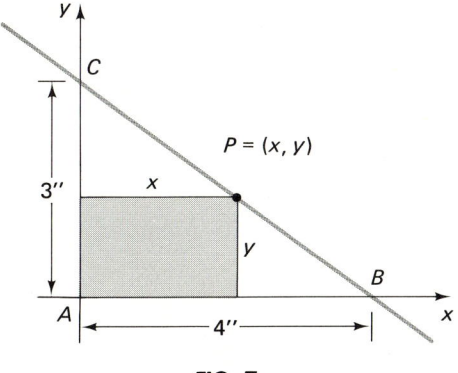

FIG. 7

37. Find the dimensions of the largest rectangle that can be inscribed in triangle ABC as shown in Fig. 7. Determine the largest area. [*Hint:* First find the equation of the line BC that the point $P = (x, y)$ lies on.]

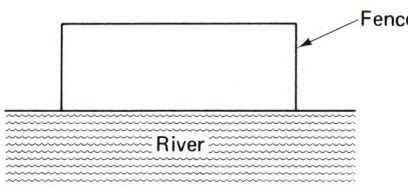

FIG. 8

38. A farmer wants to build a rectangular pen next to a river. He has 120 yards of fencing and will not fence the side next to the river. See Fig. 8.
  a. What are the dimensions of the largest area he can enclose?
  b. What is the largest area?

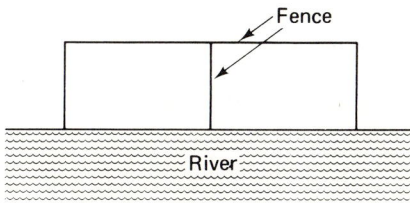

FIG. 9

39. A farmer wants to build two rectangular pens of the same size next to a river so that they are separated by one fence. See Fig. 9. If she has 240 meters of fencing and does not fence the side next to the river, what are the dimensions of the largest area she can enclose? What is that largest area?

40. In Exercise 39, if she also fences the side next to the river, what are the answers now?

In Exercises 41–43, use the formulas before Example 9 of Sec. 2.3.

41. A ball is thrown upwards at 30 miles per hour from the top of a house 30 feet high. (*Note:* 30 mph = 44 ft per sec.) Neglecting air resistance,
  a. How long will it take to reach its highest point?
  b. How high is the ball at its highest point?
  c. How long will it take to reach the ground?

42. A mortar shell is fired upward from a cliff 200 feet above the ground. Assume that the muzzle velocity is 640 feet per second and that the shell comes down missing the cliff and hitting the ground.
  a. How high does the shell go?
  b. When does it reach its highest point?
  c. When does it hit the ground?

43. A model rocket is fired upward. At the end of its burn, it has an upward velocity of 98 meters per second and is 310 meters high.
  a. What is its maximum height?
  b. How long after the end of the burn does it reach its highest point?
  c. How long until it hits the ground?

44. An orange grower finds that she gets an average yield of 50 bushels per tree when she plants 26 trees to an acre of ground. Each time she adds one tree to an acre, the yield per tree decreases by 1 bushel, due to congestion. How many trees per acre should she plant for maximum yield?

SECTION 4.2. GRAPHS OF POLYNOMIAL FUNCTIONS

In Sec. 4.1, we discussed quadratic functions. These are a special case of a class of functions called polynomial functions. A **polynomial function** is a function whose rule is a polynomial, where a **polynomial** (in x) is an algebraic expression of the form

$$a_n x^n + a_{n-1} x^{n-1} + \cdots + a_1 x + a_0$$

Here n is a nonnegative integer and the coefficients $a_0, a_1, \ldots, a_n$ are numbers. Since a polynomial is often used as a rule of a function (a polynomial function), we use notations such as $f(x)$, $g(x)$, etc., to denote polynomials. The **degree** of a polynomial is n if n is the highest power of x with a non-zero coefficient. We shall use deg $f(x)$ to denote "the degree of the polynomial (or polynomial function) $f(x)$." The polynomial with all coefficients zero is called the **zero polynomial** and is denoted by 0. The zero polynomial has no degree.

Suppose $f(x)$ is a polynomial and c is a number such that $f(c) = 0$. Then c is a **root** of the equation $f(x) = 0$ and is a **zero** of the polynomial $f(x)$. Although some mathematicians call c a "root" of the polynomial, we shall not in this text.

We now discuss graphs of polynomial functions. We begin by considering the graph of $y = ax^n$, where n is a positive integer and a is positive. Several of these graphs are sketched in Fig. 10. You can see (by plotting points, if you like) that all the graphs of $y = ax^n$, n even, are similar in shape and that all the graphs of $y = ax^n$, $n > 1, n$ odd, are similar. The larger n is, the steeper the graph is away from the origin and the flatter it is near the origin.

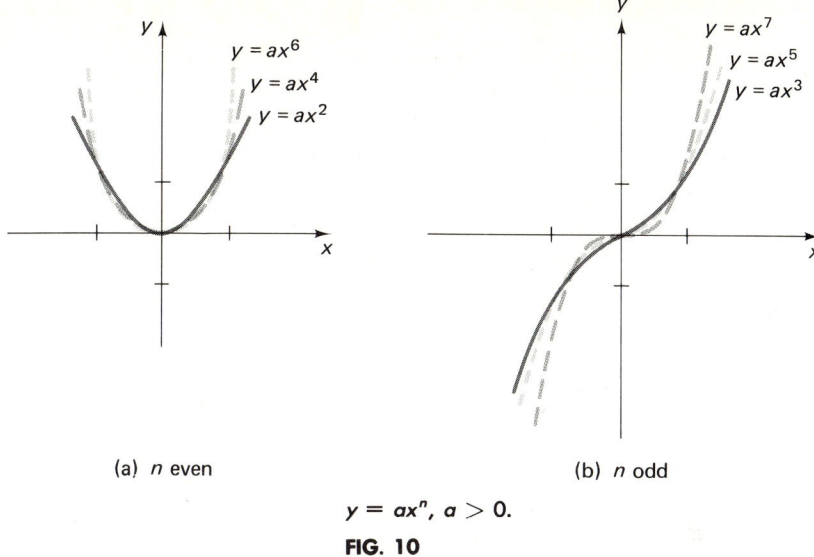

(a) *n* even (b) *n* odd

$y = ax^n$, $a > 0$.

FIG. 10

What interests us now is what happens when x gets large in absolute value. In either case, n even or n odd, we see that as x gets large and positive, $f(x)$ also gets large and positive. In mathematical terminology we say that

$f(x) = ax^n$, $a > 0$, *goes to plus infinity as x goes to plus infinity*,

which is denoted symbolically by

$$f(x) \to +\infty \quad \text{as} \quad x \to +\infty$$

Of course, $+\infty$ is not a number. The phrase "goes to plus infinity" or the notation "$\to +\infty$" is just a mathematical shorthand for "becomes large and positive."

As x gets large and negative, we see that two different things happen, since (negative)^{even} is positive and (negative)^{odd} is negative. As x gets large and negative, $y = ax^n$, $a > 0$, gets large and positive if n is even and gets large and negative if n is odd. In mathematical terminology we say that

$$f(x) = ax^n, a > 0, \text{ goes to } \begin{cases} \text{plus infinity if } n \text{ is even} \\ \text{minus infinity if } n \text{ is odd} \end{cases}$$
as x goes to minus infinity

and write

$$f(x) = ax^n \to \begin{cases} +\infty, & n \text{ even} \\ -\infty, & n \text{ odd} \end{cases} \quad \text{as} \quad x \to -\infty$$

If $a < 0$, the opposite happens with $f(x) = ax^n$, as can be seen in Fig. 11.

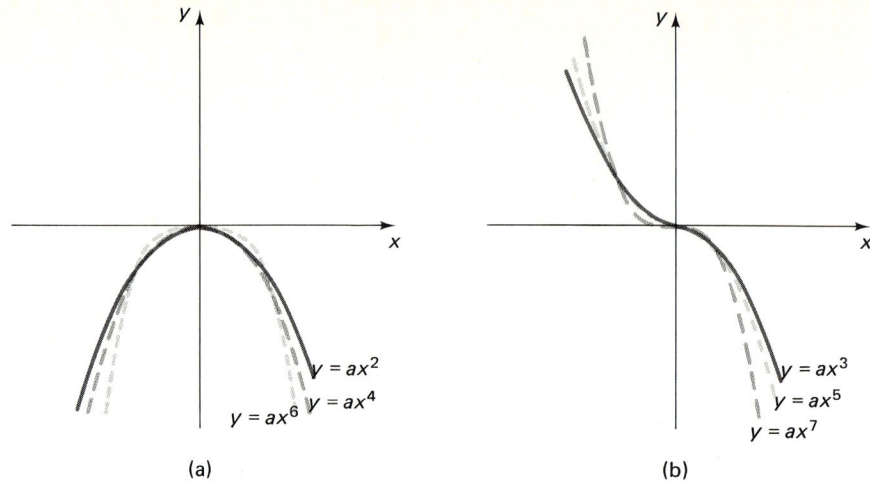

$y = ax^n$, $a < 0$.

FIG. 11.

Now consider a general polynomial of degree n, for example,

$$f(x) = a_6 x^6 + \cdots + a_1 x + a_0, \qquad a_6 \neq 0, \quad \text{if } n = 6,$$

or

$$g(x) = b_7 x^7 + \cdots + b_1 x + b_0, \qquad b_7 \neq 0, \quad \text{if } n = 7$$

To see what happens when x gets very large in absolute value, factor out the first term. In these cases, we get

$$f(x) = a_6 x^6 \left(1 + \frac{a_5}{a_6} \frac{1}{x} + \cdots + \frac{a_1}{a_6} \frac{1}{x^5} + \frac{a_0}{a_6} \frac{1}{x^6} \right)$$

or

$$g(x) = b_7 x^7 \left(1 + \frac{b_6}{b_7} \frac{1}{x} + \cdots + \frac{b_1}{b_7} \frac{1}{x^6} + \frac{b_0}{b_7} \frac{1}{x^7} \right)$$

As x grows very large in absolute value, the fractions $1/x^n$ approach 0. We write

$$\frac{1}{x^n} \to 0 \quad \text{as} \quad x \to \pm\infty$$

Thus what is inside the parentheses approaches 1 as x grows large in absolute value. Therefore the graph of $f(x)$ is close to the graph of $a_6 x^6$, and the graph of $g(x)$ is close to $b_7 x^7$ for $|x|$ *large*. However, if $|x|$ is not large, then the graphs can vastly differ, as illustrated in Fig. 12.

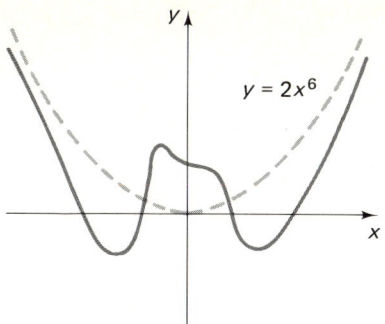
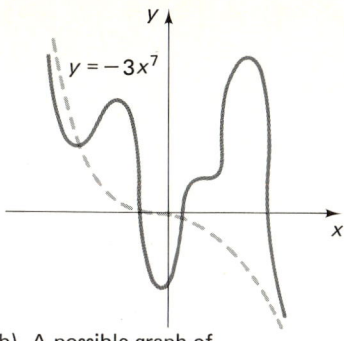

(a) A possible graph of
$y = 2x^6 + a_5x^5 + \ldots + a_1x + a_0$

(b) A possible graph of
$y = -3x^7 + a_5x^5 + \ldots + a_1x + a_0$

FIG. 12

In general, the graph of a polynomial of degree larger than 2 can have many "hills" and "valleys," as illustrated in Fig. 12. Calculus is required for a complete analysis of this. However, if we happen to have the polynomial in factored form (or if we can factor it completely), then we can use the methods of Sec. 2.7 to give us an idea of the graph.

EXAMPLE 1 Sketch the graph of

$$f(x) = -3(x-2)(x+1)^2(x+5)^3$$

SOLUTION The zeros of $f(x)$ are 2, -1, and -5. For other values of x, we determine the sign of $f(x)$ using the *graphing* method of Sec. 2.7 (but keeping the coefficient -3 with the first factor):

		-5		-1		2	
$-3(x-2)$	$+$		$+$		$+$		$-$
$(x+1)^2$	$+$		$+$		$+$		$+$
$(x+5)^3$	$-$		$+$		$+$		$+$
Sign of the product	$-$		$+$		$+$		$-$

We conclude that the graph

  Intersects the x-axis for $x = -5, -1, 2$.
  Is below the x-axis for $x < -5$ or $x > 2$.
  Is above the x-axis for $-5 < x < -1$ or $-1 < x < 2$.

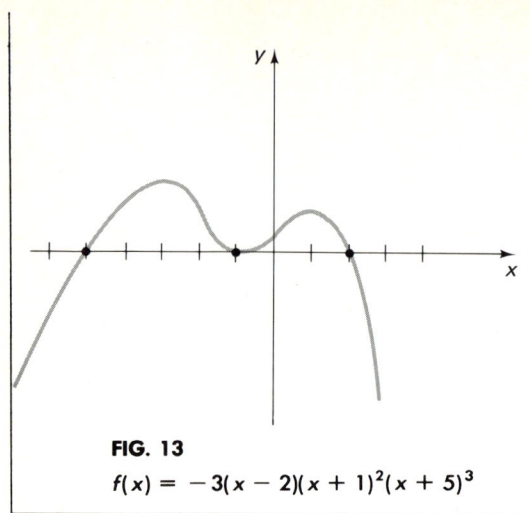

FIG. 13
$f(x) = -3(x-2)(x+1)^2(x+5)^3$

Since $f(x) = -3x^6 + a_5 x^5 + \cdots + a_1 x + a_0$ (we could determine the coefficients $a_5, \ldots, a_0$ by multiplying out, but we do not need to), $f(x) \approx -3x^6$ for $|x|$ large. Thus

$f(x) \to -\infty$ as $x \to +\infty$ or as $x \to -\infty$

Putting this information together, we sketch the graph in Fig. 13. We should note that this is only a very rough sketch. We could get a better graph by plotting more points, but without calculus we may not be able to tell how many peaks and valleys there are or exactly how high or deep the peaks and valleys are.

EXERCISES

1. If $f(x) = ax^4 + 2$, sketch the graph of f, and determine how many zeros $y = f(x)$ has if
   a. $a = 2$ **b.** $a = -2$
   c. $a = -\frac{1}{3}$ **d.** $a = 3$

2. If $f(x) = 2x^4 + a$, sketch the graph of f, and determine how many zeros $y = f(x)$ has if
   a. $a = 0$ **b.** $a = 2$
   c. $a = -2$ **d.** $a = -\frac{1}{3}$

In Exercises 3–18, sketch the graph.

3. $f(x) = \frac{1}{3}x^3 - 2$
4. $g(x) = -\frac{1}{4}x^3 - 2$
5. $h(x) = 3x^6 + 3$
6. $k(x) = -7x^4 - 4$
7. $F(x) = 2x^3 - x$
8. $G(x) = -3x^3 + x^2$
9. $H(x) = 2x^5 - 5x^7$
10. $K(x) = 3x^3 + 3x^6$
11. $f(x) = (x-1)(x+1)(x-2)$
12. $g(x) = (2x-1)(x-3)(x-5)^2$
13. $h(x) = (x-1)(2-x)(x-3)$
14. $k(x) = 2(x+7)(x+3)(x-1)(x+1)$
15. $F(x) = -3(x-2)(x+3)(x-4)(x-1)$
16. $G(x) = \frac{1}{2}(x+1)^2(x-3)^2$
17. $H(x) = -2(x-3)^2(2-3x)^3(1-x)^5$
18. $K(x) = -\frac{1}{2}(3+x)^3(x-1)^2(5-x)^4$

SECTION 4.3. RATIONAL FUNCTIONS

A function is a **rational function** if its rule is the quotient of two polynomials. For example,

$$f(x) = \frac{2x^5 - 3x^2 + 4}{-3x^2 - 2x + 5}, \quad g(x) = \frac{-3x^7 - x^3 + 13}{8x^7 - 6x^2}, \quad h(x) = \frac{5}{3x^3 - 2x + 6}$$

all determine rational functions. In this section, we study the graphs of rational functions. Just as with graphs of polynomials, a complete analysis of rational functions requires calculus. However, we can obtain a good idea of the graph if the polynomials are (or can be easily) factored completely.

We first discuss what can happen in the graph when $|x|$ is large, using the preceding three examples as illustrations.

To explain what happens, for each rational function
$$\frac{a_n x^n + \cdots + a_1 x + a_0}{b_m x^m + \cdots + b_1 x + b_0}$$
we factor $a_n x^n$ out of the numerator and $b_m x^m$ out of the denominator (similar to what we did with polynomials):

$$f(x) = \frac{2x^5}{-3x^2} \cdot \frac{1 - \frac{3}{2x^3} + \frac{2}{x^5}}{1 + \frac{2}{3x} - \frac{5}{3x^2}}, \quad g(x) = \frac{-3x^7}{8x^7} \cdot \frac{1 + \frac{1}{3x^4} - \frac{13}{3x^7}}{1 - \frac{3}{4x^5}}$$

$$h(x) = \frac{5x^0}{3x^3} \cdot \frac{1}{1 - \frac{2}{3x^2} + \frac{2}{x^3}}$$

We then have each rational function expressed as $\frac{a_n x^n}{b_m x^m}$ times a fraction, where the fraction is close to 1 when $|x|$ is large. Therefore, for $|x|$ large, the graph of the rational function is close to the graph of $a_n x^n / b_m x^m$; what this is depends on the relative sizes of n and m and on the signs of a_n and b_m. In the preceding cases,

$$f(x) \approx \frac{2x^5}{-3x^2} = -\frac{2}{3} x^3, \quad \text{so } f(x) \to -\infty \text{ as } x \to +\infty$$
$$f(x) \to +\infty \text{ as } x \to -\infty$$

$$g(x) \approx \frac{-3x^7}{8x^7} = -\frac{3}{8}, \quad \text{so } g(x) \to -\frac{3}{8} \text{ as } x \to \pm\infty$$

$$h(x) \approx \frac{5x^0}{3x^3} = \frac{5}{3} \frac{1}{x^3}, \quad \text{so } h(x) \to 0 \text{ as } x \to \pm\infty$$

Thus $y = -\frac{3}{8}$ is a horizontal asymptote of $y = g(x)$, $y = 0$ (the x-axis) is a horizontal asymptote of $y = h(x)$, and $f(x)$ has no horizontal asymptote. The general principle is as follows:

---

(4)

If
$$f(x) = \frac{a_n x^n + a_{n-1} x^{n-1} + \cdots + a_0}{b_m x^m + b_{m-1} x^{m-1} + \cdots + b_0}$$
is a rational function, then as $x \to \pm\infty$

i. $f(x) \to 0$ if $m > n$.

ii. $f(x) \to \frac{a_n}{b_m}$ if $m = n$.

iii. $f(x) \to \pm\infty$ according as $\frac{a_n}{b_m} x^{n-m} \to \pm\infty$ if $m < n$.

Section 4.3 Rational Functions

We now discuss the behavior of the graph at or near zeros of the numerator or denominator. We assume that the polynomials are factored completely and that any common factors are canceled out. First, whenever the numerator is zero, this is a zero of the whole function, so the graph intersects the *x*-axis there. Next, whenever a value is a zero of the denominator, the rational function is undefined, and we need to determine the shape of the graph near this value.

EXAMPLE 1 Graph $y = 1/x$.

SOLUTION

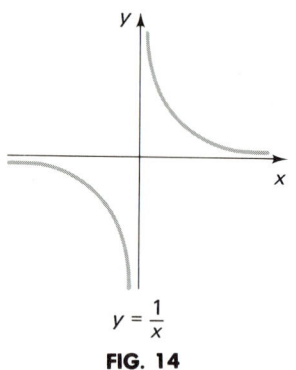

FIG. 14

We have seen this graph several times before. See Fig. 14. The degree of the denominator, which is 1, is greater than the degree of the numerator, which is 0. Thus, $y = 0$ is a horizontal asymptote. Also, $x = 0$ is a zero of the denominator, and thus the *y*-axis is a vertical asymptote.

For x near zero but greater than zero, $f(x) = 1/x$ is positive. Thus, $f(x)$ gets large and positive as x gets close to zero through numbers which are greater than zero, i.e., through numbers on the right-hand side of zero. We sometimes say that $f(x)$ *goes to plus infinity as x approaches 0 from the right-hand side*, which is denoted symbolically by

$$f(x) \to +\infty \quad \text{as} \quad x \to 0^+$$

For x near zero but less than zero, $f(x) = 1/x$ is negative. Thus, $f(x)$ gets large and negative as x gets close to zero through numbers which are less than zero, i.e., through numbers on the left-hand side of zero. We sometimes say that *$f(x)$ goes to minus infinity as x approaches 0 from the left-hand side*, which is denoted symbolically by

$$f(x) \to -\infty \quad \text{as} \quad x \to 0^-$$

The general situation is similar. Each place where the denominator is zero determines a vertical asymptote (assuming the fraction is reduced to lowest terms). On each side of an asymptote we must determine if the sign of the whole function is positive or negative. This will tell us if the function approaches $+\infty$ or $-\infty$ as x approaches that value.

EXAMPLE 2 Graph

$$f(x) = \frac{-2(x-1)(x+2)^2}{(x+3)(x+1)^2}$$

SOLUTION Since the degrees of the denominator and the numerator are the same, $f(x) \approx -2$ for $|x|$ large. That is, $y = -2$ is a horizontal asymptote.

We see that -2 and 1 are zeros of the function and that $x = -1$ and $x = -3$ are vertical asymptotes. We now determine the signs of the function as we did in Sec. 2.7:

$-2(x - 1)$	$+$	$+$	$+$	$+$	$-$
$(x + 2)^2$	$+$	$+$	$+$	$+$	$+$
$x + 3$	$-$	$+$	$+$	$+$	$+$
$(x + 1)^2$	$+$	$+$	$+$	$+$	$+$
	-3	-2		-1	1
Sign of the function	$-$	$+$	$+$	$+$	$-$

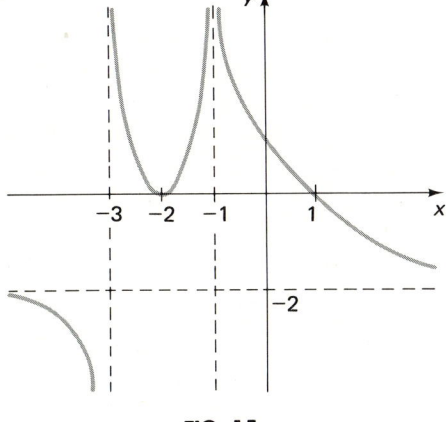

FIG. 15

Using the signs, we determine

$$f(x) \to -\infty \text{ as } x \to -3^-,$$
$$f(x) \to +\infty \text{ as } x \to -3^+$$
$$f(x) \to +\infty \text{ as } x \to -1^-$$
$$\text{or as } x \to -1^+$$

Using all the information we have, including the signs around the two zeros, $x = -2$ and $x = 1$, we graph $y = f(x)$ as in Fig. 15.

EXERCISES

Sketch the graph of the given functions.

1. $f(x) = \dfrac{1}{x - 3}$
2. $g(x) = \dfrac{1}{x + 5}$
3. $h(x) = \dfrac{-2}{x + 4}$
4. $k(x) = \dfrac{-3}{x - 3}$
5. $F(x) = \dfrac{3x - 1}{x}$
6. $G(x) = \dfrac{2x + 1}{3x}$
7. $H(x) = \dfrac{2}{(x - 3)^2}$
8. $K(x) = \dfrac{-3}{(x + 2)^2}$
9. $f(x) = \dfrac{2}{x^2 - 9}$
10. $g(x) = \dfrac{-3}{x^2 - 4}$
11. $h(x) = \dfrac{3}{x^2 + 2}$
12. $k(x) = \dfrac{2x}{x^2 + 3}$

13. $F(x) = \dfrac{2x - 1}{3x + 1}$

14. $G(x) = \dfrac{3x - 2}{4 - 3x}$

15. $H(x) = \dfrac{2x}{x^2 + 4x + 3}$

16. $K(x) = \dfrac{x - 4}{x^2 + 4x - 5}$

17. $f(x) = \dfrac{x^2}{2x^2 + x - 1}$

18. $g(x) = \dfrac{4x^2 - 1}{x^2 - 9}$

19. $h(x) = \dfrac{x^3 - x}{x^2 - 5x + 6}$

20. $k(x) = \dfrac{2x^3 - x^2}{4x^2 + 9x + 2}$

21. $F(x) = \dfrac{(x + 3)^2}{4x^5 - x^3}$

22. $G(x) = \dfrac{x^2 + x - 2}{9x - x^3}$

23. $H(x) = \dfrac{x^4 - x^2 - 12}{10x^4 + 7x^2 + 1}$

24. $K(x) = \dfrac{2x^6 - x^4 - x^2}{x^6 - 1}$

25. $f(x) = \dfrac{(x - 1)(2x + 1)^2(x + 1)^3}{x(x - 2)^3(x + 3)^2}$

26. $g(x) = \dfrac{(x + 1)^7(x - 3)^5(1 - 2x)^2}{(4x - 1)(x - 5)^6(x + 4)^7}$

27. $h(x) = \dfrac{(2x - 1)^2(3x + 1)^3(1 - x)^3}{x(x - 4)^5(2x - 3)^3}$

28. $k(x) = \dfrac{x^2(x - 4)^3(3 - x)^5}{(x - 2)^4(x + 1)^3(1 - x)}$

SECTION 4.4. CONIC SECTIONS

A **conic section**, or **conic**, is a curve in a plane that can be found by the intersection of that plane with a right circular cone as illustrated in Fig. 16. There are four (nondegenerate) types of conics, **parabola**, **circle**, **ellipse**, and **hyperbola**. This section is a survey of some of the elementary properties of conics.

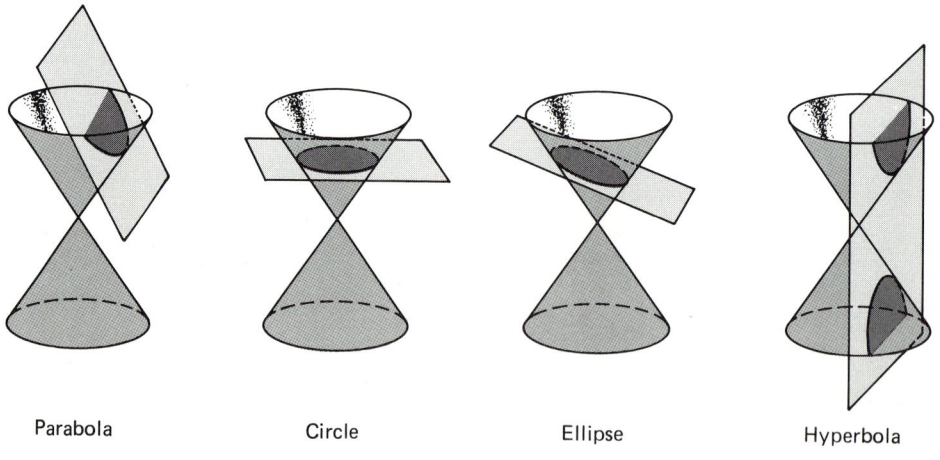

Parabola　　　Circle　　　Ellipse　　　Hyperbola

FIG. 16

In Sec. 4.1, we studied parabolas which opened upward or downward. We recall the following information from that section:

(5)

The standard form of the equation of a parabola which opens upward or downward is

$$y = ax^2, \quad a \neq 0, \qquad \text{if the vertex is the origin}$$

or

$$y - k = a(x - h)^2, \quad a \neq 0, \qquad \text{if the vertex is } (h, k)$$

The parabola opens upward if $a > 0$ and downward if $a < 0$. See Fig. 17.

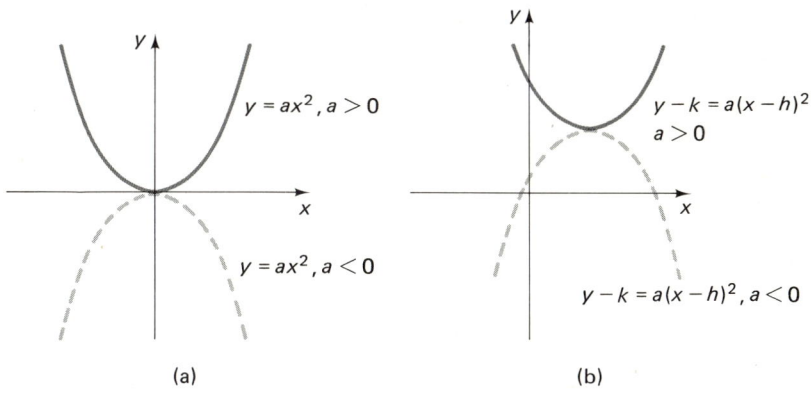

FIG. 17

If the parabola opens to the right or left, then the standard forms of the equations are the same, except that the roles of x and y are interchanged.

(6)

The standard form of the equation of a parabola which opens to the right or to the left is

$$x = ay^2, \quad a \neq 0, \qquad \text{if the vertex is the origin}$$

or

$$x - h = a(y - k)^2, \quad a \neq 0, \qquad \text{if the vertex is } (h, k)$$

The parabola opens to the right if $a > 0$ and to the left if $a < 0$. See Fig. 18.

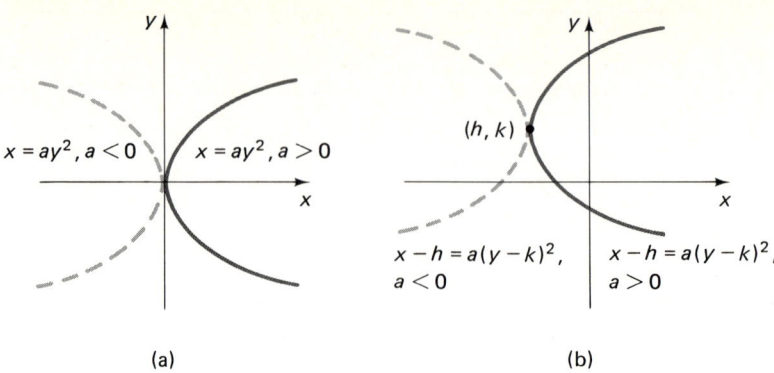

FIG. 18

The technique of completing the square to find the standard form of the equation applies to parabolas that open to the right or to the left.

EXAMPLE 1 Put the equation

$$2x + 3y^2 - 18y + 31 = 0$$

into standard form. Determine the center and which way it opens, and sketch the graph.

SOLUTION Starting with the given equation, we first divide by the coefficient of x (because in the standard form the coefficient of x is 1) and next isolate the two terms containing y or y^2 on one side:

$$2x + 3y^2 - 18y + 31 = 0$$
$$x + \tfrac{3}{2}y^2 - 9y + \tfrac{31}{2} = 0$$
$$x + \tfrac{31}{2} = -\tfrac{3}{2}y^2 + 9y$$

We next factor out the coefficient of y^2 and then complete the square:

$$x + \tfrac{31}{2} = -\tfrac{3}{2}(y^2 - 6y \qquad)$$
$$x + \tfrac{31}{2} - \tfrac{3}{2}(9) = -\tfrac{3}{2}(y^2 - 6y + 9)$$
$$x + 2 = -\tfrac{3}{2}(y - 3)^2$$

Therefore the parabola opens to the left, and its vertex is $(-2, 3)$. See Fig. 19.

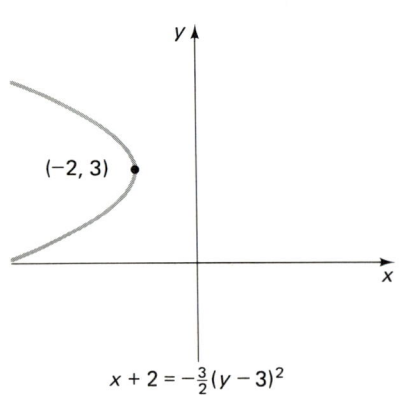

$x + 2 = -\tfrac{3}{2}(y - 3)^2$

FIG. 19

The circle is another conic section that we have discussed. We recall the following from Sec. 3.2:

> **(7)**
>
> The standard form of the equation of a circle of radius $r > 0$ is
>
> $$x^2 + y^2 = r^2 \quad \text{if the center is the origin,}$$
>
> or
>
> $$(x-h)^2 + (y-k)^2 = r^2 \quad \text{if the center is } (h, k)$$

See Fig. 20, and pay particular attention to the extreme points of the graph in the horizontal and vertical directions. This will help in graphing the next conic section, which is an ellipse.

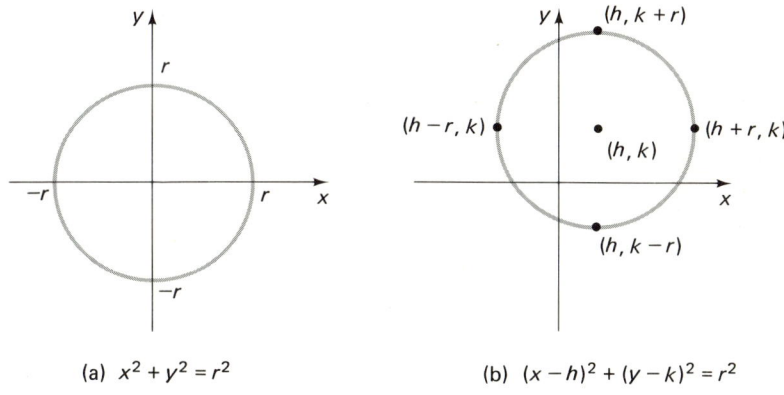

(a) $x^2 + y^2 = r^2$
(b) $(x-h)^2 + (y-k)^2 = r^2$

FIG. 20

To help understand the equation of an ellipse, we start with the standard form of the equation of a circle centered at the origin of radius $b > 0$. We divide this equation by b^2, obtaining

(8) $$\frac{x^2}{b^2} + \frac{y^2}{b^2} = 1, \quad b > 0$$

Now, if we replace x by $\dfrac{b}{a}x$, $a > 0$, and simplify, we obtain

(9) $$\frac{x^2}{a^2} + \frac{y^2}{b^2} = 1, \quad a, b > 0$$

This is an equation of an ellipse. We know from our study of graphing techniques that the effect on the graph of the circle (8) of replacing x by $(b/a)x$ is to

Contract the graph horizontally if $b > a$.
Expand the graph horizontally if $b < a$.

Section 4.4 Conic Sections

Therefore the graph of an ellipse is an elongated circle. The geometric significance of a and b can be seen by first letting one of the variables be zero and then solving for the other.

$$\text{If } y = 0, \quad \frac{x^2}{a^2} + 0 = 1, \quad x^2 = a^2, \quad x = \pm a$$

$$\text{If } x = 0, \quad 0 + \frac{y^2}{b^2} = 1, \quad y^2 = b^2, \quad y = \pm b$$

Thus, a and b give the intercepts on the x- and y-axes. See Fig. 21.

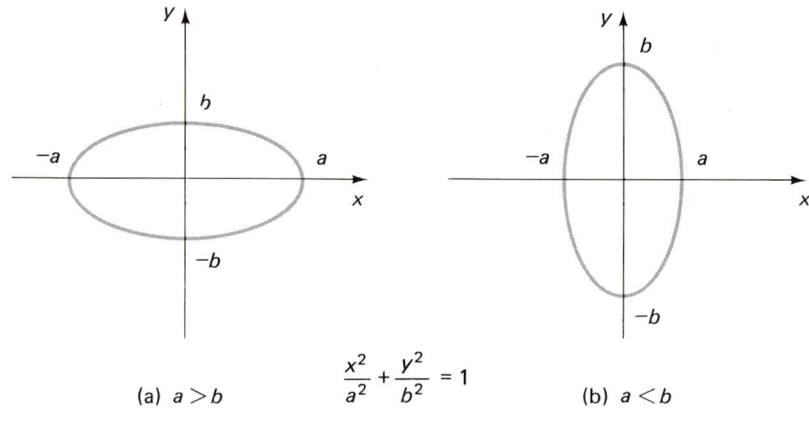

FIG. 21

If the center is translated to (h, k), then the extreme points in the x and y directions are translated also; $(\pm a, 0)$ become $(\pm a + h, k)$, and $(0, \pm b)$ become $(h, \pm b + k)$, and the standard form of the equation becomes

$$\frac{(x - h)^2}{a^2} + \frac{(y - k)^2}{b^2} = 1$$

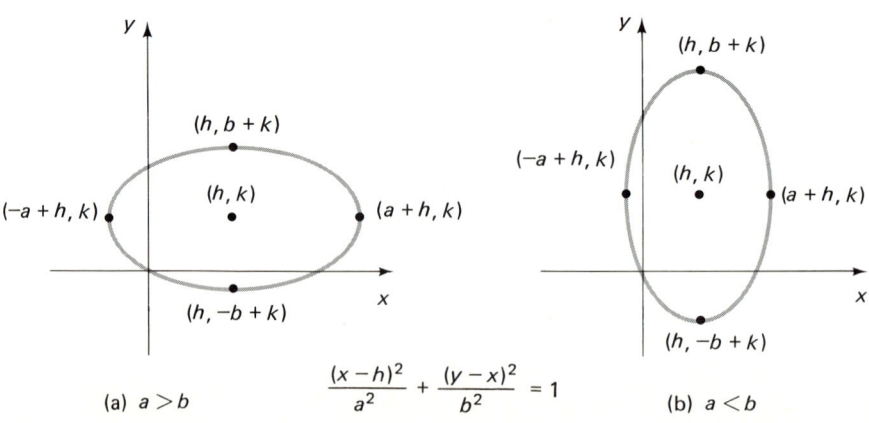

We now have the following:

> **(10)**
>
> The standard form of the equation of an ellipse is
>
> $$\frac{x^2}{a^2} + \frac{y^2}{b^2} = 1 \qquad \text{if the center is the origin}$$
>
> or
>
> $$\frac{(x-h)^2}{a^2} + \frac{(y-k)^2}{b^2} = 1 \qquad \text{if the center is } (h, k)$$
>
> where $a, b > 0$.

We now illustrate the techniques needed to put an equation for an ellipse into standard form.

EXAMPLE 2 Put the equation $9x^2 + 16y^2 = 144$ into standard form, and graph.

SOLUTION The standard form has a 1 on the right-hand side, so we divide by 144:

$$\frac{9x^2}{144} + \frac{16y^2}{144} = \frac{144}{144} \qquad \text{or} \qquad \frac{x^2}{16} + \frac{y^2}{9} = 1$$

Next, we write the denominators as a^2 and b^2, $a, b > 0$:

$$\frac{x^2}{4^2} + \frac{y^2}{3^2} = 1$$

This is now in standard form. See Fig. 22(a) for the graph.

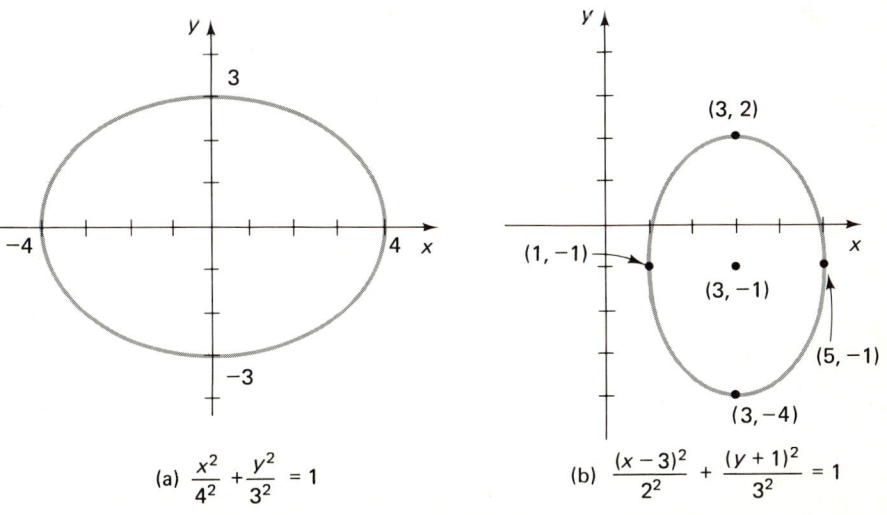

(a) $\frac{x^2}{4^2} + \frac{y^2}{3^2} = 1$

(b) $\frac{(x-3)^2}{2^2} + \frac{(y+1)^2}{3^2} = 1$

FIG. 22

EXAMPLE 3 Put the equation $9x^2 + 4y^2 - 54x + 8y + 49 = 0$ into standard form, and graph.

SOLUTION We first group the x terms and y terms together and move the constant to the other side:

$$9x^2 - 54x + 4y^2 + 8y = -49$$

We next factor out the coefficients of the square terms and then proceed to complete the square:

$$9(x^2 - 6x \quad) + 4(y^2 + 2y \quad) = -49$$
$$9(x^2 - 6x + 9) + 4(y^2 + 2y + 1) = -49 + 9(9) + 4(1)$$
$$9(x - 3)^2 + 4(y + 1)^2 = 36$$

We now divide by 36 and finally put the equation into standard form:

$$\frac{9(x-3)^2}{36} + \frac{4(y+1)^2}{36} = \frac{36}{36} \quad \text{or} \quad \frac{(x-3)^2}{2^2} + \frac{(y+1)^2}{3^2} = 1$$

We see that the center is $(3, -1)$. See Fig. 22(b) for the graph.

The remaining conic section is a hyperbola. The standard form of its equation is very similar to that of an ellipse. If the center is the origin, the standard form of the equation of a hyperbola is

$$\frac{x^2}{a^2} - \frac{y^2}{b^2} = 1 \quad \text{or} \quad \frac{y^2}{b^2} - \frac{x^2}{a^2} = 1, \quad a, b > 0$$

and the graphs are given in Fig. 23. As indicated in Fig. 23, the hyperbolas open to the right and left or up and down depending on which term has the negative sign. In addition, hyperbolas have asymptotes which serve as guidelines in sketching the graph. We can see that the asymptotes occur by rewriting the

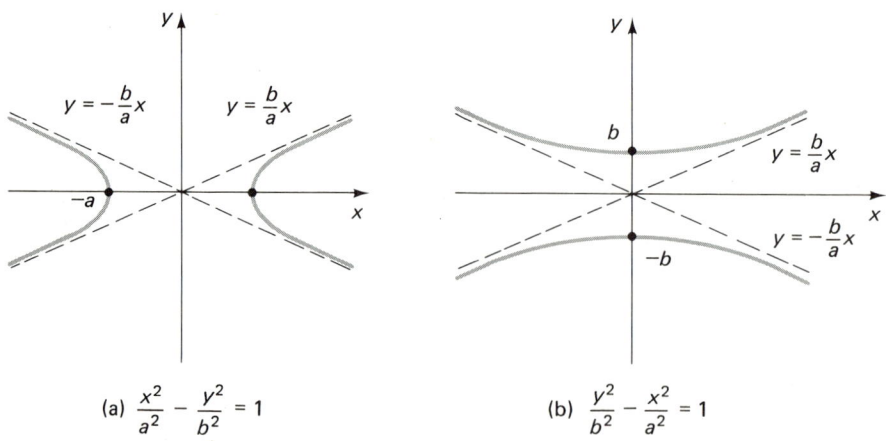

(a) $\frac{x^2}{a^2} - \frac{y^2}{b^2} = 1$ (b) $\frac{y^2}{b^2} - \frac{x^2}{a^2} = 1$

FIG. 23

equation as follows:

$$\frac{x^2}{a^2} - \frac{y^2}{b^2} = 1$$

$$b^2\left(\frac{x^2}{a^2} - 1\right) = y^2$$

$$\frac{b^2}{a^2}x^2\left(1 - \frac{a^2}{x^2}\right) = y^2$$

(11) $$y = \pm\frac{b}{a}x\sqrt{1 - \frac{a^2}{x^2}}$$

We can see that as x becomes very large, $\sqrt{1 - a^2/x^2}$ approaches 1, so that y approaches $\pm(b/a)x$. If we had started with $y^2/b^2 - x^2/a^2 = 1$, we would have obtained $y = \pm(b/a)x\sqrt{1 + a^2/x^2}$, so again y approaches $\pm(b/a)x$.

When $y = 0$ in $x^2/a^2 - y^2/b^2 = 1$, we see that $x = \pm a$. The points $(\pm a, 0)$ are called the **vertices** of the hyperbola. From equation (11) we can see that the equation is defined only for $|x| \geq a$. We could plot points to see that the general shape of the graph is as given in Fig. 23(a). In a similar manner, the equation $y^2/b^2 - x^2/a^2 = 1$ is defined only for $|y| \geq b$, and when $x = 0$, $y = \pm b$, so that the vertices are $(0, \pm b)$.

If the center of the hyperbola is (h, k), then the standard form of the equation is

$$\frac{(x-h)^2}{a^2} - \frac{(y-k)^2}{b^2} = 1 \quad \text{or} \quad \frac{(y-k)^2}{b^2} - \frac{(x-h)^2}{a^2} = 1, \quad a, b > 0$$

and the graphs are as given in Fig. 24. The asymptotes for these hyperbolas are $y - k = \pm(b/a)(x - h)$, and the vertices are indicated in Fig. 24.

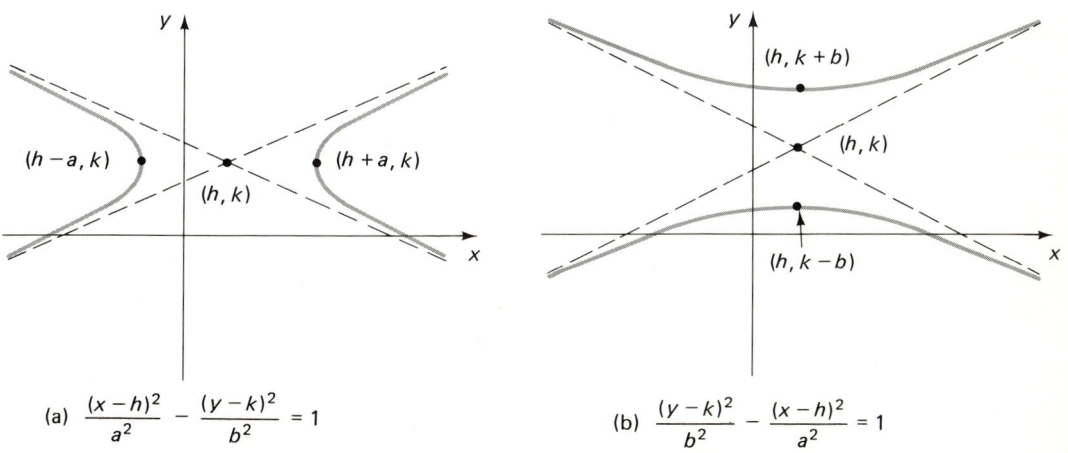

FIG. 24

Summarizing, we have the following:

> **(12)**
>
> The standard form of the equation of a hyperbola centered at the origin is
>
> $\dfrac{x^2}{a^2} - \dfrac{y^2}{b^2} = 1$ when it opens to the right and left
>
> $\dfrac{y^2}{b^2} - \dfrac{x^2}{a^2} = 1$ when it opens up and down
>
> If it is centered at (h, k), the standard form is
>
> $\dfrac{(x-h)^2}{a^2} - \dfrac{(y-k)^2}{b^2} = 1$ when it opens to the right and left
>
> $\dfrac{(y-k)^2}{b^2} - \dfrac{(x-h)^2}{a^2} = 1$ when it opens up and down
>
> The asymptotes are $y - k = \pm \dfrac{b}{a}(x - h)$.

The techniques needed to put an equation for a hyperbola into standard form are very similar to the techniques used for equations of ellipses. We give one example.

EXAMPLE 4 Put the equation $9x^2 - 16y^2 + 54x - 32y + 101 = 0$ into standard form, and graph.

SOLUTION We first group the x and y terms together and move the constant to the other side:

$$9x^2 + 54x - 16y^2 - 32y = -101$$

We next factor out the coefficient of the square terms and then proceed to complete the square:

$$9(x^2 + 6x\quad) - 16(y^2 + 2y\quad) = -101$$
$$9(x^2 + 6x + 9) - 16(y^2 + 2y + 1) = -101 + 9(9) - 16(1)$$
$$9(x+3)^2 - 16(y+1)^2 = -36$$

We now divide by -36 and put the equation into standard form:

$$\frac{9(x+3)^2}{-36} - \frac{16(y+1)^2}{-36} = \frac{-36}{-36}$$

$$-\frac{(x+3)^2}{4} + \frac{4(y+1)^2}{9} = 1$$

$$\frac{(y+1)^2}{\left(\frac{3}{2}\right)^2} - \frac{(x+3)^2}{2^2} = 1$$

We see that the center is $(-3, -1)$, it opens up and down, the vertices are $(-3, \frac{1}{2})$ and $(-3, -2\frac{1}{2})$, and the asymptotes are $y + 1 = \pm \frac{3}{4}(x + 3)$. See Fig. 25 for the graph.

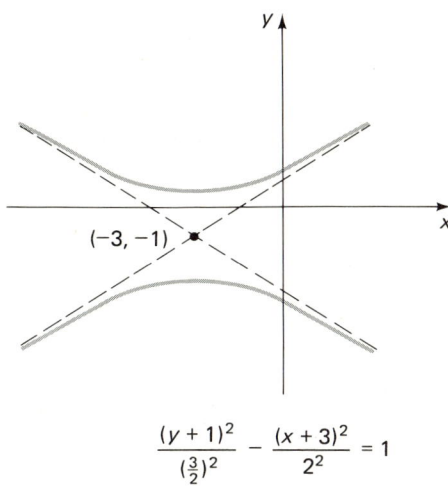

$$\frac{(y+1)^2}{(\frac{3}{2})^2} - \frac{(x+3)^2}{2^2} = 1$$

FIG. 25

EXERCISES

For the given equation, identify the conic section; put the equation into standard form; find the center, vertices, radius, asymptotes, or extreme points, whichever is appropriate; and graph.

1. $x^2 + y^2 + 4x - 2y + 1 = 0$
2. $x^2 + y^2 - 2x - 4y + 5 = 0$
3. $x = 4y^2$
4. $y = -2x^2$
5. $x = 4 - y^2$
6. $x^2 = 4 - y$
7. $x^2 + 4y^2 = 1$
8. $9x^2 + y^2 = 9$
9. $4x^2 - 9y^2 = 36$
10. $25x^2 - 4y^2 = -100$
11. $2x^2 + 4y^2 - 4x + 8y - 2 = 0$
12. $2x^2 + 3y^2 = 12$
13. $3x^2 + 2y^2 - 18 = 0$
14. $4x^2 - 8y^2 = 9$
15. $9x^2 - 8y^2 = 2$
16. $9x^2 + y^2 = 1$
17. $4x^2 + 25y^2 = 9$
18. $x^2 - 9y^2 = 36$
19. $16x^2 - 49y^2 = 25$
20. $3x^2 + 2y^2 + 12x - 8y + 38 = 0$
21. $2x^2 + 2y^2 - 4x - 10y = 0$
22. $3x^2 + 3y^2 + 3x + y + 1 = 0$
23. $3x = y^2 + 2y$
24. $x^2 - 6x + y + 2 = 0$
25. $4x^2 - 9y^2 - 8x - 36y + 4 = 0$
26. $4x^2 - 9y^2 - 8x - 36y - 68 = 0$
27. $x^2 - y^2 - x + 3y = 0$
28. $2x^2 - 2y^2 + x - 3y = 0$
29. $2x^2 - y - 6x + 1 = 0$
30. $3y^2 - 3x - 12y + 2 = 0$

REVIEW EXERCISES

In Exercises 1–4, determine the x-intercepts, y-intercepts, and vertex. Graph.

1. $f(x) = x^2 - 4x + 3$
2. $g(x) = -x^2 + 6x - 8$
3. $h(x) = 2x^2 + 3x + 4$
4. $k(x) = -3x^2 - 2x + 1$

5. Find the dimensions of the largest rectangle that can be inscribed in triangle ABC as shown.

6. Sketch the graph of $f(x) = -5x^6 + 2$.
7. Sketch the graph of $f(x) = 2x^3 - 1$.
8. Sketch the graph of $f(x) = -4(x + 5)(x - 2)(x - 7)$.
9. Sketch the graph of $f(x) = (x + 4)^3(3 - x)(2x + 5)^2$.
10. Sketch the graph of
$$f(x) = \frac{3x + 2}{2x - 1}$$
11. Sketch the graph of
$$f(x) = \frac{x^2 - 5x + 6}{x^3 + 2x^2 - 3x}$$
12. Sketch the graph of
$$f(x) = \frac{(x + 2)^4(2x - 1)^3(3 - x)}{(x - 1)^5(x + 1)^2}$$

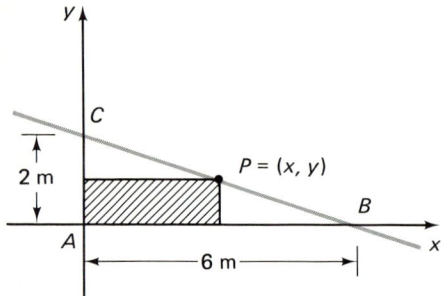

In Exercises 24–29, put the equation in the standard form of a conic section, and identify the conic section. Then find the center, vertices, radius, asymptotes, or extreme points, whichever is appropriate, and graph.

13. $9x^2 + 36x + y^2 - 2y = -28$
14. $2x^2 - 16x + 2y^2 + 20y = -10$
15. $x = 6y^2 + 6y$
16. $4x^2 - 9y^2 - 54y = 117$
17. $y^2 - x^2 + 4y + 2x = 1$
18. $16x^2 + 25y^2 = 400$

Exponential and Logarithmic Functions

CHAPTER FIVE

Knowledge of exponents goes back to the very beginnings of mathematics. Tables of exponents have been found among old Babylonian tablets, and in ancient Mesopotamia, they even knew the compound interest formula. Logarithms, on the other hand, were invented fairly recently, as a tool to transform large, tedious multiplication problems (mainly from astronomy) into relatively easy addition problems. In 1614, John Napier (1550–1617) published his *Mirifici logarithmorum canonis descriptio* ("A Description of the Marvelous Rule of Logarithms") in which he virtually developed what we would call today the system of logarithms to the base $1/e$, though he did it with a geometric description of certain ratios. Napier coined the word *logarithm* from two Greek words, *logos* ("ratio") and *arithmos* ("number"). Henry Briggs (1561–1639), first in collaboration with Napier and then on his own, modified Napier's definition and developed the common logarithm as a more easily applied computing tool. John Bürgi (1552–1632) independently developed what was almost the natural logarithm system. He probably did this earlier than Napier but only published his version in 1620. Logarithms were to remain an invaluable computing tool until the advent of calculators and computers.

The relationship between logarithms and exponentials that we find so useful for problem solving today went unsuspected for over 60 years and was not fully exploited until Leonhard Euler (1707–1783), who was also the first person to use the letter e as the base for the natural logarithm.

SECTION 5.1. EXPONENTIAL FUNCTIONS

From Chapter 1, we have various definitions of a^x, depending on the nature of x:

$a^n = a \cdots a$ (n factors)   if n is a positive integer
$a^0 = 1$ if $a \neq 0$. 0^0 is undefined
$a^{-n} = \dfrac{1}{a \cdots a}$ (n factors of a)   if n is a positive integer
$a^{1/m} = \sqrt[m]{a}$, m a positive integer
$a^{n/m} = (\sqrt[m]{a})^n$ m, n integers, $m > 0$, $\dfrac{n}{m}$ in lowest terms

Of course, a^{-n} is undefined if $a = 0$, and $a^{n/m}$ is not a real number if a is negative and m is even. It is natural to ask what is the meaning of a^x if x is irrational. For instance, what does $2^{\sqrt{2}}$ mean? We know that $\sqrt{2}$ has an infinite decimal expansion

$$\sqrt{2} = 1.4142135\ldots$$

Then each number of the sequence

$$1,\ 1.4,\ 1.41,\ 1.414,\ 1.4142\ldots$$

is a rational number (since $1.4 = 14/10$, $1.41 = 141/100$, etc.) which is an approximation to $\sqrt{2}$; the more decimal places, the better the approximation.

These rational numbers can be used to form a new sequence

$$2^1,\ 2^{1.4},\ 2^{1.41},\ 2^{1.414},\ 2^{1.4142},\ldots$$

Using some techniques in calculus, it can be shown that this second sequence approaches a unique real number; we define that real number to be $2^{\sqrt{2}}$. Whenever $a \geq 0$, we can use the same techniques to define a^x for any irrational number x; we can use it only for $a \geq 0$ because of the difficulty with even roots of negative numbers.

Under this definition, if x is an irrational number, then a^x "fits in" exactly where it should: If x is between numbers r and s, then a^x is between a^r and a^s. Moreover, it can be shown that all the laws of exponents [(3) of Sec. 1.2] still hold.

By this discussion, for each real number x there corresponds a unique real number a^x if $a > 0$. Thus we can define the exponential function as follows:

---

(1)   **DEFINITION**

If $a > 0$, the **exponential function with base a** is the function f defined by

$$f(x) = a^x$$

where x is any real number.

For different values of the base a, the exponential function $f(x) = a^x$ (and its graph) have different characteristics.

EXAMPLE 1 Let $a = 2$ or $\frac{1}{2}$, and observe the symmetry with respect to $x = 0$ (the y-axis). See Table 1 and Fig. 1.

TABLE 1

x	2^x	$(\frac{1}{2})^x$
-3	$\frac{1}{8}$	8
-2	$\frac{1}{4}$	4
-1	$\frac{1}{2}$	2
0	1	1
1	2	$\frac{1}{2}$
2	4	$\frac{1}{4}$
3	8	$\frac{1}{8}$

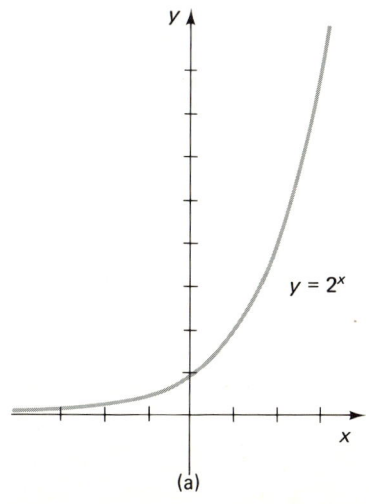

(a) $y = 2^x$

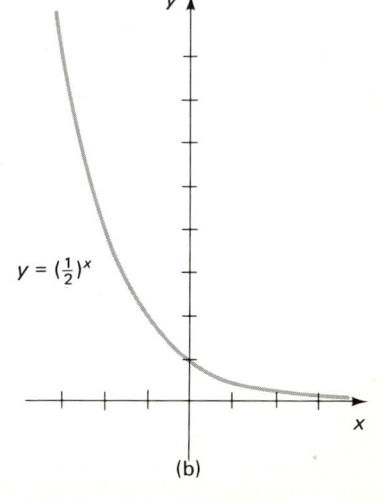

(b) $y = (\frac{1}{2})^x$

FIG. 1

Example 1 illustrates the following general situation. If $a > 1$, the function $f(x) = a^x$ is strictly increasing; to the left the x-axis is a horizontal asymptote. If $0 < a < 1$, the function $f(x) = a^x$ is strictly decreasing; to the right the x-axis is a horizontal asymptote. For different values of a, these graphs differ in *steepness*, as indicated in Fig. 2.

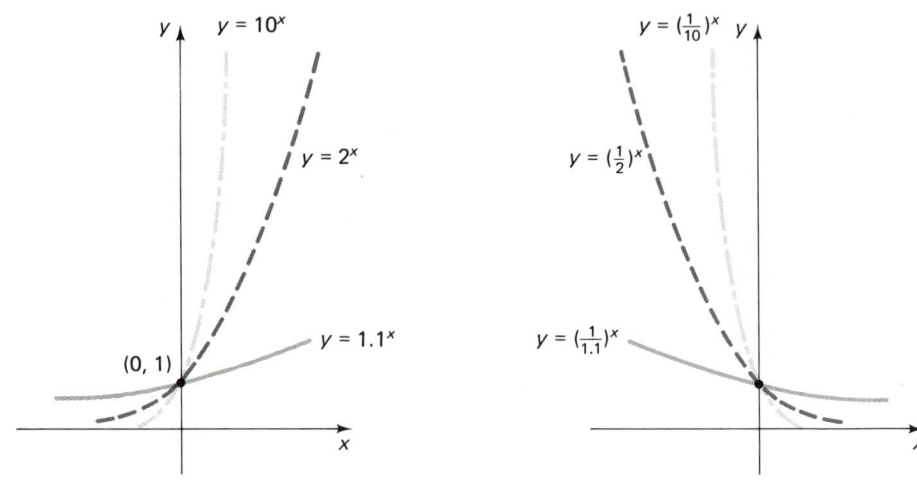

FIG. 2

Note that if $a > 1$, then $0 < a^{-1} < 1$, and the two graphs $y = a^x$ and $y = (a^{-1})^x$ are reflections of each other through the y-axis.

Using the Calculator

Most calculators have three buttons related to exponential functions:
$$[10^x], \quad [e^x], \quad [y^x]$$
On some calculators, $[e^x] = [INV][LN\,x]$ and $[10^x] = [INV][LOG\,x]$.

The $[y^x]$ button computes the number y raised to the power x. To compute $23^{1.47}$, press

Alg.: $[23][y^x][1.47][=]$
RPN: $[23][ENTER][1.47][y^x]$

You should obtain 100.401.

Reminder: The preceding computation means $23^{1.47} \approx 100.401$. In this text, unless otherwise indicated, all intermediate calculations are carried to at least eight significant figures, and the answers are rounded to six.

The $[10^x]$ button computes 10 raised to the power x. To compute $10^{3.1}$, press (on any type of calculator)

$$[3.1][10^x]$$

You should obtain 1258.93.

Note: A few calculators do not have a $[10^x]$ button. On such calculators, you can easily use the $[y^x]$ button.

The number e is irrational; $e = 2.71828\ldots$. This number turns out to be very important for both theoretical and practical reasons, and it will be discussed later. To compute $e^{-4.17}$, press

$$[-4.17][e^x]$$

You should get .0154523.

Estimating Exponentials

When using a calculator, it is extremely useful to have a rough estimate of the answer in order to check that you did not push the wrong button. (Usually, if you push the wrong button during a calculator computation, the answer is so far off that *the error is obvious if you have a rough estimate of the answer*.) In addition, learning to make rough estimates also leads to a far better understanding of the concepts involved.

In Sec. 3.3, we saw how to estimate functions through graphical estimations. We shall be applying this technique to exponential functions. First, however, we present another method of estimating, which we refer to as **rough interpolation**. This method is very useful for checking calculator answers, since, with a little practice, the estimate can be made in your head.

Suppose you have a function f whose values at some points can easily be computed, and you wish to estimate a value elsewhere, say estimate $f(c)$. Let a and b be the closest points on either side of c such that you can easily compute their function values, and make the following table:

x	f(x)
a	f(a)
c	?
b	f(b)

You then determine if c is closer to one or the other of a or b, or about the middle, and then make a corresponding estimate for $f(c)$.

For example, to estimate $\sqrt{\pi}$, we use $f(x) = \sqrt{x}$ and make the following table:

x	$y = \sqrt{x}$
1	1
π	?
4	2

As π is between 1 and 4 but closer to 4, we see that $f(\pi) = \sqrt{\pi}$ is between 1 and 2 but closer to 2. Thus we might estimate $\sqrt{\pi} \approx 1.8$ or 1.9. In fact, $\sqrt{\pi} \approx 1.77245$, but either estimate would be close enough to tell us to check our work if the answer on a calculator was vastly different.

We now illustrate how to use both graphical estimation and rough interpolation to estimate exponential values.

EXAMPLE 2 Estimate $(\sqrt[3]{2})^7$.

SOLUTION

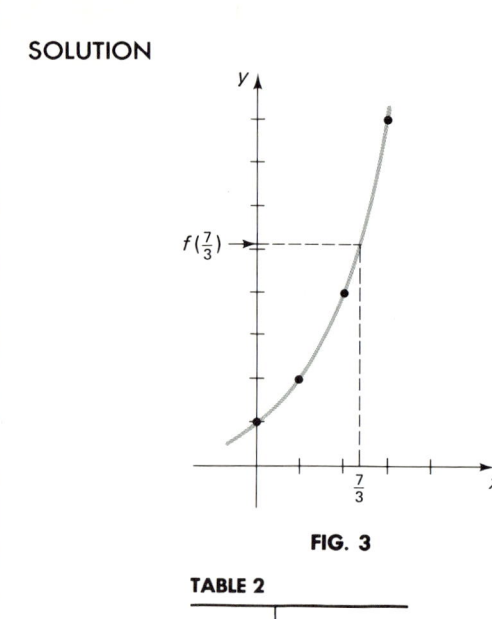

FIG. 3

TABLE 2

x	$y = 2^x$
2	4
$\frac{7}{3}$	?
3	8

We are looking for an exponential function $f(x) = a^x$ which fits this situation. Since $(\sqrt[3]{2})^7 = 2^{7/3}$ (and 2 is an easy number to work with), we choose $f(x) = 2^x$. Then the problem may be restated: Estimate $f(\frac{7}{3})$.

To estimate graphically, plot $y = f(x)$ at $x = 0, 1, 2, 3$, and then sketch the graph, obtaining Fig. 3. From this, we guess that $f(\frac{7}{3}) \approx 5.1$.

To estimate $f(7/3) = 2^{7/3}$ by rough interpolation, we use Table 2. We pick 2 and 3 because $\frac{7}{3}$ is between them (and we can easily compute 2 raised to those powers). Since $\frac{7}{3}$ is a little closer to 2, we say $2^{7/3}$ is between 4 and 8 but a little closer to 4, say $2^{7/3} \approx 5$.

Using a calculator, $2^{7/3} \approx 5.03968$.

The graphing techniques of Sec. 3.4 can be used with exponential functions. In this situation, you should indicate the asymptote and one or two points.

EXAMPLE 3 Sketch the graph of $y = 3^{2x} - 3$.

SOLUTION

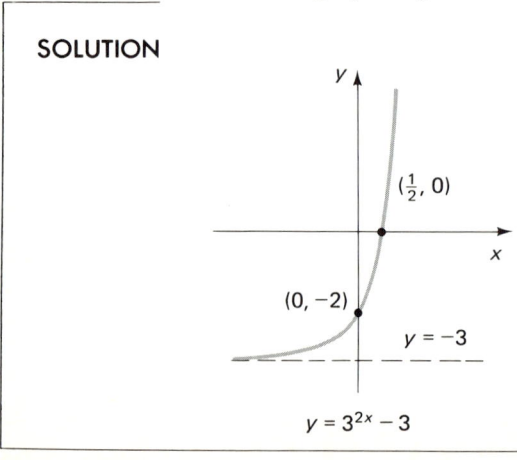

Start with the graph of $y = 3^x$. Contract it horizontally (by dividing the x-coordinates by 2), obtaining $y = 3^{2x}$. Then translate this down three units. Note that the asymptote is also translated.

EXAMPLE 4 Sketch the graph of $y = (-5)(\frac{1}{5})^x + 1$.

SOLUTION

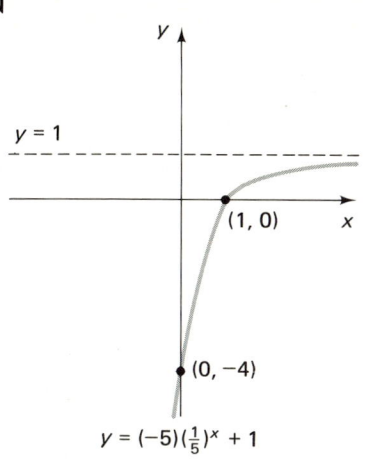

Start with the graph of $y = (\frac{1}{5})^x$. Reflect it through the x-axis, and expand it vertically (i.e., multiply all the y-coordinates by -5), obtaining the graph of $y = (-5)(\frac{1}{5})^x$. Then translate this up one unit.

The graph of $y = e^{-x^2}$ is a **normal distribution** curve. This is a very interesting and useful curve. To graph it, we first observe that it is symmetric to the y-axis. So to plot points, we need only fill in Table 3, sketch the curve for $x \geq 0$, and then use symmetry. Remember that e^u is always > 0. See Fig. 4(a). Of course, this graph can be expanded and translated, too. For example, the graph of $y = 2e^{-(x+1)^2} - 1$ is given in Fig. 4(b).

TABLE 3

x	$-x^2$	e^{-x^2}
0	0	1
.5	$-.25$	.78
1	-1	.37
1.5	-2.25	.11
2	-4	.02

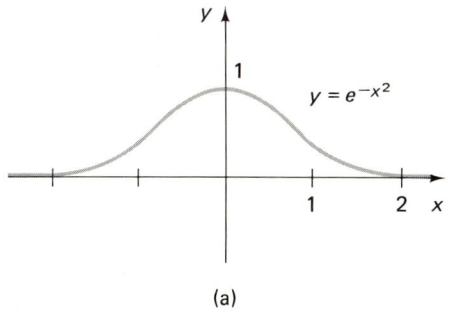

(a)

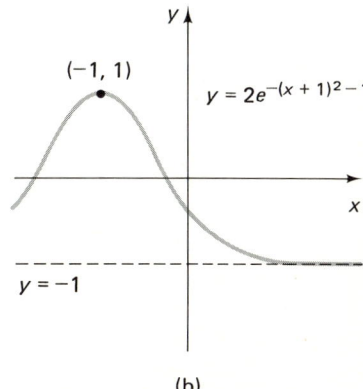

(b)

FIG. 4

EXERCISES

In Exercises 1–12, use your calculator to compute the given number.

1. $10^{3.124}$
2. $\sqrt[5]{10}$
3. $10^{-1/4}$
4. $e^{4.12}$
5. $(e^{-1.2})^{3.2}$ (Do this two ways.)
6. $20e^{-.08(3)}$
7. $\sqrt[20]{5}$
8. 53.8^{22}
9. $\left(1 + \dfrac{.08}{12}\right)^{12(5)}$
10. $(2.31 + e^{15.2})^3 14.1^{-2}$
11. $381.7 + \sqrt{10^{4.31} - 22^3}$
12. $(3.21^{4.8} + 2.07^{16.2})^{.07} 10^{30.65}$
13. Graph $y = 3^x$ and $y = (\frac{1}{3})^x$.
14. Use the graphs in Exercise 13 to estimate

   a. $\sqrt{3}$ b. $3^{\sqrt{3}}$ c. $\sqrt[4]{\frac{1}{3}}$ d. $(\frac{1}{3})^{-2.1}$

15. Use rough interpolation to estimate the numbers in Exercise 14.
16. Use your calculator to compute approximations to the numbers in Exercise 14, and compare with Exercises 14 and 15.

In Exercises 17–24, estimate the given number first graphically and then by rough interpolation. Finally, compute the number with your calculator, and compare.

17. $\sqrt{5}$
18. $5^{.2}$
19. $2^{-.2}$
20. $2^{-.9}$
21. $\sqrt[6]{2}$
22. $2^{-1/6}$
23. $\sqrt[3]{3^5}$
24. $\sqrt[5]{3^3}$

In Exercises 25–38, graph the equation, labeling at least two crucial things (a point and an asymptote or two points).

25. $y = 4^x + 1$
26. $y = -4^x + 2$
27. $y = .25^{-x} + 1$
28. $y = -.25^{-x+1}$
29. $y = 3^{-x} + 2$
30. $y = -3^{2x} - 3$
31. $y = (.1)^{x-1} - 1$
32. $y = -2e^{x-1}$
33. $y = e^{2x} - e$
34. $y = (-2)4^{2x} + 2$
35. $y = 4^{-x^2} - 1$
36. $y = -2(3^{-x^2}) + 2$
37. $y = 3^{x^2} - 3$
38. $y = 2(2^{x^2}) + 1$

39. A certain factory has 3000 light bulbs. It was determined that if all the light bulbs were replaced at essentially the same time, the number of bulbs which had burned out after t hours was $N = 3000(1 - e^{-t/500})$. (Such functions can be used to show it is more economical to replace all light bulbs at once rather than just as they burn out.)

   a. Graph this function (after appropriately labeling the axes).
   b. How many bulbs have burned out after 100 hours? 500 hours? 900 hours?
   c. If you want to replace all bulbs after 35% of them have burned out, when should you do this?

40. If a rock is dropped from a cliff or a building, then we can usually neglect air resistance, and its velocity is $V = -32t$ feet per second. However, if a sky diver jumps from an airplane with

arms and legs spread out, then air resistance has an important effect, and the velocity is given by $V = K(1 - e^{-at})$, where K and a are constants. Suppose after jumping from an airplane at 10,000 feet, the velocity of a sky diver is given by $V = 220(1 - e^{-3t})$ feet per second, t in seconds, up until the time she opens her parachute.

  a. Graph this function.
  b. What is her velocity after 3 seconds?
  c. How long does it take her to reach a velocity of 200 feet per second?
  d. What is her *terminal velocity*, i.e., the velocity she approaches but never quite reaches? Use 30 miles per hour = 44 feet per second to express this in miles per hour.

SECTION 5.2. LOGARITHMS

Logarithmic Functions

If $a = 1$, the graph of $f(x) = a^x$ is a horizontal straight line. Hence it does not have an inverse function. However if $a > 1$, the graph of $f(x) = a^x$ is strictly increasing; if $0 < a < 1$, the graph is strictly decreasing. (See Fig. 1 for the cases $a = 2$, $a = \frac{1}{2}$.) Hence, by the horizontal line test, $f(x) = a^x$ has an inverse function in these cases.

---

(2)   **DEFINITION**

If $0 < a < 1$ or $a > 1$, the inverse of the exponential function $f(x) = a^x$ is called the **logarithm function (to the base a)** and is denoted by $\log_a x$.

Warning: Although $\log_a$ contains four letters, it should be thought of as one symbol (as is f^{-1}). The symbol $\log_a$ is read "log to the base a." A related difficulty is that $\log_a x$ should be written $\log_a(x)$, as in $f^{-1}(x)$. Unfortunately, the parentheses are traditionally left out.

From the basic relationship between a function and its inverse, we have the following, which is really just another form of the definition:

---

(3)

If $0 < a < 1$ or $a > 1$, then

$$\log_a x = y \quad \text{if and only if} \quad a^y = x$$

The function $f(x) = \log_a x$ is not defined for all values of x. Since a is positive, a^y is positive. Thus with $x = a^y$, we see that

---

(4)   **THEOREM**

If $0 < a < 1$ or $a > 1$, $\log_a x$ is defined only for $x > 0$.

This restriction will be very important in some of the problems which follow.

Many logarithms can be calculated mentally using the definition in the form (3). To compute $\log_a x$, just ask "a to what power is x?"

EXAMPLE 1 Compute $\log_2 16$.

SOLUTION By (3), $\log_2 16 = y$ if and only if $2^y = 16$. Since $2^4 = 16$, $\log_2 16 = 4$. This demonstrates what we said above: To find $\log_2 16$, ask "2 to what power is 16?"

EXAMPLE 2 Compute $\log_{10} .0001$.

SOLUTION Ask "10 to what power is .0001?" Since $10^{-4} = .0001$, $\log_{10} .0001 = -4$.

EXAMPLE 3 Compute $\log_{16} 2$.

SOLUTION Ask "16 to what power is 2?" Since $2 = \sqrt[4]{16} = 16^{1/4}$, $\log_{16} 2 = \frac{1}{4} = .25$.

EXAMPLE 4 Compute $\log_3(-3)$.

SOLUTION By Theorem (4), $\log_3(-3)$ does not exist.

EXAMPLE 5 If $a > 0$, $a \neq 1$, compute $\log_a a$.

SOLUTION Ask "a to what power is a?" Since $a^1 = a$, we have $\log_a a = 1$.

EXAMPLE 6 If $a > 0$, $a \neq 1$, compute $\log_a 1$.

SOLUTION Ask "a to what power is 1?" Since $a^0 = 1$, we have $\log_a 1 = 0$.

The last two computations are special enough to emphasize.

> (5) **THEOREM**
>
> If $0 < a < 1$ or $a > 1$, then
>
> (i) $\log_a a = 1$ and   (ii) $\log_a 1 = 0$

The definition of $\log_a$ in the form (3) is also useful in solving certain equations.

EXAMPLE 7 Solve the following: **(a)** $\log_4 2 = x$; **(b)** $\log_4 x = 2$; **(c)** $\log_x 2 = 4$.

SOLUTION In all problems of this form, the first step is to use (3). What you do from there depends on the form of the resulting equation.
(a) $\log_4 2 = x$ becomes $2 = 4^x$. Since $2 = \sqrt{4} = 4^{1/2}$, $x = \frac{1}{2}$.
(b) $\log_4 x = 2$ becomes $x = 4^2$, so $x = 16$.
(c) $\log_x 2 = 4$ becomes $2 = x^4$, so $x = 2^{1/4} \approx 1.18921$. (*Note*: x, as a base of a log, must be greater than zero.)

There are two very useful identities which follow from (3). To simplify $\log_a a^x$, we ask "a to what power is a^x?" Since $a^x = a^x$, we get

> **(6) THEOREM**
> If $0 < a < 1$ or $a > 1$, then $\log_a a^x = x$ for all real x.

To simplify $a^{\log_a x}$, we use (3) and substitute $y = \log_a x$ into the expression $a^y = x$, obtaining

> **(7) THEOREM**
> If $0 < a < 1$ or $a > 1$, then $a^{\log_a x} = x$ for all $x > 0$.

EXAMPLE 8 Solve $3^{\log_3 x^2} = 25$.

SOLUTION By (7), $3^{\log_3 x^2} = x^2$. Thus the equation may be rewritten as $x^2 = 25$. The solutions are $x = \pm 5$.

EXAMPLE 9 Solve $\log_{.25}(\frac{1}{4})^{2x-1} = 5$.

SOLUTION By (6), $\log_{.25}(\frac{1}{4})^{2x-1} = 2x - 1$. Thus the equation may be rewritten as $2x - 1 = 5$. The solution is $x = 3$.

Note: Theorems (6) and (7) merely restate that $f^{-1}(f(x)) = x$ and $f(f^{-1}(x)) = x$ in the case $f(x) = a^x$ and $f^{-1}(x) = \log_a x$. [See Theorem (41) in Sec. 3.7.] There is one remaining fundamental property.

> **(8) THEOREM**
> If $0 < a < 1$ or $a > 1$, then
> i. $a^x = a^y$ if and only if $x = y$.
> ii. $\log_a x = \log_a y$ if and only if $x = y$ with $x > 0$ and $y > 0$.

In the first paragraph of this section, we observed that the function $f(x) = a^x$ is strictly increasing (if $a > 1$) or strictly decreasing (if $0 < a < 1$). Thus two different values of x give two different values of a^x, yielding part i. (This allows a^x to have an inverse.) Part ii follows similarly.

EXAMPLE 10 Solve $3^x = 81$.

SOLUTION $3^x = 81$, $3^x = 3^4$, $x = 4$ by (8i).

Section 5.2 Logarithms

EXAMPLE 11 Solve $\log_{.1}(x^2 - 10) = \log_{.1} 3x$.

SOLUTION
$$\log_{.1}(x^2 - 10) = \log_{.1} 3x$$
$$x^2 - 10 = 3x \quad [\text{by (8i)}]$$
$$x^2 - 3x - 10 = 0$$
$$(x - 5)(x + 2) = 0, \quad x = 5, -2$$

However, here we must be careful. By (4), $\log_a x$ is defined only for $x > 0$. If we substitute $x = -2$ back into the original equation, we get $\log_{.1}(-6)$, which is undefined. However, substituting $x = 5$ yields $\log_{.1}(25 - 10) = \log_{.1} 15$, which is true. Thus, $x = 5$ is the only answer for this problem. Hence it is important to remember always to check any problem which has a logarithm in its original formulation.

Estimating Logarithms by Rough Interpolation

Logarithms may be estimated by rough interpolation in essentially the same way that exponentials are.

EXAMPLE 12 Estimate $\log_2 14$ by rough interpolation.

SOLUTION

TABLE 4

x	$y = \log_2 x$
$8 (= 2^3)$	3
14	?
$16 (= 2^4)$	4

Use Table 4. We pick 8 and 16 because we can easily compute their logarithms to the base 2, and 14 is between them. Since 14 is closer to 16, we say $\log_2 14$ is between 3 and 4 but closer to 4. We might guess $\log_2 14 \approx 3.8$. (In fact, $\log_2 14 \approx 3.80736$.)

EXAMPLE 13 Estimate $\log_2 \frac{1}{5}$ by rough interpolation.

SOLUTION

TABLE 5

x	$y = \log_2 x$
$\frac{1}{4} (= 2^{-2})$	-2
$\frac{1}{5}$	?
$\frac{1}{8} (= 2^{-3})$	-3

Use Table 5. Since $\frac{1}{5}$ is between $\frac{1}{4}$ and $\frac{1}{8}$ but closer to $\frac{1}{4}$, we say $\log_2 \frac{1}{5}$ is between -2 and -3 but closer to -2. We might guess $\log_2 \frac{1}{5} \approx -2.2$. (In fact, $\log_2 \frac{1}{5} \approx -2.32193$.)

Special Cases

There are two bases that are very important for both historical and scientific reasons. Their logarithms have special names and usually have special notations.

> **(9) NOTATION**
>
> **i.** If the base is 10, the logarithm $\log_{10} x$ is called the **common logarithm** of x. Often the 10 is not written, so that $\log x$ usually means $\log_{10} x$.
>
> **ii.** If the base is e, the logarithm $\log_e x$ is called the **natural logarithm** of x. Often $\ln x$ is written for $\log_e x$.

The symbol ln is usually read "ell-en" or "lin."

Hereafter in this text, we shall usually use $\log x$ for $\log_{10} x$ and $\ln x$ for $\log_e x$. Most calculators use this notation. However, some texts use $\log x$ for $\log_e x$, so you should be careful about this abbreviation.

Using the Calculator

The functions $\log x$ and $\ln x$ are so important that most scientific calculators have separate buttons for them:

$$[\text{LOG } x] \text{ (or just } [\text{LOG}]), \quad [\text{LN } x] \text{ (or just } [\ln])$$

To compute $\log 100 = \log_{10} 100$, press (on any calculator)

$$[100][\text{LOG } x]$$

Of course, $\log 100 = \log_{10} 100 = 2$.

Computing $\ln 100 = \log_e 100$ is similar. Press

$$[100][\text{LN } x]$$

obtaining $\ln 100 = \log_e 100 \approx 4.60517$.

Since $\log_b x$ is undefined for $x \leq 0$, if you attempt to compute $\log x$ or $\ln x$ for $x \leq 0$, the calculator will give an error message.

The computation of logarithms to bases other than 10 and e usually requires a small calculation. A few calculators have a $[\log_x y]$ button. However, most calculators just have [log] and [ln] buttons. On these you must use the following:

> **(10) THEOREM**
>
> $$\log_b x = \frac{\log_a x}{\log_a b}$$
>
> where $a, b, x > 0$, $a, b \neq 1$.

This will be proved in Sec. 5.3 [see (15iii)].

By Theorem (10), to compute $\log_b x$, we can use any base a for which we can find $\log_a x$ and $\log_a b$. In particular, when using a calculator we can choose either $a = 10$ or $a = e$.

EXAMPLE 14 Compute $\log_2 3$ and $\log_2 \frac{1}{5}$.

SOLUTION

$$\log_2 3 = \frac{\log_{10} 3}{\log_{10} 2} = \frac{\log 3}{\log 2} \approx \frac{.4771212}{.3010299} \approx 1.58496$$

or

$$\log_2 3 = \frac{\log_e 3}{\log_e 2} = \frac{\ln 3}{\ln 2} \approx \frac{1.0986123}{.6931478} \approx 1.58496$$

$$\log_2 .2 = \frac{\log_{10} .2}{\log_{10} 2} = \frac{\log .2}{\log 2} \approx \frac{-.698970}{.3010299} \approx -2.32193$$

or

$$\log_2 .2 = \frac{\log_e .2}{\log_e 2} = \frac{\ln .2}{\ln 2} \approx \frac{-1.6094379}{.69314718} \approx -2.32193$$

Although on most problems you have a choice of using either the log or ln button,

For certain later problems it will be easier to use the ln button,

so you should get used to using it now. To do the computation on a calculator is not difficult. For example, to compute $\ln 3 / \ln 2$,

Alg.: [3][LN x][÷][2][LN x][=]
RPN: [3][LN x][ENT][2][LN x][÷]

EXERCISES

In Exercises 1–14, find the given number without using a calculator.

1. $\log_2 8$
2. $\log_8 2$
3. $\log_8 1$
4. $\log_8 8$
5. $\log_3(\log_3 27)$
6. $\log_2[\log_2(\log_2 4)]$
7. $\log_9 \frac{1}{81}$
8. $\log_9 3$
9. $\log_9 27$
10. $\log_{10} 10{,}000{,}000$
11. $\log_{10} .000000000001$
12. $\log_{.01} 10$
13. $\log_{1/2} 4$
14. $\log_{.2} 125$

In Exercises 15–28, estimate the given number by rough interpolation. Then use your calculator to find that number, and compare.

15. $\log_3 4$
16. $\log_3 11$
17. $\log_3 21$
18. $\log_2 11$
19. $\log_2 39$
20. $\log_5 31$
21. $\log_7 14$
22. $\log_3 60$
23. $\log_3 .1$
24. $\log_3 .01$
25. $\log_5 .3$
26. $\log_7 150$
27. $\log_7 .2$
28. $\log_{12} 140$

Use (3), (6), (7), and (8) to solve Exercises 29–48.

29. $x = \log_3 3^{15}$
30. $x = \frac{1}{2}^{\log_{.5} 23}$
31. $\log_2 2^x = 4$
32. $(\frac{1}{4})^{\log_{.25} x^2} = 9$
33. $\log_5 5^{2x-3} = 4$
34. $(\frac{3}{5})^{\log_{.6}(2x)} = x^2 - 8$
35. $\log_3 9 = x$
36. $\log_x 9 = 3$
37. $\log_3 x = 9$
38. $\log_2 2 = x$
39. $\log_x 2 = 2$
40. $\log_2 x = 2$
41. $\log_9 3 = x$
42. $\log_x 3 = 9$
43. $\log_9 x = 3$
44. $2^x = 16$
45. $3^{2x-1} = 3^{21}$
46. $\log_3(x+2) = \log_3 8$
47. $\log x = \log(x^2 - 12)$
48. $\ln x^2 = \ln(3x + 10)$

49. This is very important and should be done before Sec. 5.3. Make a copy of the following table and fill it in using the [LOG] button on your calculator. Then
   a. Find three relationships that you would guess would always be true (for $x, y > 0$).
   b. Find at least two relationships that look like they might be true but that the table shows are not [for example, $\log(x - y) \neq \log x - \log y$].

x	y	log x	log y	log(xy)	log(x+y)	log x + log y	(log x)(log y)
2	3						
5.1	.21						
.01	.3						
1	5						

x^y	log x^y	(log x)^y	y log x	log $\frac{x}{y}$	$\frac{\log x}{\log y}$	log x − log y

SECTION 5.3. GRAPHS AND PROPERTIES OF LOGARITHMS

Recall that if you have the graph of $y = f(x)$, the graph of $y = f^{-1}(x)$ is obtained by reflection through the line $y = x$. Applying this to $f(x) = a^x$ so that $f^{-1}(x) = \log_a x$, we easily get the graph of $y = \log_a x$, shown in Fig. 5.

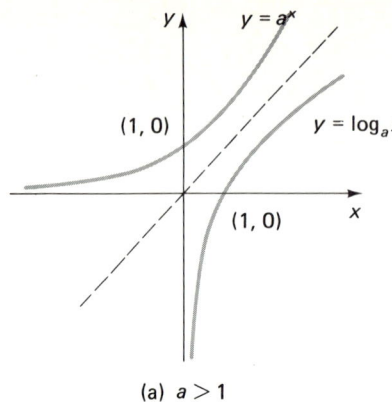
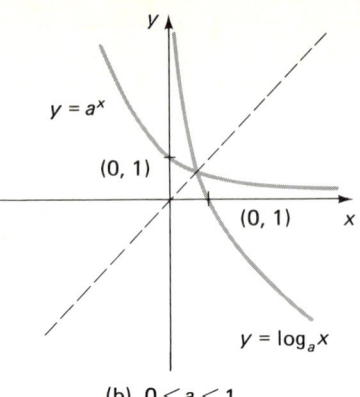

(a) $a > 1$ (b) $0 < a < 1$

FIG. 5

From these graphs, we see that the graph of $y = \log_a x$ is strictly increasing if $a > 1$ and strictly decreasing if $0 < a < 1$.

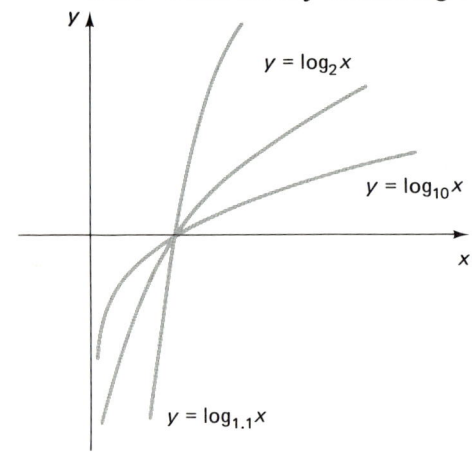

Note that the graph of $y = \log_a x$ crosses the x-axis at $x = 1$ (since $\log_a 1 = 0$) and that the y-axis is a vertical asymptote. For different values of a, the graphs of $y = \log_a x$ differ in *steepness*, as indicated in Fig. 6. (The case $0 < a < 1$ is analogous. Compare with Fig. 2, Sec. 5.1.)

Graphs of $y = \log_a x$ various $a > 1$.

FIG. 6

We can also observe what the domains and ranges of $f(x) = a^x$ and $f^{-1}(x) = \log_a x$ are, namely,

(11)
$$f: \mathbb{R} \to (0, \infty); \qquad f^{-1} = \log_a : (0, \infty) \to \mathbb{R}.$$

In other words, the domain of $f(x) = a^x$ is the set of all real numbers, and the range is the set of all positive real numbers. The domain of $f^{-1}(x) = \log_a x$ is the set of all positive real numbers, and the range is the set of all real numbers. This illustrates the earlier discussion of inverse functions where we observed that the domain of a function is the range of its inverse and vice versa.

Estimation

The techniques of graphical estimation can be applied to logarithmic functions.

EXAMPLE 1 Graphically estimate $\log_2 3$ and $\log_2 \frac{1}{5}$.

SOLUTION In Table 6, we list a few easy-to-find values near the numbers we are interested in. Then we plot them to get Fig. 7. From the graph, $\log_2 3 \approx 1.6$ and $\log_2 \frac{1}{5} \approx -2.3$.

TABLE 6

x	$y = \log_2 x$
$\frac{1}{8}$	-3
$\frac{1}{4}$	-2
$\frac{1}{2}$	-1
1	0
2	1
4	2

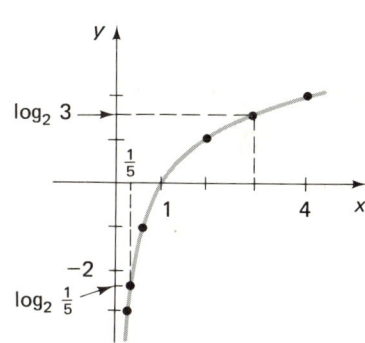

FIG. 7

Just as with exponential functions, the graphing techniques can be used with logarithmic functions. Again, you should indicate the asymptote and one or two points.

EXAMPLE 2 Sketch the graph of $y = \log_3(x - 2)$.

SOLUTION

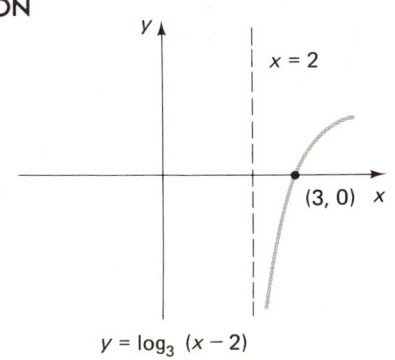

Start with the graph of $y = \log_3 x$ and translate it two units to the right.

EXAMPLE 3 Sketch the graph of $y = 3\log_{1/4}(x + 4)$.

SOLUTION

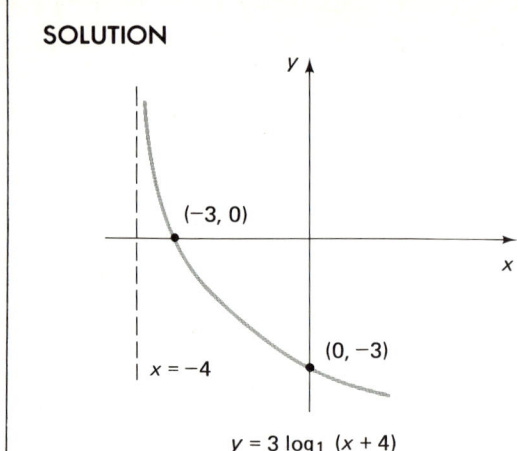

$y = 3\log_{\frac{1}{4}}(x+4)$

Start with the graph of $y = \log_{1/4} x$. Expand it vertically (by multiplying all y-coordinates by 3), obtaining the graph of $y = 3\log_{1/4} x$. Then translate this four units to the left.

There are many situations that arise naturally which involve exponential or logarithmic relations. To solve some of the equations that arise, we must first study some of the basic properties of logarithms.

There are three fundamental properties of logarithms.

---

(12)   **THEOREM**

If $0 < a < 1$ or $a > 1$, then

i. $\log_a(uw) = \log_a u + \log_a w, \quad u, w > 0.$

ii. $\log_a\left(\dfrac{u}{w}\right) = \log_a u - \log_a w, \quad u, w > 0.$

iii. $\log_a(u^c) = c\log_a u, \quad u > 0, c$ any real number.

The properties of logarithms listed in (12) are the reasons logarithms were invented in the early seventeenth century. Exercise 66 illustrates how they were used.

The proof of (12i–iii) uses, several times, the second form of the definition of logarithm, (3). We restate it here for convenience:

---

(13)

If $0 < a < 1$ or $a > 1$, then

$\log_a x = y \quad$ if and only if $\quad a^y = x.$

Proof of (12i): Let $r = \log_a u$ and $s = \log_a w$. Then $a^r = u$ and $a^s = w$, by (13). Multiplying, we obtain $a^r a^s = uw$, so $a^{r+s} = uw$ by the laws of exponents. By (13), $\log_a(uw) = r + s$. Substituting $r = \log_a u$ and $s = \log_a w$ yields $\log_a(uw) = \log_a u + \log_a w$.

Since the proof of (12ii) uses (12iii), we prove (12iii) first.

Proof of (12iii): Let $r = \log_a u$. By (13), $a^r = u$. Raising both sides to the power c, we obtain $(a^r)^c = u^c$. By the laws of exponents, $a^{cr} = u^c$. By (13) again, $cr = \log_a u^c$. Substituting $r = \log_a u$, we obtain $c \log_a u = \log_a u^c$.

Proof of (12ii): This follows from (12i) and (12iii):

$$\log_a\left(\frac{u}{w}\right) = \log_a(uw^{-1})$$
$$= \log_a u + \log_a w^{-1} \quad [\text{by (12i)}]$$
$$= \log_a u + (-1)\log_a w \quad [\text{by (12iii)}]$$
$$= \log_a u - \log_a w$$

Example 4 is very helpful in understanding the properties of logarithms given in (12).

EXAMPLE 4 Suppose $\log_7 2 \approx .36$ and $\log_7 3 \approx .56$. Using only this and fundamental properties, find **(a)** $\log_7 6$ and **(b)** $\log_7 \frac{72}{7}$.

SOLUTION For **(a)**, $\log_7 6 = \log_7(3 \cdot 2) = \log_7 3 + \log_7 2 \approx .56 + .36 = .92$. For **(b)**, $\log_7 \frac{72}{7} = \log_7(8 \cdot 9) - \log_7 7 = \log_7 2^3 + \log_7 3^2 - 1 = 3\log_7 2 + 2\log_7 3 - 1 \approx 3(.36) + 2(.56) - 1 = 1.2$.

EXAMPLE 5 Simplify $9^{2\log_9 7 + 3\log_9 2}$.

SOLUTION
$$9^{2\log_9 7 + 3\log_9 2} = 9^{\log_9 7^2 + \log_9 2^3} \quad [\text{by (12iii)}]$$
$$= 9^{\log_9(7^2 2^3)} \quad [\text{by (12i)}]$$
$$= 7^2 2^3 = 49(8) = 392 \quad [\text{by (7)}]$$

EXAMPLE 6 Simplify $\log_3(x^2 - 9) - \log_3(x - 3)$.

SOLUTION
$$\log_3(x^2 - 9) - \log_3(x - 3) = \log_3 \frac{x^2 - 9}{x - 3} \quad [\text{by (12ii)}]$$
$$= \log_3(x + 3)$$

Frequently it is necessary to use the properties of logarithms when solving equations involving exponentials or logarithms.

EXAMPLE 7 Solve $30e^{.12x} = 6$.

SOLUTION Divide by 30 first:
$$e^{.12x} = \tfrac{6}{30} = .2$$

Next take ln's, using (8ii):
$$\ln e^{.12x} = \ln .2$$

Now use $\ln e^u = u$, by (6):
$$.12x = \ln .2$$

Finally, solve for x:
$$x = \frac{\ln .2}{.12} \approx -13.412$$

There are problems that appear similar where the solutions require that you take either ln's or roots. The key to deciding what to do is the following:

(14)

If the unknown is in the exponent, take ln's.

If the unknown is in the base, take roots.

EXAMPLE 8 Solve for x: **(a)** $12^x = 30$ and **(b)** $x^{12} = 30$.

SOLUTION

(a) $12^x = 30$
$\ln 12^x = \ln 30$ [by (8ii)]
$x \ln 12 = \ln 30$ [by (12iii)]
$x = \dfrac{\ln 30}{\ln 12} \approx 1.36874$

(b) $x^{12} = 30$
$(x^{12})^{1/12} = \pm 30^{1/12}$
$x = \pm 30^{1/12}$
$x \approx \pm 1.32768$

EXAMPLE 9 If $5000 = 2000(1 + r^n)$, find **(a)** r if $n = 13$ and **(b)** n if $r = .06$.

SOLUTION

(a) $5000 = 2000(1 + r)^{13}$
$2.5 = (1 + r)^{13}$
$(2.5)^{1/13} = 1 + r$
$r = (2.5)^{1/13} - 1$
$r \approx .0730273$

(b) $5000 = 2000(1 + .06)^n$
$2.5 = 1.06^n$
$\ln 2.5 = \ln 1.06^n$
$n = \dfrac{\ln 2.5}{\ln 1.06}$
$n \approx 15.7252$

There are several further properties of logarithms in addition to the fundamental properties given in (12). Some of the more important are the following [the fundamental properties (12) are used in their proofs]:

---

(15)   **THEOREM**

If $a, b > 0$ and $a, b \neq 1$, then

i. $\log_a \dfrac{1}{x} = -\log_a x, \quad x > 0.$

ii. $y^x = e^{x \ln y}, \quad y > 0, \text{ all } x.$

iii. $\log_b x = \dfrac{\log_a x}{\log_a b}, \quad x > 0.$

Proof of (15i):

$$\log_a \frac{1}{x} = \log_a x^{-1}$$
$$= (-1) \log_a x \qquad [\text{by (12iii)}]$$
$$= -\log_a x$$

Proof of (15ii):

$$y^x = e^{\ln(y^x)} \qquad [\text{by (7)}]$$
$$= e^{x \ln y} \qquad [\text{by (12iii)}]$$

Proof of (15iii):

$$x = b^{\log_b x} \qquad [\text{by (7)}]$$
$$\log_a x = \log_a(b^{\log_b x}) \qquad [\text{by (8ii)}]$$
$$\log_a x = (\log_b x)(\log_a b) \qquad [\text{by (12iii)}]$$
$$\log_b x = (\log_a x)/\log_a b \qquad [\text{divide by } \log_a b]$$

You have already been using (15iii) to compute with logarithms. [See Theorem (10).] It is most frequently used with either $a = e$ or $a = 10$. However, the most important of the relationships in Theorem (15) is probably (15ii). This is used by calculators to compute y^x when the $[y^x]$ button is pressed. Since $\ln y$ is defined only for $y > 0$, this is why calculators do not compute y^x for $y \leq 0$. In fact, on some algebraic calculators if you want to compute $(3.1)^{5.2}$ and press $[3.1][y^x]$, the number 1.131402 appears in the display. This is just $\ln 3.1$; the machine has just started the computation $e^{5.2 \ln 3.1}$.

EXAMPLE 10   Use (15ii) (and your calculator) to compute 2^3.

SOLUTION   By (15ii), $2^3 = e^{3 \ln 2}$, so press

                Alg.: $[2][\text{LN}][\times][3][=][e^x]$

                RPN: $[2][\text{LN}][3][\times][e^x]$

You should, of course, get 8, but you might get something like 7.9999998 due to calculator error.

EXERCISES

1. Graph $y = \log_3 x$ and $y = \log_{1/3} x$.

In Exercises 2–7, use Exercise 1 to estimate graphically the given numbers. Then use your calculator to compute the numbers, and compare.

2. $\log_3 8$, $\log_{1/3} 8$
3. $\log_3 4$, $\log_{1/3} 4$
4. $\log_3 1.2$, $\log_{1/3} 1.2$
5. $\log_3 .8$, $\log_{1/3} .8$
6. $\log_3 .2$, $\log_{1/3} .2$
7. $\log_3 \frac{1}{20}$, $\log_{1/3} \frac{1}{20}$

In Exercises 8–30, graph the equation, labeling at least two crucial things (a point and an asymptote or two points).

8. $y = \log_4 x$
9. $y = \log_4(-x)$
10. $y = -\log_4 x$
11. $y = -\log_4(-x)$
12. $y = \log_3(x + 3)$
13. $y = \log_{1/3}(x - 1)$
14. $y = 2\log_{.3}(x + 2)$
15. $y = -\log(x - 2)$
16. $y = 3\log(x + 3)$
17. $y = \log_{.2}(2x)$
18. $y = \ln x - 1$
19. $y = \ln(x + e)$
20. $y = \log_{.5}(x - 1) + 2$
21. $y = \log_5(4x - 8)$
22. $y = \log_{1/3}(2x - 2)$
23. $y = 3\log_4(3 - x)$
24. $y = -2\log_2(4 - 2x)$
25. $y = \log_{.5}(1 - x) + 2$
26. $y = 2\log_4(x + 2) + 2$
27. $y = \log_7 |x|$
28. $y = |\log_7 x|$
29. $y = e^{\ln x}$
30. $y = \log 10^x$

31. a. Compute $\ln \frac{3}{4}$, $\ln 3/\ln 4$, $\ln 2^3$, and $(\ln 2)^3$.

b. Does
$$\log_b \frac{x}{y} = \frac{\log_b x}{\log_b y} \quad \text{or} \quad \log_b x^y = (\log_b x)^y ?$$

c. Prove:
$$\frac{\log_b x}{\log_b y} = \frac{\ln x}{\ln y}, \quad x, y, b > 0 \text{ and } y, b \neq 1.$$

In Exercises 32–40, assume that $\log_5 2 \approx .43$ and $\log_5 3 \approx .68$. Using only this and the fundamental properties, find the given numbers.

32. $\log_5 6$
33. $\log_5 10$
34. $\log_5 12$
35. $\log_5 \frac{3}{2}$
36. $\log_5 \frac{2}{5}$
37. $\log_5 \frac{4}{15}$
38. $\log_5 125$
39. $\log_5 \frac{9}{50}$
40. $\log_5 .1$

In Exercises 41–55, simplify. Do not use your calculator.

41. $3^{\log_3 5}$
42. $4^{3\log_4 x}$
43. $7^{\log_7 5 + \log_7 4}$
44. $23^{4\log_{23} 2 - 2\log_{23} 4}$
45. $\log_7 7^x$
46. $\log_{.3} .3^x - \log_{.3} .3^y$ (Do two ways.)

47. $(\log_3 9^{-1})^{-1}$ (Do two ways.)
48. $\log_2(2 \cdot 8^x)$
49. $\{\log_{2/3}[(\frac{2}{3})^2]^3\}^2$
50. $\log_{.1} x^4 - \log_{.1} \sqrt{x}$
51. $3\log_4 x^2 - \frac{1}{2}\log_4 \sqrt{x}$
52. $\log_9(x^2 - 1) - \log_9(x + 1)$
53. $\log_{12}(x^3 + 1) - \log_{12}(x + 1)$
54. $\log_a[a(a + 1)]$
55. $\log_a \dfrac{a}{\sqrt{x}} + \log_a \sqrt{\dfrac{x}{a}}$

In Exercises 56–63, solve for x.

56. $2^x = 8$
57. $2^x = 9$
58. $x^{20} = 8$
59. $(1 + x)^5 = 4$
60. $5^{1+x} = 4$
61. $e^{.423x} = 16.1$
62. $\log_5(1 + x) = 2$
63. $\ln(51.7x) = 3.2$

64. Suppose $A = P(1 + r)^n$.
   a. If $P = 200$, $r = .02$, and $n = 12$, find A.
   b. If $A = 4000$, $r = .04$, and $n = 8$, find P.
   c. If $A = 5000$, $P = 2000$, and $n = 12$, find r.
   d. If $A = 3000$, $P = 1000$, and $r = .07$, find n.

65. For the following, compute y^x by $e^{x \ln y}$ (on your calculator) after first computing the answer by hand:
   a. 2^4
   b. 4^{-1}
   c. 10^{16}
   d. $(.01)^{-23}$

66. Compute in two ways:
$$x = \frac{14.38\sqrt{72.79}\,(3.91)^4}{37.91}$$
   a. Using your calculator, the way it is.
   b. First show that
$$\ln x = \ln 14.38 + \tfrac{1}{2}\ln 72.79 + 4\ln 3.91 - \ln 37.91$$

Then use your calculator to compute the right-hand side. Denote the result by y. (Before calculators, people had to look up the logarithms in tables and then do the computations by hand.) Finally, compute $e^y = e^{\ln x} = x$, and compare with part a. (Before calculators, the last step would be done using tables.)

SECTION 5.4. APPLICATIONS

In this section we shall discuss some applications of exponentials and logarithms. In doing so, we shall see one place where the number e arises naturally. We begin by examining compound interest.

Suppose you invest $1000 for 2 years in a savings account that pays 6% per year compounded quarterly. What this means is that each quarter, interest is earned at the rate of 1/4 of 6% or 1.5% = .015, and this interest is accumulated with the principal to earn interest in subsequent quarters. Thus, during the first 3 months, the $1000 earns $1000(.015) = $15 interest, so you have $1000 + $15 = $1015 after 3 months. During the second 3 months, the $1015 earns $1015(0.15)

≈ $15.23 interest, so you have $1015 + $15.23 = $1030.23 after 6 months. During the third 3 months, the $1030.23 earns $1030.23(.015) ≈ $15.45, so after 9 months you have $1030.23 + $15.45 = $1045.68. It keeps going like this five more times until you have $1126.49 after 2 years.

If it had been compounded monthly, then each month interest is earned at the rate of 1/12 of 6% or .5% = .005, and it would take 24 computations like the above to compute that you would have $1127.16 after 2 years. Fortunately, there is an easy formula.

First we observe how much money we have after one quarter or month, etc., if our money is invested in an account which compounds interest quarterly or monthly, etc.

---

(16)

Let Q be an amount of money invested for a period of time at a rate of interest r in decimal form. Then Q earns rQ for the time period. Thus at the end of the time period, the amount A that the investment is worth is

$$A = Q + rQ = Q(1 + r)$$

The expression $Q(1 + r)$ is more useful, as will be seen below. We now derive the formula.

Suppose that we invest P dollars for n time periods and that r is the rate of interest for each period. In the first example $n = 8$; in the second, $n = 24$. Using (16), we then have $P(1 + r)$ after the first period. Now for the second period $Q = P(1 + r)$, so we have $P(1 + r)(1 + r) = P(1 + r)^2$ at the end of the second period. For the third period, $Q = P(1 + r)^2$, so we have $P(1 + r)^2(1 + r) = P(1 + r)^3$ at the end of the third period. After four periods, we have $P(1 + r)^4$, after five, $P(1 + r)^5$, and so on. In general,

---

(17)   THEOREM

If P dollars is invested at a rate of interest of r per period, then the amount A that the investment is worth after n time periods is

$$A = P(1 + r)^n$$

EXAMPLE 1 If $1000 is invested at 6% compounded quarterly for 2 years, then $r = .06/4 = .015$, $n = 2(4) = 8$. By (17), $A = 1000(1 + .015)^8 ≈ 1126.49$. If it is compounded monthly for 2 years, then $r = .06/12 = .005$, $n = 2(12) = 24$, and $A = 1000(1 + .005)^{24} ≈ 1127.16$. If it is compounded daily, then $A = 1000(1 + .06/365)^{730} ≈ 1127.49$.

Comparing this answer with the previous one shows that very little is gained by increasing the number of times it was compounded. Monthly to daily increased the earnings only by 33 cents over 2 years.

This leads to a very interesting question: What exactly does happen as we shorten the time interval when the interest is compounded. As an example, we examine how much interest $100,000 earns for 1 year at 8% compounded at different time intervals:

COMPOUNDED	AMOUNT AFTER 1 YEAR	INTEREST EARNED FOR 1 YEAR
Annually	$100,000(1 + .08) = 108,000$	$8000
Semiannually	$100,000\left(1 + \frac{.08}{2}\right)^2 = 108,160$	8160
Quarterly	$100,000\left(1 + \frac{.08}{4}\right)^4 \approx 108,243.22$	8243.22
Monthly	$100,000\left(1 + \frac{.08}{12}\right)^{12} \approx 108,299.99$	8299.99
Daily	$100,000\left(1 + \frac{.08}{365}\right)^{365} \approx 108,328.39$	8328.39
Hourly	$100,000\left(1 + \frac{.08}{8760}\right)^{8760} \approx 108,328.69$	8328.69
Each minute	$100,000\left(1 + \frac{.08}{525,600}\right)^{525,600} \approx 108,328.7067$	8328.7067

These numbers appear to be getting closer and closer to some particular number. This is in fact the case, and that number is very interesting: It is $100,000 e^{.08} \approx 108,328.7068$. Using calculus, one can prove the following:

---

(18)   **THEOREM**

As k gets larger and larger, $\left(1 + \frac{R}{k}\right)^k$ approaches e^R.

This is one of the reasons the number e is so important and interesting (other reasons arise in calculus).

Notation: We shall use R for the stated annual interest and r for the interest per time period (month, quarter, etc.).

The preceding discussion shows how compounding more often leads to exponential growth. The limit is what is sometimes called **compounded continuously**.

---

(19)   **DEFINITION**

If P dollars is invested at an annual rate of interest R compounded continuously, then after n years it is worth

$$A = P(e^R)^n = Pe^{Rn}$$

Section 5.4 Applications

EXAMPLE 2 The $100,000 compounded continuously for 1 year at 8% gives
$$\$100{,}000 e^{.08} = \$108{,}328.71$$
So the most interest that can be earned at 8% is $8328.71, just 32 cents more than that earned when compounded daily.

EXAMPLE 3 Investing $1000 for 5 years at 6% compounded continuously gives $A = \$1000(e^{.06})^5 = \$1000 e^{.30} \approx \$1349.86$.

EXAMPLE 4 (a) If $2000 is invested at 8% compounded quarterly, how long does it take to double?
(b) If it is compounded continuously, how long then?
(c) At what rate should we invest it to get it to double in $7\frac{1}{2}$ years if the interest is compounded monthly?

Before solving this example, it may be good to refer back to the "key" in Sec. 5.3, (14), and the examples which followed.

SOLUTION (a) Let n = the number of years. After n years, the 2000 is worth $2000[1 + .08/4]^{4n}$, so we want to solve
$$2000(1.02)^{4n} = 4000$$
Then $(1.02)^{4n} = 2$, $4n \ln(1.02) = \ln 2$, so $n = \ln 2 / 4 \ln 1.02 \approx 8.7507$, or just over $8\frac{3}{4}$ years.
(b) We want to solve $2000 e^{.08n} = 4000$. Then $e^{.08n} = 2$, $.08n = \ln 2$, $n = (\ln 2)/.08 \approx 8.664$ years.
(c) Let R = rate (per year). Since $7\frac{1}{2}$ years is 90 months, we want to solve $2000(1 + \frac{1}{12} R)^{90} = 4000$. Then $(1 + \frac{1}{12} R)^{90} = 2$, $1 + \frac{1}{12} R = 2^{1/90}$, $\frac{1}{12} R = 2^{1/90} - 1$, $R = 12(2^{1/90} - 1) \approx .09276$, so 9.276%.

There are many other situations in which exponential (and hence logarithmic) relations arise.

EXAMPLE 5 **Depreciation.** Suppose a new car depreciates at the rate of 30% a year. Thus each year the car loses in value 30% of what it was worth at the beginning of that year. Then the arithmetic works just like interest compounded annually except the rate is negative. For example, a $4000 car after 5 years is worth
$$\$4000(1 - .30)^5 = \$4000(.7)^5 = \$672.30$$

EXAMPLE 6 **Exponential growth.** Under favorable conditions many things such as populations grow exponentially. This means the following:

(20)

If P is the population after time t and P is growing exponentially, then
$$P = P_0 e^{kt}$$
where $k > 0$ is a proportionality constant and P_0 is the population at time $t = 0$.

For example, suppose that the population of a certain kind of rabbit is growing exponentially and that it doubles every 9 months (if you start with enough rabbits of both sexes). If you start with 100 rabbits (50 of each sex), how many do you have after 4 years?

SOLUTION Since the population is growing exponentially, we use the formula $P = P_0 e^{kt}$. We know the initial population is $P_0 = 100$. To find k, we use the fact that the population doubles after 9 months($= 3/4$ year). So $P = 200$ when $t = 3/4$ year. Solving for k, we obtain $200 = 100 e^{3k/4}$, $\ln 2 = 3k/4$, $k = (4/3)\ln 2 \approx .924196$. Thus $P = 100 e^{.924196 t}$. When $t = 4$, $P \approx 4032$ rabbits.

EXAMPLE 7 **Exponential decay.** Many things decay exponentially: radioactive substances, some chemicals during certain reactions, air pressure as a function of altitude. This means the following:

---

(21)

If A is the amount of a substance after time t and A is decaying exponentially, then

$$A = A_0 e^{-kt}$$

where $k > 0$ is a proportionality constant and A_0 is the amount at time $t = 0$.

For example, if you start with 10 grams of a certain radioactive isotope of iodine, you have 8 grams after 10 days. **(a)** What is its half-life? **(b)** How much do you have after a year?

SOLUTION First we must find the constants in the formula $A = A_0 e^{-kt}$. We know $A_0 = 10$, so we only need to determine k. We are given that when $t = 10$ (days), $A = 8$, so $8 = 10 e^{-10k}$, $k = -\frac{1}{10}\ln .8 \approx .0223144$.

(a) The **half-life** is the time that it takes for half of the isotope to decay. Thus we need to find the time t when we only have 5 grams left, i.e., when $A = 5$. Hence we solve the following for t:

$$5 = 10 e^{-.0223144 t}$$
$$t = \frac{\ln .5}{-.0223144} \approx 31.0633$$

so the half-life is about 31 days.

(b) $A \approx 10 e^{-.0223144(365)} \approx .00290296$ gram.

Note: The only essential difference between the formulas for exponential growth and decay is the sign in the exponent. The sign is positive for growth and negative for decay.

EXERCISES

1. Compute $\left(1 + \dfrac{2}{n}\right)^n$ for $n = 1, 10, 100, 1000, 10^6$. Then compute e^2 and compare.

2. Compute $\left(1 + \dfrac{-2}{n}\right)^n$ for $n = 1, 10, 100, 1000, 10^6$. Then compute e^{-2} and compare.

3. Compute $\left(1 + \dfrac{-20}{n}\right)^n$ for $n = 1, 10, 100, 1000, 10^6$. Then compute e^{-20} and compare.

4. Compute $\left(1 + \dfrac{.03}{n}\right)^n$ for $n = 1, 10, 100, 1000, 10^6$. Then compute $e^{.03}$ and compare.

5. Suppose $5000 is invested in an account which pays 8%. How much is this worth after 10 years if it is compounded
   a. Annually?   b. Quarterly?   c. Daily?   d. Continuously?

6. Suppose you had $1000 to invest, and you wanted to double it in 8 years. At what rate of interest would you have to invest if it was compounded
   a. Annually?   b. Quarterly?   c. Daily?   d. Continuously?

7. Suppose you had $10,000 to invest, and you wanted to triple it in 14 years. At what interest rate would you have to invest if it was compounded
   a. Annually?   b. Semiannually?   c. Monthly?   d. Continuously?

8. Suppose $10,000 is invested in an account which pays 9.25%. How much is this worth after 5 years if it is compounded
   a. Annually?   b. Monthly?   c. Daily?   d. Continuously?

9. What is the best way to invest $1000 for 3 years: **(a)** $9\frac{1}{2}\%$ compounded annually, **(b)** $9\frac{3}{8}\%$ compounded quarterly, or **(c)** $9\frac{1}{4}\%$ compounded continuously?

10. What is the best way to invest $5000 for 2 years: **(a)** 10% compounded annually, **(b)** $9\frac{7}{8}\%$ compounded monthly, or **(c)** $9\frac{3}{4}\%$ compounded continuously?

11. Suppose you are going to need $20,000 in 15 years. How much do you have to invest now in an account that pays 8.5% compounded
    a. Annually?   b. Quarterly?   c. Weekly?   d. Continuously?

12. A company has to pay off a loan of $2 million in 5 years. To cover this loan, how much would it have to set aside right now if it could earn 10.5% compounded
    a. Annually?   b. Monthly?   c. Semimonthly?   d. Continuously?

13. Suppose a $100,000 piece of machinery is depreciating at 10% a year.
    a. How much is it worth after 3 years?
    b. How long will it take to be worth only $20,000?

14. Suppose a Thunderbird depreciates at 20% a year and an Edsel depreciates at 60% a year. How long does it take for each one to be worth half of its original price? How much of their original price are they worth after 3 years?

15. A woman buys an apartment house for $5 million as a tax shelter. She wants to depreciate the building at such a rate that it will be worth only $1 million after 7 years, when she will sell it. What rate of depreciation should she claim on her income tax form? If she wanted to claim 15% depreciation per year, how long would it take to depreciate to $1 million?

16. Suppose 100 mice (of both sexes) got free in a psychology lab and the population began to grow exponentially.
    a. If there were 110 mice after 1 week, in how many weeks would the population double?

b. If the building could hold at most 2000 mice, how long would it take before they began falling out the windows?

17. Suppose a population of ants is growing exponentially. On June 1, there are 800 ants, and on July 4 of the same year, there are 1000.
  a. How many ants will there be on September 15 of the same year?
  b. On what day will the population have doubled?

18. If you start with 1 gram of a certain isotope, you have .9 gram after a week.
  a. Find the formula that gives you how much is left after t days.
  b. How much is left after a year (365 days)?
  c. What is the half-life of this isotope?
  d. How long will it be until there is 1/10 gram left?

19. Suppose a certain radioactive substance has a half-life of 21 days. Suppose you start with 50 grams of this substance.
  a. Find a formula that gives you how much is left after t days.
  b. How long will it take until there are 5 grams left?
  c. How much is left after a year (365 days)?

20. Suppose the atmospheric pressure at sea level is 15 pounds per square inch and the pressure is halved every 3 miles of vertical ascent.
  a. Find the exponential equation which expresses this.
  b. Find the pressure at 10 miles above sea level.

21. A certain satellite has a power supply whose output in watts is given by the equation $P = 40e^{-t/900}$, where t is the number of days the battery has operated.
  a. If it is operated continuously after the satellite is placed into orbit, how many watts is the battery putting out after 1 year?
  b. If it takes at least 10 watts to operate the satellite, how many days can the satellite be used?

22. After a person is about 25 years old, his basal metabolism slowly decreases. Let N be the number of calories per day that a person needs and $M = N - 1200$. For the average person, M decreases exponentially at the rate of 10% every 7 years.
  a. Find a formula expressing N as a function of time t in years.
  b. Suppose a certain person requires about 2500 calories per day when he is 25 years' old. If he does not increase his physical activity as he gets older, how many calories will he require per day when he is 40 years old? 60 years old?
  c. Graph the function obtained in part a as applied to the person in part b.

23. The simple circuit in Fig. 8 consists of a constant voltage E, an induction of L henries, and a resistance of R ohms. The current I at any time t is given by

$$I = \frac{E}{R}(1 - e^{-(R/L)t})$$

Suppose for a given circuit that $E = 120$, $R = 50$, and $L = 40$.
  a. Find the current I when $t = 2, 10,$ and 100.
  b. Graph this function.
  c. For what time t will $I = 2$? $I = 4$?

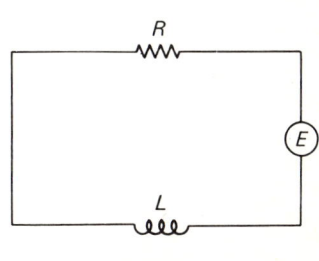

FIG. 8

24. The pH of a solution is defined by
$$\text{pH} = -\log[\text{H}^+]$$
where $[\text{H}^+]$ is the concentration of hydrogen ions in solution (in moles per liter).
   a. Find the pH of a solution for which $[\text{H}^+] = 8.321 \times 10^{-7}$.
   b. Find $[\text{H}^+]$ when the pH of a solution is 4.61.

25. The magnitude M of an earthquake is usually measured on the Richter scale. The intensity I of an earthquake (i.e., how strong the earthquake is) and M are related by the formula $M = \log(I/I_0)$, where I_0 was the intensity of an arbitrarily chosen earthquake. If one earthquake which measured 4.2 on the Richter scale hit San Francisco and a day later a second one hit which measured 5.7, how much stronger was the second than the first? [*Hint*: Translate $M = \log(I/I_0)$ to the equivalent exponential equation.]

26. The *logistic function*
$$P = \frac{aP_0 e^{at}}{bP_0 e^{at} + (a - bP_0)}$$
is another population growth function (which also occurs in the study of the spread of information).
   a. Show that $P = P_0$ when $t = 0$.
   b. If $a = 3$, $b = .002$, and $P_0 = 1000$, find P when $t = .1$.
   c. Find t when $P = 1200$. (*Hint*: You might substitute $u = e^{at}$ and solve for u first.)

SECTION 5.5. TWO IMPORTANT FORMULAS AND THEIR APPLICATIONS*

There are two very important formulas relating interest and making deposits or payments over a period of time. Their derivations are indicated in Sec. 12.2, Exercises 45 and 46.

Suppose that you invest P dollars at the beginning of each time period (each month, each year, etc.) and that r is the rate of interest per period. Then after n time periods the total amount A of money invested plus interest earned is

(22)   FORMULA 1
$$A = P\left(1 + \frac{1}{r}\right)[(1 + r)^n - 1].$$

EXAMPLE 1 Suppose that your local savings bank pays 5% per year compounded monthly and that at the beginning of each month you deposit $100. How much do you have after 5 years? How much did you deposit? How much interest did your money earn?

SOLUTION The time period is a month. Since .05 (i.e., 5%) is the interest rate per year, the interest rate per month is $r = .05/12$. The number of time periods is

*This section is optional.

$n = 12(5) = 60$. By Formula 1, the amount you have after 5 years is

$$A = \$100\left(1 + \frac{12}{.05}\right)\left[\left(1 + \frac{.05}{12}\right)^{60} - 1\right] \approx \$6828.98$$

You deposited $100 per month for 60 months, or $6000, so your money earned $828.98 in interest.

For the second formula, suppose that an amount A of money is borrowed and that it will be paid back by installment payments, one at the end of each time period for n periods. Let P = the amount of each payment per period, let r = the interest rate per period, and assume the interest is compounded at the end of each period.

(23) **FORMULA 2**

$$A = P\frac{1 - (1 + r)^{-n}}{r}.$$

Note: The A in Formula 1 represents the amount of money *accumulated* at the end; the A in Formula 2 represents the amount of money at the *beginning*.

Note: With an installment plan payment, the interest due is taken out first, and the remainder reduces the amount borrowed. This has the effect of compounding the interest each period.

EXAMPLE 2 Suppose you borrow $3800 for a new car at 11% interest for 36 months with monthly payments (hence compounded monthly). **(a)** How much are your payments? **(b)** How much do you pay altogether? **(c)** How much interest do you pay?

SOLUTION
(a) Here, $r = .11/12$, $n = 36$, and you wish to solve

$$\$3800 = P\frac{1 - (1 + .11/12)^{-36}}{.11/12}$$

so

$$P = \$3800\frac{.11/12}{1 - (1 + .11/12)^{-36}} \approx \$124.41$$

(b) You pay $124.41(36) = $4478.76 altogether.
(c) You pay $4478.76 − $3800 = $678.76 in interest.

Note: If you made payments for 48 months instead, the answers would be **(a)** $98.21 **(c)** $914.22. *Be aware of the differences.*

Section 5.5 Two Important Formulas and Their Applications

EXAMPLE 3 Suppose you buy a new washer and dryer for $441.35. You can afford to pay $20 a month. How many payments would you make if you **(a)** Charged it on a charge card at 18% per year = 1.5% per month? **(b)** Borrowed the money from a credit union at 12% per year = 1% per month?

SOLUTION
(a) Here $A = \$441.35$, $P = \$20$, $r = 1.5\%$, and you want to solve for n in

$$441.35 = 20 \frac{1 - (1 + .015)^{-n}}{.015}$$

So

$$441.35(.015)/20 = 1 - (1.015)^{-n}$$

$$(1.015)^{-n} = 1 - 441.35(.015)/20$$

$$n = \frac{-\ln[1 - 441.35(.015)/20]}{\ln 1.015} \approx 26.9998 \approx 27$$

Thus there are 27 payments. **(b)** Just change all the .015s to .01, and you get

$$n = \frac{-\ln[1 - 441.35(.01)/20]}{\ln 1.01} = 25.057 \approx 25$$

So there are 25 payments (your twenty-fifth payment will be slightly more).

EXERCISES

1. In Example 2, suppose you made payments semimonthly for 36 months. What are your answers to parts (a)–(c) then?

2. Suppose you deposit $10 each week in a savings account that pays 5% annually compounded weekly. How much do you have at the end of a year? Two years? How much interest did your money earn the first year?

3. Suppose you buy a home and take out a mortgage for $20,000 at 9% per year, with monthly payments (compounded monthly). Answer the following questions if it is a 20-year, 25-year, or 30-year mortgage.
  a. How much is each monthly payment?
  b. How much do you pay altogether?
  c. How much interest do you pay?

4. As in Example 3, suppose that you buy a color TV for $300 and that you make payments of $20 per month. How many payments would you make
  a. If you charge it on a charge card at 18% per year?
  b. If you borrow the money from your credit union at 12% per year?

5. Suppose that you put an amount A into an account that compounds interest at a rate $= r$ each time period and that you wish this account to pay you P dollars at the end of each period for n periods (at which time A will be all used up).
   a. Explain why the formula relating A, P, r, and n is exactly the same as Formula 1. (*Hint*: Change places with the bank.)
   b. If you put $10,000 into a 6% annual account, compounding monthly, how much could it pay you at the end of each month for the next 30 years (and be all used up then)?
   c. Could this account pay out a fixed amount at the end of each month forever? (If that is too long, say 1000 years.)
   d. If in part b you wanted $100 a month, how long would it last?

6. Suppose you win $100,000 in the lottery, but the way the state pays you is $10,000 now and $10,000 at the end of each year for the next 9 years. Suppose that the way the state does this, after making that first payment, is to put a fixed sum into a savings account paying $7\frac{1}{2}\%$ compounded annually and that then that savings account pays you the $10,000 at the end of each year (and that the account is empty after the 9 years).
   a. How much does the state have to set aside?
   b. How much money does the state save (and you lose) doing it this way rather than all at once?

7. I have just received a $50,000 inheritance. I wish to use this to increase my income for the next 25 years by giving me a check at the end of each month. Suppose I can invest the money in an account which pays 7% per year, compounded monthly (and which will send me a monthly check).
   a. If I want the maximum amount I can get each month and have the $50,000 all used up at the end of the 25 years, how much will I get each month?
   b. If instead I only want $300 per month, how much is left at the end of the 25 years? (*Hint*: First compute how much of the $50,000 I would have to set aside just to give me the $300 per month.)

8. A large corporation sells bonds worth $20 million. To buy back these bonds in 30 years, it sets up an account (called a **sinking fund**) into which it will make a payment at the end of each quarter, starting 3 months from now. (Of course, it pays interest on the bonds for the 30 years, too.)
   a. If the account pays 8.5% compounded quarterly, how much does the corporation have to deposit each quarter to have the $20 million in 30 years?
   b. How much will it actually have paid into the sinking fund?
   c. How much interest will the fund have earned?

9. A company wants to **float** (i.e., to sell) a 20-year bond issue. It estimates it can afford to pay into a sinking fund (see Exercise 8) $20,000 at the end of each quarter. Suppose the sinking fund pays 9.2% compounded quarterly.
   a. What is the largest amount of money the company can raise?
   b. If the company wanted to raise $8 million but still only pay $20,000 per month into the sinking fund, how long do the bonds have to be issued for?

10. A town is about to put in a sewer system at a cost of $1.5 million. They will pay for it by selling 10-year bonds and then assessing all property owners a special semiannual sewer tax to be deposited into a sinking fund. Suppose the sinking fund will pay 8.5% compounded semiannually.
   a. How much will the total semiannual assessment for the whole town have to be?
   b. How much will the property owners have to pay over the 10-year period?
   c. How much interest will the sinking fund earn altogether?

11. Suppose you buy a new bedroom suite for $600. You put $100 down and borrow the balance, paying off the loan in monthly installments for 3 years.
   a. If you borrow the balance from the furniture store which charges 18% annual interest, compounded monthly, what are your payments?
   b. If you borrow the balance from a credit union which charges 12% annual interest, compounded monthly, what are your payments?
12. Suppose in Exercise 11 that instead of paying off the loan in exactly 3 years, you decide to pay exactly $30 at the end of each month until the loan is paid off. How many months will it take to pay off the loan if you borrow the $500 from
   a. The furniture store?   b. The credit union?
13. Suppose that you and your spouse are living in an apartment and that you decide you want to save $8000 for a down payment on a house in 3 years. If you make monthly payments into a savings account, how much do you have to save each month if the account pays
   a. 5% compounded monthly?   b. 7% compounded monthly?
14. If in Exercise 13 you decide you can save $500 each month, how many months does it take you to save the $8000 if the account pays
   a. 5% compounded monthly?   b. 7% compounded monthly?

SECTION 5.6. EXPONENTIAL AND LOGARITHMIC EQUATIONS

In the preceding sections, we have been solving equations involving exponentials and logarithms. In this section we shall expand on those techniques.

Exponential Equations

When the unknown is in the exponent, we have already seen that taking logarithms of both sides often leads to a solution.

EXAMPLE 1 Solve for x: $4^{2x-1} = 7^{x+2}$.

SOLUTION
$$4^{2x-1} = 7^{x+2}$$
$$\ln 4^{2x-1} = \ln 7^{x+2}$$
$$(2x - 1) \ln 4 = (x + 2) \ln 7$$
$$2(\ln 4)x - \ln 4 = (\ln 7)x + 2 \ln 7$$
$$(2 \ln 4 - \ln 7)x = 2 \ln 7 + \ln 4$$
$$x = \frac{2 \ln 7 + \ln 4}{2 \ln 4 - \ln 7} \approx 6.38478$$

Sometimes it is necessary to do some algebra first.

EXAMPLE 2 Solve for x: $2 \cdot 3^{2x} = 4 \cdot 5^{3x}$.

SOLUTION
$$2 \cdot 3^{2x} = 4 \cdot 5^{3x}; \quad 2 \cdot 9^x = 4 \cdot 125^x;$$
$$\frac{9^x}{125^x} = \frac{4}{2}; \quad \left(\frac{9}{125}\right)^x = 2; \quad \ln\left(\frac{9}{125}\right)^x = \ln 2;$$
$$x = \frac{\ln 2}{\ln(9/125)} \approx -.263445$$

Substitution is a basic mathematical tool. Its use is to simplify problems, making them more manageable.

EXAMPLE 3 Solve for x: $(3^x + 3^{-x})/2 = 7$.

SOLUTION Let $u = 3^x$. Then the equation becomes $(u + u^{-1})/2 = 7$. We now solve this equation for u and then use the values obtained for u to solve $3^x = u$ for x:

$$u + \frac{1}{u} = 14, \qquad u^2 + 1 = 14u$$

$$u^2 - 14u + 1 = 0, \qquad u = \frac{14 \pm \sqrt{14^2 - 4}}{2} = 7 \pm \tfrac{1}{2}\sqrt{192}$$

Then, since $3^x = u$, we have

$$3^x = 7 + \tfrac{1}{2}\sqrt{192}, \qquad 3^x = 7 - \tfrac{1}{2}\sqrt{192}$$

Solving these for x yields

$$x = \frac{\ln\left(7 + \tfrac{1}{2}\sqrt{192}\right)}{\ln 3} \approx 2.39749$$

$$x = \frac{\ln\left(7 - \tfrac{1}{2}\sqrt{192}\right)}{\ln 3} \approx -2.39749$$

Substitution is also a powerful tool in logarithmic equations. The other tools used to solve logarithmic equations are the properties of Secs. 5.2, 5.3, and 5.4. Sometimes it is difficult to decide whether to use the properties or substitution in a particular problem. The key is the following:

(24)

1. If all the logarithms are functions of the same algebraic expression, try substitution $[u = \log_b(\ )]$.

2. If the logarithms are functions of different expressions, try the properties.

EXAMPLE 4 Solve for x: $[\log_3(x-1)]^2 - 3\log_3(x-1) = 4$.

SOLUTION Here, both $\log_3$'s are functions of $x - 1$, so we try substituting $u = \log_3(x-1)$. The equation becomes

$$u^2 - 3u = 4 \quad \text{or} \quad u^2 - 3u - 4 = 0$$

Factoring gives

$$(u - 4)(u + 1) = 0, \quad \text{so } u = 4, -1$$

Substituting back $\log_3(x - 1) = u$, we obtain

$$\log_3(x - 1) = 4 \quad \text{or} \quad \log_3(x - 1) = -1$$

Solving for x, we obtain

$$x - 1 = 3^4 \quad \text{or} \quad x - 1 = 3^{-1}$$
$$x = 3^4 + 1 = 82 \quad \text{or} \quad x = 3^{-1} + 1 = \tfrac{1}{3} + 1 = 4/3$$

Both answers $x = 82$ and $4/3$ check.

EXAMPLE 5 Solve for x: $2\log_3(x+1) - \log_3(x+4) = 2\log_3 2$.

SOLUTION Here, the $\log_3$'s are functions of different algebraic expressions, so we use the properties:

$$2\log_3(x+1) - \log_3(x+4) = 2\log_3 2$$
$$\log_3(x+1)^2 - \log_3(x+4) = \log_3 2^2$$
$$\log_3 \frac{(x+1)^2}{x+4} = \log_3 4$$
$$\frac{(x+1)^2}{x+4} = 4$$
$$x^2 + 2x + 1 = 4(x+4) = 4x + 16$$
$$x^2 - 2x - 15 = 0$$
$$(x-5)(x+3) = 0, \quad x = 5, -3$$

But $x = -3$ does not work in the original equation because $\log_3(-2)$ is undefined. Since $x = 5$ does check, it is the only solution.

EXERCISES

In Exercises 1–37, solve for x.

1. $2^x = 16$
2. $2^{x-1} + 2 = 34$
3. $3^x = 16$
4. $3^{x-1} + 1 = 34$
5. $\log_4(2x - 1) = 2$
6. $[\log_5(x + 1)]^2 = 4$
7. $\log x^2 = 1$
8. $\log_2 x = \log_2(x^2 - 2)$
9. $3^{2x} \cdot 3^{x^2} = 27$
10. $7^{-x} = 5 \cdot 7^x$
11. $4^{x+2} = 7^{x-1}$
12. $(10^x)^2 = 6$

13. $\log x + \log(x - 9) = 1$
14. $\log x - \log(x - 9) = 1$
15. $(\log_4 x)^2 + 15 \log_4 x = 16$
16. $[\log_2(x - 1)]^2 = 4 \log_2(x - 1)$
17. $3 \cdot 4^{2x} + 5 \cdot 4^x - 2 = 0$
18. $3^{2x} - 4 \cdot 3^{x+1} + 27 = 0$
19. $[\log_4(x + 1)]^2 - 3 \log_4(x + 1) - 4 = 0$
20. $9(\log_3 x^2)^2 + 17 \log_3 x^2 = 2$
21. $\log(x + 1) - \log(x - 1) = 1$
22. $\log(2x + 1) - \log(x - 2) = 1$
23. $\log x^2 = (\log x)^2$
24. $(\log_2 x)^2 - \log_2 x^2 = 3$
25. $\log \sqrt{x} = \sqrt{\log x}$
26. $2[\log_4(x - 1)]^2 - 2 \log_4 \sqrt{x - 1} = 1$
27. $3^x + 8 \cdot 3^{-x} = 9$
28. $2 \cdot 4^x + 4^{-x} = 9/2$
29. $\dfrac{4^x + 4^{-x}}{2} = 8$
30. $2^x + 2^{-x} = 0$
31. $4(2^x + 2^{-x}) = 17$
32. $4^{1+x} + 4^{1-x} = 10$
33. $\dfrac{2^x - 2^{-x}}{2^x + 2^{-x}} = \dfrac{1}{2}$
34. $\dfrac{3^x + 3^{-x}}{3^x - 3^{-x}} = 9$
35. $5^{2x} 7^{-3x} = 6$
36. $(2x)^5 (7x)^{-3} = 6$
37. $\log x^3 = (\log x)^3$

SECTION 5.7. LINEAR INTERPOLATION*

In the past when logarithms were taught without calculators, extensive logarithm tables were provided so that people could compute with logarithms. Thus, traditionally, table reading and linear interpolation have been part of the course. Although calculators now replace tables, some instructors still feel that interpolation is a useful skill that students should learn. This section fills that need.

A table of mortgage rates was chosen for this, because it is in a context that we have already discussed and because it is similar to tables that many people actually use. See Table 7.

TABLE 7
Table of Monthly Mortgage Payments Per $1000 of Loan

ANNUAL PERCENT	LENGTH OF MORTGAGE		
	20 YR	25 YR	30 YR
8%	$8.3644	$7.7182	$7.3376
9%	$8.9973	$8.3920	$8.0462
10%	$9.6502	$9.0870	$8.7757
11%	$10.3219	$9.8011	$9.5232
12%	$11.0109	$10.5322	$10.2861
13%	$11.7158	$11.2784	$11.0620

EXAMPLE 1 What would the payments be for a $40,000 mortgage for 20 years at 10%?

SOLUTION In Table 7, you look in the 20-year column and the 10% row and find the figure $9.6502. This means for each $1000 of mortgage, you pay $9.6502 each

*This section is optional.

month. Since the mortgage is to be for $40,000, you multiply,
$$40(\$9.6502) = \$386.008$$
so you will pay $386.01 each month.

Often when we have a table of data, the information we want is in between the entries of the table. We then use interpolation to approximate the information we want.

EXAMPLE 2 For the mortgage in Example 1, what would be the payments if you are given a 25-year mortgage at 11.3%?

SOLUTION *By Rough Interpolation*: The idea of rough interpolation is to pick numbers that are roughly accurate but that are nice enough so that you can do the computations *in your head*. We use linear interpolation, which we describe next, to obtain more accurate approximations.

TABLE 8

ANNUAL PERCENT	LENGTH OF MORTGAGE, 25 YR
11%	$9.8011
11.3%	?
12%	$10.5322

To estimate the rate per $1000 of loan by rough interpolation we use Table 7 to form Table 8. We pick 11% and 12% because 11.3% is between them, and these figures are listed in Table 7. Since 11.3% is a little closer to 11% than 12%, we guess that the rate per $1000 is a little closer to $9.8011 than to $10.5322, say estimate that the rate is roughly $10 per $1000. To solve the problem, we multiply this by 40, obtaining

$$40(\$10) = \$400$$

for the mortgage payment each month.

SOLUTION *By Linear Interpolation*: Linear interpolation is very similar to rough interpolation, but it uses ratios to obtain an estimate that is usually more accurate than a rough guess. We start by forming Table 9, which is essentially identical to Table 8.

TABLE 9

	PERCENT	25 YEARS	
1% { .3% { 11%	$9.8011 } d		
11.3%	? }	$.7311	
12%	$10.5322 }		

We next form ratios of differences from each side of the table and set the ratios

236 EXPONENTIAL AND LOGARITHMIC FUNCTIONS

equal:

$$\frac{\text{Smaller difference}}{\text{Larger difference}} = \frac{\text{Smaller difference}}{\text{Larger difference}}$$

$$\frac{.3\%}{1\%} = \frac{d}{\$.7311}$$

Solving for d, we obtain

$$d = .3(\$.7311) = \$.21933$$

which we round off to $d \approx \$.2193$. The value for the rate per \$1000 at 11.3% is then approximately

$$\$9.8011 + d = \$9.8011 + \$.2193 = \$10.0204$$

To solve the problem, we multiply this by 40, obtaining

$$\$10.0204(40) = \$400.816$$

so this gives an estimate of \$400.82 for the monthly mortgage payment.

It is interesting to compare the above estimates with the actual rate, which is \$10.0188 per \$1000 of loan, or \$400.75 per month for the mortgage payment. Thus the linear interpolation is accurate to the nearest dime.

It is helpful to have a geometric picture of linear interpolation. Suppose you have a function f whose values you know at some points; say they are given in a table. However, you wish to approximate a value between those you know. Then linear interpolation pretends that between the values you know the graph of the function is a straight line, and it uses this line to estimate the value. See Fig. 9.

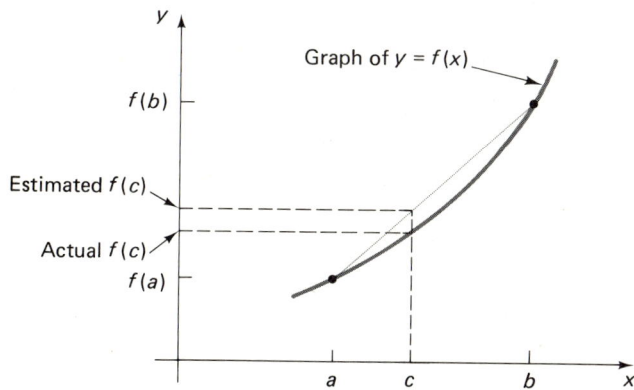

Linear interpolation: You know $f(a)$ and $f(b)$; estimate $f(c)$.

FIG. 9

EXERCISES

In Exercises 1–12, use Table 7. If interpolation is required, use both rough and linear interpolation. (You may use Formula 2 of Sec. 4.5 to compute the actual value if you wish.)

1. What would be the monthly mortgage payments for a $60,000 loan for 30 years at 8%?
2. What would be the monthly mortgage payments for a $45,000 loan for 20 years at 12%?
3. What would be the monthly mortgage payments for a $30,000 loan for 25 years at 13%?
4. What would be the monthly mortgage payments for a $70,000 loan for 25 years at 9%?
5. What would be the monthly mortgage payments for a $55,000 loan for 20 years at 12.2%?
6. What would be the monthly mortgage payments for a $15,000 loan for 30 years at 10.9%?
7. What would be the monthly mortgage payments for a $38,000 loan for 25 years at 8.6%?
8. What would be the monthly mortgage payments for a $110,000 loan for 30 years at 9.1%?
9. What would be the monthly mortgage payments for a $85,000 loan for 22 years at 8%?
10. What would be the monthly mortgage payments for a $62,000 loan for 26 years at 11%?
11. What would be the monthly mortgage payments for a $35,000 loan for 29 years at 9%?
12. What would be the monthly mortgage payments for a $74,000 loan for 24 years at 13%?
13. A 25-year mortgage for $32,000 has monthly payments of $281.89. What is the annual interest rate?
14. A 30-year mortgage for $60,000 has monthly payments of $582.84. What is the annual interest rate?
15. A 20-year mortgage for $47,000 has monthly payments of $478.82. What is the annual interest rate?
16. A 25-year mortgage for $26,000 has monthly payments of $276.75. What is the annual interest rate?

REVIEW EXERCISES

1. Compute $3.68^{-4.1}$.
2. Compute $11.5\sqrt{5.9^3 - e^{2.3}}$.
3. Use the graph of $y = 5^x$ to estimate $5^{.3}$, and then compute the number on your calculator and compare.
4. Use rough interpolation to estimate $2^{4.6}$, and then compute the number on your calculator and compare.
5. Without using your calculator, find $\log_{81} 3$ and $\log_3 81$.
6. Use rough interpolation to estimate $\log_5 87$, and then compute the number on your calculator and compare.

In Exercises 7–12, solve for x without using a calculator.

7. $\log_2 32 = x$
8. $\log_3 x = -3$
9. $\log_x 8 = 3$
10. $\log_6 6^{4x-1} = 19$
11. $3.5^{\log_{3.5} 2x^2} = 98$
12. $\log_7(2x + 1) = \log_7(3x - 8)$

In Exercises 13–20, sketch the graph.

13. $y = 5^x$
14. $y = (\frac{1}{5})^x$
15. $y = 5^{x-2}$
16. $y = 5^x - 2$

238 EXPONENTIAL AND LOGARITHMIC FUNCTIONS

17. $y = \log_5 x$
18. $y = \log_5(-x)$
19. $y = \log_5(x - 2)$
20. $y = (\log_5 x) - 2$

In Exercises 21–24, simplify without using your calculator.

21. $\log_2 68 - \log_2 17$
22. $\log_9 \dfrac{3}{\sqrt{2}} + \log_9 \sqrt{2}$
23. $\log_4 4^{x^2-4x}$
24. $11^{2\log_{11} 3 + \log_{11} 2}$

25. Using $\log_5 2 \approx .43$ and $\log_5 3 \approx .68$, compute $\log_5 \frac{2}{3}$, $\log_5 4$, and $\log_5 60$.

In Exercises 26–33, solve for x.

26. $\log_5(x - 2) + \log_5(x + 2) = 1$
27. $3^{2x+1} = 5.6$
28. $\ln(4.85x) = 5.981$
29. $(2x + 1)^3 = 5.6$
30. $[\log_3(x + 2)]^2 - 7\log_3(x + 2) = -12$
31. $\dfrac{5^x + 5^{-x}}{5^x - 3(5^{-x})} = 3$
32. $3^{5x} 4^{-2x} = 7$
33. $4^{2x} - 4^{x+1} - 21 = 0$

34. If $A = P(1 + r)^n$, $P = 1000$, $r = .01$, and $n = 24$, find A.

35. Suppose $5000 is invested in an account which pays 10.5%. How much is it worth after 4 years if interest is compounded
   a. Annually? b. Monthly? c. Daily? d. Continuously?

36. If you have $10,000 to invest and you want to triple it in 16 years, at what interest rate would you have to invest if interest is compounded
   a. Annually? b. Quarterly? c. Daily?

37. You have purchased a car for $9600. It depreciates 18% each year. How much is it worth after 4 years? How long does it take to be worth $4800?

38. Suppose a population of flies is growing exponentially. On May 1, there are 500 flies, and on May 18 there are 600 flies.
   a. How many flies will there be on May 31?
   b. On what day will the population have doubled?

39. Suppose you start with 100 grams of a radioactive substance which has a half-life of 25 days.
   a. How many grams are left after 40 days?
   b. How long will it take until there are 5 grams left?

40. Suppose you deposit $60 into a savings account at the beginning of each month. The account earns interest at a rate of 7.5% annually compounded monthly. How much do you have after 6 years? How much did you deposit? How much interest did your money earn?

41. Suppose you purchase a lot for $12,000. You put $1000 down and borrow the rest at 13% for 5 years with monthly payments (compounded monthly).

42. Use Table 7 in Sec. 5.7 and linear interpolation to find the monthly mortgage payments for a $40,000 loan for 25 years at 12.6%.
   a. How much is each monthly payment?
   b. How much do you pay altogether?
   c. How much interest do you pay?

43. Use Table 7 in Sec. 5.7 and linear interpolation to find the monthly mortgage payments for a $50,000 loan for 21 years at 11%.

CHAPTER SIX
Systems of Equations and Inequalities

The solving of simultaneous linear equations occurs in the earliest of recorded mathematical history, but matrices and determinants are fairly modern inventions of Western mathematics.

In the Far East, however, a Chinese text written around 250 B.C., *Chui-chang suan-shu*, used matrices to solve linear systems in essentially the same way we do in Sec. 6.4. This is one of many examples of the great intellectual promise of early Eastern culture. However, Eastern intellectual development suffered several serious setbacks, such as the ordering of the burning of books in 213 B.C. by the Chinese emperor.

In the West, it was not until 1858 that Arthur Cayley (1821–1895) invented matrices and started the study of the algebra of matrices. It was 60 years later in 1925 that Heisenberg recognized that this abstract mathematics was exactly the tool he needed to describe his revolutionary work in quantum mechanics.

Determinants were first suggested by Leibniz in 1693, but their properties were discovered later by several mathematicians. In 1750, Gabriel Cramer (1704–1752) published *Cramer's Rule*, but it was probably known to Colin Maclaurin (1698–1746) in 1729. Cramer's notation was better, however, and this is probably why the method was associated with him.

SECTION 6.1. SYSTEMS OF TWO EQUATIONS IN TWO VARIABLES

In this section, we develop techniques for solving two equations in two variables. By a system of two equations in x and y we mean any pair of equations having x and y as variables. A **solution** to such a system is an ordered pair of numbers (a, b) such that when a and b are substituted for x and y, respectively, a true statement is obtained for each equation. Two systems are **equivalent** if they have the same solutions.

Recall that the graph of an equation in x and y is the set of all points in the plane whose coordinates satisfy the equation. Hence a pair (a, b) is a solution to a system of two equations if and only if the point (a, b) is in the intersection of the two graphs. Sometimes it is helpful to examine the graphs to help us understand the nature of the solutions.

EXAMPLE 1 Solve the systems of equations:

(a) $y = 4 - x^2$
$y = 2x + 1$

(b) $y = 4 - x^2$
$y = 2x + 5$

(c) $y = 4 - x^2$
$y = 2x + 6$

SOLUTION Although these three systems are similar, we shall see that the nature of their solutions is quite different. We first solve each system algebraically by substituting $y = 4 - x^2$ from the first equation in for y in the second equation.

(a) $y = 4 - x^2$
$y = 2x + 1$

(b) $y = 4 - x^2$
$y = 2x + 5$

(c) $y = 4 - x^2$
$y = 2x + 6$

$$4 - x^2 = 2x + 1$$
$$0 = x^2 + 2x - 3$$
$$0 = (x + 3)(x - 1)$$
$$x = 1, -3$$

When $x = 1$,
$$y = 4 - x^2 = 3$$
When $x = -3$,
$$y = 4 - x^2 = -5$$

Ans.: $(x, y) = (1, 3), (-3, -5)$.

$$4 - x^2 = 2x + 5$$
$$0 = x^2 + 2x + 1$$
$$0 = (x + 1)^2$$
$$x = -1$$

When $x = -1$,
$$y = 4 - x^2 = 3$$

Ans.: $(x, y) = (-1, 3)$.

$$4 - x^2 = 2x + 6$$
$$0 = x^2 + 2x + 2$$
No real solution exists.

These three systems are graphed in Fig. 1.

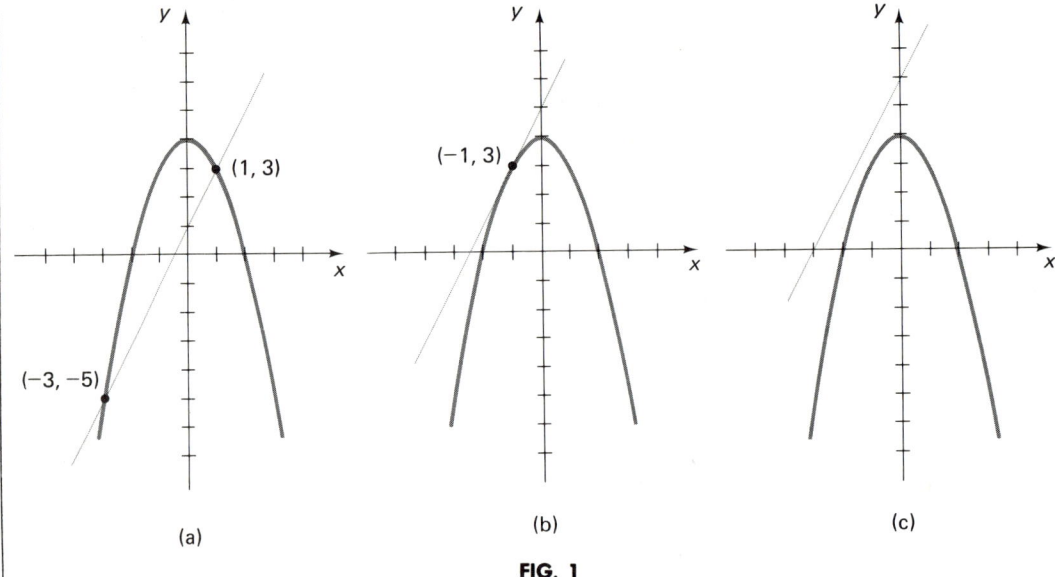

FIG. 1

The geometry helps explain why the algebraic solutions are different. For part (a), the graphs intersect at two points, so there are two solutions for the system. For part (b), the graphs intersect at one point, so the system has one solution. For part (c), the graphs do not intersect, so there are no real solutions.

We now turn to the problem of finding the solutions to a system of two equations in two variables. The primary method is called the **method of substitution**. The process is as follows:

1. Solve one of the equations for one variable in terms of the other. You may pick either variable in either equation, so you should look over both equations and pick the easiest one to work with.

2. Substitute the expression obtained in step 1 into the other equation. You now have an equation in one variable.
3. Solve the equation obtained in step 2.
4. One at a time, substitute each solution obtained in step 3 back into the expression obtained in step 1 to obtain the corresponding value for the other variable.

EXAMPLE 2 Solve the system

$$2x - 3y = 14$$
$$2x^2 + y^2 = 36$$

SOLUTION When one equation is linear and the other is of degree 2 in both variables, pick one of the variables in the linear equation. In this case, we solve the first equation for x:

$$x = \tfrac{3}{2}y + 7$$

$$2(\tfrac{3}{2}y + 7)^2 + y^2 = 36$$

$$2(\tfrac{9}{4}y^2 + 21y + 49) + y^2 = 36$$

$$11y^2 + 84y + 124 = 0$$

$$(y + 2)(11y + 62) = 0, \quad y = -2, -62/11$$

When $y = -2$,

$$x = \tfrac{3}{2}(-2) + 7 = 4$$

When $y = -62/11$,

$$x = \tfrac{3}{2}(-\tfrac{62}{11}) + 7 = -\tfrac{16}{11}$$

Thus the answers are $(x, y) = (4, -2), (-16/11, -62/11)$, both of which check.

Warning: In Example 3, we did *not* write $x = 4, -16/11, y = -2, -62/11$, as this suggests that there are four answers: $(x, y) = (4, -2), (4, -62/11), (-16/11, -2)$, and $(-16/11, -62/11)$.

EXAMPLE 3 Solve the system

$$x^2 + y^2 = 4$$
$$y - x^2 = -4$$

SOLUTION Since there is an x^2 in both equations, one way to solve this system is to solve one of the equations for x^2 and substitute the answer into the other equation. We pick the second equation, but they are both equally easy to use:

$$y + 4 = x^2$$

$$(y + 4) + y^2 = 4$$

$$y^2 + y = 0$$

$$y(y + 1) = 0, \quad y = 0, -1$$

When $y = 0$, $x^2 = 4 + 0 = 4$, so $x = \pm 2$.
When $y = -1$, $x^2 = 4 + (-1) = 3$, so $x = \pm \sqrt{3}$.
Thus there are four answers: $(x, y) = (2, 0), (-2, 0), (\sqrt{3}, -1), (-\sqrt{3}, -1)$, which all check.

Some systems are more easily solved by first letting u represent some quantity.

EXAMPLE 4 Solve the system
$$y = 4^{2x} + 19$$
$$y = 17(4^x) + 3$$

SOLUTION First let $u = 4^x$. The system becomes
$$y = u^2 + 19$$
$$y = 17u + 3$$

The first equation is already solved for y, so we substitute that into the second equation, obtaining
$$u^2 + 19 = 17u + 3$$

Solving for u, we obtain
$$u^2 - 17u + 16 = 0$$
$$(u - 16)(u - 1) = 0$$
$$u = 1, 16$$

Since $4^x = u$, $x = \log_4 u$. We also know that $y = 17u + 3$.
When $u = 1$, $x = \log_4 1 = 0$, and $y = 17(1) + 3 = 20$.
When $u = 16$, $x = \log_4 16 = 2$, and $y = 17(16) + 3 = 275$.
Thus the two answers are $(x, y) = (0, 20), (2, 275)$.

Sometimes the graphs will easily show that a system has no solution, whereas it is difficult to determine that fact algebraically.

EXAMPLE 5 Solve the system
$$y = 3^x$$
$$x = 3^y$$

SOLUTION This system is equivalent to
$$y = 3^x$$
$$y = \log_3 x$$

The two graphs are sketched in Fig. 2(a). It is clear that the graphs do not intersect, so the system has no solution.

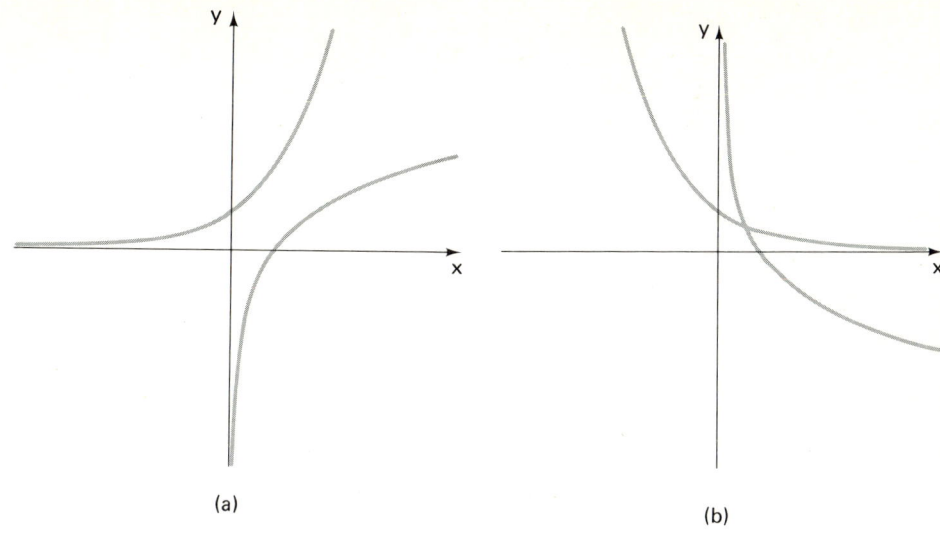

FIG. 2

Sometimes the graphs will show that there is a solution and its approximate value, even when it is difficult to determine the solutions algebraically.

EXAMPLE 6 Determine the nature of the solutions to the system

$$y = \left(\tfrac{1}{3}\right)^x$$
$$x = \left(\tfrac{1}{3}\right)^y$$

SOLUTION This system is equivalent to

$$y = \left(\tfrac{1}{3}\right)^x$$
$$y = \log_{1/3} x$$

Their graphs are sketched in Fig. 2(b). By the graphs, it is clear that there is exactly one solution, at approximately $x = .5$. By the symmetry of these equations, a solution (x, y) is on the line $y = x$.

EXERCISES

In Exercises 1–20, solve the given system, if possible.

1. $y = x^2 + 1$
   $y = x + 1$
2. $y = x^2 - x$
   $x + y = 9$
3. $x + y^2 = 2$
   $2x + 3y = -5$
4. $2x - y^2 = 2$
   $y + 4x = 14$
5. $x = y^2$
   $-x = y + 2$
6. $y + x^2 = 0$
   $y - 4x = 1$

7. $2x - y = 5$
$3x - 2y = 7$

8. $2x + 3y = 9$
$2x - 5y = -31$

9. $xy = 6$
$x + 3y = 11$

10. $xy = -4$
$2x + y = -7$

11. $x + 2y = 3$
$x^2 + y^2 = 9$

12. $2x - 3y = 0$
$x^2 + y^2 = 4$

13. $x^2 + y^2 = 20$
$x + y = 1$

14. $x^2 + y^2 = 10$
$x - y = 2$

15. $x^2 + y^2 = 4$
$2x^2 + y^2 = 5$

16. $3x^2 + 4y^2 = 3$
$x^2 + y^2 = 1$

17. $x^2 + y^2 = 9$
$3x^2 + y^2 = 3$

18. $x^2 + y^2 = 1$
$9x^2 + 2y^2 = 10$

19. $2x^2 + y^2 = 8$
$y - x^2 = 0$

20. $x^2 - y^2 = 6$
$x - y^2 + 6 = 0$

21. Find the values of k when the system
$$y - 2x = k$$
$$x^2 - y = 0$$
has
  a. No real solution
  b. One real solution
  c. Two real solutions
Interpret the answer geometrically.

22. Find the values of k when the system
$$x^2 + y^2 = 9$$
$$y = x + k$$
has
  a. No real solution
  b. One real solution
  c. Two real solutions
Interpret the answer geometrically.

In Exercises 23–32, solve the systems of equations, if algebraically possible. If not, try graphing to determine the nature of the solutions.

23. $y = 3^x + 1$
$y = 2 \cdot 3^x - 8$

24. $y = 3^{2x}$
$y = x^3 + 2$

25. $y = 2^x$
$x = 2^y$

26. $y = \log x$
$x = \log y$

27. $y = 2^x$
$x = 2^{y-2}$

28. $y = 2^x - 2$
$x = 2^y$

29. $y = (\frac{1}{2})^x$
$x = (\frac{1}{2})^y$

30. $y = (\frac{1}{2})^x$
$x = (\frac{1}{2})^y + 1$

31. $2^x = 3^y$
$9 \cdot 2^x = 6^y$

32. $2^x + y = 1$
$x + 2^y = 1$

33. a. Show that if $x = a, y = b$ is the solution to Exercise 29, $a = b$. That is, $a = (\frac{1}{2})^a$ and $a = \log_{1/2} a$.
  b. Devise a method to find a to three decimal places on your calculator.

SECTION 6.2. LINEAR EQUATIONS IN TWO VARIABLES

In Sec. 6.1, we studied systems of two equations in two unknowns and used substitution as a method of solving them. In this section, we restrict our attention to linear equations and develop a method for this special type that is usually quicker and easier than substitution. This method will be extended to solve three or more linear equations (in three or more variables) in the following sections. The method is called **Gaussian elimination**.

The two systems

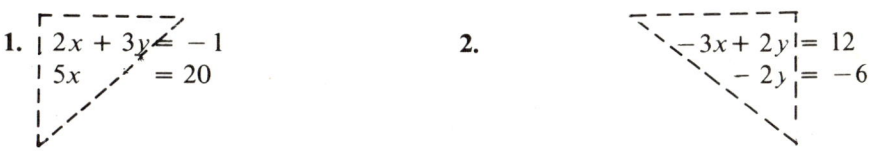

are said to be in **triangular form**, because of the shape of the system on the left-hand side of the equal signs (as indicated by the dashed lines). It is very easy to solve systems in triangular form. You solve the second equation for the variable which appears and then substitute that value back into the first equation and solve for the other variable. This process is called **backsubstitution**. For example, to solve system 2,

$$-2y = -6, \quad y = 3$$
$$-3x + 2(3) = 12, \quad -3x = 6, \quad x = -2$$

Suppose, now, we are given an arbitrary system of linear equations to solve. Our method is first to transform the system to an equivalent triangular system which we can then easily solve by backsubstitution. Unfortunately, it takes longer to describe and justify the process than to actually do it. So we first give some examples and then formalize the process.

EXAMPLE 1 Solve the system

$$x + 4y = -10$$
$$2x - 3y = 13$$

SOLUTION Multiply the first equation by -2:

$$-2x - 8y = 20$$
$$2x - 3y = 13$$

Replace the second equation with the sum of the two equations:

$$-2x - 8y = 20$$
$$-11y = 33$$

This system is now in triangular form, so we solve it as before:

$$-11y = 33, \quad y = -3$$
$$-2x - 8(-3) = 20, \quad -2x = -4, \quad x = 2$$

We check our solution $(x,y) = (2, -3)$ by substituting back into the original equations:

$$x + 4y = -10: \quad 2 + 4(-3) = 2 - 12 = -10$$
$$2x - 3y = 13: \quad 2(2) - 3(-3) = 4 + 9 = 13$$

The solution checks.

Section 6.2 Linear Equations in Two Variables

Note: It only took two steps to reach triangular form. But we could have reached triangular form in one step. This is accomplished by replacing the second equation by "the second equation plus -2 times the first equation." (This multiplying and adding is done term by term. The first equation is left alone.) Thus the one step would be

$$x + 4y = -10 \qquad x + 4y = -10$$
$$2x - 3y = 13 \qquad -11y = 33$$

Now solve as before.

EXAMPLE 2 Solve the system

$$2x - 9y = 7$$
$$5x + 6y = 8$$

SOLUTION Multiply the first equation by 5 and the second equation by -2:

$$10x - 45y = 35$$
$$-10x - 12y = -16$$

Replace the second equation with the sum of the two equations:

$$10x - 45y = 35$$
$$ -57y = 19$$

This system is in triangular form, so we solve it:

$$y = -\tfrac{1}{3}, \quad 10x - 45(-\tfrac{1}{3}) = 35, \quad 10x = 20, \quad x = 2$$

Checking the solution $(x, y) = (2, -\tfrac{1}{3})$ in the original problem, we find

$$2(2) - 9(-\tfrac{1}{3}) = 4 + 3 = 7$$
$$5(2) + 6(-\tfrac{1}{3}) = 10 - 2 = 8$$

It checks.

We now formalize what we have been doing. Recall that two systems are equivalent if they have exactly the same solutions. Our method of solution is to use certain operations in order to transform the original system through a sequence of systems, ending with a triangular system. At each step the operations used are such that the system obtained is equivalent to the previous system. Consequently the final system is equivalent to the first, and hence solving the final system (by backsubstitution) gives all the solutions to the original system. The operations which may be used are described in the following theorem:

(1)   **THEOREM**

The following operations transform a system of equations into an equivalent system:
 i. Interchanging the position of any two equations

ii. Multiplying both sides of an equation by a nonzero number

iii. Replacing an equation with that equation plus a multiple of another equation of the system

We leave a formal proof to a course in linear algebra. The main idea is to observe that each step is reversible so that no solutions are introduced or lost.

We now examine geometrically the nature of the solutions to a system of two linear equations in two variables. As shown in Sec. 3.5, the graph of a linear equation is a straight line. We know that if two different lines intersect, they intersect in exactly one point. Thus if the two linear equations represent different lines which intersect, the system of those two equations has exactly one solution. When a system of linear equations has one or more solutions, it is called **consistent**. When it has exactly one solution, the system is called **consistent and independent**, which we usually shall shorten to **independent**. Therefore, if the two linear equations represent different lines which intersect, the system is (consistent and) *independent*.

EXAMPLE 3 Graph the system

$$x + 4y = -10$$
$$2x - 3y = 13$$

SOLUTION We rewrite the equations in slope-intercept form,

$$y = -\tfrac{1}{4}x - \tfrac{5}{2}$$
$$y = \tfrac{2}{3}x - \tfrac{13}{3}$$

and then graph them in Fig. 3(a). The solution is $(2, -3)$. This system is independent.

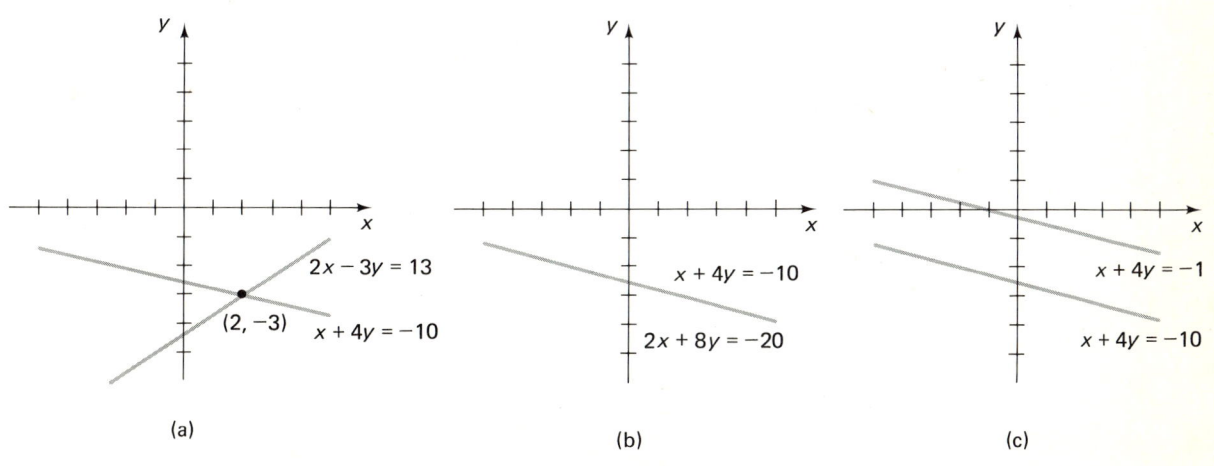

FIG. 3

If the graphs of two equations are the same line, then every point on that line represents a solution to the system. Therefore the system has an infinite number of solutions and is called **consistent and dependent**, which we usually shall shorten to **dependent**.

EXAMPLE 4 Solve the system
$$x + 4y = -10$$
$$2x + 8y = -20$$

SOLUTION These are graphed in Fig. 3(b). They are exactly the same lines, so the system is dependent and has infinitely many solutions. We examine what happens algebraically in the system. Replace the second equation with the sum of the second equation plus -2 times the first equation:
$$x + 4y = -10$$
$$0 = 0$$

We see that the original system is equivalent to a system where there is just one equation in two variables and hence there are infinitely many solutions. The equation $0 = 0$ is the trademark of this situation.

To solve this system, we let $y = s$, where s is any number, and then substitute s for y in the one equation. Thus
$$y = s, \qquad x + 4s = -10, \qquad x = -4s - 10$$

The complete solution is
$$(x, y) = (-4s - 10, s), \qquad s \text{ any number}$$

A third thing that can happen with the two linear equations is that they represent two different lines which are parallel. Then the lines do not intersect at all, and the corresponding system of equations does not have a solution. In this situation, the system is called **inconsistent**.

EXAMPLE 5 Solve the system
$$x + 4y = -10$$
$$x + 4y = -4$$

SOLUTION These lines are graphed in Fig. 3(c). The lines are different and parallel, so there is no solution. We examine what happens algebraically in the system. Replace the second equation with the second equation minus the first:
$$x + 4y = -10$$
$$0 = 6$$

Since $0 \neq 6$, the triangular system can have no solution, and hence the original system has none. This type of obviously false statement, that $0 = 6$, is the trademark of an inconsistent system.

We summarize the three types of linear systems of two equations in two variables in the following table:

TYPE OF SYSTEM	NUMBER OF SOLUTIONS	GEOMETRIC INTERPRETATION
Consistent		
Independent	One	Lines are different but intersect
Dependent	Infinite number	Lines are coincident
Inconsistent	None	Lines are different and parallel

Many word problems can be translated into a system of linear equations and then solved by the preceding techniques.

EXAMPLE 6 There are 30 coins in a bag. There are only dimes and quarters, and they are worth $6.30. How many of each coin are there?

SOLUTION Since we are looking for the number of each coin, we let

$$d = \text{number of dimes}$$
$$q = \text{number of quarters}$$

(*Note*: It is important to write "*number* of dimes," etc., not just an ambiguous phrase such as "dimes.") Since altogether there are 30 coins, one equation is

$$d + q = 30$$

Since each dime is worth $.10, the d dimes are worth $.10d$. Similarly, the q quarters are worth $.25q$. Since altogether the bag holds $6.30, the second equation is

$$.10d + .25q = 6.30$$

Therefore, the system to solve is

$$d + q = 30$$
$$.10d + .25q = 6.30$$

Solving, we replace the second equation with that equation plus $-.10$ times the first:

$$d + q = 30$$
$$.15q = 3.30$$

Solving this, we obtain $q = 22$ and $d = 8$, so there are 22 quarters and 8 dimes in the bag.

EXERCISES

In Exercises 1–20, solve the given system of equations, if possible, stating whether they are independent, inconsistent, or dependent.

1. $2x + 4y = 10$
   $x - 3y = -5$

2. $3w - 2z = -3$
   $2w - 4z = 2$

Section 6.2 Linear Equations in Two Variables

3. $4p - 6q = 4$
   $6p - 9q = -4$
4. $8m + 4n = -2$
   $6m + 3n = 6$
5. $3a - b = 3$
   $6a - 3b = 10$
6. $-3c - 3d = -5$
   $2c + 12d = 0$
7. $18x + 3y = 13$
   $-2x - 9y = -65$
8. $2x + 3y = -30$
   $6x + 2y = 6$
9. $2u - v = 4$
   $-8u + 4v = -12$
10. $3p + 2q = 5$
    $-6p - 4q = -10$
11. $12c - 8d = 4$
    $9c - 6d = 3$
12. $20m - 10n = 15$
    $-12m + 6n = -9$
13. $.02x - .03y = 0$
    $.04x + .09y = .05$
14. $.11x + .09y = -.007$
    $-.51x - .27y = .003$
15. $31.2x - 41.7y = 201.03$
    $62.4x + 59.6y = -41.24$
16. $.051w - .021z = .36897$
    $.017w + .064z = -.44346$
17. $-2p + q = 4$
    $6p - 3q = 10$
18. $15c - 21d = -6$
    $-20c + 28d = 9$
19. $15m - 20n = 10$
    $-21m + 28n = -14$
20. $-14a - 20b = 10$
    $21a + 30b = -15$

In Exercises 21–30, solve. (You might first make a substitution such as $u = 1/x$, $v = 1/y$ or $u = x^2$, $v = y^2$, etc. which converts the given system into a system of linear equations.)

21. $\dfrac{2}{x} + \dfrac{1}{y} = 2$
    $\dfrac{6}{x} - \dfrac{3}{y} = 4$
22. $\dfrac{4}{x} - \dfrac{3}{y} = -1$
    $\dfrac{1}{x} + \dfrac{3}{y} = 2$
23. $\dfrac{3}{x} - \dfrac{2}{y} = 1$
    $\dfrac{1}{x} - \dfrac{6}{y} = -1$
24. $\dfrac{2}{x} - \dfrac{9}{y} = 2$
    $\dfrac{-4}{x} - \dfrac{3}{y} = 3$
25. $5x^2 - 3y^2 = -7$
    $-2x^2 + 3y^2 = 10$
26. $3x^2 - 7y^2 = -1$
    $x^2 - y^2 = 5$
27. $4|x| + 3|y| = 13$
    $5|x| - |y| = 2$
28. $2|x| + 6|y| = 5$
    $-4|x| + 10|y| = 1$
29. $3x^2 + 2y^2 = 1$
    $6x^2 - 4y^2 = 5$
30. $2|x| - 3|y| = -10$
    $|x| + |y| = 0$

In Exercises 31–34: **(a)** Determine the constant k such that the system is dependent. **(b)** For other values of k, is the system independent or inconsistent?

31. $6x - 4y = 2$
    $-9x + 6y = k$
32. $12x + 3y = -9$
    $-8x - 2y = k$
33. $9x + 6y = 12$
    $kx + 2y = 4$
34. $2x - 6y = 4$
    $-3x + ky = -6$

35. Find two numbers whose sum is -11 and difference is 3.
36. Find two numbers whose sum is -3 and difference is -11.
37. A bag contains $2.90 in nickels and dimes. If there are 40 coins in all, how many of each coin are there?

38. Type *A* beer is 5% alcohol and Type *B* beer is 2% alcohol. How much of each type should be mixed to get 100 liters of beer which is 3.2% alcohol?

39. A woman can row 6 kilometers downstream in 1 hour and row back the 6 kilometers in 2 hours. What is the rate of the stream, and how fast can she row in still water?

40. Jane and Jim are 2 miles apart. If Jim starts walking directly away from Jane and she starts walking at the same time after him, she will overtake him in 1 hour. If they start walking at the same time toward each other, they will meet in 12 minutes. How fast do they each walk?

41. The price of admission to a certain movie was $2.50 for adults and $1.10 for children. If 540 tickets were sold for a total of $888, how many of each kind were sold?

42. A woman receives income from two investments totalling $15,000, one at $7\frac{1}{2}\%$ and the other at $8\frac{3}{4}\%$ simple interest per year. If altogether each year she earns $1175, how much is invested at each rate?

SECTION 6.3. SYSTEMS OF LINEAR EQUATIONS IN THREE OR MORE VARIABLES

A solution to an equation in x, y, and z is an ordered triple of numbers (a, b, c) such that when a is substituted for x, b for y, and c for z, the equation is true. A solution to a system of equations in x, y, and z is a triple which is simultaneously a solution to all the equations of the system. Similar definitions hold for equations in four or more variables.

It is our purpose in this section to introduce an efficient method for solving systems of three or more linear equations. The method presented here is the basis for most computer and programmable calculator programs designed to solve linear systems.

As in the previous section, the system:

$$2x + y - z = 3$$
$$-3y + 5z = -1$$
$$2z = 8$$

is said to be in **triangular form** because of the shape of the system on the left-hand side of the equal signs (as indicated by the dashed lines), and it is easily solved by backsubstitution.

EXAMPLE 1 Solve the system

$$2x + y - z = 3$$
$$-3y + 5z = -1$$
$$2z = 8$$

SOLUTION We solve the last equation and then work backwards, each time substituting the values we know into the previous equation:

$$2z = 8, \quad \text{so } z = 4$$
$$-3y + 5(4) = -1, \quad \text{so } y = 7$$
$$2x + 7 - 4 = 3, \quad \text{so } x = 0$$

Our method for solving an arbitrary system of linear equations is to transform it to an equivalent system in triangular form. This in turn can be easily solved by backsubstitution as in Example 1. We do this in much the same way as in Sec. 6.2. The elementary algebraic operations listed in Theorem (1) apply to systems of equations in any number of variables. To aid in the description of the operations, we label the positions of the equations.

EXAMPLE 2 Solve the system

$$2x - 4y - 2z = 10 \quad [1]$$
$$2x - 3y - 3z = 5 \quad [2]$$
$$-3x - y - z = -13 \quad [3]$$

These numbers indicate the equations in the first, second, and third positions, respectively.

SOLUTION We begin by dividing equation [1] by 2. This makes the coefficient of the x term equal to 1, which will be very helpful in the next step.

$$x - 2y - z = 5 \quad [1]$$
$$2x - 3y - 3z = 5 \quad [2]$$
$$-3x - y - z = -13 \quad [3]$$

The next step to transforming the system into triangular form is to make the coefficient of x equal to zero in equations [2] and [3]. We do this using the first equation. Specifically, we replace equation [2] with equation [2] plus -2 times equation [1] and replace equation [3] with equation [3] plus 3 times equation [1].

$$x - 2y - z = 5 \quad [1]$$
$$y - z = -5 \quad [2]$$
$$-7y - 4z = 2 \quad [3]$$

Now replace equation [3] with equation [3] plus 7 times equation [2]. (Note that we must not use equation [1] again, as this would wreck the zeros already obtained.)

$$x - 2y - z = 5$$
$$y - z = -5$$
$$-11z = -33$$

This is now in triangular form, so we solve it as before:

$$z = 3; \quad y - 3 = -5, \; y = -2; \quad x - 2(-2) - (3) = 5, \; x = 4$$

The answer is $(x, y, z) = (4, -2, 3)$. We check this using the original equations:

$$2x - 4y - 2z = 10: \quad 2(4) - 4(-2) - 2(3) = \quad 8 + 8 - 6 = 10$$
$$2x - 3y - 3z = 5: \quad 2(4) - 3(-2) - 3(3) = \quad 8 + 6 - 9 = 5$$
$$-3x - y - z = -13: \quad -3(4) - (-2) - (3) = -12 + 2 - 3 = -13$$

The answer checks.

Sometimes the form of a system can be confusing if the variables are not properly spaced to allow for zero coefficients.

EXAMPLE 3 Solve the system

$$4x + y = -7$$
$$x - 2z = 4$$
$$3y + 2z = -3$$

SOLUTION First, realign the variables. Then interchange equations [1] and [2] in order to get a coefficient of 1 for x in the first equation.

$$4x + y = -7 \quad [1]$$
$$x - 2z = 4 \quad [2]$$
$$ 3y + 2z = -3 \quad [3]$$

$$x - 2z = 4 \quad [1]$$
$$4x + y = -7 \quad [2]$$
$$ 3y + 2z = -3 \quad [3]$$

Now replace [2] with [2] + (−4)[1]. In the resulting system, replace [3] with [3] + (−3)[2].

$$x - 2z = 4 \quad [1]$$
$$y + 8z = -23 \quad [2]$$
$$3y + 2z = -3 \quad [3]$$

$$x - 2z = 4$$
$$y + 8z = -23$$
$$-22z = 66$$

Solving, we get $(x, y, z) = (-2, 1, -3)$, which checks.

Equations in three (or more) variables can be inconsistent or dependent also.

EXAMPLE 4 Solve the system

$$x + 2y - 3z = 1 \quad [1]$$
$$2x - y + z = 3 \quad [2]$$
$$3x + 11y - 16z = 1 \quad [3]$$

SOLUTION Replace [2] with [2] + (−2)[1] and [3] with [3] + (−3)[1]:
$$x + 2y - 3z = 1$$
$$-5y + 7z = 1$$
$$5y - 7z = -2$$

Replacing [3] with [3] + [2] yields
$$x + 2y - 3z = 1$$
$$-5y + 7z = 1$$
$$0 = -1$$

Since $0 = -1$ is not true, the system is inconsistent and has no solutions.

EXAMPLE 5 Solve the system

$$x + 2y - 3z = 1 \quad [1]$$
$$2x - y + z = 3 \quad [2]$$
$$3x + 11y - 16z = 2 \quad [3]$$

SOLUTION The coefficients on the left are the same as in Example 4. As done there, replace [2] with [2] + (−2)[1] and [3] with (−3)[1]. Then replace the resulting [3] with [3] + [2].

$$\begin{array}{ll} x + 2y - 3z = 1 & \qquad x + 2y - 3z = 1 \\ -5y + 7z = 1 & \qquad -5y + 7z = 1 \\ 5y - 7z = -1 & \qquad 0 = 0 \end{array}$$

To solve this, we let $z = s$, substitute this in, and solve first for y and then for x as if s were a specific number:

$$-5y + 7s = 1, \qquad -5y = -7s + 1, \qquad y = \tfrac{7}{5}s - \tfrac{1}{5}$$
$$x + 2(\tfrac{7}{5}s - \tfrac{1}{5}) - 3s = 1, \qquad x = \tfrac{1}{5}s + \tfrac{7}{5}$$

Thus the system is dependent, there are an infinite number of solutions, given by

$$(x, y, z) = \left(\tfrac{1}{5}s + \tfrac{7}{5}, \tfrac{7}{5}s - \tfrac{1}{5}, s\right), \qquad s \text{ any number}$$

The techniques in this section can be used to solve a variety of word problems.

EXAMPLE 6 Find the quadratic function whose graph goes through the points $(-1, 4)$, $(1, -2)$, and $(2, 1)$.

SOLUTION We wish to find a function of the form

$$f(x) = ax^2 + bx + c$$

such that the three given points satisfy the equation $y = ax^2 + bx + c$. Substituting in, we get

For $(-1, 4)$: $\quad 4 = a(-1)^2 + b(-1) + c$
For $(1, -2)$: $\quad -2 = a(1)^2 + b(1) + c$
For $(2, 1)$: $\quad 1 = a(2)^2 + b(2) + c$

Rewriting, we get the system

$$\begin{array}{l} a - b + c = 4 \\ a + b + c = -2 \\ 4a + 2b + c = 1 \end{array}$$

This is solved by the methods in this section to get $(a, b, c) = (2, -3, -1)$. Thus the function is $f(x) = 2x^2 - 3x - 1$.

EXAMPLE 7 A merchant mixes three grades of coffee costing $2.50, $3.50, and $3.90 per pound. If he makes 200 pounds of mixture which costs $3.38 per pound and he uses 35 more pounds of the $3.50 grade than the $2.50 grade, how many pounds of each grade does he use?

SOLUTION We first carefully define the variables. Let

x = number of pounds of the $2.50 grade
y = number of pounds of the $3.50 grade
z = number of pounds of the $3.90 grade

Since there are three variables, we need three equations. One equation is

$$\underbrace{\text{No. lb}}_{x} \text{ plus } \underbrace{\text{no. lb}}_{y} \text{ plus } \underbrace{\text{no. lb}}_{z} \text{ is } \underbrace{\text{total no. lb}}_{200}$$
$$x + y + z = 200$$

The value of a grade is the price per pound times the total number of pounds used. A second equation is

$$\underbrace{\text{Value of}}_{2.50x} \text{ plus } \underbrace{\text{value of}}_{3.50y} \text{ plus } \underbrace{\text{value of}}_{3.90z} \text{ is } \underbrace{\text{value of mixture}}_{3.38(200)}$$
$$2.50x + 3.50y + 3.90z = 3.38(200)$$

Since he has 35 more pounds of the $3.50 grade than the $2.50 grade, a third equation is $y = x + 35$. We now solve the system

$$x + y + z = 200$$
$$25x + 35y + 39z = 6760$$
$$-x + y = 35$$

using the methods described, obtaining 50 pounds of $2.50 grade, 85 pounds of $3.50 grade, and 65 pounds of $3.90 grade.

We now consider one type of difficulty that arises; namely, what do you do when the coefficients are not easy to work with? For example, suppose we wanted to solve

$$5x + 3y + 2z = 1 \quad [1]$$
$$2x + y + z = 0 \quad [2]$$
$$3x - 2y + 3z = 1 \quad [3]$$

Then for the first step we could do one of the following:

1. Make one of the coefficients of x equal to 1 by adding a multiple of another equation, for example, replacing [3] with [3] + (−1)[2]
2. Multiply equations [1] and [3] by 2, making all the coefficients of x multiples of 2
3. Make one of the coefficients of x equal to 1 by choosing an equation whose coefficient of x is nonzero and dividing that equation by the coefficient of x

When working by hand, usually method 1 is the best choice. However, it is usually method 3 (or a variation of it) which is used when programming a computer.

EXERCISES

In Exercises 1–28, solve the given system, if possible. State whether it is independent, inconsistent, or dependent.

1. $\begin{aligned} x + y + z &= 2 \\ 2x + y - 3z &= -12 \\ 3x + 4y + z &= 1 \end{aligned}$

2. $\begin{aligned} x - 2y + 2z &= -1 \\ 5x - 4y + 3z &= 11 \\ 2x + 5y - z &= 0 \end{aligned}$

3. $\begin{aligned} 3x - 2y + 3z &= 10 \\ -x + 2y - 3z &= -2 \\ -3x - 4y + 3z &= -15 \end{aligned}$

4. $\begin{aligned} 4x + 6y + 7z &= -9 \\ -6x - 3y + z &= 21 \\ 2x + 3y - z &= -9 \end{aligned}$

5. $\begin{aligned} 4x - 3y - 2z &= 4 \\ 3x + 2y + z &= -2 \\ -2x - 7y - 4z &= 1 \end{aligned}$

6. $\begin{aligned} 2x + 3y - 6z &= 4 \\ 5x + 7y - 8z &= 1 \\ 3x + 4y - 2z &= 3 \end{aligned}$

7. $\begin{aligned} 2x - y + z &= 2 \\ x + 2y + z &= 4 \\ 3x + y + 2z &= 6 \end{aligned}$

8. $\begin{aligned} 5x + y + z &= 4 \\ 2x - y - z &= 3 \\ 3x + 2y + 2z &= 1 \end{aligned}$

9. $\begin{aligned} 2x - 2y + 3z &= -7 \\ 3x + y + z &= 3 \\ -2x + y - z &= -2 \end{aligned}$

10. $\begin{aligned} 3x + 4y + 2z &= 6 \\ -2x - 2y + 3z &= -8 \\ x + 2y - z &= 4 \end{aligned}$

11. $\begin{aligned} 2x - y - z &= 1 \\ 3x + 2y + z &= 3 \\ 4x - y - z &= 3 \end{aligned}$

12. $\begin{aligned} 5x + 2y - 2z &= 17 \\ -3x - y + z &= -10 \\ 2x + 2y - z &= 7 \end{aligned}$

13. $\begin{aligned} 3x - y + 3z &= 2 \\ 2x + y - 2z &= -3 \\ x + 3y - 7z &= 1 \end{aligned}$

14. $\begin{aligned} 2x + y - z &= 1 \\ 3x - y &= 4 \\ x - 2y + z &= -2 \end{aligned}$

15. $\begin{aligned} 2x + y - 8z &= 1 \\ x - 2y + z &= -2 \\ 4x - 3y - 6z &= -3 \end{aligned}$

16. $\begin{aligned} 4x - 5y - z &= 3 \\ 6x - 7y - 4z &= 5 \\ 2x - 3y + 2z &= 1 \end{aligned}$

17. $\begin{aligned} x + y + z - 2w &= 3 \\ 2x + y - z + 2w &= -2 \\ x - y + z - 4w &= 6 \\ 3x + y + z - 2w &= 5 \end{aligned}$

18. $\begin{aligned} x - y + 2z - w &= 7 \\ 3x - 2y - 2z + 2w &= 9 \\ -2x + y + 4z + 3w &= -8 \\ -x - y + 2z + 2w &= 0 \end{aligned}$

Note: The systems in Exercises 19–22 are called **homogeneous**, because of the zeros on the right-hand side of the equal signs. Homogeneous equations are never inconsistent (why?), but they may be either dependent or independent.

19. $\begin{aligned} 2x - 3y + z &= 0 \\ x + 2y - z &= 0 \\ x - 5y + 2z &= 0 \end{aligned}$

20. $\begin{aligned} 4x - 3y - 2z &= 0 \\ -2x + 2y - 5z &= 0 \\ 8x - 7y + 8z &= 0 \end{aligned}$

21. $\begin{aligned} 3x - 2y + 2z &= 0 \\ x + y - z &= 0 \\ 2x - y + 3z &= 0 \end{aligned}$

22. $\begin{aligned} 3x - 4y + 5z &= 0 \\ 6x - 3y + 10z &= 0 \\ -2x - y - 5z &= 0 \end{aligned}$

258 SYSTEMS OF EQUATIONS AND INEQUALITIES

23. $\begin{aligned} x + y &= -3 \\ 2x + z &= -1 \\ y - z &= -4 \end{aligned}$

24. $\begin{aligned} 2x - y &= 9 \\ x + z &= 7 \\ y + 2z &= 5 \end{aligned}$

25. $\begin{aligned} 3x - y &= -1 \\ 6x + 2z &= 1 \\ 2y - 4z &= 6 \end{aligned}$

26. $\begin{aligned} 4x - 3y &= 1 \\ 6x - 8z &= 1 \\ 6y - 4z &= 1 \end{aligned}$

27. $\begin{aligned} 2x + y &= 1 \\ 4x - z &= 3 \\ -2y - z &= 1 \end{aligned}$

28. $\begin{aligned} 10x - 5y &= 15 \\ 4x + z &= 5 \\ 2y + z &= -1 \end{aligned}$

In Exercises 29 and 30, solve.

29. $\begin{aligned} \frac{1}{x} + \frac{2}{y} - \frac{3}{z} &= 1 \\ \frac{2}{x} - \frac{4}{y} + \frac{6}{z} &= 2 \\ -\frac{1}{x} + \frac{2}{y} + \frac{6}{z} &= 2 \end{aligned}$

30. $\begin{aligned} \frac{6}{x} + \frac{2}{y} - \frac{4}{z} &= 0 \\ \frac{3}{x} + \frac{1}{y} + \frac{4}{z} &= 9 \\ \frac{3}{x} - \frac{8}{y} - \frac{6}{z} &= 3 \end{aligned}$

31. Find the quadratic function $f(x) = ax^2 + bx + c$ whose graph goes through the points $(-2, -1)$, $(-1, 2)$, and $(1, -4)$.

32. A woman starts a business and sells shares to pay for the building and equipment. The first 2 years it loses $1.50 per share each year, but the third year it loses only $.50 per share. She plots the points $(1, -\frac{3}{2})$, $(2, -\frac{3}{2})$, $(3, -\frac{1}{2})$ and feels that a quadratic function might fit the data.

   a. Find that quadratic function.

   b. Use that function to predict how the business will do the fourth year.

33. A bag contains 40 coins in nickels, dimes, and quarters. If there are $6 in the bag and twice as many dimes as nickels, how many of each coin are there in the bag?

34. A merchant wishes to blend 50 pounds of tobacco from three types of tobacco: type A, which costs $4 per pound; type B, which costs $3.50 per pound; and type C, which costs $2.50 per pound. If he wants to use twice as much type C as type A and have the resulting mixture cost $3.20 per pound, how many pounds of each type should he use?

35. A box contains beetles (which have 6 legs), spiders (which have 8 legs) and centipedes (which have 100 legs). The box contains 78 heads, 1656 feet, and 6 more spiders than twice the number of centipedes. How many of each species are in the box?

36. A swimming pool is being filled by three pipes A, B, and C. Pipes A and B by themselves could fill it in $1\frac{1}{5}$ hours, pipes A and C by themselves could fill it in $1\frac{1}{3}$ hours, and pipes B and C by themselves could fill it in $1\frac{5}{7}$ hours. How long will it take them working together to fill the pool?

37. A large petroleum storage tank can be filled by three different pipes A, B, and C. When all three pipes are running, they can fill the tank in 3 hours. Pipes A and B by themselves can fill the tank in 4 hours. Pipes A and C can fill the tank in 8 hours by themselves. How long would it take each running alone to fill the tank?

38. A merchant mixes peanuts costing $.75 per kilogram, Brazil nuts costing $1.25 per kilogram, and cashews costing $2.00 per kilogram to obtain 130 kilograms of a nut mixture costing $1.50 per kilogram. If she mixes twice as many kilograms of peanuts as Brazil nuts, how many pounds of each nut does she use?

39. A chemist mixes a 10% HCL (hydrochloric acid) solution, a 30% HCL solution, and a 50% HCL solution to obtain 100 liters of 32% HCL. If she uses twice as much of the 50% solution as the 30% solution, how many liters of each solution does she use?

40. A collection of nickels, dimes, and quarters is worth $4.35. If there are 41 coins in all and the number of nickels is two more than twice the number of dimes, how many of each coin is in the collection?

41. A grocer mixes candy worth $.50 per kilogram with nuts worth $1.10 per kilogram and raisins worth $.90 per kilogram to obtain a 200-kilogram mixture worth $.89 per kilogram. If he uses as many kilograms of nuts as the sum of the other two, how many kilograms of each item does he use?

42. A woman invested different amounts at 8%, $8\frac{3}{4}$% and 9%, all simple annual interest. Altogether she has $40,000 invested and earns $2455 per year. How much does she have invested at each rate if she has $4000 more invested at 9% than at 8%?

SECTION 6.4. MATRIX SOLUTIONS OF LINEAR SYSTEMS

Suppose we had the following system of linear equations to solve:

$$2x + y + 6z = 4$$
$$x - y + 2z = 0$$
$$-3x - 3y - 8z = -9$$

Using the method of Sec. 6.3, we would reduce the system to triangular form and then solve. If you consider carefully the process of reducing the system to triangular form, you realize that once they are properly aligned, the variables x, y, and z and the equal signs in the equations play no role in the operations other than "place holders."

To make the reduction process quicker and easier, we devise a method for eliminating the variables and equal signs while reducing the system to triangular form. This method begins by associating with a system of equations a *rectangular array* made up only of the constants in the system:

$$\begin{array}{r} 2x + y + 6z = 4 \\ x - y + 2z = 0 \\ -3x - 3y - 8z = -9 \end{array} \leftrightarrow \begin{bmatrix} 2 & 1 & 6 & 4 \\ 1 & -1 & 2 & 0 \\ -3 & -3 & -8 & -9 \end{bmatrix}$$

Such an array is called a **matrix**. (The plural of matrix is **matrices**.) Not only are the numbers in a matrix important, but their position in the matrix is also important. This is because the numbers are the constants of the linear system with which it is associated.

The **rows** of a matrix are the numbers, in order, which appear horizontally. The preceding example has three rows; the first row is 2 1 6 4, the second row is 1 −1 2 0, and the third row is −3 −3 −8 −9. Matrices also have **columns**, which are the numbers, in order, which appear vertically. The preceding matrix has four columns, and it is called a 3 × 4 matrix (read "3 by 4") because it has three rows and four columns. Matrices will be formally defined and some of their properties discussed in the following sections. In this section, we shall just be using them to help reduce linear systems to triangular form. For this purpose, it is important to be able to go back and forth between a linear system and its associated matrix.

EXAMPLE 1 Find the linear system in x, y, and z associated with the matrix
$$\begin{bmatrix} 2 & 3 & -1 & 2 \\ 0 & -3 & 1 & -4 \\ 0 & 0 & -4 & 2 \end{bmatrix}$$

SOLUTION The three rows give three equations. The first column gives the coefficients of the x's, the second column gives the coefficients of the y's, etc. We get

$$\begin{aligned} 2x + 3y - z &= 2 \\ 0x - 3y + z &= -4 \\ 0x + 0y - 4z &= 2 \end{aligned} \quad \text{or} \quad \begin{aligned} 2x + 3y - z &= 2 \\ -3y + z &= -4 \\ -4z &= 2 \end{aligned}$$

The matrix in Example 1 corresponds to a system in triangular form, so we say the matrix is in **triangular form**.

The process of using matrices to reduce a system to triangular form is to reduce the matrix to triangular form using the same types of operations. We illustrate this using the system given at the beginning of this section.

System in equation form:

$$\begin{aligned} 2x + y + 6z &= 4 \\ x - y + 2z &= 0 \\ -3x - 3y - 8z &= -9 \end{aligned}$$

System in matrix form:

$$\begin{bmatrix} 2 & 1 & 6 & 4 \\ 1 & -1 & 2 & 0 \\ -3 & -3 & -8 & -9 \end{bmatrix}$$

Interchange equations [1] and [2]:

$$\begin{aligned} x - y + 2z &= 0 \\ 2x + y + 6z &= 4 \\ -3x - 3y - 8z &= -9 \end{aligned}$$

Interchange rows [1] and [2]:

$$\begin{bmatrix} 1 & -1 & 2 & 0 \\ 2 & 1 & 6 & 4 \\ -3 & -3 & -8 & -9 \end{bmatrix}$$

Replace equation [2] with [2] + (−2)[1], and replace equation [3] with [3] + 3[1]:

$$\begin{aligned} x - y + 2z &= 0 \\ 3y + 2z &= 4 \\ -6y - 2z &= -9 \end{aligned}$$

Replace row [2] with [2] + (−2)[1], and replace row [3] with [3] + 3[1]:

$$\begin{bmatrix} 1 & -1 & 2 & 0 \\ 0 & 3 & 2 & 4 \\ 0 & -6 & -2 & -9 \end{bmatrix}$$

Replace equation [3] with [3] + 2[2]:

$$\begin{aligned} x - y + 2z &= 0 \\ 3y + 2z &= 4 \\ 2z &= -1 \end{aligned}$$

Replace row [3] with [3] + 2[2]:

$$\begin{bmatrix} 1 & -1 & 2 & 0 \\ 0 & 3 & 2 & 4 \\ 0 & 0 & 2 & -1 \end{bmatrix}$$

We would now translate the matrix on the right back to the system on the left and solve that system by backsubstitution, obtaining $(x, y, z) = (\frac{8}{3}, \frac{5}{3}, -\frac{1}{2})$.

By translating a system of three or more equations to a matrix, we reduce the amount of work needed to solve the system.

We now formalize the operations which we use on matrices.

> **(2) THEOREM**
>
> Suppose we are given a matrix associated with a system of linear equations. Then each of the following operations on the rows of that matrix results in a matrix which is associated with a system equivalent to the original system:
>
> i. Interchanging two rows
>
> ii. Multiplying a row by a nonzero number (where the number is multiplied times each term in the row)
>
> iii. Replacing a row with that row plus a multiple of another row (two rows are added by adding the terms in the same column together)

For the proof, refer back to Theorem (1) and observe that these operations correspond to the operations given there.

Let us look at another example.

EXAMPLE 2 Solve the system

$$2x + y = 1$$
$$8x - z = 12$$
$$y + z = 1$$

SOLUTION We first find the corresponding matrix, *being careful to put in the necessary zeros*:

$$\begin{bmatrix} 2 & 1 & 0 & 1 \\ 8 & 0 & -1 & 12 \\ 0 & 1 & 1 & 1 \end{bmatrix}$$

Replace row [2] with [2] + (-4)[1]:

$$\begin{bmatrix} 2 & 1 & 0 & 1 \\ 0 & -4 & -1 & 8 \\ 0 & 1 & 1 & 1 \end{bmatrix}$$

Interchange rows [2] and [3], and then replace [3] with [3] + 4[2]:

$$\begin{bmatrix} 2 & 1 & 0 & 1 \\ 0 & 1 & 1 & 1 \\ 0 & -4 & -1 & 8 \end{bmatrix}, \quad \begin{bmatrix} 2 & 1 & 0 & 1 \\ 0 & 1 & 1 & 1 \\ 0 & 0 & 3 & 12 \end{bmatrix}$$

We now translate back to a triangular system in $x, y,$ and z:

$$2x + y = 1$$
$$y + z = 1$$
$$3z = 12$$

We solve this by backsubstitution, as in Sec. 6.3. The answer is $(x, y, z) = (2, -3, 4)$. This checks in the original system.

If a system is either dependent or inconsistent, it can still be handled with matrices.

EXAMPLE 3 Solve the system
$$2x - 3y - 3z = 1$$
$$-4x + 2y + z = 5$$
$$-2x - y - 2z = 6$$

SOLUTION Find the corresponding matrix:
$$\begin{bmatrix} 2 & -3 & -3 & 1 \\ -4 & 2 & 1 & 5 \\ -2 & -1 & -2 & 6 \end{bmatrix}$$

Replace row [2] with [2] + 2[1], and replace row [3] with [3] + [1]:
$$\begin{bmatrix} 2 & -3 & -3 & 1 \\ 0 & -4 & -5 & 7 \\ 0 & -4 & -5 & 7 \end{bmatrix}$$

Replace [3] with [3] + (−1)[2], and then translate back:
$$\begin{bmatrix} 2 & -3 & -3 & 1 \\ 0 & -4 & -5 & 7 \\ 0 & 0 & 0 & 0 \end{bmatrix} \qquad \begin{array}{r} 2x - 3y - 3z = 1 \\ -4y - 5z = 7 \\ 0 = 0 \end{array}$$

This system is dependent. We solve it as in Sec. 6.3, getting
$$(x, y, z) = \left(-\tfrac{3}{8}s - \tfrac{17}{8},\ -\tfrac{5}{4}s - \tfrac{7}{4},\ s\right)$$

Note: If the last row of the matrix had been 0 0 0 1, then the system would have been inconsistent.

EXERCISES

Solve, using matrices.

1. $3x - 4y = 10$
   $6x + y = 11$

2. $4x - y = -11$
   $3x + 2y = 0$

3. $x + 2y - z = 3$
   $2x + 3y + 2z = -1$
   $-3x - y + 3z = -14$

4. $2x - 2y + 3z = -5$
   $4x + 5y - 4z = -2$
   $6x - y + z = -13$

5. $2p - 3q + r = -19$
   $3p + 4q - 2r = 21$
   $p - 2q + r = -13$

6. $4a - 2b + c = -3$
   $8a + 3b - c = 11$
   $-2a - 5b + 3c = -14$

7. $3a - b + c = 0$
   $9a + 3b - 2c = 9$
   $-6a - 4b + 3c = -9$

8. $2c - 3d + e = -6$
   $c + d - 2e = 7$
   $-2c - 7d + 9e = -34$

9. $3x - y + 2z = 2$
   $2x + 3y - z = -1$
   $x - 4y + 3z = 2$

10. $4x - 3y - 2z = 3$
    $2x + 4y - z = 1$
    $10x - 2y - 5z = 8$

11. $2x + 3y = -8$
    $x - 3z = -10$
    $2y + z = -1$

12. $6x - 2y = 6$
    $-3x + z = 0$
    $3y - z = -5$

13. $\begin{aligned} 2x - y + 2z - w &= 5 \\ x + y - z + w &= 0 \\ -x - y + 2z - w &= 2 \\ 3x + 2y + z - w &= 1 \end{aligned}$

14. $\begin{aligned} x - 2y + z - w &= -2 \\ 2x - y + 2z - w &= 0 \\ -x + y + z - 2w &= -6 \\ 3x - y - z + w &= 10 \end{aligned}$

15. $\begin{aligned} a - 2b + c + s - t &= 4 \\ a - c - s &= 1 \\ b + c + t &= -2 \\ 2a - c + 2s &= 12 \\ a + 2b + s + t &= 4 \end{aligned}$

For additional exercises, use matrices to solve Exercises 1–42 in Sec. 6.3.

SECTION 6.5. DETERMINANTS AND CRAMER'S RULE

We have seen that any linear system can be solved by reducing it to triangular form. In this section, we introduce the concept of a determinant and then show how determinants can be used for solving linear systems; the method that does this is called Cramer's rule. Determinants have other very important applications (but we leave a thorough treatment of determinants to a higher course in algebra).

A determinant is a number which is associated with each square matrix, where by a **square matrix** we mean a matrix with the same number of rows as columns (i.e., a square matrix is an $n \times n$ matrix). We define the determinants of 2×2 and 3×3 matrices; the pattern is similar for larger matrices.

---

(3) **DEFINITION**

The determinant of a 2×2 matrix $A = \begin{bmatrix} a_1 & b_1 \\ a_2 & b_2 \end{bmatrix}$

is denoted by $|A|$ or $\begin{vmatrix} a_1 & b_1 \\ a_2 & b_2 \end{vmatrix}$ and is defined by

$$|A| = \begin{vmatrix} a_1 & b_1 \\ a_2 & b_2 \end{vmatrix} = a_1 b_2 - a_2 b_1$$

EXAMPLE 1 If $A = \begin{bmatrix} 2 & -4 \\ -5 & -\frac{3}{2} \end{bmatrix}$

then the determinant of A is

$|A| = \begin{vmatrix} 2 & -4 \\ -5 & -\frac{3}{2} \end{vmatrix}$ (The arrows indicate the products.)

$= 2(-\frac{3}{2}) - (-5)(-4) = -3 - 20 = -23$

We now state Cramer's rule for two equations in two unknowns. Suppose we have the following system to solve for x and y:
$$a_1 x + b_1 y = c_1$$
$$a_2 x + b_2 y = c_2$$
We can solve this using the methods of Sec. 6.2, getting the unique solution

(4) $$x = \frac{c_1 b_2 - c_2 b_1}{a_1 b_2 - a_2 b_1}, \quad y = \frac{a_1 c_2 - a_2 c_1}{a_1 b_2 - a_2 b_1}$$

provided $a_1 b_2 - a_2 b_1 \neq 0$. (If $a_1 b_2 - a_2 b_1 = 0$, the system is either dependent or inconsistent. See Exercises 23 and 24.)

We now rewrite the solutions given in (4) using three determinants. Let

$$D = \begin{vmatrix} a_1 & b_1 \\ a_2 & b_2 \end{vmatrix}$$

In other words, D is the determinant of the matrix of coefficients of the system. Note that $D = a_1 b_2 - a_2 b_1$ is the denominator of the fractions given in (4). In D, first replace the a's (which are the coefficients of x) with the c's, getting

$$D_x = \begin{vmatrix} c_1 & b_1 \\ c_2 & b_2 \end{vmatrix} \quad \text{(The shading emphasizes the column which is replaced.)}$$

Next in D, replace the b's (which are the coefficients of y) with the c's, getting

$$D_y = \begin{vmatrix} a_1 & c_1 \\ a_2 & c_2 \end{vmatrix}$$

Then in equations (4), $D_x = c_1 b_2 - c_2 b_1$ and $D_y = a_1 c_2 - a_2 c_1$ are the numerators of the answers for x and y, respectively. Thus, we have shown the following:

---

(5) **CRAMER'S RULE FOR TWO UNKNOWNS**

In the system
$$a_1 x + b_1 y = c_1$$
$$a_2 x + b_2 y = c_2$$
if $D = a_1 b_2 - a_2 b_1 \neq 0$, then
$$x = \frac{D_x}{D}, \quad y = \frac{D_y}{D}$$
where D, D_x, and D_y are as already given.

We shall see that this is part of a general pattern.

EXAMPLE 2 Solve using Cramer's rule:
$$3x + 4y = -2$$
$$-2x - 5y = 3$$

SOLUTION

$$x = \frac{\begin{vmatrix} -2 & 4 \\ 3 & -5 \end{vmatrix}}{\begin{vmatrix} 3 & 4 \\ -2 & -5 \end{vmatrix}} = \frac{10 - 12}{-15 + 8} = \frac{-2}{-7} = \frac{2}{7}$$

$$y = \frac{\begin{vmatrix} 3 & -2 \\ -2 & 3 \end{vmatrix}}{\begin{vmatrix} 3 & 4 \\ -2 & -5 \end{vmatrix}} = \frac{9 - 4}{-15 + 8} = \frac{5}{-7} = -\frac{5}{7}$$

These answers check.

We now turn to determinants of 3×3 matrices, which are defined using determinants of *submatrices*.

---

(6) **DEFINITION**

The determinant of a 3×3 matrix

$$A = \begin{bmatrix} a_1 & b_1 & c_1 \\ a_2 & b_2 & c_2 \\ a_3 & b_3 & c_3 \end{bmatrix}$$

is denoted by $|A| = \begin{vmatrix} a_1 & b_1 & c_1 \\ a_2 & b_2 & c_2 \\ a_3 & b_3 & c_3 \end{vmatrix}$ and is defined by

$$|A| = \begin{vmatrix} a_1 & b_1 & c_1 \\ a_2 & b_2 & c_2 \\ a_3 & b_3 & c_3 \end{vmatrix} = a_1 \begin{vmatrix} b_2 & c_2 \\ b_3 & c_3 \end{vmatrix} - a_2 \begin{vmatrix} b_1 & c_1 \\ b_3 & c_3 \end{vmatrix} + a_3 \begin{vmatrix} b_1 & c_1 \\ b_2 & c_2 \end{vmatrix}$$

Fortunately, this formula for $|A|$ does not have to be memorized. Instead, all that you have to remember is that $|A|$ is the sum of three terms, the signs in front of the terms alternate (plus, minus, plus), and the ith term is just a_i times the determinant of the 2×2 matrix obtained from A by deleting the row and column containing a_i.

EXAMPLE 3 If $A = \begin{bmatrix} 2 & 0 & -3 \\ -4 & 2 & 1 \\ -3 & -1 & 5 \end{bmatrix}$ then the determinant of A is

$$|A| = \begin{vmatrix} 2 & 0 & -3 \\ -4 & 2 & 1 \\ -3 & -1 & 5 \end{vmatrix} = 2 \begin{vmatrix} 2 & 1 \\ -1 & 5 \end{vmatrix} - (-4) \begin{vmatrix} 0 & -3 \\ -1 & 5 \end{vmatrix} + (-3) \begin{vmatrix} 0 & -3 \\ 2 & 1 \end{vmatrix}$$

$$= 2[10 - (-1)] + 4(0 - 3) - 3(0 + 6) = 22 - 12 - 18 = -8$$

We are now ready to describe Cramer's rule for three linear equations in three unknowns. Consider the system

$$a_1x + b_1y + c_1z = d_1$$
$$a_2x + b_2y + c_2z = d_2$$
$$a_3x + b_3y + c_3z = d_3$$

Define the following determinants:

$$D = \begin{vmatrix} a_1 & b_1 & c_1 \\ a_2 & b_2 & c_2 \\ a_3 & b_3 & c_3 \end{vmatrix} \qquad D_x = \begin{vmatrix} d_1 & b_1 & c_1 \\ d_2 & b_2 & c_2 \\ d_3 & b_3 & c_3 \end{vmatrix}$$

$$D_y = \begin{vmatrix} a_1 & d_1 & c_1 \\ a_2 & d_2 & c_2 \\ a_3 & d_3 & c_3 \end{vmatrix} \qquad D_z = \begin{vmatrix} a_1 & b_1 & d_1 \\ a_2 & b_2 & d_2 \\ a_3 & b_3 & d_3 \end{vmatrix}$$

Note that D is made up of the coefficients of the variables and that D_x, D_y, and D_z are obtained from D by replacing the corresponding coefficients with the d's, as indicated.

---

(7)   **CRAMER'S RULE FOR THREE UNKNOWNS**

In the system

$$a_1x + b_1y + c_1z = d_1$$
$$a_2x + b_2y + c_2z = d_2$$
$$a_3x + b_3y + c_3z = d_3$$

if $D \neq 0$, then the system has a unique solution given by

$$x = \frac{D_x}{D}, \qquad y = \frac{D_y}{D}, \qquad z = \frac{D_z}{D}$$

where D, D_x, D_y, and D_z are as already given.

We shall not give the proof. If $D = 0$, then the system is either dependent or inconsistent, and Cramer's rule cannot be used.

EXAMPLE 4   Solve using Cramer's rule:

$$3x + 2y + 4z = 4$$
$$3x + 2y - 2z = -\tfrac{1}{2}$$
$$-3x - y + 2z = 0$$

SOLUTION   Computing these determinants by (6), we obtain

$$D = \begin{vmatrix} 3 & 2 & 4 \\ 3 & 2 & -2 \\ -3 & -1 & 2 \end{vmatrix} = 18, \qquad D_x = \begin{vmatrix} 4 & 2 & 4 \\ -\tfrac{1}{2} & 2 & -2 \\ 0 & -1 & 2 \end{vmatrix} = 12,$$

Section 6.5   Determinants and Cramer's Rule

$$D_y = \begin{vmatrix} 3 & 4 & 4 \\ 3 & -\frac{1}{2} & -2 \\ -3 & 0 & 2 \end{vmatrix} = -9, \quad D_z = \begin{vmatrix} 3 & 2 & 4 \\ 3 & 2 & -\frac{1}{2} \\ -3 & -1 & 0 \end{vmatrix} = \frac{27}{2}$$

Therefore,

$$x = \frac{D_x}{D} = \frac{12}{18} = \frac{2}{3}, \quad y = \frac{D_y}{D} = \frac{-9}{18} = -\frac{1}{2}, \quad z = \frac{D_z}{D} = \frac{27/2}{18} = \frac{3}{4}$$

These answers check.

EXERCISES

In Exercises 1–10, evaluate the given determinants.

1. $\begin{vmatrix} 2 & 3 \\ -1 & -2 \end{vmatrix}$

2. $\begin{vmatrix} 3 & 2 \\ \frac{1}{2} & 1 \end{vmatrix}$

3. $\begin{vmatrix} 2\sqrt{2} & \sqrt{3} \\ -2\sqrt{3} & -\sqrt{2} \end{vmatrix}$

4. $\begin{vmatrix} 3 & -2 \\ -6 & 4 \end{vmatrix}$

5. $\begin{vmatrix} 2 & 1 & 3 \\ -3 & -1 & 2 \\ 3 & 4 & 6 \end{vmatrix}$

6. $\begin{vmatrix} 4 & 0 & -1 \\ 0 & 2 & 1 \\ 3 & 1 & -2 \end{vmatrix}$

7. $\begin{vmatrix} 5 & -1 & -2 \\ 4 & 1 & -1 \\ 1 & 2 & 3 \end{vmatrix}$

8. $\begin{vmatrix} -1 & 2 & 1 \\ 3 & 1 & -1 \\ 4 & 1 & 4 \end{vmatrix}$

9. $\begin{vmatrix} -1 & 1 & 2 \\ -2 & 1 & 1 \\ 5 & 4 & 7 \end{vmatrix}$

10. $\begin{vmatrix} 2 & -1 & 1 \\ -1 & 3 & 1 \\ 0 & 5 & 3 \end{vmatrix}$

In Exercises 11–18, solve the given system using Cramer's rule.

11. $2x - y = -4$
    $3x + 2y = 1$

12. $3x - 2y = 13$
    $3x + 3y = 0$

13. $3x + 4y = -1$
    $6x - 8y = 8$

14. $4x - 3y = \frac{3}{2}$
    $5x + 9y = 14$

15. $2x - y + 3z = -3$
    $3x + 3y - z = 10$
    $-x - y + z = -4$

16. $2x - 3y + 3z = 13$
    $-3x - y + 4z = -3$
    $4x + 3y + z = 7$

17. $3x + 2y + z = 1$
    $6x - 2y - 4z = -7$
    $3x + 4y + z = 2$

18. $6x + y + 4z = -4$
    $3x - 3y + 6z = -1$
    $-3x + 2y - 4z = 0$

We defined a 3×3 determinant by *expanding* along the first column. Exercises 19–22 illustrate that determinants can be found by expanding along any column; expanding along rows is done similarly. In Exercises 19–22, evaluate the given determinants using the definition in the text and also by the rules

(a) $\begin{vmatrix} a_1 & b_1 & c_1 \\ a_2 & b_2 & c_2 \\ a_3 & b_3 & c_3 \end{vmatrix} = -b_1 \begin{vmatrix} a_2 & c_2 \\ a_3 & c_3 \end{vmatrix} + b_2 \begin{vmatrix} a_1 & c_1 \\ a_3 & c_3 \end{vmatrix} - b_3 \begin{vmatrix} a_1 & c_1 \\ a_2 & c_2 \end{vmatrix}$

and

(b) $\begin{vmatrix} a_1 & b_1 & c_1 \\ a_2 & b_2 & c_2 \\ a_3 & b_3 & c_3 \end{vmatrix} = c_1 \begin{vmatrix} a_2 & b_2 \\ a_3 & b_3 \end{vmatrix} - c_2 \begin{vmatrix} a_1 & b_1 \\ a_3 & b_3 \end{vmatrix} + c_3 \begin{vmatrix} a_1 & b_1 \\ a_2 & b_2 \end{vmatrix}$

(Note that each 2 × 2 determinant is obtained from the 3 × 3 by deleting the row and column in which the element multiplied by it lies.)

19. $\begin{vmatrix} 2 & -1 & 1 \\ 3 & 0 & 2 \\ -1 & 0 & 1 \end{vmatrix}$
20. $\begin{vmatrix} 4 & 0 & 4 \\ -1 & 3 & 1 \\ -2 & 0 & 4 \end{vmatrix}$

21. $\begin{vmatrix} 3 & 2 & 0 \\ -1 & 1 & 2 \\ -3 & -1 & 0 \end{vmatrix}$
22. $\begin{vmatrix} -1 & 0 & 4 \\ 2 & 0 & 1 \\ -3 & 0 & 4 \end{vmatrix}$

23. Prove that if

$$\begin{vmatrix} a & c \\ b & d \end{vmatrix} = 0, \quad a \neq 0$$

then there is a number k such that $b = ka$, $d = kc$.

24. Use Exercise 23 to show that in the system $\begin{array}{l} a_1 x + b_1 y = c_1 \\ a_2 x + b_2 y = c_2 \end{array}$ if $\begin{vmatrix} a_1 & b_1 \\ a_2 & b_2 \end{vmatrix} = 0$ then either the system is dependent or inconsistent. Assume $a_1 \neq 0$. (*Hint*: It depends on whether $c_2 = kc_1$ or not.)

When variables represent numbers, you can evaluate determinants with such variables in the entries exactly as discussed in this section. In Exercises 25–28, evaluate the determinant.

25. $\begin{vmatrix} x^2 & 9 \\ x & x \end{vmatrix}$
26. $\begin{vmatrix} y^2 & -y \\ 2 & 3 \end{vmatrix}$

27. $\begin{vmatrix} z & 3 & 1 \\ 0 & z & z^2 \\ -2 & z & 2 \end{vmatrix}$
28. $\begin{vmatrix} w^2 & 1 & w \\ w & w & w \\ 0 & -1 & 1 \end{vmatrix}$

SECTION 6.6. OPERATIONS ON MATRICES

In Sec. 6.4, we introduced the concept of a matrix and associated matrices with systems of linear equations. In this section, we describe a few operations on matrices (for example, how to add and multiply them) and relate the operations to certain manipulations of linear systems. These operations lead to a whole theory for matrices which has many mathematical, scientific and business applications.

We begin by developing a slightly different notation for the entries of a matrix and formally defining a matrix. We have already described specific 2 × 3 matrices, for example,

$$A = \begin{bmatrix} 1 & -2 & \frac{3}{2} \\ -4 & 0 & 14.2 \end{bmatrix}$$

A general 2×3 matrix is denoted by

$$A = \begin{bmatrix} a_{11} & a_{12} & a_{13} \\ a_{21} & a_{22} & a_{23} \end{bmatrix}$$

---

(8)   **DEFINITION**

If m and n are positive integers, then an **$m \times n$ matrix** (over the real numbers) is a rectangular array of m rows and n columns of the form

$$\begin{bmatrix} a_{11} & a_{12} & a_{13} & \cdots & a_{1n} \\ a_{21} & a_{22} & a_{23} & \cdots & a_{2n} \\ \vdots & \vdots & \vdots & & \vdots \\ a_{m1} & a_{m2} & a_{m3} & \cdots & a_{mn} \end{bmatrix}$$

where each a_{ij} is a real number. The numbers m and n are called the **dimensions** of the matrix.

This way of denoting elements a_{ij} of a matrix is called **double-subscript notation**. It is a very convenient way to denote where in the matrix an element is located. The first subscript tells what row the element is in, while the second subscript tells what column the element is in. Sometimes instead of writing out the matrix as was done in the definition, we simply write "the $m \times n$ matrix $A = (a_{ij})$." This indicates that the entries of A are denoted by a's, that we are using double-subscript notation, and that A has m rows and n columns.

If $A = (a_{ij})$ and $B = (b_{ij})$ are $m \times n$ matrices, we define A to be **equal** to B if their corresponding entries are equal term by term; i.e.,

---

(9) $A = B$ if and only if $a_{ij} = b_{ij}$ for every i and j.

EXAMPLE 1

$$\begin{bmatrix} (\sqrt{2})^4 & 0^3 & 3^2 \\ (-2)^3 & (-1)^2 & \sqrt[3]{27} \end{bmatrix} = \begin{bmatrix} 4 & 0 & 9 \\ -8 & 1 & 3 \end{bmatrix}$$

If $A = (a_{ij})$ and $B = (b_{ij})$ are $m \times n$ matrices, we define their **sum** $A + B$ to be the $m \times n$ matrix obtained by adding the corresponding entries; i.e.,

---

$$(a_{ij}) + (b_{ij}) = (a_{ij} + b_{ij})$$

EXAMPLE 2

$$\begin{bmatrix} 2 & -3 & -15.1 \\ 8 & -2 & .01 \end{bmatrix} + \begin{bmatrix} -3 & 3 & 4.2 \\ 6 & 0 & .24 \end{bmatrix} = \begin{bmatrix} -1 & 0 & -10.9 \\ 14 & -2 & .25 \end{bmatrix}$$

If two matrices are of different dimensions, their sum is not defined.

The commutative and associative laws of addition hold (when defined). That is, if A, B, and C are $m \times n$ matrices, then

$$A + B = B + A, \qquad A + (B + C) = (A + B) + C$$

The **$m \times n$ zero matrix**, denoted by 0, is the $m \times n$ matrix whose entries are all zero. If A is an $m \times n$ matrix, then

$$A + 0 = A = 0 + A$$

so that the zero matrix is the **additive identity**.

If A is an $m \times n$ matrix, its **negative** is denoted by $-A$ and is the $m \times n$ matrix obtained from A by replacing each entry with its negative; i.e.,

$$-(a_{ij}) = (-a_{ij})$$

It is easy to see that

$$A + (-A) = 0 = (-A) + A$$

We are now ready to define the product of two matrices. The definition may seem a little strange at first, but it is justified by its application. If A and B are two matrices, the only restriction for their product AB to be defined is that *the number of columns of A must equal the number of rows of B*. Their product C will then have the same number of rows as A and the same number of columns as B. Thus if A is an $m \times n$ matrix and B is an $n \times p$ matrix, then $C = AB$ is an $m \times p$ matrix. To compute the element c_{ij} in the ith row and jth column of C, we first single out the ith row of A and the jth column of B:

$$\begin{bmatrix} a_{11} & a_{12} & \cdots & a_{1n} \\ \vdots & \vdots & & \vdots \\ a_{i1} & a_{i2} & \cdots & a_{in} \\ \vdots & \vdots & & \vdots \\ a_{m1} & a_{m2} & \cdots & a_{mn} \end{bmatrix} \begin{bmatrix} b_{11} & \cdots & b_{1j} & \cdots & b_{1p} \\ b_{21} & \cdots & b_{2j} & \cdots & b_{2p} \\ \vdots & & \vdots & & \vdots \\ b_{n1} & \cdots & b_{nj} & \cdots & b_{np} \end{bmatrix}$$

We now multiply the elements pairwise and then add the products, using the formula

(10)
$$c_{ij} = a_{i1}b_{1j} + a_{i2}b_{2j} + \cdots + a_{in}b_{nj}$$

EXAMPLE 3 Compute the following product of a 2×3 and a 3×4 matrix (which will give us a 2×4 matrix):

$$\begin{bmatrix} 1 & 2 & -1 \\ -3 & 0 & 4 \end{bmatrix} \begin{bmatrix} 3 & 4 & -1 & 3 \\ 1 & -1 & 0 & 1 \\ 2 & -1 & 0 & 1 \end{bmatrix}$$

SOLUTION We compute the eight entries as follows:

$$c_{11} = 1(3) + 2(1) + (-1)(2) = 3$$
$$c_{12} = 1(4) + 2(-1) + (-1)(-1) = 3$$
$$c_{13} = 1(-1) + 2(0) + (-1)(0) = -1$$
$$c_{14} = 1(3) + 2(1) + (-1)(1) = 4$$
$$c_{21} = (-3)(3) + 0(1) + 4(2) = -1$$
$$c_{22} = (-3)(4) + 0(-1) + 4(-1) = -16$$
$$c_{23} = (-3)(-1) + 0(0) + 4(0) = 3$$
$$c_{24} = (-3)(3) + 0(1) + 4(1) = -5$$

Thus the product matrix is the 2×4 matrix

$$\begin{bmatrix} 3 & 3 & -1 & 4 \\ -1 & -16 & 3 & -5 \end{bmatrix}$$

Example 4 illustrates how matrices and matrix multiplication are used in inventory control and cost analysis.

EXAMPLE 4 A certain store sells brand X TV's and radios. The following matrix on the left gives the sales of these items for 3 months; the matrix on the right gives the sales price and dealer's cost of these items. Use matrix multiplication to generate a matrix which has as its entries the total dollar sales and the total dollar costs of brand X items for each of the 3 months.

	Jan.	Jul.	Dec.			TV's	radio
TV's	22	32	30	retail price		245	23
Radios	13	15	11	dealer cost		170	18

SOLUTION For January, the total dollar sales price of brand X items is

$$\underbrace{245}_{\text{Retail price of a TV}} \times \underbrace{22}_{\substack{\text{no. of}\\\text{TV's}\\\text{sold}}} + \underbrace{23}_{\substack{\text{retail price}\\\text{of a radio}}} \times \underbrace{13}_{\substack{\text{no. of}\\\text{radios}\\\text{sold}}}$$

and the total dollar cost is

$$\underbrace{170}_{\substack{\text{Cost of}\\\text{a TV}}} \times \underbrace{22}_{\substack{\text{no. of}\\\text{TV's}\\\text{sold}}} + \underbrace{18}_{\substack{\text{cost of}\\\text{a radio}}} \times \underbrace{13}_{\substack{\text{no. of}\\\text{radios}\\\text{sold}}}$$

Similar computations hold for July and December. However, these are exactly the computations performed in multiplying the matrices in the proper order:

$$\begin{bmatrix} 245 & 23 \\ 170 & 18 \end{bmatrix} \begin{bmatrix} 22 & 32 & 30 \\ 13 & 15 & 11 \end{bmatrix} = \begin{bmatrix} 5689 & 8185 & 7603 \\ 3974 & 5710 & 5298 \end{bmatrix}$$

Therefore, the requested matrix is

	Jan.	July	Dec.
Total dollar sales	5689	8185	7603
Total retail cost	3974	5710	5298

One thing to note is that multiplication is not commutative.

EXAMPLE 5

$$\begin{bmatrix} 1 & 3 \\ 2 & 6 \end{bmatrix} \begin{bmatrix} 3 & -6 \\ -1 & 2 \end{bmatrix} = \begin{bmatrix} 0 & 0 \\ 0 & 0 \end{bmatrix}, \quad \begin{bmatrix} 3 & -6 \\ -1 & 2 \end{bmatrix} \begin{bmatrix} 1 & 3 \\ 2 & 6 \end{bmatrix} = \begin{bmatrix} -9 & -27 \\ 3 & 9 \end{bmatrix}$$

In fact, for many pairs of matrices, AB is defined, whereas BA is not, as in Example 3. There are several laws that do hold. For example, the associative law of multiplication

$$A(BC) = (AB)C$$

and the two distributive laws

$$A(B + C) = AB + AC, \quad (A + B)C = AC + BC$$

hold whenever everything is defined. Let I_n be the square $n \times n$ matrix with 1's along the diagonal and zeros elsewhere, so that

$$I_2 = \begin{bmatrix} 1 & 0 \\ 0 & 1 \end{bmatrix}, \quad I_3 = \begin{bmatrix} 1 & 0 & 0 \\ 0 & 1 & 0 \\ 0 & 0 & 1 \end{bmatrix}, \quad \text{etc.}$$

Then I_n is called the **$n \times n$ identity matrix** and is the **multiplicative identity**; i.e.,

$$AI_n = A, \quad I_n B = B$$

if A is any $m \times n$ matrix and B is any $n \times p$ matrix.

We now relate some of the operations discussed in terms of manipulating systems of linear equations. Consider the following system of m linear equations in n variables:

(11)
$$a_{11}x_1 + a_{12}x_2 + \cdots + a_{1n}x_n = b_1$$
$$a_{21}x_1 + a_{22}x_2 + \cdots + a_{2n}x_n = b_2$$
$$\cdots$$
$$a_{m1}x_1 + a_{m2}x_2 + \cdots + a_{mn}x_n = b_m$$

Let

$$A = \begin{bmatrix} a_{11} & a_{12} & \cdots & a_{1n} \\ a_{21} & a_{22} & \cdots & a_{2n} \\ \vdots & \vdots & & \vdots \\ a_{m1} & a_{m2} & \cdots & a_{mn} \end{bmatrix}, \quad X = \begin{bmatrix} x_1 \\ x_2 \\ \vdots \\ x_n \end{bmatrix}, \quad B = \begin{bmatrix} b_1 \\ b_2 \\ \vdots \\ b_m \end{bmatrix}$$

Using the definition of matrix multiplication, we see that the product AX is an $m \times 1$ matrix,

$$AX = \begin{bmatrix} a_{11}x_1 + a_{12}x_2 + \cdots + a_{1n}x_n \\ a_{21}x_1 + a_{22}x_2 + \cdots + a_{2n}x_n \\ \vdots & \vdots & & \vdots \\ a_{m1}x_1 + a_{m2}x_2 + \cdots + a_{mn}x_n \end{bmatrix}$$

Using the definition of equality of matrices, the system of m equations of (11) is equivalent to the single equation of matrices

(12) $$AX = B$$

Equation (12) is called the **matrix equation associated with the system** (11). The matrix A is called the **coefficient matrix**, and B is called the **matrix of constants**.

EXAMPLE 6 Find the matrix equation associated with the system
$$2x - 3y + 4z = 5$$
$$4x \quad\quad - z = -2$$

SOLUTION

$$\begin{bmatrix} 2 & -3 & 4 \\ 4 & 0 & -1 \end{bmatrix} \begin{bmatrix} x \\ y \\ z \end{bmatrix} = \begin{bmatrix} 5 \\ -2 \end{bmatrix}$$

EXERCISES

In Exercises 1–6, find $A + B$.

1. $A = \begin{bmatrix} 2 & 3 & 1 \\ -1 & 0 & -5 \end{bmatrix}, B = \begin{bmatrix} 4 & 1 & -1 \\ -2 & 3 & 4 \end{bmatrix}$

2. $A = \begin{bmatrix} 8 & 1 \\ 0 & -2 \\ 3 & 1 \end{bmatrix}, B = \begin{bmatrix} 4 & -2 \\ -1 & 1 \\ -3 & 0 \end{bmatrix}$

3. $A = \begin{bmatrix} 1 & 2 \\ 8 & -1 \end{bmatrix}, B = \begin{bmatrix} 3 & -1 \\ -1 & 2 \end{bmatrix}$

4. $A = \begin{bmatrix} 4 & 1 & 2 \\ 2 & -1 & -2 \\ 0 & 0 & 1 \end{bmatrix}, B = \begin{bmatrix} -1 & -2 & -3 \\ -1 & 0 & 0 \\ 8 & -1 & -4 \end{bmatrix}$

5. $A = \begin{bmatrix} 1 & 2 & 0 & -3 \\ -2 & 0 & -1 & 2 \end{bmatrix}, B = \begin{bmatrix} 3 & -1 & -1 & 0 \\ 4 & 1 & 0 & -8 \end{bmatrix}$

6. $A = \begin{bmatrix} 1 & 2 & 0 & 3 & 4 \end{bmatrix}, B = \begin{bmatrix} -3 & -2 & 1 & 0 & -1 \end{bmatrix}$

In Exercises 7–16, find AB and BA, if possible, or state "undefined."

7. $A = \begin{bmatrix} 1 & 2 \\ -3 & -1 \end{bmatrix}, B = \begin{bmatrix} 4 & 1 \\ -1 & -2 \end{bmatrix}$

8. $A = \begin{bmatrix} -1 & 0 & 1 \\ 2 & 1 & 3 \\ 4 & 0 & 0 \end{bmatrix}, B = \begin{bmatrix} 4 & 0 & -1 \\ 2 & 2 & 0 \\ 0 & 1 & -1 \end{bmatrix}$

9. $A = \begin{bmatrix} 2 & 1 & -3 \\ 0 & 1 & 0 \end{bmatrix}, B = \begin{bmatrix} 3 & -1 \\ 0 & 1 \\ 4 & 0 \end{bmatrix}$

10. $A = \begin{bmatrix} 1 & 2 & -3 \end{bmatrix}, B = \begin{bmatrix} 0 \\ 1 \\ -2 \end{bmatrix}$

11. $A = \begin{bmatrix} 2 & -1 \\ 0 & 3 \end{bmatrix}, B = \begin{bmatrix} 3 & -1 & 0 \\ 1 & 0 & -1 \end{bmatrix}$

12. $A = \begin{bmatrix} 3 & 5 \\ -1 & 0 \end{bmatrix}, B = \begin{bmatrix} 4 & -1 \\ -2 & 0 \\ 8 & 1 \end{bmatrix}$

13. $A = \begin{bmatrix} 1 & -2 \\ 0 & 3 \end{bmatrix}, B = \begin{bmatrix} 1 & -8 \\ 0 & 9 \end{bmatrix}$

14. $A = \begin{bmatrix} 0 & 0 & 1 \\ 2 & -1 & 3 \\ 0 & 1 & 0 \end{bmatrix}, B = \begin{bmatrix} 4 \\ 3 \\ -1 \end{bmatrix}$

15. $A = \begin{bmatrix} 2 & -1 \\ 1 & 0 \\ 3 & 1 \end{bmatrix}, B = \begin{bmatrix} 3 \\ 1 \\ 2 \end{bmatrix}$

16. $A = \begin{bmatrix} 1 \\ 2 \end{bmatrix}, B = \begin{bmatrix} 3 & 0 & -1 \\ 0 & 1 & 2 \end{bmatrix}$

In Exercises 17 and 18, let

$$A = \begin{bmatrix} 4 & 1 \\ -2 & 0 \end{bmatrix}, \quad B = \begin{bmatrix} 1 & -1 \\ 0 & 1 \end{bmatrix}$$

17. Show that $(A - B)(A + B) \neq A^2 - B^2$, where $A^2 = AA$ and $B^2 = BB$.

18. Show that $(A + B)^2 \neq A^2 + 2AB + B^2$, where $2AB = AB + AB$.

In Exercises 19–22, for the two systems given, **(a)** add the systems term by term and then find the associated matrix equation and **(b)** find the associated matrix equations and then add the corresponding matrices.

19. $\begin{array}{r} 2x - 3y = -2 \\ 4x - y = 3 \end{array}$ $\begin{array}{r} 5x + 2y = -7 \\ 2x - 8y = -1 \end{array}$

20. $3x + y = -5$
$2x + y - 2z = -1$
$5x + z = 6$

$2y + z = 4$
$4x - y - z = 1$
$-5x + y = -2$

21. $2x - y - z = 1$
$x + y + 5z = -6$

$3x + 2y - 8z = 4$
$2x + 4y + 3z = -2$

22. $2x - y = 1$
$3x + 5y = 2$
$x + 6y = 1$

$3x + 2y = 4$
$x - y = -2$
$6x - y = -2$

In Exercises 23–28, for the two systems given, **(a)** substitute the second system in the first, simplify, and then find the associated matrix equation and **(b)** find the associated matrix equation and then substitute in and multiply the coefficient matrices.

23. $3x - 4y = 2$
$4x + 5y = 3$

$x = 2s - t$
$y = 3s + 4t$

24. $2x + 3y - z = 4$
$-x - 2y + 3z = 1$

$x = s - 3t$
$y = 2s + 3t$
$z = s + t$

25. $7x - 2y = 10$
$3x + 4y = -3$

$x = 2p + 3q - 2r$
$y = 4p - q + r$

26. $4x - 3y - 2z = 4$
$2x + 3y - z = 2$
$x + z = 1$

$x = p - q$
$y = 2p + 3q$
$z = -p + 2q$

27. $3x - y - z = 2$
$y + 2z = -5$
$2x - z = 1$

$x = p - q - r$
$y = p + r$
$z = 2p - 3q$

28. $3x - 2y + z = 1$
$4x - y + 3z = -4$

$x = 2p - q$
$y = p + 2r$
$z = q - r$

29. A store sells brand X and brand Y refrigerators. The following matrices give the sales figures and costs of these items for 3 months. Use matrix multiplication to determine the total dollar sales and total costs of these items for the 3 months.

	Dec.	Apr.	Aug.			X	Y
Brand X	15	10	12	retail price		300	250
Brand Y	17	11	15	dealer cost		210	180

30. A cycle shop sells two grades of bicycles, Easy Roller (E.R.) and Super Rider (S.R.), manufactured by the same company. The following matrices give the sales of these items for 4 months and the sale price and dealer's cost of these items. Use matrix multiplication to determine the total dollar sales and the total costs of this company's items for the 4 months.

	Feb.	Mar.	Apr.	May			E.R.	S.R.
Easy Roller	6	10	13	12	retail price		120	180
Super Rider	4	7	8	8	dealer cost		70	100

SECTION 6.7. SYSTEMS OF INEQUALITIES

We have already begun the study of inequalities in earlier sections. Inequalities in one variable were studied in Chapter 2, and simple inequalities in two variables were examined as relations in Chapter 3. We shall now examine inequalities in two variables, such as

$$2x + 3y < 4y + 5 \quad \text{or} \quad 3y \geq 2 - x^2$$

in more detail. We then use the knowledge of their solutions to find solutions to systems of inequalities.

A **solution** to an inequality in two variables is an ordered pair of real numbers (a, b) which, when a is substituted for x and b for y in the inequality, produces a true statement. To **solve** an inequality means to find all solutions. The **graph of an inequality** is the graph of the set of all solutions as a subset of the plane.

EXAMPLE 1 Solve and sketch the graph of the inequality $4x - 3 \leq 6x + y$.

SOLUTION Subtracting $6x$ from both sides yields $-2x - 3 \leq y$. Thus the solution is the set of all ordered pairs (x, y) such that $y \geq -2x - 3$. In set notation, the solution is

$$\{(x, y) | y \geq -2x - 3\}$$

To graph the inequality, we first graph the equality $y = -2x - 3$. This is a straight line l with slope -2 and y-intercept -3. See Fig. 4(a). Since $y \geq -2x - 3$ means either $y = -2x - 3$ or $y > -2x - 3$, every point on l is a solution to $y \geq -2x - 3$. Now suppose (a, b) is on l, so that $b = -2a - 3$. Let us look at any other point (a, c) with the same x coordinate, a. If $c > b$, then since $b = -2a - 3$, we have $c > -2a - 3$, which means that (a, c) is a solution to the inequality. Note that since $c > b$, (a, c) is directly above (a, b). On the other hand, if $c < b$, then $c < -2a - 3$, and (a, c) is not a solution to the inequality. Thus every point on l or above l is a solution to the inequality, while every point below l is not a solution. See Fig. 4(b).

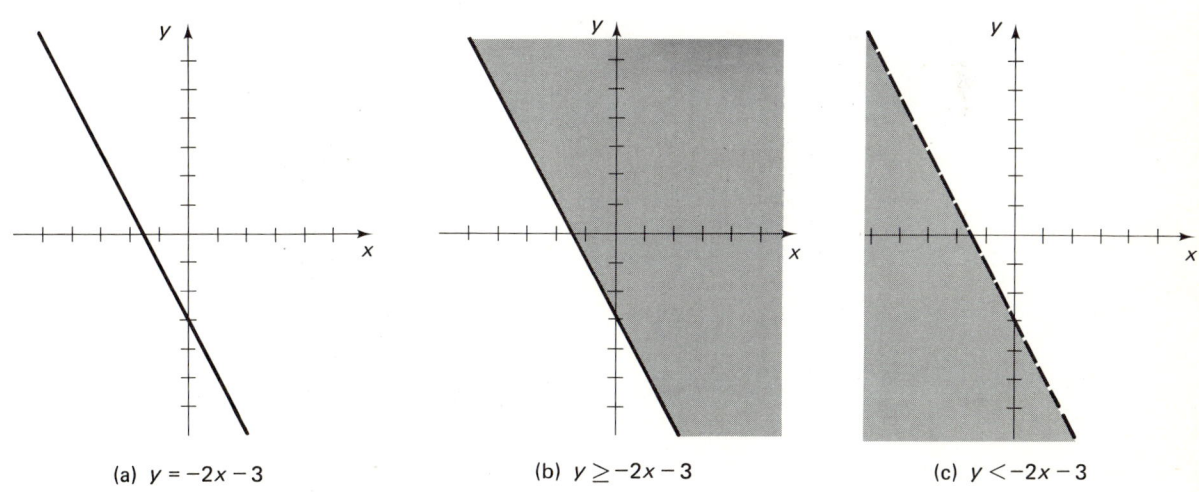

(a) $y = -2x - 3$ (b) $y \geq -2x - 3$ (c) $y < -2x - 3$

FIG. 4

The region sketched in Fig. 4(b) is called a **half plane**. Since the boundary is included, the region is called a **closed half plane**, and we draw the boundary with a solid line. Similarly, the graph of the inequality $y < -2x - 3$ in Fig. 4(c) is the **open half plane** below the line $y = -2x - 3$. The line is dashed to indicate it is not part of the graph.

This illustrates the following general principle:

> If f is a function, then the graph of the inequality $y > f(x)$ is the set of all points in the plane which lie *above* the graph of $y = f(x)$. Similarly, the graph of $y < f(x)$ is the set of all points which lie *below* the graph.

Of course, corresponding statements hold for the inequalities $y \geq f(x)$ and $y \leq f(x)$.

Consequently, when graphing inequalities, it is beneficial to work them into the form $y > f(x)$ or $y < f(x)$.

EXAMPLE 2 Solve and graph $y - x(x - 1) > 5x$.

SOLUTION

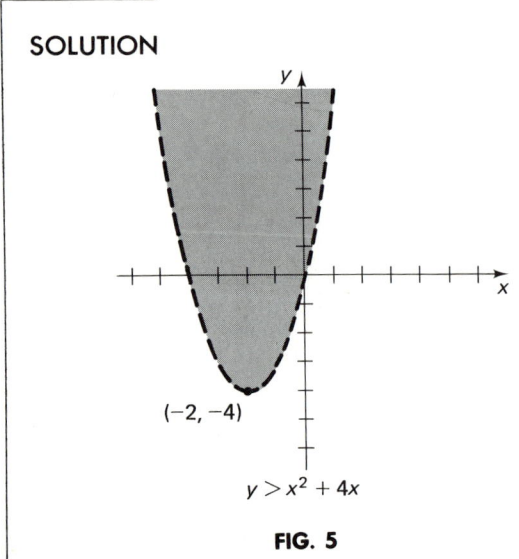

(−2, −4)

$y > x^2 + 4x$

FIG. 5

We clear parentheses and then rewrite this in the form $y > f(x)$:

$$y - x^2 + x > 5x, \qquad y > x^2 + 4x$$

The solution set is $\{(x,y) | y > x^2 + 4x\}$. To graph this, we first graph $y = x^2 + 4x$, with a dashed line. This equation is the parabola $y = (x + 2)^2 - 4$. We then shade in the region above the parabola. See Fig. 5.

Sometimes inequalities are more easily put in the form $x > f(y)$, $x < f(y)$, etc. The corresponding principles apply.

> If f is a function, then the graph of the inequality $x > f(y)$ is the set of all points in the plane which lie to the *right* of the graph of $x = f(y)$. Similarly, the graph of $x < f(y)$ is the set of all points which lie to the *left* of the graph.

EXAMPLE 3 Solve and graph $0 \leq x + y^2 + 1$.

SOLUTION

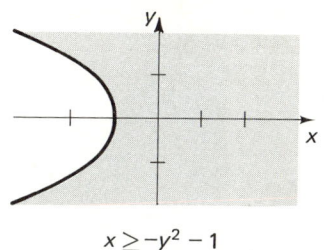

$x \geq -y^2 - 1$

FIG. 6

We rewrite this in the form $f(y) \leq x$,
$$-y^2 - 1 \leq x$$
so that the solution set is $\{(x, y) | x \geq -y^2 - 1\}$. We first graph $x = -y^2 - 1$ with a heavy line, since this is part of the solution. We then shade in the area to the right of this graph. See Fig. 6.

In Sec. 6.1, we defined the concept of a system of equations. **Systems of inequalities** are defined analogously, as are the **solution** and the **graph** of such a system and the concept of equivalent systems. We consider several examples.

EXAMPLE 4 Solve and graph the system
$$x + y \leq 2$$
$$2x - y \leq 7$$

SOLUTION

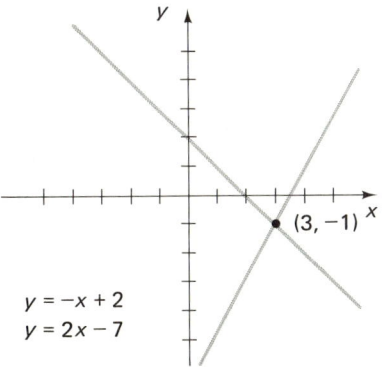

$y = -x + 2$
$y = 2x - 7$

FIG. 7

The given system is equivalent to
$$y \leq -x + 2$$
$$y \geq 2x - 7$$
We begin by sketching the lines $y = -x + 2$ and $y = 2x - 7$. They intersect at the point $(3, -1)$. See Fig. 7. A point where boundary lines intersect is called a **vertex** (the plural is **vertices**).

One way to graph the solution to the system is to shade in the solution to $y \leq -x + 2$ and then shade the solution to $y \geq 2x - 7$ with a different shading. See Fig. 8(a). The solution to the system is the set of all points which satisfy *both* inequalities. This is the region with the double (darkest) shading in Fig. 8(a), including its boundary. The solution is graphed in Fig. 8(b).

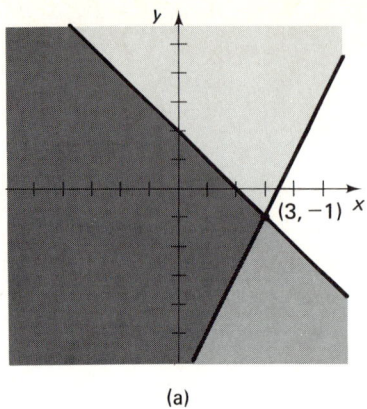
(a)

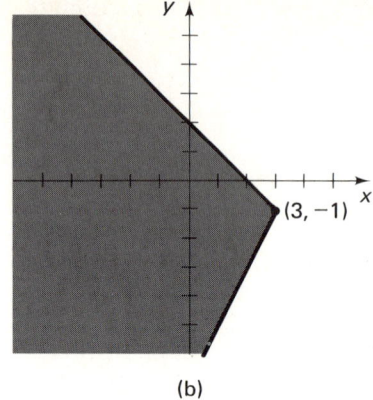
(b)

FIG. 8

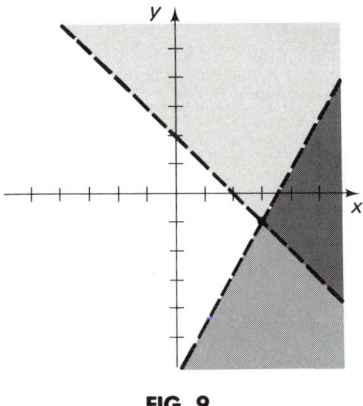

FIG. 9

This method can become messy and difficult to interpret when there are three or more inequalities involved. Hence we introduce a second method for solving systems of inequalities, which we shall call the **complement method**. In this method, we shade in the regions which are *not* the solution to each inequality. On the boundary, we draw a line dashed if it was solid before and vice versa. See Fig. 9. Thus a point is in the overall shaded area exactly when it is not in the solution to at least one of the inequalities of the system. So a point is in the *unshaded* area exactly when it is in the solution. In this way, the solution can be determined in Fig. 9 and then graphed separately, if it is so desired, as in Fig. 8(b).

We now give a more complicated example to illustrate the simplicity of the complement method.

EXAMPLE 5 Graph the system

$$y \leq 2x + 2$$
$$y \geq 2x - 1$$
$$x \geq -1$$
$$y \leq -x - 2$$

Find the coordinates of all vertices.

280 SYSTEMS OF EQUATIONS AND INEQUALITIES

SOLUTION For each inequality, we draw the boundary line and shade in the region *not* in the solution. See Fig. 10(a). Note that the boundary lines are dashed. The solution to the system is the *unshaded area*, which is graphed in Fig. 10(b).

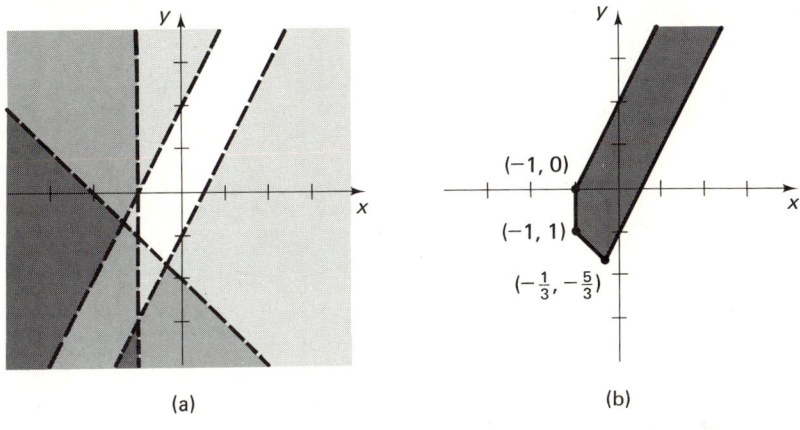

(a)　　　(b)

FIG. 10

The graph of a system which contains nonlinear inequalities is found in exactly the same way.

EXAMPLE 6 Sketch the graph of the system

$$x^2 + (y + 1)^2 \leq 2$$
$$y \geq x^2 - 1$$

Find the coordinates of all vertices.

SOLUTION The graph of $x^2 + (y + 1)^2 = 2$ is a circle centered at $(0, -1)$ and of radius $\sqrt{2}$. The graph of $y = x^2 - 1$ is a parabola with vertex at $(0, -1)$. These are graphed in Fig. 11(a). The solution is the region above the parabola and inside the circle (recall that these inequalities were discussed in Sec. 3.2) and is sketched in Fig. 11(b).

Of course we could graph each inequality individually or use the complement method to arrive at the answer. But when systems are simple, such as this one, you can do either method in your head and just write down the answer.

To find the points of intersection, we substitute $y = x^2 - 1$ in the other equation and then solve:

$$x^2 + (y + 1)^2 = 2, \quad y = x^2 - 1$$
$$x^2 + [(x^2 - 1) + 1]^2 = 2$$
$$x^2 + x^4 = 2$$
$$x^4 + x^2 - 2 = 0$$
$$(x^2 + 2)(x^2 - 1) = 0$$

$$x^2 = -2, \quad \text{no real solution}$$
$$x^2 = 1, \quad x = \pm 1$$

Using $y = x^2 - 1$, we obtain the vertices $(-1, 0)$, $(1, 0)$.

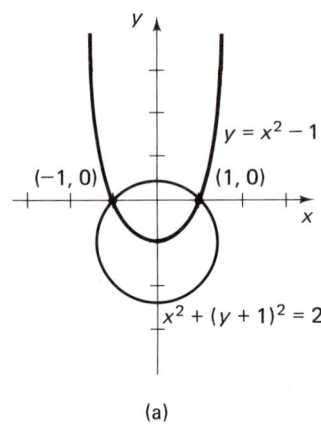
(a)

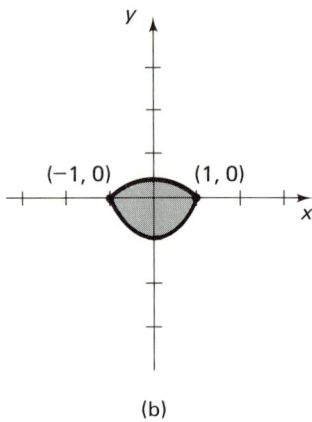
(b)

FIG. 11

EXERCISES

In Exercises 1–12, find the solution, and sketch the graph of the given inequality.

1. $x + y \geq 0$
2. $x + y < 1$
3. $2x - y < 3 + 2y$
4. $x + 5y \geq 10 - x$
5. $x + 2 < y^2$
6. $x - y^2 \leq 0$
7. $y^2 + 1 \leq x^2$
8. $(y - 1)^2 + (x + 2)^2 \geq 4$
9. $y \leq x^3$
10. $y > x^{1/3}$
11. $xy \geq 1$
12. $y < \dfrac{1}{x - 1}$

In Exercises 13–30, sketch the graph of the given system, labeling all vertices.

13. $2x - y < 3$
    $x > y + 1$
14. $y \geq -2x + 1$
    $2x \leq y - 5$
15. $2x + 3y \geq -13$
    $-4x + y \geq -2$
16. $2x + 3y < 3$
    $5x - 6y < 12$
17. $2x + 3y \leq 12$
    $2x - y \geq 4$
    $x \geq 0$
    $y \geq 1$
18. $3x - y > -9$
    $2x + 3y < 5$
    $x > -10$
    $y > -1$
19. $x + y < 4$
    $x - y > 6$
    $3x + y > -4$
20. $3x - y \geq 2$
    $2x + y \leq 1$
    $3x - y \leq 4$
    $x + y \geq -1$

21. $x^2 + y^2 \leq 4$
    $x - y \geq -1$
22. $x^2 + y^2 > 1$
    $x^2 + y^2 < 9$
23. $(x - 1)^2 + y^2 > 1$
    $x^2 + y^2 < 1$
24. $x^2 - y \leq 1$
    $x + y \geq 1$
25. $x^2 + y^2 > 1$
    $y > x^2 - 1$
26. $y - x^2 \geq 0$
    $x - y^2 \geq 0$
27. $y > 2^x$
    $y < 3^x$
    $x < 1$
28. $y \geq \ln x$
    $y + x - e \leq 1$
    $x \geq 1$
29. $y \geq \log x$
    $y + x \geq 1$
    $x^2 + (y - 1)^2 \leq 10^2$
30. $y \geq 3^{-x}$
    $y \leq 2^{-x}$
    $y \geq \frac{1}{2}$

31. A truck traveling from Chicago to New York is to be loaded with two types of cargo. Each carton of cargo A weighs 50 kilograms, is $1\frac{1}{2}$ cubic meters in volume, and earns \$4 for the driver. Each carton of cargo B weighs 40 kilograms, is 2 cubic meters in volume, and earns \$5.50 for the driver. The truck can carry no more than 5000 kilograms and no more than 200 cubic meters of cargo. Find a system of inequalities that describes all possibilities, and sketch the graph.

32. A tennis racket manufacturer makes two types of rackets, Big Ace and Super Server. Because of demand, she wants to make at least twice as many Big Aces as Super Servers. It takes $1\frac{1}{2}$ hours to make a Big Ace and 2 hours to make a Super Server. Altogether, she wants to make at least 25 rackets and work no more than 40 hours. Find a system of inequalities that describes all possibilities, and sketch the graph.

33. A theater contains 500 seats. For a certain show, the management wants to sell \$4 and \$5 tickets. Thy have to sell at least 200 \$4 tickets and 100 \$5 tickets, but they want to take in at least \$2000. Find a system of inequalities that describes all possibilities, and sketch the graph.

34. An appliance store sells two brands of washers. Because of demand, the management wants at least twice as many of brand B as brand A. They make \$40 profit on each brand A washer and \$30 on each brand B washer. They can stock at most 40 washers, and they want to make at least \$1700 from selling all machines in stock. Find a system of inequalities that describes all possibilities, and sketch the graph.

35. A store owner wishes to stock two types of felt-tipped pens, pen A which costs him 30 cents a pen and pen B which costs him 45 cents a pen. He can spend at most \$500 for the pens, and he wants an inventory of at least 400 of type A and 500 of type B. Find a system of inequalities that describes all possibilities, and sketch the graph.

SECTION 6.8. LINEAR PROGRAMMING

We are now ready to discuss a technique used to solve certain very important types of applied problems. In these problems, there are various constraints which give rise to systems of inequalities such as we studied in Sec. 6.7. In addition, there is a function which we shall be trying to maximize or minimize on the set of solutions to the system of inequalities. The technique for doing this is called **linear programming**. The term *linear* refers to the fact that the constraints of the problem lead to a system of linear inequalities and that the function to be maximized or minimized is linear. This technique has actually been used by many companies to help make inventory, shipping routes, and warehouse placement decisions, by the Army Corps of Engineers to help locate dams, and by public transit companies in drawing up schedules. These are just a few of the diverse applications of the technique.

We shall first illustrate the technique with an example and then discuss it in more detail.

EXAMPLE 1 Several women form a small company to make two styles of women's custom bathing suits, bikini and one-piece. To make a bikini, it takes 10 minutes of cutting time and 30 minutes of sewing time. To make a one-piece suit, it takes 30 minutes of cutting time and 15 minutes of sewing time. Altogether they can spend at most 20 hours a day cutting and at most 15 hours a day sewing. Suppose they earn $10 profit for each bikini they make and $8 profit for each one-piece suit they make. If the market is such that they can sell all the suits they can produce, how many of each type of suit should they make in order to earn the most profit each day?

SOLUTION We start off by determining what it is we want to do. From the last sentence, we see we want to maximize the profit P made each day. We see that P depends on how many bikinis and how many one-piece suits are made each day. So we let

x = number of bikinis made each day
y = number of one-piece suits made each day

Then since the profit on each bikini is $10 while on each one-piece it is $8, the total profit is

$$P = 10x + 8y$$

We now turn to the constraints. To help organize the data, we form a table. This is almost always a good practice for this type of problem in order to help translate the statement of the constraints into mathematical equations and inequalities:

PROCESS	BIKINI	ONE-PIECE	TOTAL TIME AVAILABLE
Cutting	10 min	30 min	1200 min
Sewing	30 min	15 min	900 min

We get two inequalities:

$$\underbrace{\text{Time cutting bikinis}}_{10x} + \underbrace{\text{time cutting one-piece suits}}_{30y} \leq \underbrace{\text{time available for cutting}}_{1200}$$

$$\underbrace{\text{Time sewing bikinis}}_{30x} + \underbrace{\text{time sewing one-piece suits}}_{15y} \leq \underbrace{\text{time available for sewing}}_{900}$$

Finally, there is an implicit but unstated restriction, namely, that you do not make a negative number of suits. Thus

$$x \geq 0 \quad \text{and} \quad y \geq 0$$

So altogether the system of constraints is

$$\begin{array}{l} 10x + 30y \leq 1200 \\ 30x + 15y \leq 900 \\ x \geq 0 \\ y \geq 0 \end{array} \quad \text{or} \quad \begin{array}{l} x + 3y \leq 120 \\ 2x + y \leq 60 \\ x \geq 0 \\ y \geq 0 \end{array}$$

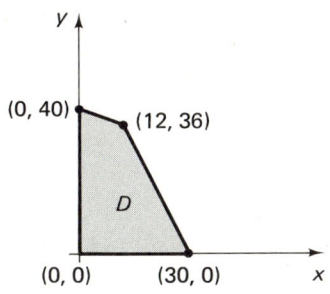

We graph this system in Fig. 12 (using the techniques from Sec. 6.7).

We are now ready to examine more closely the expression we wish to maximize, $P = 10x + 8y$. In fact, P is a *function of two variables*. What this means is that to every ordered pair of real numbers (x,y), P assigns a unique real number. In this case, we do not want to consider all ordered pairs but just those in the region D, sketched in Fig. 12. Thus, P exactly fulfills the conditions to be a function with domain D, with range the reals $\mathbb{R}$, and with rule $P(x, y) = 10x + 8y$. (See the definition of function in Sec. 3.3.)

Since x and y must be integers, there are only a finite number of ordered pairs (x,y) we need examine in the domain D. So we could solve this problem by computing all the corresponding values (there are only 827 of them) and picking out the largest. However, there is a theorem which cuts down considerably on the number of values we need to compute.

(13) THEOREM OF LINEAR PROGRAMMING

Suppose F is a function of the form $F(x, y) = ax + by + c$. If the domain of F is a polygon together with its interior, then F has a maximum and a minimum value, and these values occur at a vertex of the domain.

TABLE 1

(x, y)	$P = 10x + 8y$
(0, 0)	0
(30, 0)	300
(12, 36)	408
(0, 40)	320

This theorem tells us that in this case we need check only four values since the domain D has only four vertices. (And if this were a situation in which we were considering all points (x, y) in D, not just those with x and y being integers, we would still only have to check the same four values.) The four vertices are listed in Table 1, together with corresponding values of P. From the table we see that they should make 12 bikinis and 36 one-piece suits a day, for a maximum profit per day of $408.

Now that we have seen how linear programming can be used to solve a problem, let us review the steps involved. The first thing we always have to do is decide what we are trying to maximize or minimize. This can usually be determined from the last sentence or two of the problem. The next step is to define *carefully* the variables involved in this function. Always write this out in detail, as we do in the examples. Next, use these variables to write the constraints as inequalities. (The inequalities must be linear.) If your variables are x and y, do not forget the inequalities $x \geq 0, y \geq 0$ which are usually inherent in the problem but not mentioned. Once you have the system of inequalities, you graph the inequalities and determine the coordinates of the vertices of the solution region. Finally, you find the value of the function to be maximized or minimized at each vertex; the largest such value is the maximum of the function on the whole solution region, and the smallest such value is the minimum. The coordinates of the corresponding vertex are the values of the variables where the maximum or minimum occurs.

EXAMPLE 2 A company has two warehouses W_1 and W_2. The company receives an order for 50 units of one of its products from customer A and an order for 85 units of the same product from customer B. At that time, the company has 90 units of this product stored at W_1 and 80 units at W_2. If the shipping costs from the warehouses to the customers are as given in Table 2, how should the order be filled to minimize the total shipping cost? What is the minimal cost?

TABLE 2

WAREHOUSE	CUSTOMER	SHIPPING COST PER UNIT
W_1	A	$10
W_2	A	$14
W_1	B	$13
W_2	B	$15

SOLUTION The thing we want to do is to minimize the total shipping cost C. This depends on how much is shipped from each warehouse to each customer. So we let

x = number of units shipped from W_1 to A
y = number of units shipped from W_1 to B

Since altogether A ordered 50 units and B ordered 85 units,

$50 - x$ = number of units shipped from W_2 to A
$85 - x$ = number of units shipped from W_2 to B

Using Table 1, we see that

the cost of shipping 50 units to A is $10x + 14(50 - x)$
the cost of shipping 85 units to B is $13y + 15(85 - y)$

Therefore the total cost C is
$$C = 10x + 14(50 - x) + 13y + 15(85 - y)$$
or
$$C = 1975 - 4x - 2y$$

We now consider the constraints. Since A ordered 50 units, the amount shipped from W_1 to A is no more than 50. Of course a negative number of units cannot be shipped, so the amount shipped is at least zero. Thus
$$0 \leq x \quad \text{and} \quad x \leq 50$$
Similarly, since B ordered 85 units,
$$0 \leq y \quad \text{and} \quad y \leq 85$$
Since W_1 has 90 units and W_2 has 80 units, the total amounts shipped from W_1 and W_2 are no more than 90 and 80 units, respectively. Therefore
$$\begin{array}{cc} x + y \leq 90 & x + y \leq 90 \\ (50 - x) + (85 - y) \leq 80 & \text{or} \quad x + y \geq 55 \end{array}$$

Therefore the total system of inequalities is
$$\begin{array}{ll} 0 \leq x, & x \leq 50 \\ 0 \leq y, & y \leq 85 \\ 55 \leq x + y, & x + y \leq 90 \end{array}$$

We graph the complement of these inequalities in Fig. 13(a) and then read off the graph of the system in Fig. 13(b).

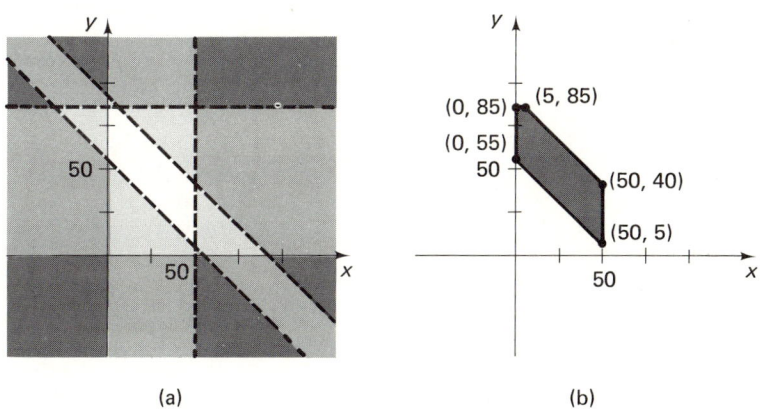

(a)   (b)

FIG. 13

Thus the domain of $C = 1975 - 4x - 2y$ has a boundary with five vertices. By the theorem of linear programming, we need only compute the value of C at these five vertices, which is done in Table 3.

TABLE 3

(x, y)	$C = 1975 - 4x - 2y$
(0, 85)	1805
(0, 55)	1865
(50, 5)	1765
(50, 40)	1695
(5, 85)	1785

We see from Table 3 that we should use up the complete inventory of W_1, shipping 50 units to customer A and 40 units to customer B. The remaining 45 units for customer B should be shipped from W_2. The total shipping cost is $1695.

The problems we have done here involve just two variables. Linear programming problems that involve more variables generally can be solved by essentially the same techniques, but they require more sophisticated mathematics. In fact, this is one of the situations in which matrices play a very important role.

EXERCISES

In Exercises 1–8, find both the maximum and the minimum values and the points at which these values are attained for the given function subject to the given constraints.

1. $F = 2x - 3y + 2$
   $x + y \leq 3$
   $2x + y \leq 4$
   $x \geq 0, y \geq 0$

2. $G = 3x + 2y + 1$
   $x + y \leq 10$
   $x + 2y \leq 6$
   $2x + y \geq 2$
   $x \geq 0, y \geq 0$

3. $P = 4x + 3y - 1$

   Same constraints as Exercise 1

4. $Q = 2x - 5y + 7$

   Same constraints as Exercise 2

5. $M = -8x - 2y + 12$
   $y - 2x \leq -1$
   $3y - x \geq 2$
   $2y - x \leq 2$
   $x \leq 10$

6. $N = -5x + 4y - 3$
   $y + 3x \geq 6$
   $3y + 2x \geq 11$
   $4y + x \geq 2$
   $4y + 3x \leq 6$

7. $S = 5x - 2y - 3$

   Same constraints as Exercise 5

8. $T = -4x - 7y - 2$

   Same constraints as Exercise 6

9. A truck traveling from Chicago to New York is to be loaded with two types of cargo. Each carton of cargo A weighs 50 kilograms, is 5 cubic meters in volume, and earns $4 for the driver. Each carton of cargo B weighs 40 kilograms, is 6 cubic meters in volume, and earns $3.50 for the driver. The truck can carry no more than 5000 kilograms and no more than 600 cubic meters of cargo. How many cartons of each cargo should the driver carry to earn the most money? How much money does the driver earn?

10. A tennis racket manufacturer makes two types of rackets, Big Ace and Super Server. Because of demand, she wants to make at least twice as many Big Aces as Super Servers. It takes $1\frac{1}{2}$ hours to make a Big Ace and 2 hours to make a Super Server. Altogether each week she wants to make at least 20 rackets and work no more than 40 hours. If she earns $12 for each Big Ace she makes and $18 for each Super Server, how many of each racket should she make a week to earn the most money? How much does she then earn?

11. A trucker is driving from San Francisco to Los Angeles. He will take a load of two types of cargo. Each crate of cargo A is 5 cubic feet in volume, weighs 100 pounds, and earns $15 for the driver. Each crate of cargo B is 4 cubic feet in volume, weighs 150 pounds, and earns $13 for the driver. If the truck can carry no more than 400 cubic feet of cargo and no more than 9,750 pounds, how many crates of each type should the trucker take to earn the most money? How much money will he earn?

TABLE 4

CUSTOMER	W_1	W_2
A	15	18
B	12	10

TABLE 5

MAN	J_1	J_2
A	23	25
B	19	17

TABLE 6

CONTRACTOR	D_1	D_2
A	9	12
B	8	13

12. A company has two warehouses W_1 and W_2. At the same time it receives an order for 150 units of one of its products from customer A and 200 units of the same item from customer B. The company has 250 units of that item at each of its warehouses. If the shipping cost per unit from each warehouse to each customer is as given in Table 4, how should the company fill the orders in order to minimize shipping costs? What is this cost?

13. A company has two different jobs, a first job J_1 which requires at most 30 hours a week and a second job J_2 which requires at most 60 hours a week. The company has two men A and B to work on these jobs. They each work 40 hours a week. Because of different training and abilities, the company determines that the value to the company per hour of each man doing each job is as given in Table 5. How should the company make the job assignments to get the most value? What is this value?

14. A government agency must resupply two depots D_1 and D_2 with a certain article. D_1 needs 15,000, and D_2 needs 30,000. The agency receives bids which include varying manufacturing and shipping costs from two contractors A and B. Each contractor is willing to supply up to 25,000 articles at a cost per article to the different depots as given in Table 6. If the agency must award contracts in such a way that the total dollar cost to the government is at a minimum, how should it award the contracts? What is the total cost?

15. A manufacturer of ball-point pens makes a profit of 8 cents each on a deluxe model and 5 cents each on a standard model. The manufacturing process requires three main machines A, B, and C. To manufacture 1000 deluxe pens, machine A must be used for 4 hours; machine B, for 2 hours; and machine C, for 1 hour. To manufacture 1000 standard pens, machine A must be used for 1 hour; machine B, for 2 hours; and machine C, for 2 hours. Machine A is available for 20

hours a day; machine B, for 16 hours; and machine C, for 14 hours. If the machines are otherwise available when needed and the manufacturer can sell as many pens as can be produced, how many of each type should be produced a day to earn the most profit? What is this profit?

16. A farmer has at most 200 acres available for planting two crops A and B. The seed for crop A costs \$6 per acre, and the seed for crop B costs \$10 per acre. Labor costs \$40 per acre for A and \$60 per acre for B. The expected income is \$100 per acre for A and \$160 per acre for B. If the farmer wants to spend at most \$1640 for seed and at most \$10,000 for labor, how many acres of each crop should be planted to make the most money? How much does the farmer make?

17. The government of a developing nation has a source of two types of food supplements A and B. One cup of A has 4 grams of protein, 1/8 milligram of riboflavin, and 100 calories and costs 10 cents. One cup of B has 16 grams of protein, 100 calories, and 1/4 milligram of riboflavin and costs 15 cents. The government wants to supplement each adult's diet with at least 48 grams of protein, 500 calories, and 1 milligram of riboflavin, and it can spend at most 75 cents per day per adult. How much of each supplement should they distribute to each adult per day in order to meet the above requirements and minimize the cost per person? What is the cost per person?

Note: Exercises 18–20 have three variables to find. However, in each problem there is a condition given that allows the problem to be reduced to two variables. For example, in Exercise 18, if you let x, y, and z be the amounts (in millions) in each type of investment, then $z = 50 - x - y$.

18. A pension fund has \$50 million to invest in stocks, bonds, and treasury notes. The rules of the fund require at least \$5 million in each type, that at least half be invested in bonds and treasury notes, and that the amount invested in bonds be at most four times the amount invested in notes. The annual yields for the three investments are 6% for treasury notes, 8% for bonds, and 8.5% for stocks. How much should the fund invest in each type of security to bring the highest yield? What is that yield?

19. A farmer wants to plant 200 acres with crops A, B, and C. The seed for crop A costs \$6 per acre; for B, \$10 per acre; and for C, \$8 per acre. Labor costs \$40 per acre for A, \$60 per acre for B, and \$80 per acre for C. She wants to spend at most \$1700 for seed and \$12,000 for labor. If the yield per acre for each crop is \$120 for A, \$150 for B, and \$170 for C, how many acres of each crop should she plant to earn the most money? How much does she earn?

20. A trucking company ships grapefruits, oranges, and tangerines from Florida to Chicago. Each truck carries 170 crates and must include at least 10 but not more than 50 crates of tangerines, at least 20 crates of grapefruits, and at least four times as many crates of oranges as crates of grapefruits. If the profit per crate is \$5 for oranges, \$6 for tangerines, and \$7 for grapefruits, how many crates of each type should be put on each truck in order to maximize profit? What is that profit?

REVIEW EXERCISES

In Exercises 1–14, solve the given system, if possible. For linear systems, state whether the equations are independent, dependent, or inconsistent.

1. $y = x^2 - 1$
   $y = 2x + 2$
2. $2x^2 + y^2 = 3$
   $x + y = 1$
3. $xy = -2$
   $y + x = 1$
4. $3x + 4y = 10$
   $5x - 2y = 8$

5. $2x - 5y = 4$
   $x - 4y = 9$

6. $-\dfrac{2}{x} + \dfrac{3}{y} = 4$
   $\dfrac{2}{x} - \dfrac{1}{y} = -8$

7. $1.65x + 3.89y = 2.47$
   $5.04x - 2.55y = -3.6$

8. $.6x + .9y = 1.7$
   $1.8x + 2.7y = 5.1$

9. $x + y + 2z = 7$
   $x - y - 3z = -6$
   $2x + 3y + z = 4$

10. $3x - 2y = 3$
    $5y - 7z = 4$
    $8z - 5x = 19$

11. $\dfrac{1}{x} + \dfrac{2}{y} + \dfrac{3}{z} = 4$
    $\dfrac{2}{x} + \dfrac{3}{y} - \dfrac{5}{z} = -5$
    $\dfrac{3}{x} + \dfrac{4}{y} - \dfrac{6}{z} = -7$

12. $x - y - 2w = 7$
    $ 2y + z + 3w = -9$
    $x + 2z + 3w = -11$
    $x + 4y + z = 4$

13. $x + 3y - z = 5$
    $3x + 7y = 19$
    $-2x - 8y + 5z = -5$

14. $x - 2y - 4z = 8$
    $2x - 3y - 5z = 18$
    $3x - 5y - 9z = 26$

15. Evaluate the determinant: $\begin{vmatrix} 4 & 5 \\ -2 & 3 \end{vmatrix}$

16. Evaluate the determinant: $\begin{vmatrix} 1 & -1 & 2 \\ 3 & 0 & 2 \\ -2 & 1 & -3 \end{vmatrix}$

17. Determine which of $A + B$, $A + C$, and $B + C$ are defined, and find those sums:

$A = \begin{bmatrix} 5 & 0 & 1 \\ 2 & -3 & -4 \end{bmatrix}$, $B = \begin{bmatrix} -1 & -2 & -3 \\ 4 & 5 & 6 \end{bmatrix}$, $C = \begin{bmatrix} 6 & -7 \\ -7 & 6 \end{bmatrix}$

18. Determine which of AB, AC, BC, BA, CA, and CB are defined, and find those products:

$A = \begin{bmatrix} 1 & 2 & 0 \\ -1 & 0 & 3 \end{bmatrix}$, $B = \begin{bmatrix} 3 & -1 \\ 0 & 4 \end{bmatrix}$, $C = \begin{bmatrix} 1 & 0 \\ 1 & -2 \\ 1 & -1 \end{bmatrix}$

In Exercises 19–20, sketch the graph of the given systems, labeling all vertices.

19. $y \leq 3x - 1$
    $y \leq x + 7$
    $y \geq x - 1$

20. $y - x \leq 3$
    $2y + 3x \leq 16$
    $3y + 2x \geq 14$
    $y \geq 0$

21. A company manufactures two products. They sell one product for $5 per unit and the other for $8.60 per unit. If they sold a total of 30 units for $214.80, how many units of each type did they sell?

22. You have decided to sell cans of mixed nuts. You are using pecans, cashews, and walnuts. You have spent a total of $340 while purchasing pecans at $3 per pound, cashews at $2 per pound, and walnuts at $1 per pound. In all, you have purchased 160 pounds of nuts. If, instead, you had purchased the same amount of pecans, twice as many cashews, and four times as many walnuts, you would have 330 pounds of nuts. How many pounds of each kind of nut did you purchase?

23. Find a system of inequalities, and sketch a graph for the following information: A merchant is buying two brands of a product. Brand A costs $600, brand B costs $200, but the merchant can spend at most $3600. Brand A weighs 80 pounds, brand B weighs 160 pounds, and each requires 3 cubic feet of space. However, the delivery vehicle can hold at most 1040 pounds and at most 24 cubic feet of cargo.

24. A farmer is going to buy bags of soybeans and corn for seed. There are only six bags of soybeans available. The soybeans cost $25 per bag, the corn costs $10 per bag, and the farmer can afford to spend at most $190. One bag of soybeans can be used to plant 1 acre, while one bag of corn plants 2 acres. The farmer wants to plant at most 22 acres. The profit per bag is estimated to be $250 for soybeans and $125 for corn. How many bags of each should the farmer purchase in order to maximize profit?

Right Triangle Trigonometry

CHAPTER SEVEN

The exact origins of trigonometry are lost in prehistory. All that we know is that early man became interested in astronomy for several reasons: for its relation to religion (and astrology), to predict the seasons and planting time, and as an aid to navigation and geography. The mathematics that people developed to describe their observations in astronomy formed the beginnings of trigonometry. Indeed, as trigonometry developed over the centuries, it remained an appendage to astronomy until 1200–1400 A.D.

The early Babylonians divided the circle into 360 equal parts, giving us degrees, perhaps because they thought there were 360 days in a year. The number system they developed was based on the number 60 (a sexagesimal system rather than our decimal system), and because of this, they divided a degree into 60 minutes and a minute into 60 seconds. (Also, the Babylonians divided the day into hours and divided an hour into 60 minutes and a minute into 60 seconds.)

Ancient Egyptian and Babylonian scholars knew theorems about the ratios of sides of similar triangles. Both cultures used rudimentary trigonometric computations by 2000 B.C. However, the Greeks were the first to systematically study trigonometric relationships. The first significant trigonometry book was written by Ptolemy of Alexandria around the second century A.D. This book was later called the *Almagest* by the Arabs for "the greatest" and was to be the definitive text on astronomy and trigonometry until the sixteenth century.

The sine function was invented in India, perhaps around 300 to 400 A.D. However, it was really a modification of a slightly different function used by the Greeks. By the end of the ninth century A.D., all six trigonometric functions, and the identities relating them, were known to the Arabs.

George Rheticus (1514–1577) was the first to define trigonometric functions completely in terms of right triangles. Before him, they were defined in terms of half chords in circles.

SECTION 7.1. ANGLES

Most people reading this already have an intuitive feeling for an angle. For completeness, we give a definition of an angle. It is one of many possible definitions of an angle and is more general than is needed for this chapter, but it will be useful to have this definition for later work.

> **(1)**   **DEFINITION**
> Suppose *l* and *m* are two half lines with a common endpoint *O*,
>
>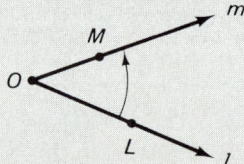
>
> **FIG. 1**
>
> *L* is a point on *l*, and *M* is a point on *m*. See Fig. 1. Then an **angle** from *OL* to *OM* (denoted ∠*LOM*) is the amount of rotation of *l* about *O* (in the plane containing *l* and *m*) required so that *l* will coincide with *m*. The side being rotated, *OL*, is called the **initial side**, the other side, *OM*, is the **terminal side**, and the common point *O* is the **vertex**.

Note that the definition says "an" angle. This is because for a given configuration, the initial side may be rotated clockwise or counterclockwise, and it may be rotated completely around several times before stopping at the terminal side. See Fig. 2 for a few of the angles determined by the same configuration.

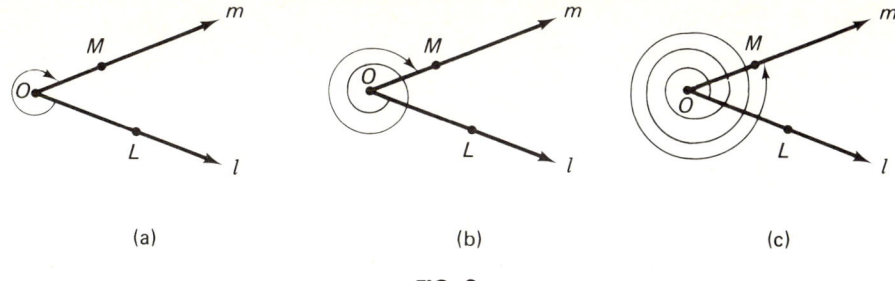

FIG. 2

If the rotation is counterclockwise as in Fig. 2(c), we say that the angle is **positive**, while if the rotation is clockwise as in Fig. 2(a) and (b), we say that the angle is **negative**.

Given an angle, we often find it convenient to measure it, so that we can compare it with other angles or do computations involving the angle. There are two widely used units for measuring angles, called degrees and radians.* People have been measuring angles in degrees since the Babylonians, but in many modern applications the radian is a more useful unit.

> **(2)**   **DEFINITION**
>
> One **degree**, denoted $1°$, is $\frac{1}{360}$ of a complete rotation counterclockwise. If $x \geq 0$ and θ is $x/360$ of a complete rotation counterclockwise, we say that the degree measurement of θ is $x°$ and write $\theta = x°$. If the rotation is clockwise, $\theta = -x°$.

In other words, if a complete rotation is divided into 360 equal parts, each measuring $1°$, the degree measurement of θ is the number of these parts (or fractions thereof) in θ, with the sign assigned according to our sign convention: plus for counterclockwise and minus for clockwise rotations.

EXAMPLE 1

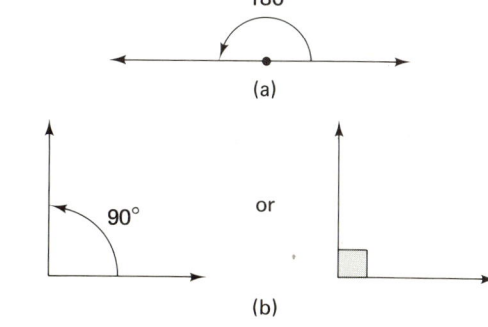

FIG. 3 (continued on next page)

(Rotations are counterclockwise unless stated otherwise.)

(a) If θ is $\frac{1}{2} = \frac{180}{360}$ of a complete rotation, $\theta = 180°$. This is called a **straight angle**.

(b) If θ is $\frac{1}{4} = \frac{90}{360}$ of a complete rotation, $\theta = 90°$. This is called a **right angle**.

(c) If θ is $63.21/360$ of a complete rotation, $\theta = 63.21°$.

(d) If θ is $\frac{6}{5} = \frac{432}{360}$ of a complete rotation, $\theta = 432°$.

(e) If θ is $\frac{142}{360}$ of a complete rotation clockwise, $\theta = -142°$.

---

*There are other units for measuring angles. For example, another unit which is used in Europe, mainly in surveying, is a **gradian** or **grad**. One grad is $1/100$ of a right angle.

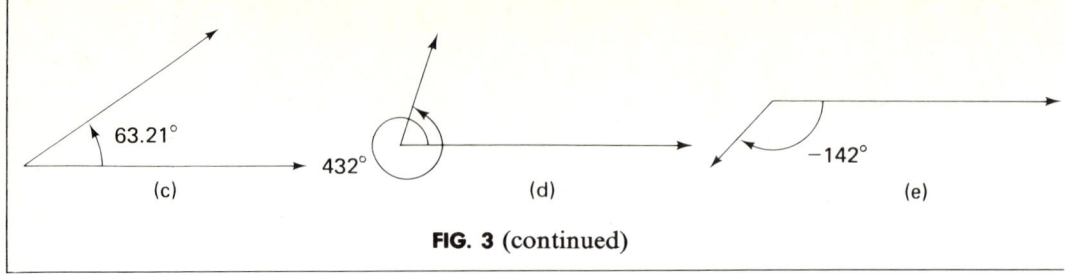

FIG. 3 (continued)

The degree can be further divided into minutes and seconds. One **minute**, denoted 1′, is $\frac{1}{60}$ degree and one **second**, denoted 1″, is $\frac{1}{60}$ minute. Thus 1° is broken up into minutes and seconds in exactly the same way that an hour is. With the advent of computers and hand-held calculators which work with decimal numbers, it is usually best to have degrees in decimal notation. If an angle is given in degrees, minutes, and seconds, this can easily be converted to degrees in decimal notation and vice versa.

EXAMPLE 2   Convert 6°13′18″ to degrees in decimal notation.

SOLUTION   Since 1′ = 60″, we have 1 = 1 min/60 sec. Thus

$$18'' = 18 \text{ sec} \frac{1 \text{ min}}{60 \text{ sec}} = .3'$$

At this stage, 6°13′18″ may be written 6°13.3′. Now, since 1° = 60′, we have $1 = \frac{1°}{60 \text{ min}}.$ Thus

$$13.3' = 13.3 \text{ min} \frac{1°}{60 \text{ min}} \approx .221667°$$

Thus 6°13′18″ ≈ 6.22167°.

EXAMPLE 3   Convert 87.4129° to degrees, minutes, and seconds.

SOLUTION   The 87.4129° is 87° plus a part of a degree, namely, .4129°. We must convert .4129° to minutes and seconds. Since 1° = 60′, we have 1 = 60 min/1 deg, so

$$.4129° = .4129 \text{ deg} \frac{60 \text{ min}}{1 \text{ deg}} = 27.774'$$

The 24.774′ is 24′ plus a part of a minute, namely, .774′. Since 1′ = 60″, we have 1 = 60 sec/1 min, so

$$.774' = .774 \text{ min} \frac{60 \text{ sec}}{1 \text{ min}} = 46.44''$$

Thus 87.4129° = 87°24′46.44″.

A second unit of measurement of angles is radians (rad).

> **(3)** **DEFINITION**
>
> On the initial side *l* of an angle θ, let *P* be the point one unit from the vertex. If θ is positive or 0, let *s* be the length of the circular arc traced out by *P* as *l* rotates through θ; if θ is negative, let *s* be the negative of that length. Then the **radian measure** of θ is *s*, and we write
>
> $$\theta = s \text{ rad}$$

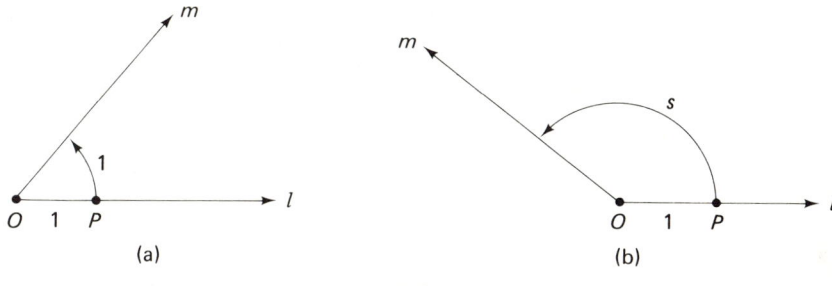

FIG. 4

In Fig. 4(a) and (b), the arc lengths along the circle of radius 1 are 1 and *s*, respectively. Thus Fig. 4(a) illustrates an angle of 1 rad, while Fig. 4(b) illustrates an angle of *s* rad.

Since a circle of radius *r* has a circumference of $2\pi r$, a circle of radius 1 (which is referred to as a unit circle) has circumference 2π. When θ is a complete rotation, *P* travels all around the circumference of a unit circle. This yields the following:

> If θ is a complete rotation (counterclockwise), $\theta = 2\pi$ rad.

On the other hand, we already know that one complete rotation (counterclockwise) is 360°. Consequently, 360° = 2π rad, or

> **(4)**
> $$180° = \pi \text{ rad}$$

From (4), it follows that

> **(5)**
> $$\frac{180°}{\pi \text{ rad}} = 1 \quad \text{and} \quad \frac{\pi \text{ rad}}{180°} = 1$$

Section 7.1 Angles

There are what we use to change from degrees to radians or radians to degrees. The quantity to be changed is simply multiplied by whichever fraction allows you to cancel out the unit you started with and leave the one you desire. When doing conversions or any other calculations involving π, it is important to realize that π is an irrational number and hence may be expressed as an infinite nonrepeating decimal number. The [π] button on your calculator gives π rounded off, usually to 8 or 10 significant digits and displaying 3.1415926. You must remember that this is only an approximation.

We can use the conversion process to find out what 1 rad is in degrees and vice versa:

(6)
$$1 \text{ rad} = 1 \text{ rad} \cdot \frac{180°}{\pi \text{ rad}} \approx 57.2958°$$

and

$$1° = 1 \text{ deg} \cdot \frac{\pi \text{ rad}}{180 \text{ deg}} \approx .0174532 \text{ rad}$$

From this, we see that as a rough estimate 1 rad $\approx$ 60°. Hence

(7)
For rough estimates for conversion, multiply by either

$$1 \approx \frac{60°}{1 \text{ rad}} \quad \text{or} \quad 1 \approx \frac{1 \text{ rad}}{60°}$$

EXAMPLE 4 Convert **(a)** 8.4 rad to degrees and **(b)** 37° to radians by first (i) estimating, using (7) and any other reasonable approximations that would allow mental arithmetic and then (ii) using (5) and your calculator.

SOLUTION

(a) (i) $8.3 \text{ rad} \approx 8 \text{ rad} \cdot \dfrac{60°}{1 \text{ rad}} = 480°$

(ii) $8.3 \text{ rad} = 8.3 \text{ rad} \cdot \dfrac{180°}{\pi \text{ rad}} \approx 475.555°$

(b) (i) $37° \approx 36 \text{ deg} \cdot \dfrac{1 \text{ rad}}{60 \text{ deg}} = \dfrac{6}{10} \text{ rad} = .6 \text{ rad}$

(ii) $37° = 37 \text{ deg} \cdot \dfrac{\pi \text{ rad}}{180 \text{ deg}} \approx .645772 \text{ rad}$

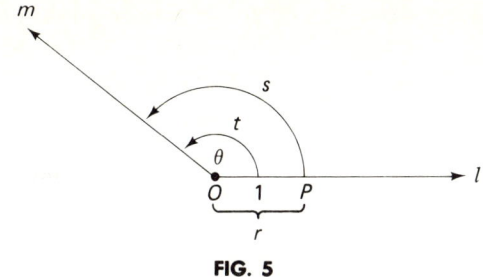

FIG. 5

Let us examine radian measure more closely. A theorem in geometry says that for a fixed angle θ, the ratio of arc length to radius is constant. Thus in Fig. 5, $\frac{t}{1} = \frac{s}{r}$. Since θ has measure t rad, we see the following:

---

(8) **THEOREM**

Let P be a point on the initial side of an angle θ. Suppose P is r units from the vertex O. As the initial side of θ is rotated to the terminal side, suppose P traces out an arc of length s. Then

$$\theta = \frac{s}{r} \text{ rad}$$

Theorem (8) leads to an interesting observation about radians. Suppose, for instance, that $s = 8$ centimeters and $r = 2$ centimeters. Then

$$\frac{s}{r} = \frac{8 \text{ cm}}{2 \text{ cm}} = 4$$

That is, the units cancel, leaving no units whatsoever. Consequently the radian measure of an angle is usually regarded as a real number. For this reason, henceforth we shall use the following:

CONVENTION

We shall write $\theta = s$ instead of $\theta = s$ rad except in places where "rad" is needed for clarity or emphasis.

The formula $\theta = s/r$ from Theorem (8) is very useful in many applied problems.

EXAMPLE 5 Find the approximate diameter of the moon if an observer on the earth measures an angle of $.5°$ from the top to the bottom of the disk when the moon is 240,000 miles away.

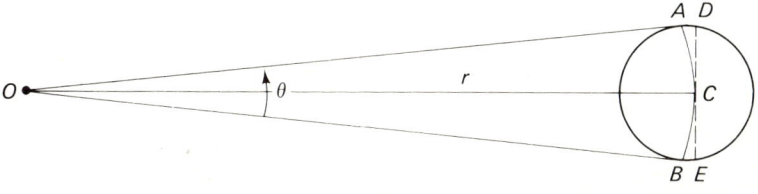

FIG. 6

Section 7.1 Angles 299

SOLUTION Referring to Fig. 6, let O be the eye of the observer, C be the center of the moon, r be the distance from O to C, and DE be a diameter of the moon perpendicular to OC. Finally, let A and B be the points where the circle of radius r centered at O intersects the moon. Then r is approximately the distance from the earth to the moon, and the length of DE is approximately s, where s is the length of the arc ACB. From $s/r = \theta$ (rad) and since $\theta = .5° = .5(\pi/180)$,

$$s = r\theta = 240{,}000(.5)\left(\frac{\pi}{180}\right) \text{ mi}$$

Estimating

$$s = 240{,}000(.5)\left(\frac{\pi}{180}\right) \approx 240{,}000\left(\frac{1}{2}\right)\left(\frac{1}{60}\right) = 2000 \text{ mi}$$

With a calculator, $s \approx 2094.395$ miles, but the distance $r = 240{,}000$ is probably accurate to only two significant figures. Thus the most we can say is that the diameter ≈ 2100 miles.

EXERCISES

In Exercises 1–8, find the degree and radian measure of θ.

1. θ is $\frac{4}{3}$ of a complete rotation (counterclockwise).
2. θ is $\frac{3}{4}$ of a complete rotation (counterclockwise).
3. θ is $\frac{3}{2}$ of a complete rotation (counterclockwise).
4. θ is $\frac{5}{6}$ of a complete rotation (counterclockwise).
5. θ is $\frac{1}{3}$ of a complete rotation (clockwise).
6. θ is $\frac{1}{5}$ of a complete rotation (clockwise).
7. θ is $\frac{8}{9}$ of a complete rotation (clockwise).
8. θ is $\frac{8}{3}$ of a complete rotation (clockwise).

In Exercises 9–20, convert the given angle to degrees in decimal notation.

9. $14°25'12''$
10. $434°43'48''$
11. $-24°15'$
12. $27°43'15''$
13. $231°32'57''$
14. $-38°11'19''$
15. $84°6'7''$
16. $5°5'5''$
17. $-132°41'40''$
18. $189°17''$
19. $1227°4'38''$
20. $-625°10'53''$

In Exercises 21–26, convert the given angle to degrees, minutes, and seconds.

21. $9.468°$
22. $106.53°$
23. $-75.6135°$
24. $804.06°$
25. $431.98°$
26. $-147.317°$

In Exercises 27–42, convert the angle from degrees to radians. First estimate your answer.

27. $135°$
28. $15°$

29. 75°
30. −30°
31. 120°
32. 270°
33. 180°
34. −45°
35. −85.4°
36. 18°
37. 311°14′19″
38. −160.41°
39. −46°12′
40. 52°28′48″
41. 861.428°
42. 74°26′24″

In Exercises 43–58, convert the angle from radians to degrees in decimal notation. Estimate first.

43. $\frac{7\pi}{6}$ rad
44. $\frac{\pi}{6}$ rad
45. $\frac{5\pi}{4}$ rad
46. $-\frac{\pi}{4}$ rad
47. 9π rad
48. $\frac{\pi}{3}$ rad
49. $\frac{5\pi}{2}$ rad
50. $-\frac{\pi}{2}$ rad
51. 3 rad
52. 2 rad
53. .8 rad
54. −1 rad
55. 8.2 rad
56. .4 rad
57. .349 rad
58. $-\frac{2\pi}{3}$ rad

In Exercises 59–66, convert the angle from radians to degrees, minutes, and seconds.

59. 2.5 rad
60. 4.13 rad
61. .682 rad
62. 13.9 rad
63. −1.6 rad
64. −3.07 rad
65. −.091 rad
66. −22 rad

67. The length of an arc of a circle is 7.00 centimeters. Find the angle (in degrees and radians) it makes at the center of the circle if the radius is 3.491 centimeters.

68. The arc of a circle 10.00 centimeters long makes an angle of 35.81° at the center. Find the radius.

69. An arc of a circle makes an angle of 173.91° at the center. If the radius of the circle is 15.72 meters, find the length of the arc.

70. A wheel makes 120 revolutions in 1 minute. In 1 second, through how many radians and how many degrees does it turn?

71. The large hand on a wall clock is 1 foot, 7 inches long. In 20 minutes
   a. Through what angle (in both degrees and radians) does the hand move?
   b. How many inches does its tip move?

72. The hour hand on a clock is 10 inches long. In a 50-minute class period,
   a. Through what angle (in both degrees and radians) does the hand move?
   b. How far does its tip move?

73. The second hand on a wristwatch is $\frac{3}{4}$ inch long. In one second,
   a. Through what angle (in both degrees and radians) does the hand move?
   b. How far does its tip move?

74. Your horse on a merry-go-round is 18.3 meters from the center. Suppose the merry-go-round makes 12.4 revolutions per ride. If you take three rides,
  a. Through what angle do you move?
  b. How many meters do you travel?

75. The height of a mountain 10 miles away makes an angle of 3.2° at the eye of an observer. How high is the mountain to the nearest 10 feet?

76. Suppose that the sun is 93 million miles away from the earth and that its diameter makes an angle of 32' at a point on the earth. What is the approximate diameter of the sun?

77. A satellite is in a circular orbit 1000 miles above the surface of the earth. The radius of the earth is about 4000 miles. How many miles (to two significant figures) does the satellite travel while sweeping an angle of 40° with the center of the earth?

78. If the satellite in Exercise 77 makes 1 revolution every 80 minutes, how fast (in miles per hour) is it traveling?

79. Suppose a bike wheel has a radius of 27 inches. If the bike is rolling at 10 miles per hour, through what angle does a spoke turn in 1 minute?

80. If the bike wheel in Exercise 79 made 200 revolutions in a minute, how fast was the bike traveling?

SECTION 7.2. THE TRIGONOMETRIC FUNCTIONS

The trigonometric functions were originally defined as functions of angles in right triangles and then extended to more general situations. To follow this approach, we need to review briefly some facts about triangles.

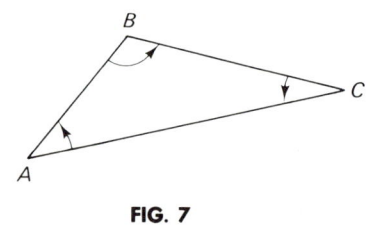

FIG. 7

First we must clarify what we mean by an angle in a triangle. See Fig. 7. An angle in a triangle is obtained by rotating the initial side counterclockwise through the interior of the triangle until the first time it coincides with the terminal side. Thus the three angles are each positive. Moreover, the following is true (see Exercise 35 for a proof):

The sum of the three angles of any triangle is $180° = \pi$ rad.

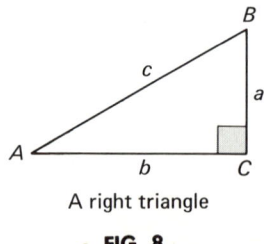

A right triangle

FIG. 8

If one of the angles of a triangle is a right angle, the triangle is called a **right triangle**. See Fig. 8. The sides of a right triangle adjacent to the right angle are called **legs** of the triangle, while the side opposite the right angle is called the **hypotenuse** of the triangle. In any right triangle, the Pythagorean theorem holds.*

*See Exercise 34 for a proof. Also, see *The Pythagorean Proposition* by E. S. Loomis, NCTM, Ann Arbor, MI, 1968, which contains 256 different proofs of the Pythagorean theorem.

PYTHAGOREAN THEOREM

In a right triangle with legs of lengths a and b and hypotenuse of length c (as in Fig. 8),

$$a^2 + b^2 = c^2$$

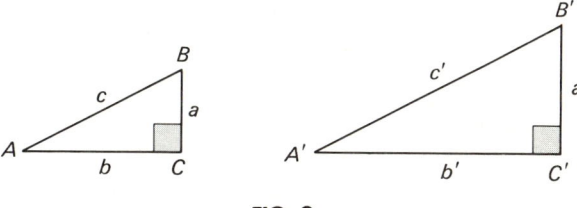

FIG. 9

Now let us consider the two right triangles in Fig. 9. Suppose $\angle A = \angle A'$. Then since $\angle C = \angle C' = \pi/2$ (rad) and $\angle A + \angle B + \angle C = \pi = \angle A' + \angle B' + \angle C'$ (the sum of the angles of any triangle is π), it follows that $\angle B = \angle B'$. Thus if $\angle A = \angle A'$, the two right triangles have precisely the same three angles. Two triangles with the same three angles are called **similar**, and it is known that in similar triangles the ratios of the lengths of corresponding sides are the same. Thus in our example

$$\frac{a}{c} = \frac{a'}{c'}, \qquad \frac{b}{c} = \frac{b'}{c'}, \qquad \frac{a}{b} = \frac{a'}{b'}, \qquad \text{etc.}$$

If we know the sides of one triangle and one side of a second triangle which is similar to the first, we can use the ratios of corresponding sides to find the remaining sides of the second triangle.

EXAMPLE 1 Find the coordinates of P if P is one unit from the origin along the line from the origin to $(-4, 3)$. See Fig. 10(a).

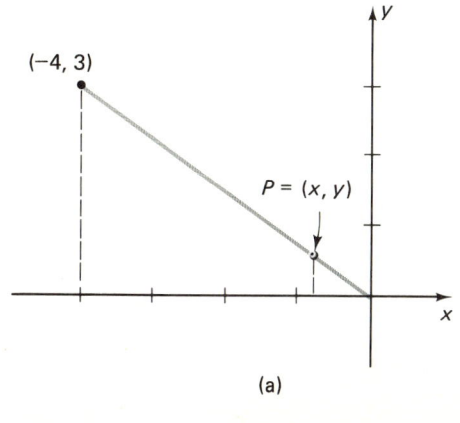

(a)

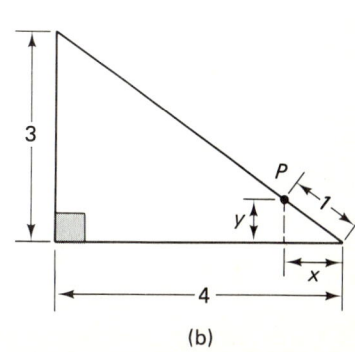

(b)

FIG. 10

Section 7.2 The Trigonometric Functions

SOLUTION If we drop perpendiculars from $(-4, 3)$ and P to the x-axis, we obtain right triangles with the *lengths* of the sides indicated in Fig. 10(b). By the Pythagorean theorem, the larger triangle has hypotenuse of length 5. Since the two triangles are similar,

$$\frac{|y|}{1} = \frac{3}{5} \quad \text{or} \quad |y| = \frac{3}{5} \quad \text{and} \quad \frac{|x|}{1} = \frac{4}{5} \quad \text{or} \quad |x| = \frac{4}{5}$$

Since (x, y) is in the second quadrant, $(x, y) = (-\frac{4}{5}, \frac{3}{5})$.

We are now ready to define the trigonometric functions of any acute angle θ. Let A be the vertex of θ. Pick any point B other than A on the terminal side of θ and drop a perpendicular from B to the initial side of θ, labeling the point of intersection as C. See Fig. 11(a). We now have a right triangle ABC with sides a, b, c. By the discussion preceding Example 1, we know that the six ratios a/c, b/c, a/b, b/a, c/b and c/a are independent of what point B we picked on the terminal side of θ. Thus the ratios depend only on the size of θ; i.e., they are a function of θ alone. These six functions are the **trigonometric functions** and are called the **sine** of θ (denoted by $\sin \theta$), the **cosine** of θ (denoted by $\cos \theta$), the **tangent** of θ (denoted by $\tan \theta$), the **cotangent** of θ (denoted by $\cot \theta$), the **secant** of θ (denoted by $\sec \theta$), and the **cosecant** of θ (denoted by $\csc \theta$).

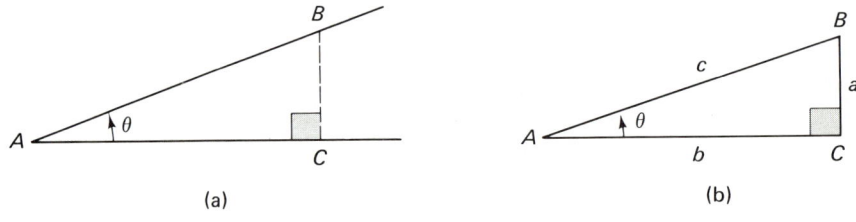

FIG. 11

The trigonometric functions are defined as follows [referring to Fig. 11(b)]:

(9)	DEFINITION		
$\sin \theta = \dfrac{a}{c} = \dfrac{\text{opposite side}}{\text{hypotenuse}}$		$\csc \theta = \dfrac{c}{a} = \dfrac{\text{hypotenuse}}{\text{opposite side}}$	
$\cos \theta = \dfrac{b}{c} = \dfrac{\text{adjacent side}}{\text{hypotenuse}}$		$\sec \theta = \dfrac{c}{b} = \dfrac{\text{hypotenuse}}{\text{adjacent side}}$	
$\tan \theta = \dfrac{a}{b} = \dfrac{\text{opposite side}}{\text{adjacent side}}$		$\cot \theta = \dfrac{b}{a} = \dfrac{\text{adjacent side}}{\text{opposite side}}$	

EXAMPLE 2 In each of the following, find the six trigonometric functions of θ:

SOLUTION We apply the definitions (9):

(a) $\sin\theta = \frac{3}{5}$ $\csc\theta = \frac{5}{3}$ (b) $\sin\theta = \frac{12}{13}$ $\csc\theta = \frac{13}{12}$

$\cos\theta = \frac{4}{5}$ $\sec\theta = \frac{5}{4}$ $\cos\theta = \frac{5}{13}$ $\sec\theta = \frac{13}{5}$

$\tan\theta = \frac{3}{4}$ $\cot\theta = \frac{4}{3}$ $\tan\theta = \frac{12}{5}$ $\cot\theta = \frac{5}{12}$

From the definitions (9), it is immediate that there are relationships among the trigonometric functions. For instance, each of the functions defined on the same line of (9) is clearly the reciprocal of the other. Thus

(10)
$$\csc\theta = \frac{1}{\sin\theta}, \quad \sec\theta = \frac{1}{\cos\theta}, \quad \text{and} \quad \cot\theta = \frac{1}{\tan\theta}$$

Moreover, $\frac{\sin\theta}{\cos\theta} = \frac{a/c}{b/c} = \frac{a}{b} = \tan\theta$. Since in addition $\cot\theta = \frac{1}{\tan\theta}$, we have the following:

(11)
$$\tan\theta = \frac{\sin\theta}{\cos\theta} \quad \text{and} \quad \cot\theta = \frac{\cos\theta}{\sin\theta}$$

One further relation follows directly from the Pythagorean theorem: $a^2 + b^2 = c^2$. If both sides are divided by c^2, one obtains $(a/c)^2 + (b/c)^2 = 1$. By using $\sin\theta = a/c$ and $\cos\theta = b/c$, the following holds:

(12)
$$\sin^2\theta + \cos^2\theta = 1$$

Note: In the last equation we have used a convention regarding powers of trigonometric functions. When raising a trigonometric function, say $\sin\theta$, to a power n, we write $\sin^n\theta$ instead of $(\sin\theta)^n$ whenever $n \neq -1$. [We shall see later that the notations $\sin^{-1}\theta$, $\cos^{-1}\theta$, etc., have a special meaning. To denote $1/\sin\theta$, $1/\cos\theta$, etc., using exponents, write $(\sin\theta)^{-1}$, $(\cos\theta)^{-1}$, etc.]

Equations (10), (11), and (12) are called the **elementary trigonometric identities**. You can sometimes use these identities to find the remaining trigonometric functions from one of them.

EXAMPLE 3 If θ is an angle in a right triangle and $\cos \theta = \frac{3}{5}$, find the remaining trigonometric functions of θ.

SOLUTION Since $\cos \theta = \frac{3}{5}$, from (10) we have

$$\sec \theta = \frac{1}{\cos \theta} = \frac{5}{3}$$

From $\sin^2 \theta + \cos^2 \theta = 1$, identity (12), we have

$$\sin^2 \theta + \tfrac{9}{25} = 1 \quad \text{or} \quad \sin^2 \theta = \tfrac{16}{25}, \quad \sin \theta = \pm \tfrac{4}{5}$$

Since θ is in a right triangle, $\sin \theta$ is positive, so $\sin \theta = \frac{4}{5}$. Then from (10),

$$\csc \theta = \frac{1}{\sin \theta} = \frac{5}{4}$$

Finally, from (11),

$$\tan \theta = \frac{\sin \theta}{\cos \theta} = \frac{4/5}{3/5} = \frac{4}{3} \quad \text{and} \quad \cot \theta = \frac{\cos \theta}{\sin \theta} = \frac{3}{4}$$

ALTERNATE SOLUTION

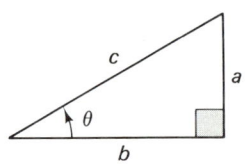

The same problem may be done directly from the definitions of the trigonometric functions and the Pythagorean theorem. Since θ is an angle in a right triangle and $\cos \theta = \frac{3}{5}$, we can look at any right triangle with one angle θ in which the ratio $b/c = \frac{3}{5}$. The easiest is to choose $b = 3$ and $c = 5$. Then $a^2 + b^2 = c^2$, or $a^2 + 9 = 25$, or $a^2 = 16$. Thus, $a = 4$ (not $a = \pm 4$ since a length is positive). Consequently,

$$\sin \theta = \frac{a}{c} = \frac{4}{5}, \; \tan \theta = \frac{a}{b} = \frac{4}{3}, \; \csc \theta = \frac{c}{a} = \frac{5}{4},$$

$$\sec \theta = \frac{c}{b} = \frac{5}{3}, \text{ and } \cot \theta = \frac{b}{a} = \frac{3}{4}$$

Trigonometry can be extremely useful in solving word problems which involve triangles. Let us look at an example.

EXAMPLE 4 From a point 80 feet from the base of a smokestack along level ground, it is observed that the angle between the ground and a line drawn from the point to the top of the smokestack has tangent .635. How tall is the smokestack?

SOLUTION

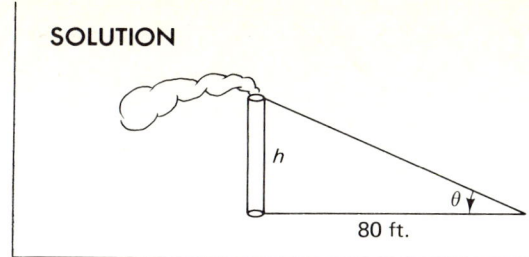

First we draw a picture as shown. We are given that $\tan \theta = .635$. However, since $\tan \theta = h/80$,

$$h = 80 \tan \theta = 80(.635) = 50.8 \text{ feet}$$

EXERCISES

In Exercises 1–6, assume the following triangles are similar:

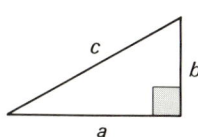 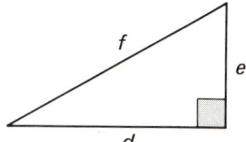

1. If $a = 4$, $b = 3$, and $d = 8$, find c, e, f, and then find a/c and d/f.
2. If $a = 3$, $c = 5$, and $e = 10$, find b, d, f, and then find b/c and e/f.
3. If $d = 36$, $e = 15$, and $a = 12$, find f, b, c, and then find b/a and e/d.
4. If $a = 24$, $c = 25$, and $d = 36$, find b, e, f, and then find a/c and d/f.
5. If $a = 2$, $b = 3$, and $f = 12$, find c, d, e, and then find a/b and d/e.
6. If $e = 3$, $f = 4$, and $c = 96$, find a, b, d, and then find f/e and c/b.
7. Find the coordinates of a point P if P is one unit from the origin along the line from the origin to $Q = (-3, 4)$.
8. Find the coordinates of a point P if P is one unit from $Q = (-5, -12)$ along the line from the origin to Q.
9. Find the coordinates of a point P if P is along the line from the origin through $Q = (2, -4)$ and P is twice as far from the origin as Q is.
10. If P has x-coordinate -7 and is along the line from the origin through $Q = (-\frac{4}{5}, \frac{3}{5})$, how far is P from the origin?

In Exercises 11–18, find the six trigonometric functions of θ.

11.

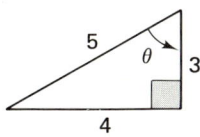

12.

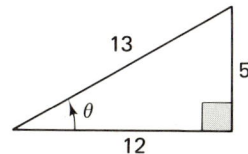

13.

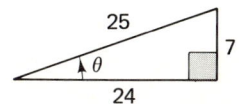

14.

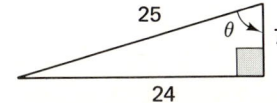

15. 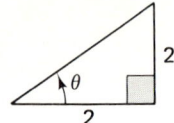   **16.**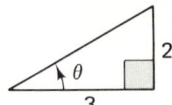

17. 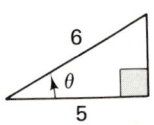   **18.**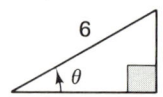

Find the trigonometric functions of θ in Exercises 19–26. θ is an angle in a right triangle.

19. $\sin \theta = \frac{12}{13}$ **20.** $\tan \theta = \frac{3}{4}$

21. $\cot \theta = 1$ **22.** $\cos \theta = \frac{3}{4}$

23. $\cos \theta = \frac{1}{2}$ **24.** $\sin \theta = \frac{1}{2}$

25. $\sin \theta = \frac{2}{3}$ **26.** $\tan \theta = \frac{3}{2}$

27. Find two further relationships like equation (12) among the trigonometric functions by dividing the Pythagorean relation $a^2 + b^2 = c^2$ first by a^2 and then by b^2.

28. Suppose a 12-meter ladder is resting against a wall with the top of the ladder at the top of the wall. Suppose the ground, the wall, and the ladder form a right triangle. If θ is the angle between the ground and the ladder and $\sin \theta = .623$, how tall is the wall, and how far from the wall is the base of the ladder?

29. One leg of a triangle is 9.2 feet long. The tangent of the angle between the other leg and the hypotenuse is 1.56. How long is the other leg?

30. John and Diane are on the edges of opposite sides of a canyon. Nancy is on the same edge as Diane, 20 meters from her. A line from John to Diane would be perpendicular to a line from Diane to Nancy. The angle between the line from Nancy to Diane and the line from Nancy to John has tangent 1.437. How far apart are John and Diane?

31. A telephone pole is perpendicular to the ground. A taut anchoring cable 60 feet long runs from the ground to the top of the pole (so that the ground, cable, and pole form a right triangle). If the tangent of the angle between the ground and the cable is 2.6, how tall is the pole (to the nearest foot)?

32. A 3-meter ramp is put from a level driveway to a loading dock. If the cosine of the angle between the ramp and the driveway is .95, how high is the loading dock?

33. A 6-foot electric cord is stretched taut from the bottom of a clock on a wall to a socket on the same wall. The socket is 1 foot above the floor. If you drew a line from the clock straight down to the floor, it would make an angle with the cord whose cosine is .9. How high above the floor is the bottom of the clock?

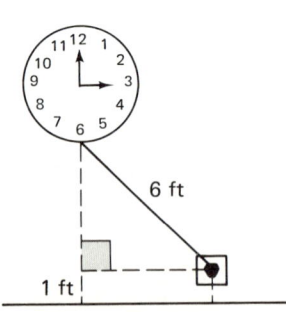

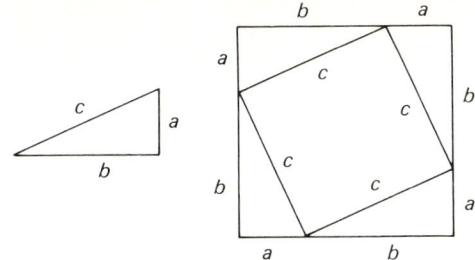

34. Prove the Pythagorean theorem which says that in a right triangle labeled as shown, $a^2 + b^2 = c^2$. Do so by computing the area of the accompanying large square in two ways: first in the normal way and then as the sum of the areas of the four copies of the triangle and the inner square. Set the two equal, and see what happens.

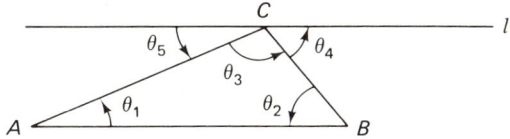

35. Give a proof that the sum of the angles in a triangle is 180° as follows. Given any triangle ABC, draw a line through the vertex C parallel to side AB and label as shown. State the sum of $\theta_5 + \theta_3 + \theta_4$, giving a reason. Give a reason why $\theta_5 = \theta_1$. Give a reason why $\theta_4 = \theta_2$. What can you conclude about $\theta_1 + \theta_2 + \theta_3$?

SECTION 7.3. EVALUATION OF TRIGONOMETRIC FUNCTIONS

In this section we consider how to compute the values of the trigonometric functions of an acute angle θ when you are given just the angle. This is relatively easy if $\theta = \pi/6 = 30°$, $\theta = \pi/4 = 45°$, or $\theta = \pi/3 = 60°$. Because of this and the fact that they arise fairly often, these angles are called **special angles**. (Observe that we are following our convention of not writing "rad" if radians is the measure but *always writing the symbol* ° if degrees is the measure.)

Trigonometric Functions of $\pi/4 = 45°$

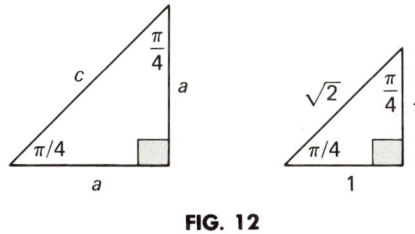

FIG. 12

If one acute angle of a right triangle has measure $\pi/4 = 45°$, then so does the other since the sum of the angles in any triangle is $\pi = 180°$. Any triangle having two equal angles is called **isosceles**; the sides opposite the equal angles are also equal. In the present case, denote the equal sides by a. Then the hypotenuse c can be found by the Pythagorean theorem, $c^2 = a^2 + a^2 = 2a^2$, or $c = \sqrt{2}\,a$. We could now compute the trigonometric ratios, but for any two choices of a positive number for a, the resulting triangles would be similar and the corresponding trigonometric ratios the same. Consequently, for convenience we choose $a = 1$ (which means $c = \sqrt{2}$). See Fig. 12. Then

$$\sin \frac{\pi}{4} = \cos \frac{\pi}{4} = \frac{1}{\sqrt{2}} \approx .707107$$

$$\csc \frac{\pi}{4} = \sec \frac{\pi}{4} = \frac{\sqrt{2}}{1} \approx 1.41421$$

$$\tan \frac{\pi}{4} = \cot \frac{\pi}{4} = \frac{1}{1} = 1$$

Remark: Although most of you have been taught to rationalize denominators, we shall not emphasize that here. This is because when evaluating on the calculator, it is quicker to evaluate $1/\sqrt{2}$ (press [2][√][1/x]) than to evaluate $\sqrt{2}/2$ (press on alg.: [2][√][÷][2][=]).

Trigonometric Functions of $\pi/6 = 30°$ and $\pi/3 = 60°$

If one acute angle of a right triangle has measure $\pi/6$, then the other acute angle has measure $\pi/3$. Consequently, we can use the same right triangle to compute the trigonometric functions of both angles. To see how the sides of such a triangle are related, we start with an equilateral triangle. See Fig. 13. Each of the angles of triangle ABC measures $\pi/3 = 60°$, and each of the sides has equal length s. The altitude CD bisects the side AB and the angle $\angle ACB$. Thus triangle CAD is a $\pi/6, \pi/3, \pi/2$ right triangle ($30°, 60°, 90°$ right triangle). To compute the trigonometric functions, we may assign any nonzero value to s, since for all such values the triangles are similar, and hence corresponding ratios are the same. For convenience, we assume $s = 2$. Then the length of AD, which we denote by $|AD|$, is $|AD| = s/2 = 1$ (the side opposite the hypotenuse in a $30°, 60°, 90°$ right triangle is half as long as the hypotenuse), and $|CD| = \sqrt{2^2 - 1^2} = \sqrt{3}$. See Fig. 14.

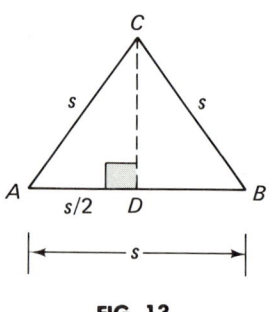

FIG. 13

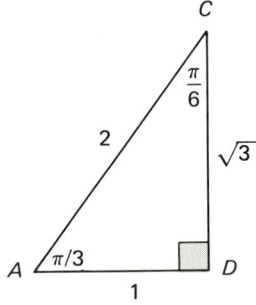

FIG. 14

Thus

$$\sin \frac{\pi}{3} = \frac{\sqrt{3}}{2} \approx .866025 \qquad \csc \frac{\pi}{3} = \frac{2}{\sqrt{3}} \approx 1.15470$$

$$\cos \frac{\pi}{3} = \frac{1}{2} = .5 \qquad\qquad \sec \frac{\pi}{3} = 2$$

$$\tan \frac{\pi}{3} = \sqrt{3} \approx 1.73205 \qquad\qquad \cot \frac{\pi}{3} = \frac{1}{\sqrt{3}} \approx .577350$$

and

$$\sin \frac{\pi}{6} = \frac{1}{2} = .5 \qquad\qquad \csc \frac{\pi}{6} = 2$$

$$\cos \frac{\pi}{6} = \frac{\sqrt{3}}{2} \approx .866025 \qquad\qquad \sec \frac{\pi}{6} = \frac{2}{\sqrt{3}} \approx 1.15470$$

$$\tan \frac{\pi}{6} = \frac{1}{\sqrt{3}} \approx .577350 \qquad\qquad \cot \frac{\pi}{6} = \sqrt{3} \approx 1.73205$$

We now know how to compute the trigonometric functions of the special angles $\pi/4$, $\pi/3$, and $\pi/6$. These occur so frequently that you should either memorize them or, better yet, *memorize the triangles from which they are obtained* (Figs. 12 and 14).

Unfortunately, in general if you are given an angle θ, there is no easy way to compute the trigonometric functions of θ directly, but there are sophisticated techniques for doing this that require many calculations. In the past, extensive tables of the trigonometric functions of various angles were provided so that it would be practical to use the functions. Nowadays these functions are built into scientific calculators, making it much easier to use trigonometric functions than it was in the past.

Most scientific calculators have a switch or a button which allows you to place the calculator in degree or radian mode. If the calculator is in degree mode, all calculations involving trigonometric functions are performed with the assumption all angles are in degrees. Similarly, in radian mode, the same calculations would be performed with the assumption all angles are in radians. *Make sure you have read your manual and know how to put your calculator into both radian and degree mode.*

WARNING

A very common error which occurs when using calculators in trigonometry is to have the calculator in the wrong mode. If you are getting incorrect answers, the first thing to check is that your calculator is in the correct mode.

Once the calculator is in the appropriate mode, the value of $\sin \theta$, $\cos \theta$, or $\tan \theta$ is found in exactly the same way on *all* machines, whether RPN or algebraic: You enter θ *first*, and then press the [sin], [cos], or [tan] button, *not* the other way around.

EXAMPLE 1 In (a)–(c), the angles are measured in degrees, so your calculator should be in degree mode.
(a) Find sin 23°. Press [23][sin], obtaining .390731.
(b) Find tan 54°. Press [54][tan], obtaining 1.37638.
(c) Find $\sin^2 18° + 2 \tan 37° \cos 8°$. Using algebraic without parentheses, press [8][cos][×][37][tan][×][2][+][18][sin][x^2][=], obtaining 1.58793.

In (d) and (e), the angles are measured in radians, so your calculator should be in radian mode.
(d) Find sin 1. Press [1][sin], obtaining .841471.
(e) Find $3 \cos^4 .83$. Using algebraic, press [.83][cos][x^y][4][×][3][=], obtaining .622326.

Calculators do not have buttons for all the trigonometric functions. Instead they have only [sin], [cos], and [tan] buttons. This is because these are the most commonly used trigonometric functions, and we have seen that the remaining trigonometric functions can be obtained by using the identities (10): $\csc \theta = 1/\sin \theta$, $\sec \theta = 1/\cos \theta$, and $\cot \theta = 1/\tan \theta$.

EXAMPLE 2 Find csc 31°. Since $\csc 31° = 1/\sin 31°$, in degree mode press [31][sin][1/x], obtaining 1.94160.

Let us look at an example to begin to see how knowledge of the trigonometric functions of the angles of a right triangle can be useful. First we define some terms which occur frequently. The **angle of elevation** of point B from point A is the angle θ between a horizontal line through A and the line through A and B. See Fig. 15. The word *elevation* indicates that B is higher than A. The **angle of depression** of B from A is the same, except that B is below A.

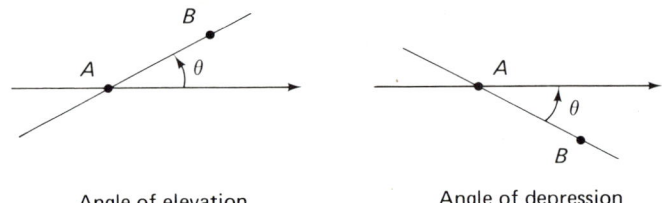

Angle of elevation   Angle of depression

FIG. 15

EXAMPLE 3 The angle of elevation of the top of a 40-foot tree from a point on the ground is 30° (the ground is level). How far is the point from the base of the tree?

SOLUTION

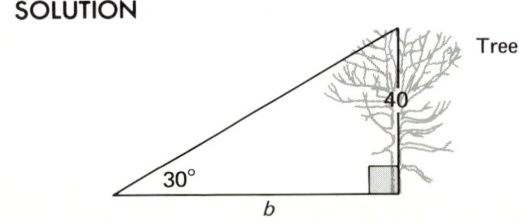

We have the accompanying picture. Since $\tan 30° = 40/b$,

$$b = \frac{40}{\tan 30°} = \frac{40}{1/\sqrt{3}}$$

$$= 40\sqrt{3} \text{ ft} \approx 69.2820 \text{ ft}$$

EXERCISES

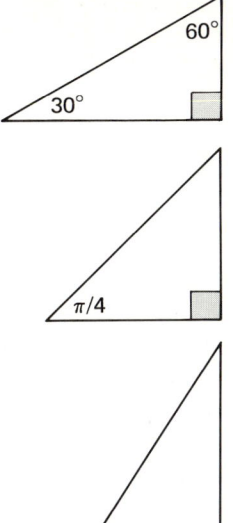

1. Copy the triangle shown on a separate piece of paper. *Without* looking back in the text, label the sides with appropriate lengths, and then compute the six trigonometric functions of 30° and of 60°.

2. Copy the triangle shown on a separate piece of paper. *Without* looking back in the text, label the sides with appropriate lengths, and then compute the six trigonometric functions of $\pi/4$.

3. Copy the triangle shown on a separate piece of paper. *Without* looking back in the text, label the sides with appropriate lengths, and then compute the six trigonometric functions of $\pi/3$ and $\pi/6$.

4. Fill in the following chart without using a calculator:

	$\sin\theta$	$\cos\theta$	$\tan\theta$	$\csc\theta$	$\sec\theta$	$\cot\theta$
30°						
45°						
60°						

In Exercises 5–31, evaluate the trigonometric functions, using a calculator only where necessary.

5. $\tan 60°$
6. $\cos 12.3°$
7. $\tan 57°$
8. $\csc 22°$
9. $\cos 34°15'$
10. $\cot 71.36°$
11. $\csc 63.5°$
12. $\sec 82°$
13. $\sin 45°$
14. $\tan 30°$
15. $\cos 30°$
16. $\sin 60°$
17. $\sin 1$
18. $\cos .5$
19. $\cos \dfrac{\sqrt{3}}{2}$
20. $\tan \dfrac{\pi}{4}$
21. $\tan \dfrac{\pi}{5}$
22. $\sin \dfrac{\pi}{6}$
23. $\csc 1.1$
24. $\sec .75$
25. $\cot \dfrac{2}{5}$
26. $\csc \dfrac{\pi}{3}$
27. $\sec \dfrac{\pi}{5}$
28. $\cot \dfrac{\pi}{5}$
29. $\cos 12°17'31''$
30. $\sin 58°46'6''$
31. $\cot 26°32''$

In Exercises 32–39, evaluate the given expressions.

32. $3.6 \sin 26.2° \tan 74°$

33. $8 \sin^3(\pi/4)$

34. $\sec 14° \cos 41° + 5 \csc 19.42°$

35. $\cot^2 .86 + \dfrac{\sin .48}{1 + \sec .3}$

36. $6.753 + 4.2 \tan 1.14 - \dfrac{3}{\csc .75}$

37. $\sin^3 32° + \sin 24° \tan 14°$

38. $\dfrac{\cos .42 + \tan 1.03}{\sin .17 - \cos .68}$

39. $(\tan 32° + 1)(\cos 84° - \sin 38°)\csc^3 51°$

40. Verify the identities $\sin^2 \theta + \cos^2 \theta = 1$ and $1 + \tan^2 \theta = \sec^2 \theta$ for $\theta = 15°, 42°, \pi/3, 1$. (If they do not work, explain why.)

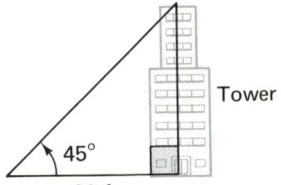

41. From a point 36 feet from the base of a tower along level ground, the angle of elevation of the top of the tower is 45°. How tall is the tower?

42. The angle of elevation of the top of a tower from the tip of its shadow (along level ground) is $\pi/6$. If the tower is 60 feet tall, how long is its shadow?

43. A field is in the shape of a right triangle. One leg is 521 yards long, and that leg makes an angle of 60° with the hypotenuse. Find the area of the field. [Recall that the area of a triangle is $\frac{1}{2}$(base)(height).]

44. A plane is 6 miles directly over city A. If the plane flew on a path that makes an angle of 30° with the ground, it would land at airport B. How far is it from A to B?

45. On a baseball diamond, what is the sine of the angle between a line from home to first and a line from first to third?

46. The angle of depression of a car from a balloon is 45°. If the balloon is 2000 feet from the car, how high is the balloon?

47. A building (with straight sides) is 43.6 meters high. From a point on the edge of the roof it is observed that if you look directly across the street, the angle of depression to the near edge of the base of the building there is 60°. How far apart are the buildings?

48. A partially full oil storage tank is a cylinder 50 feet tall and 30 feet in diameter. From a point on the top edge, the angle of depression of the surface of the oil on the opposite side of the tank is 30°. What is the height of the oil in the tank? How many cubic feet of oil are in the tank?

SECTION 7.4. INVERSE TRIGONOMETRIC FUNCTIONS

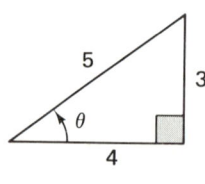

FIG. 16

Suppose the right triangle shown in Fig. 16 is given. We would like to determine the angle θ. It seems likely that we could use our knowledge of the trigonometric functions of θ, say the fact that $\sin \theta = \frac{3}{5}$, to find θ. This is the case, for whenever θ is an acute angle in a right triangle ($0 < \theta < \pi/2$), the trigonometric functions each have an inverse function.

We first show that sine and cosine have inverse functions. In Fig. 17, let OA have length 1, let l be a half line originating at O, let AB be perpendicular to l, and let θ be angle $\angle BOA$. Then since $|OA| = 1$, $\sin \theta = |AB|$ and $\cos \theta = |OB|$.

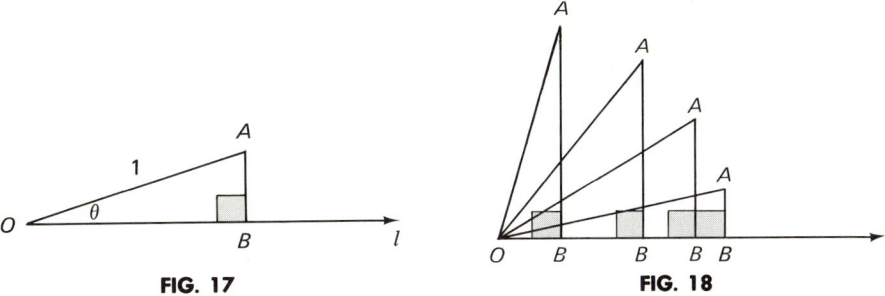

FIG. 17

FIG. 18

Now consider what happens when θ (in radians) increases from "nearly" 0 to "nearly" $\pi/2$. We see from Fig. 18 that $|AB| = \sin \theta$ increases from "nearly" 0 to "nearly" 1, and for no two different values of θ is $|AB|$ the same length.

Thus we have seen the following:

(13)

i. If $0 < \theta < \pi/2$ and $u = \sin \theta$, then $0 < u < 1$.

ii. For each u, $0 < u < 1$, there is a unique θ, $0 < \theta < \pi/2$, such that $\sin \theta = u$.

Similarly, the corresponding result holds for $\cos \theta$. For as θ increases from "nearly" 0 to "nearly" $\pi/2$, we see from Fig. 18 that $|OB| = \cos \theta$ decreases from "nearly" 1 to "nearly" 0, and for no two different values of θ is $|OB|$ the same length.

(14)

i. If $0 < \theta < \pi/2$ and $u = \cos \theta$, then $0 < u < 1$.

ii. For each u, $0 < u < 1$, there is a unique θ, $0 < \theta < \pi/2$, such that $\cos \theta = u$.

The statements (13) and (14) are exactly what we need for the sine and cosine functions to have inverse functions. The inverse sine function is usually denoted by either $\sin^{-1}$ (like f^{-1}) or arcsin. The other inverse trigonometric functions are denoted similarly.

(15)   DEFINITION

i. Suppose $0 < u < 1$. Then $\sin^{-1} u = \theta$ (or $\arcsin u = \theta$) if and only if $0 < \theta < \pi/2$ and $\sin \theta = u$.

ii. Suppose $0 < u < 1$. Then $\cos^{-1} u = \theta$ (or $\arccos u = \theta$) if and only if $0 < \theta < \pi/2$ and $\cos \theta = u$.

Of course θ may be measured in degrees; then the restriction on θ in (15i) and (15ii) would be $0° < \theta < 90°$. (However, if we want $\theta = \sin^{-1} u$ or $\theta = \cos^{-1} u$ in degrees, we shall explicitly state this. Otherwise they will be measured in radians.)

You can think of the equation $\theta = \sin^{-1} u$ as "θ is the angle, $0 < \theta < \pi/2$, whose sine is u." In particular, note the following:

Warning: $\sin^{-1} u$ *does not mean* $1/\sin u$. For example, $\sin^{-1} .5 = \pi/6 \approx .523599$ (since $\sin \pi/6 = .5$), whereas $1/\sin .5 \approx 1/.47942554 \approx 2.08583$.

EXAMPLE 1 Find $\sin^{-1}(\sqrt{2}/2)$ in both radians and degrees.

SOLUTION Since $\sin(\pi/4) = \sqrt{2}/2$, we have $\sin^{-1}(\sqrt{2}/2) = \pi/4 = 45°$.

Scientific calculators easily compute $\sin^{-1}$ and $\cos^{-1}$.

First make sure the calculator is in the proper mode.

Then the process is basically the same on most machines, whether RPN or algebraic. There usually is a button marked [INV] or [ARC] or [F]. To find $\sin^{-1} u$, you enter u *first*, next press that button marked [INV] or [ARC] or [F], and then press [sin]. The computation of $\cos^{-1}$ is similar. However, *there are variations. Consult your manual* if you have any problems.

EXAMPLE 2 Find $\cos^{-1} .23$ in both degree and radian modes.

SOLUTION Put your calculator into the proper mode and then press [.23][INV][cos], obtaining $76.7029°$ or 1.33872 (rad).

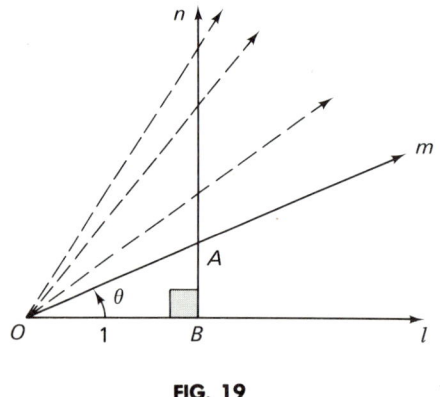

FIG. 19

We now turn to the tangent function, which also has an inverse if $0 < \theta < \pi/2$. In Fig. 19, let θ be an angle with vertex O, let l be the half line originating at O which is the initial side of θ, and let m be the terminal side of θ. Further, let B be the point on l one unit from O, let n be the line perpendicular to l at B, and let A be the intersection of m and n. Then θ is angle $\angle BOA$. Since AOB is a right triangle,

$$\tan \theta = \frac{|AB|}{|OB|} = \frac{|AB|}{1} = |AB|$$

Consider what happens as θ increases from "nearly" 0 to "nearly" $\pi/2$. We see that m gets rotated and that $|AB|$ increases from "nearly" 0 and becomes arbitrarily large. In particular, for no two different values of θ is $|AB|$ the same length, and $|AB|$ may be any positive real number. Thus since $|AB| = \tan \theta$, we have seen the following:

> 1. If $0 < \theta < \pi/2$ and $u = \tan\theta$, then $u > 0$.
> 2. For each $u > 0$, there is a unique θ, $0 < \theta < \pi/2$, such that $\tan\theta = u$.

As with sine and cosine, this is exactly what we need for tangent to have an inverse function.

> (16)   **DEFINITION**
>
> Suppose $u > 0$. Then $\tan^{-1} u = \theta$ (or $\arctan u = \theta$) if and only if $0 < \theta < \pi/2$ and $\tan\theta = u$.

Again, θ may be in degrees, in which case the restriction on θ is $0° < \theta < 90°$, and you can think of the equation $\theta = \tan^{-1} u$ as "θ is the angle, $0 < \theta < \pi/2$, whose tangent is u."

EXAMPLE 3 Find $\tan^{-1} 1$ in both radians and degrees.

SOLUTION Since $\tan(\pi/4) = 1$, we have $\tan^{-1} 1 = \pi/4 = 45°$.

Your calculator computes $\tan^{-1}$ similar to the way it computes $\sin^{-1}$ and $\cos^{-1}$.

EXAMPLE 4 Use your calculator to find $\arctan 10$ in both radian and degree modes.

SOLUTION Put your calculator into the proper mode and then press [10][INV][tan], obtaining 1.47113 (rad) or 84.2894°.

The remaining trigonometric functions have inverses, but we shall not use them here.

EXERCISES

In Exercises 1–6, evaluate in both degrees and radians without using your calculator.

1. $\sin^{-1} .5$
2. $\cos^{-1} \dfrac{1}{\sqrt{2}}$
3. $\cos^{-1} .5$
4. $\tan^{-1} \dfrac{1}{\sqrt{3}}$
5. $\sin^{-1} \dfrac{\sqrt{3}}{2}$
6. $\tan^{-1} \sqrt{3}$

In Exercises 7–30, evaluate in both degrees and radians, using your calculator only where necessary.

7. $\sin^{-1} \frac{\sqrt{2}}{2}$
8. $\cos^{-1} .5$
9. $\tan^{-1} 1$
10. $\tan^{-1} 5.3$
11. $\sin^{-1} \frac{\pi}{4}$
12. $\cos^{-1} \frac{\sqrt{3}}{2}$
13. $\cos^{-1} .1$
14. $\sin^{-1} .9$
15. $\sin^{-1} .8$
16. $\sin^{-1} .7$
17. $\sin^{-1} .6$
18. $\sin^{-1} .5$
19. $\sin^{-1} .4$
20. $\sin^{-1} .3$
21. $\sin^{-1} .2$
22. $\sin^{-1} .1$
23. $\tan^{-1} .2$
24. $\tan^{-1} .5$
25. $\tan^{-1} .8$
26. $\tan^{-1} \frac{3}{2}$
27. $\tan^{-1} 3$
28. $\tan^{-1} 10$
29. $\tan^{-1} 100$
30. $\tan^{-1} 1000$

Exercises 31–36 refer to the following triangle:

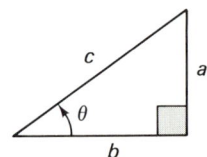

31. If $a = 3$ and $c = 5$, find $\sin \theta$, and then find θ (in degrees).
32. If $b = 11$ and $c = 15.3$, find $\cos \theta$, and then find θ (in radians).
33. If $a = .603$ and $b = .128$, find $\tan \theta$, and then find θ (in degrees).
34. If $a = 643$ and $c = 751$, find $\sin \theta$, and then find θ (in radians).
35. If $b = 1.89$ and $c = 10.4$, find $\cos \theta$, and then find θ (in degrees).
36. If $a = 37.2$ and $b = 129.48$, find $\tan \theta$, and then find θ (in radians).
37. The roads joining three villages form a right triangle, with the right angle at village X. It is 4 kilometers from village X to village Y and 9 kilometers from village Y to village Z. At what angle (in degrees) do the roads meet at village Z?
38. A small rocket is to be launched from the ground aimed toward a big tree on the other side of a brick wall. If it is 12 feet from the launching position to the base of the wall and the wall is 10 feet tall, what must be the minimum angle in *whole* degrees between the rocket and the ground so that when launched, the rocket will clear the wall?
39. A 12-foot ladder is leaning against the wall in a room. The top of the ladder is 7 feet, 4 inches from the floor. What (acute) angle (in degrees) does the ladder make with the floor?
40. A tree is 17.2 meters tall. What is the angle of inclination (in radians) of the top of the tree from a point which is 41.3 meters from the base of the tree along level ground?
41. Two flagpoles are 50 feet apart. One pole is 80 feet tall, while the other is 60 feet tall. Find the angle of depression (in radians) of the top of the shorter pole from the top of the taller pole.

42. A golfer is about to hit his golf ball. It is 183 yards straight west from his ball to the hole. He hits the ball straight but slightly off-line. It ends up 24 yards straight north of the hole. What is the angle (in degrees) between the intended line of flight and the actual line of flight of the ball?

43. It is 153 meters from an observation point on the ground to a hang-glider flying in the air. It is 47 meters along level ground from that point to a point directly under the glider. What is the angle of depression of the observation point from the glider?

44. The cross section of a roof (and the beams across at the base) is in the shape of an isosceles triangle. It is 36 feet across at the base, and each of the slant edges is 26 feet long. What is the pitch of the roof? (The pitch is the angle between the base and a slanted side of the triangle).

SECTION 7.5. APPLICATIONS OF RIGHT TRIANGLES

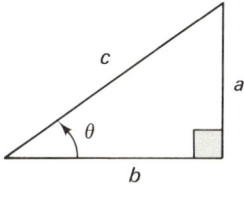

FIG. 20

There are many instances where you can use right triangles to find the length or width or height of some object which would otherwise be very difficult to measure. The principle involved is that if you are given one acute angle and one side of a right triangle or two sides of a right triangle, you can use trigonometry to find the remaining sides and angles.

EXAMPLE 1 In Fig. 20, suppose $\theta = 1$ and $a = 3.2$ centimeters. Find b and c.

SOLUTION Since $\tan \theta = \dfrac{a}{b}$, we have $b = \dfrac{a}{\tan \theta} = \dfrac{3.2}{\tan 1} \approx 2.05469$ cm. Since $\sin \theta = \dfrac{a}{c}$, we have $c = \dfrac{a}{\sin \theta} = \dfrac{3.2}{\sin 1} \approx 3.80286$ cm.

EXAMPLE 2 In Fig. 20, suppose $b = 4.69$ kilometers and $c = 16.38$ kilometers. Find a and θ.

SOLUTION Since $\cos \theta = b/c = 4.69/16.38$, $\theta = \cos^{-1}(4.69/16.38) \approx 1.28041$. To find a, you can either use the Pythagorean theorem,

$$a = \sqrt{c^2 - b^2} = \sqrt{16.38^2 - 4.69^2} \approx 15.6942$$

or trigonometry, say the fact that $\sin \theta = a/c$, to obtain

$$a = c \sin \theta = 16.38 \sin\left[\cos^{-1}\left(\dfrac{4.69}{16.38}\right)\right] \approx 15.6942$$

Examples 1 and 2 illustrate how, given sufficient information about a right triangle, you can find "missing" information. When doing word problems, you should draw pictures if at all possible and look for triangles, particularly right triangles. If a right triangle is involved, then after you label the parts, you can proceed as in Examples 1 and 2 to find the desired information.

EXAMPLE 3 Suppose you are at the base of a vertical cliff and you wish to know its height. You walk out along level ground to a point 100 feet from the base of the cliff, and from that point you measure the angle of elevation of the top of the cliff to be 68.1°. What is its height?

SOLUTION

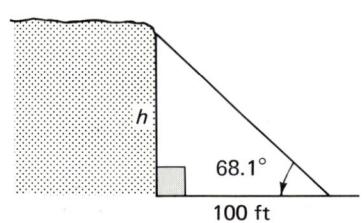

First draw a picture, letting h be the unknown height. Then $\tan 68.1° = h/100$, or $h = 100 \tan 68.1° \approx 248.758$ feet.

If in the original problem we had stated that the data are accurate to three significant figures, we would now round the answer to $h \approx 249$ feet.

In many cases the information given makes it natural to draw a triangle, but the triangle is not a right triangle. In that case, the best thing to do in order to use the techniques we have developed is to drop an altitude from one of the vertices, thereby forming two right triangles. In Fig. 21, we illustrate how dropping an altitude from vertex C of triangle ABC forms two right triangles ACD and BCD whether the base D of the altitude falls within the original triangle or not.

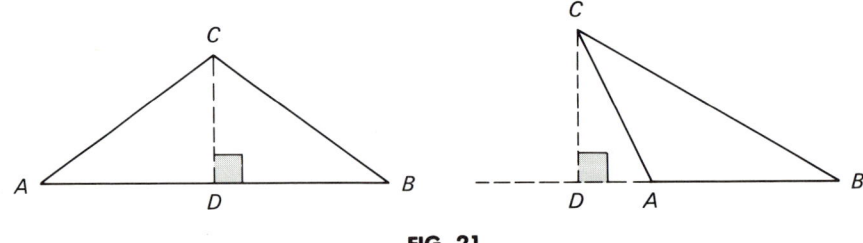

FIG. 21

EXAMPLE 4 Suppose you are on level ground and you wish to measure the height of a mountain above ground level. From some point A along the level ground you determine that the angle of elevation of the peak is .95, while from a point B which is 200 feet farther from the mountain, the angle of elevation of the peak is .92. Find the height.

SOLUTION First we draw a picture:

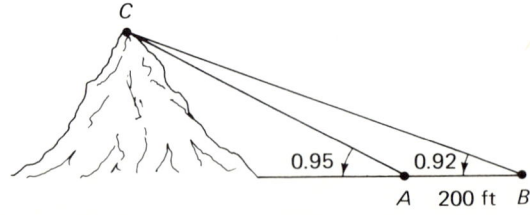

RIGHT TRIANGLE TRIGONOMETRY

We see triangle ABC, but the height h is the length of an altitude from vertex C. We drop this altitude, forming right triangles ACD and BCD. In each right triangle we know an acute angle, but we do not know any of the sides.

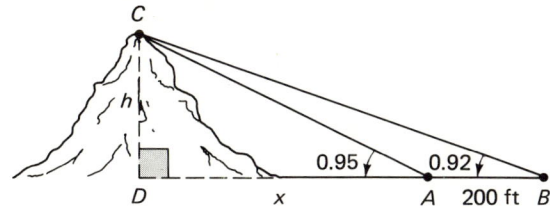

We must use some algebra to solve this problem. Let $x = |AD|$. Then $|BD| = x + 200$. From triangle ACD, $\tan .95 = h/x$, while from triangle BCD, $\tan .92 = h/(x + 200)$. Thus we have two equations in two unknowns:

$$\tan .95 = \frac{h}{x} \quad \text{and} \quad \tan .92 = \frac{h}{x + 200}$$

We can solve these equations for h as follows: From the first equation, $x \tan .95 = h$, or $x = h/\tan .95$. From the second equation,

$$h = (x + 200) \tan .92 = x \tan .92 + 200 \tan .92$$

Substituting $x = h/\tan .95$ into this equation yields

$$h = \frac{h}{\tan .95} \tan .92 + 200 \tan .92$$

$$h - h\frac{\tan .92}{\tan .95} = 200 \tan .92$$

$$h\left(1 - \frac{\tan .92}{\tan .95}\right) = 200 \tan .92$$

$$h = \frac{200 \tan .92}{1 - \dfrac{\tan .92}{\tan .95}} \approx 4314.94 \text{ ft}$$

If the problem had stated that the data are accurate to two significant figures, we would now round the answer to $h \approx 4300$ feet.

EXERCISES

Remember that unless otherwise stated, we assume all data are accurate to eight significant figures and that the answers are given correct to six significant figures.

In Exercises 1–19, use the given quantities to find the quantities indicated. (See the triangle on the next page.)

1. $\theta = 15°$, $b = 12.4$ feet. Find a, c.
2. $\theta = .6$, $a = 38$ feet. Find b, c.

3. $\theta = .432$, $c = 29.2$ meters. Find a, b.
4. $\theta = 78°24'$, $a = 6.1$ kilometers. Find b, c.
5. $\varphi = 1.12$, $c = 3.6$ inches. Find a, b.
6. $\varphi = .58$, $a = 13.1$ centimeters. Find b, c.
7. $\tan \theta = 1.1$, $c = 19.3$ feet. Find a, b.
8. $\tan \theta = .43$, $a = 6.5$ inches. Find b, c.
9. $\cos \theta = .125$, $a = 48.7$ centimeters. Find b, c.
10. $\cos \theta = .64$, $b = 183$ inches. Find a, c.
11. $\sin \theta = .312$, $b = 245$ inches. Find a, c.
12. $\sin \theta = .8743$, $c = 12$ inches. Find a, b.
13. $\sin \theta = .56$, $a = 93$ feet. Find b, c.
14. $a = 3$ meters, $b = 4$ meters. Find c, θ, φ.
15. $a = 5.23$ feet, $c = 12.81$ feet. find θ, φ.
16. $b = 4.635$ kilometers, $c = 21.289$ kilometers. Find θ, φ.
17. $a = 976$ miles, $b = 1329$ miles. Find θ, φ.
18. $a = .0115$ millimeter, $c = .0925$ millimeter. Find θ, φ.
19. $b = 47,531$ yards, $c = 51,982$ yards. Find θ, φ.

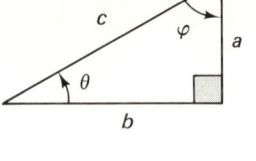

20. A tree casts a shadow (along level ground) which is 82 feet long. If the angle of elevation of the top of the tree from the tip of the shadow is 42°, how tall is the tree?
21. Steps to the entrance to a building rise a total of 2 feet. They are to be torn out and replaced by a ramp for wheelchairs, inclined at 10°. How long (along the slant) is the ramp?
22. A mountain on the moon makes a shadow, which we can measure fairly accurately from earth to be 5 kilometers. We can also measure the angle of the sun's rays to be .23 rad. Approximately how tall is the mountain (in meters)? Assume two significant figure accuracy.

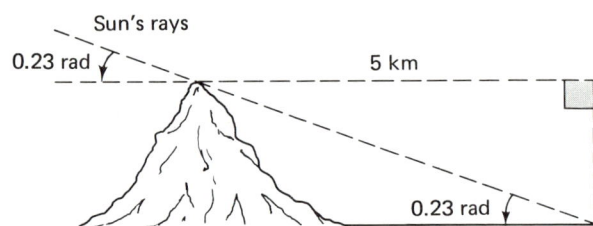

23. A plane flying level at 10,000 feet flies directly over an observer on the ground. Two minutes later the angle of elevation of the plane from the observer is .1411 rad. How many miles has the plane flown in those 2 minutes? How fast was the plane flying (in miles per hour)? Assume four significant figure accuracy.
24. A seaplane is on an approach path for landing which is 12° below horizontal. It passes over a ship at an altitude of 1626 feet. How far from the ship will the plane touch down? Assume four significant figure accuracy.
25. A field is in the shape of a right triangle, with a fence along the two legs and a river (with a straight bank) along the hypotenuse. If the side along the river is 436.9 meters long and that side makes an angle of .53627 with one of the legs, how long is the fence?
26. A 15-foot guy wire runs from the ground (which is level) to the middle of a telephone pole. The guy wire makes an angle of 39°18'12" with the pole. How tall is the pole?

27. An oil well driller has just struck oil, and a geyser is spouting forth. From a point along level ground 62 feet from the base of the geyser the driller observes that the angle of elevation of the top is about 37°. To the nearest foot, how tall is the geyser?

28. A straight railroad track is rising at a constant rate. At the end of 1 mile, it is 138.3 feet higher than at the start. What is the angle between the rail and a horizontal?

29. A laser is on level ground 6.570 meters from a mark at the base of a wall. At what angle with the ground should the laser be inclined in order to point at an object 8.831 meters up the wall directly over the mark? Assume four significant figure accuracy.

30. The ground around a smokestack is level. From the ground at a distance of 126 feet from the base, the angle of elevation of the top is 34.31509°. What is the angle of inclination of the top from a point on the ground 205 feet from the base?

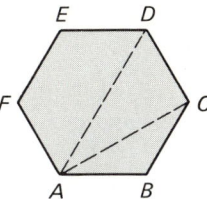

31. In the regular hexagon shown, find the length of AC and of AD if each side has length 6.23. Use the fact that in a regular n-gon, each angle is $[(n-2)/n]180°$.

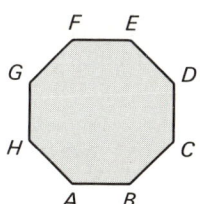

32. In the regular octagon shown, find the length of AC if each side has length 7.

33. The angle of elevation of a mountain peak from a point on level ground is 65°, while from a point 100 feet farther from the peak it is 61°. How high is the mountain to the nearest foot?

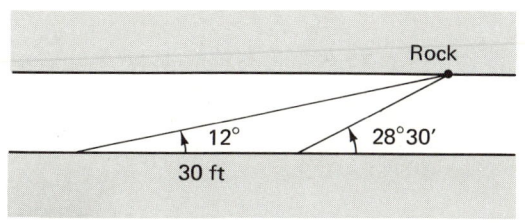

34. You wish to determine how far it is across a river with straight, parallel banks. As you stand on one bank, you notice that your line of sight to a rock farther up on the opposite bank makes an angle of 12° (with your bank). After you walk 30 feet along the bank, you notice the angle is now 28°30′. How wide is the river to the nearest foot?

35. As you stand on a bridge which is 100 feet above the water, you are looking at an approaching barge. If the angle of depression of the front of the barge is .507 rad and the angle of depression of the rear of the barge is .303 rad, find the length of the barge to the nearest foot.

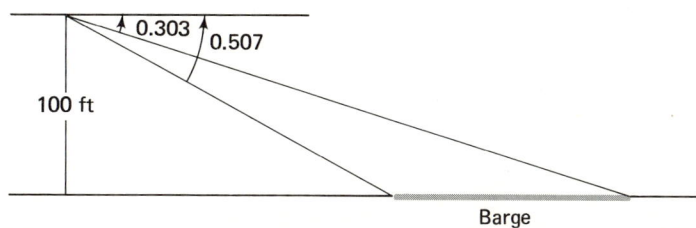

36. You are at the window of a building looking out from a point which is 93.071 feet above ground level (the ground nearby is flat). As you look at a nearby building, you notice that the angle of depression to its base is 72° and the angle of elevation to its peak is 81.62°. How far apart are the buildings, and how tall is the second building?

37. A plane is flying at 480 miles per hour on a level course directly away from you. At one time you observe that its angle of elevation is 72°, while 5 minutes later it is 5°. How much above the ground level where you are is the plane flying? Assume two significant figure accuracy.

38. Three stars (label them A, B, and C) are the vertices of a right triangle with right angle at A. The angle $\angle ACB$ is known to be .7364. A spaceship traveling along the line from A to C observes that the angle between its path and a line from it to B is .8946. If it is 1.64 light-years from A to B (a light-year is the distance a ray of light would travel in a year), how far is it from the spaceship to C?

39. A field is in the shape of a right triangle with hypotenuse of length 7250 yards and one angle .632. Suppose we label the vertices A, B, and C, where the right angle is at A and $\angle ABC = .632$. Then BC is the hypotenuse, so $|BC| = 7250$. A path starts at vertex C and proceeds in a straight line through the field until it meets side AB in a point D and then proceeds from there along side AB to vertex B. The angle $\angle ACD = .237$. How long is the path to the nearest yard? (*Hint*: First find $|AC|$ and $|AB|$, and then find $|AD|$ and $|CD|$.)

40. A ship is sailing on a straight course. A lighthouse is sighted ahead, and it is noted that the angle between a line to the lighthouse and the ship's course is 27°. After traveling 2.65 kilometers, the ship has gone past the lighthouse. The angle between the ship's course and a line to the lighthouse is now 42°. How close (to the nearest meter) did the ship come to the lighthouse?

41. Two radio towers are standing in a field. One is 123.2 feet tall. From a point part way up the second, it is observed that the angle of depression of the base of the first tower is 34°13′, while the angle of elevation of its top is 51°35′. How far apart are the towers?

42. The roads between cities A, B, and C form a right triangle, with right angle at city A. City D is located along the road from A to B and is 6.4 kilometers from A. There is a straight road from C to D, and road CD meets road AD in an angle of 53°. Road CB meets road AB in an angle of 39°. Find the remaining distances between the towns.

43. A plane is flying directly away from you at a height of 6996 feet. It is flying at a constant speed on a level course. At one time, you observe that the angle of elevation of the plane is 47°, while 2 minutes later it is 10°. How fast is the plane flying (in miles per hour)?

REVIEW EXERCISES

In Exercises 1–6, convert to degrees in decimal notation.

1. 62°18′18″
2. −41°13′27″
3. 4817°52′7″
4. 3°36″
5. −187°35′43″
6. 95°42′

In Exercises 7–12, convert to radians.

7. 60°
8. −240°
9. 84°
10. 154.286°
11. −471.32°
12. 321°14′51″

In Exercises 13–18, convert to degrees in decimal notation.

13. $\dfrac{2\pi}{3}$
14. $-\dfrac{\pi}{6}$
15. 2.3
16. -3.4172
17. $-.047$
18. 631.2

19. A right triangle has legs of lengths 1 and 2. Find the six trigonometric functions of the angle θ opposite the side of length 1.

20. A right triangle has hypotenuse of length 3 and one leg of length 1. Find the six trigonometric functions of the angle θ opposite the side of length 1.

21. If θ is an angle in a right triangle and $\sin\theta = \tfrac{1}{4}$, find the remaining trigonometric functions of θ.

22. If θ is an angle in a right triangle and $\cot\theta = 2$, find the remaining trigonometric functions of θ.

In Exercises 23–32, evaluate using a calculator.

23. $\sin 1°$
24. $\sin 1$
25. $\csc 61.4°$
26. $\cos 40°10'20''$
27. $\sec 27.23°$
28. $\sec .912$
29. $\tan 14.0279°$
30. $\cot 1.0279°$
31. $\cot 1.0279$
32. $\cos .8206$

33. Find $(\sin .294 - 3\cos .82)^3$.

34. Find $(\cot^4 18° - \tan 41.2°)/(3\sec^2 52° + 1)$.

35. A balloon is rising straight up above a point A on the ground. An observation point B is 600 feet from A along the (level) ground. At one time, it is noticed that the angle of elevation of the balloon is 30°. A short while later the angle of elevation is 60°. How much did the balloon rise in the meantime?

36. Three small islands, A, B, and C, are the vertices of a right triangle with right angle at C. If it is 28.2 miles from A to B and the angle between AB and AC is 30°, how far is it from A to C and from B to C? If a plane went down somewhere inside the triangle, how much area would have to be searched?

In Exercises 37–44, evaluate in both degrees and radians. Use your calculator only when necessary.

37. $\sin^{-1} .5$
38. $\cos^{-1} \dfrac{\sqrt{2}}{2}$
39. $\sin^{-1} .48$
40. $\tan^{-1} .5$
41. $\tan^{-1} 1$
42. $\cos^{-1} .91$
43. $\sin^{-1} \dfrac{\sqrt{3}}{2}$
44. $\tan^{-1} \sqrt{2}$

45. In a right triangle, one angle is 39.2°, and the hypotenuse has length 6.3. Find the length of the legs.

46. In a right triangle, one angle is .517 rad, and the adjacent leg has length 345.2. Find the length of the other leg.

47. One angle θ in a right triangle has $\sin\theta = .234$. If the side opposite θ has length .2897, find the length of the adjacent side.

48. One angle θ in a right triangle has $\cos\theta = .234$. If the side opposite θ has length 31.2, find the length of the adjacent side.

49. The side of a hill makes an angle of 12° with a horizontal. A horizontal mine shaft goes into the hill. A vertical air shaft is to be drilled into the mine shaft starting from a point 150 feet up along the side of the hill from the mine entrance. How long will the air shaft be?

50. A ship is proceeding on a straight course. At one time, the angle between the ship's course and a line from the ship to a small island is 18.2°. After the ship proceeds a mile, the angle has changed to 37.43°. How close will the ship pass to the island?

51. You are at the top of a building 432 feet tall. As you look at a nearby smaller building, you observe that the angle of depression of the base is 81° and the angle of depression of the top is 49.26°. Find the height of the nearby building.

General Trigonometry

CHAPTER EIGHT

We mentioned in Chapter 1 that the Arabs assimilated mathematics from the neighbors they conquered, organized and expanded it somewhat, and then passed it along to the Europeans.

The origin of the word *sine* illustrates this. When the Hindus invented the sine function, they called it *jiva* for "half chord," which the Arabs adopted as *jiba*. However, around 1150, an Englishman, Robert of Chester, translated the Arabic into Latin (the scientific language of Europe until the nineteenth century). He confused *jiba* with the Arabic word *jaib*, which means "bay" or "inlet," and so he used the Latin word meaning "bay" or "inlet," which is *sinus*. Later, sinus was shortened to sine.

As mathematics spread from Arabia to Europe, it very slowly evolved. An Arab, Nasir Eddin al-Tusi (1201–1274), began to treat trigonometry as a field of study separate from astronomy. But this approach was not adopted in Europe until Regiomontanus (1436–1476), who also began to use decimals in trigonometric tables, rather than sexagesimal fractions.

Francois Viete (1540–1603) developed a generalized approach to trigonometry. However, he was hesitant to use negative numbers, and he did not know about periodicity.

The Bernoulli brothers Jean (1667–1748) and Jacques (1654–1705) were the first to consider trigonometric functions as functions of numbers, not just angles.

Gilles Persone de Roberval (1602–1675), a contemporary of Descartes, was the first to graph the sine curve, though he only did it for $0° \leq x \leq 90°$.

SECTION 8.1. THE TRIGONOMETRIC FUNCTIONS

In this section, we shall extend the definitions of the trigonometric functions to arbitrary angles.

First, suppose θ is any angle. We say θ is in **standard position** if its vertex is the origin and its initial side is the positive x-axis. Figure 1 illustrates some angles in standard position.

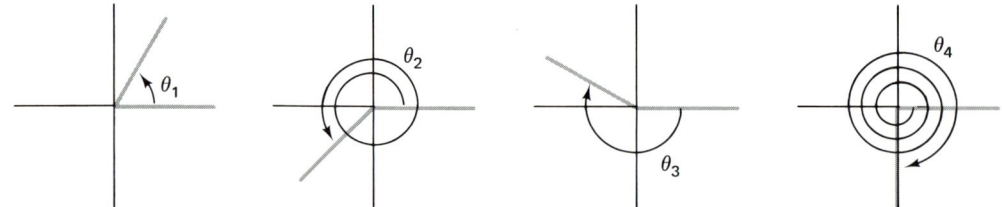

FIG. 1

We shall use the phrase "θ is in Q_3" to mean "θ is an angle in standard position, and its terminal side is in the third quadrant." We shall use the corresponding phrase when the terminal side is in any other quadrant. For example, in Fig. 1, θ_1 is in Q_1, θ_2 is in Q_3, and θ_3 is in Q_2. An angle in standard position with its terminal side on an axis is not in a quadrant; such an angle is called a **quadrantal angle** or a **between quadrant angle**. In Fig. 1, θ_4 is a quadrantal angle.

If θ is in Q_1 and is acute, we can apply the techniques of the previous chapter to find the trigonometric functions of θ. First, let $P = (x,y)$ be any point (different from the origin) on the terminal side of θ. Then $r = \sqrt{x^2 + y^2}$ is the distance from P to the origin. See Fig. 2(a). The definitions of the trigonometric functions state that

$$\sin\theta = \frac{y}{r} \qquad \csc\theta = \frac{r}{y}$$
$$\cos\theta = \frac{x}{r} \qquad \sec\theta = \frac{r}{x}$$
$$\tan\theta = \frac{y}{x} \qquad \cot\theta = \frac{x}{y}$$

In this form, the six trigonometric functions generalize immediately to arbitrary angles.

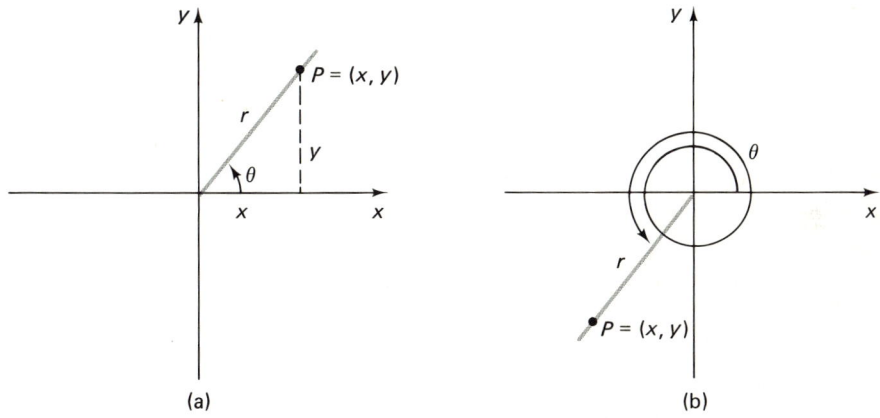

FIG. 2

(1)   **DEFINITION**

Let θ be an angle in standard position, let $P = (x, y)$ be any point (different from the origin) on the terminal side of θ, let $r = \sqrt{x^2 + y^2}$ be the distance from P to the origin. See Fig. 2(b). Then

$$\sin\theta = \frac{y}{r} \qquad\qquad \csc\theta = \frac{r}{y}, \quad y \neq 0$$
$$\cos\theta = \frac{x}{r} \qquad\qquad \sec\theta = \frac{r}{x}, \quad x \neq 0$$
$$\tan\theta = \frac{y}{x}, \quad x \neq 0 \qquad \cot\theta = \frac{x}{y}, \quad y \neq 0$$

It can easily be shown that the trigonometric functions of θ as defined in (1) do not depend on which point P (different from the origin) is chosen on the terminal side. We express this by saying that the trigonometric functions are **well defined**. The proof uses straight calculations if θ is a quadrantal angle and similar triangles otherwise. (See Exercises 48–51.)

From Definition (1), observe that $\tan\theta$ and $\sec\theta$ are undefined if $x = 0$, while $\cot\theta$ and $\csc\theta$ are undefined if $y = 0$.

Let us see how we can use the definitions.

EXAMPLE 1

Suppose $(-2, -3)$ is a point on the terminal side of θ. See Fig. 3. Find the trigonometric functions of θ.

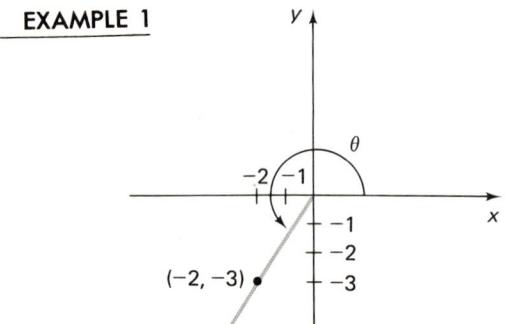

FIG. 3

SOLUTION Set $r = \sqrt{(-2)^2 + (-3)^2} = \sqrt{4 + 9} = \sqrt{13}$. From Definition (1),

$$\sin\theta = \frac{y}{r} = \frac{-3}{\sqrt{13}} \approx -.832050 \qquad \csc\theta = \frac{r}{y} = \frac{\sqrt{13}}{-3} \approx -1.20185$$

$$\cos\theta = \frac{x}{r} = \frac{-2}{\sqrt{13}} \approx -.554700 \qquad \sec\theta = \frac{r}{x} = \frac{\sqrt{13}}{-2} \approx -1.80278$$

$$\tan\theta = \frac{y}{x} = \frac{-3}{-2} = \frac{3}{2} \qquad \cot\theta = \frac{x}{y} = \frac{-2}{-3} = \frac{2}{3}$$

As Example 1 illustrates, trigonometric functions may be negative. For instance, since r is always positive, $\sin\theta = y/r$ and $\csc\theta = r/y$ have the same sign as y. Thus $\sin\theta$ and $\csc\theta$ are positive when θ is in Q_1 or Q_2 and negative when θ is in Q_3 or Q_4. The signs of the other trigonmetric functions can be found similarly. The charts in Fig. 4 indicate the signs according to where θ lies. These can be combined as in the single chart in Fig. 5. This is remembered by saying "<u>A</u>ll <u>S</u>tudents <u>T</u>ake <u>C</u>ourses" (or something similar). In this chart the letters indicate the positive functions: A denotes all, S denotes sine (and its reciprocal cosecant), T denotes tangent (and cotangent), while C denotes cosine (and secant).

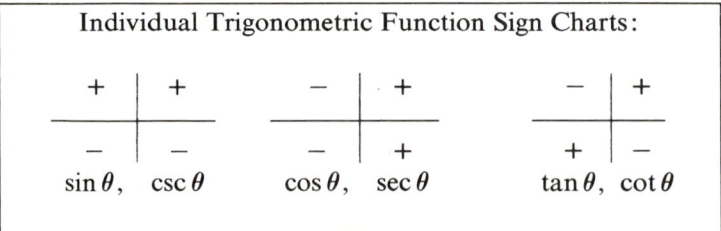

FIG. 4

Combined Trigonometric Function Sign Chart:

$$\begin{array}{c|c} S & A \\ \hline T & C \end{array}$$

FIG. 5

In the previous chapter, we always assumed that θ was an acute angle, which meant that the terminal side would be in the first quadrant. Thus all the trigonometric functions were positive (we just saw this in our sign charts). Now when we are given one of the trigonometric functions and asked to find the others, we may not be able to determine the signs unless we have further information.

EXAMPLE 2 Suppose $\cos \theta = \dfrac{3}{5}$. Find the remaining trigonometric functions of θ.

SOLUTION

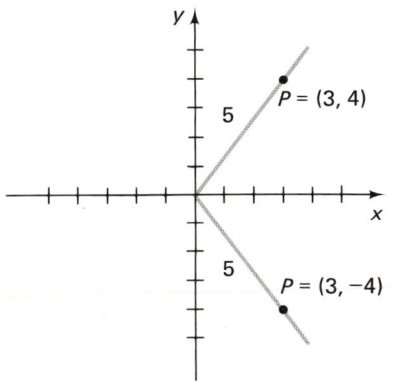

FIG. 6

First, observe from the sign charts that the terminal side of θ is in Q_1 or Q_4 since $\cos \theta$ is positive. Next we know that we may choose any convenient point $P = (x,y)$ on the terminal side when computing the trigonometric functions. Since $\cos \theta = 3/5 = x/r$, we choose the point P with $x = 3$ and y such that $r = \sqrt{x^2 + y^2} = 5$. In other words, $y^2 = r^2 - x^2 = 25 - 9 = 16$, or $y = \pm 4$. See Fig. 6. Then from the definitions,

$$\sin \theta = \pm \frac{4}{5}, \quad \csc \theta = \pm \frac{5}{4}, \quad \sec \theta = \frac{5}{3}, \quad \tan \theta = \pm \frac{4}{3}, \quad \text{and} \quad \cot \theta = \pm \frac{3}{4}$$

EXAMPLE 3

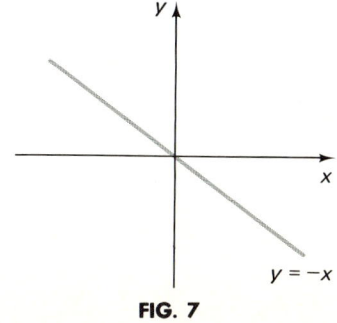

FIG. 7

Suppose that for every point on the terminal side of θ, $y = -x$. See Fig. 7. Find the trigonometric functions of θ.

SOLUTION The terminal side of θ is in either Q_2 or Q_4. We pick a point on the terminal side in each quadrant and find the trigonometric functions by the definitions: $x = -1$, $y = 1$ satisfies the equation and is in Q_2 ($r = \sqrt{2}$), and $x = 1$, $y = -1$ satisfies the equation and is in Q_4 ($r = \sqrt{2}$). Hence

$$\sin\theta = \pm\frac{1}{\sqrt{2}}, \qquad \cos\theta = \mp\frac{1}{\sqrt{2}}, \qquad \tan\theta = -1$$

$$\csc\theta = \pm\sqrt{2}, \qquad \sec\theta = \mp\sqrt{2}, \qquad \cot\theta = -1$$

EXERCISES

In Exercises 1–15, use the information given to find the values of six trigonometric functions of θ, where θ is an angle in standard position.

1. $(-4, 3)$ is on the terminal side of θ.
2. $(12, -5)$ is on the terminal side of θ.
3. $(2, 3)$ is on the terminal side of θ.
4. $(-2, -3)$ is on the terminal side of θ.
5. $(-1, -3)$ is on the terminal side of θ.
6. $(1, -2)$ is on the terminal side of θ.
7. $(3, -2)$ is on the terminal side of θ.
8. $(6, 4)$ is on the terminal side of θ.
9. For some point on the terminal side of θ, $r = 7$, $x = -2$.
10. For some point on the terminal side of θ, $r = 6$, $y = -3$.
11. For some point on the terminal side of θ, $r = 4$, $x = 2$.
12. For some point on the terminal side of θ, $r = 3$, $y = 1$.
13. For every point on the terminal side of θ, $y = 3x$.
14. For every point on the terminal side of θ, $y = -2x$.
15. For every point on the terminal side of θ, $y = -4x$.

In Exercises 16–23, determine what quadrant θ is in and the signs of $\sin\theta$, $\cos\theta$, and $\tan\theta$.

16. $756°$
17. $3467°$
18. $-255.6°$
19. $-1106.53°$
20. $\frac{11\pi}{7}$
21. 2.67
22. -8.0325
23. $-\frac{97\pi}{3}$

In Exercises 24–43, find the values of the trigonometric functions of θ.

24. $\sin\theta = \frac{5}{13}$
25. $\cos\theta = -\frac{7}{25}$
26. $\tan\theta = -\frac{3}{4}$
27. $\cot\theta = \frac{12}{5}$
28. $\csc\theta = -\frac{25}{7}$
29. $\sec\theta = \frac{13}{5}$
30. $\sin\theta = \frac{3}{4}$, $\cos\theta$ negative
31. $\cos\theta = -\frac{2}{3}$, $\sin\theta$ negative
32. $\sin\theta = -\frac{1}{2}$, $\tan\theta$ positive
33. $\tan\theta = 2$, $\cos\theta$ negative
34. $\cos\theta = \frac{1}{3}$, $\cot\theta$ positive
35. $\tan\theta = -\frac{3}{4}$, $\sin\theta$ negative
36. $\csc\theta = 3$, $\sec\theta$ negative
37. $\sin\theta = -\frac{1}{4}$, $\sec\theta$ positive
38. $\cos\theta = \frac{1}{5}$, θ in Q_4
39. $\sin\theta = -\frac{2}{5}$, θ in Q_3
40. $\tan\theta = \sqrt{3}$, θ in Q_3
41. $\tan\theta = \sqrt{3}$, θ in Q_1
42. $\sec\theta = -2$, θ in Q_2
43. $\sin\theta = \frac{3}{5}$, θ in Q_2

In Exercises 44–47, find the unknown coordinate of B and the distance of B from the origin if A and B are on the terminal side of the same angle.

44. $A = (4, -12)$, $B = (x, -24)$
45. $A = (-7, -24)$, $B = (-3, y)$
46. $A = (-2, 3)$, $B = (-5, y)$
47. $A = (1, 4)$, $B = (x, 27)$

48. Assume θ is in Q_3. Show that Definition (1) does not depend on which point (different from the origin) is chosen on the terminal side as follows: Let $P = (x, y)$ and $P' = (x', y')$ be any two such points, and drop perpendiculars to the x-axis. Use similar triangles and the signs of x, y, x', y' to show the values of the trigonometric functions are the same whether you use P or P'.

49. Do Exercise 48 under the assumption θ is in Q_2.

50. Do Exercise 48 under the assumption θ is in Q_1.

51. Do Exercise 48 under the assumption θ is in Q_4.

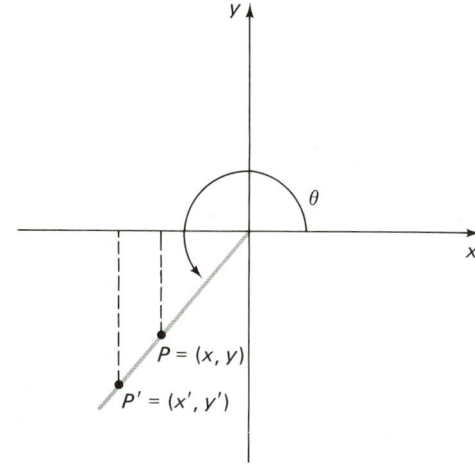

SECTION 8.2. EVALUATING TRIGONOMETRIC FUNCTIONS

If we are given an angle θ and we wish to use Definition (1) to compute one or more of the trigonometric functions of θ, we must first find the coordinates of a point P on the terminal side of θ. In a few cases, this is easy.

EXAMPLE 1 Find the trigonometric functions of $\theta = -540°$.

SOLUTION

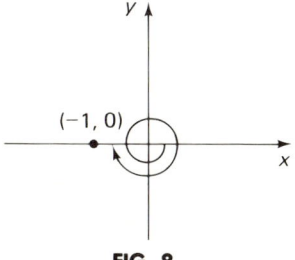

FIG. 8

The points on the terminal side all have the form $(x, 0)$, $x \leq 0$. We may pick any such point except $(0, 0)$ to compute the values of the trigonometric functions. The calculations are easiest if we choose $(-1, 0)$. (See Fig. 8.) Then $r = \sqrt{(-1)^2 + 0^2} = 1$, and from Definition (1),

$$\sin \theta = \frac{y}{r} = \frac{0}{1} = 0 \qquad \csc \theta = \frac{r}{y} \text{ is undefined}$$

$$\cos \theta = \frac{x}{r} = \frac{-1}{1} = -1 \qquad \sec \theta = \frac{r}{x} = \frac{1}{-1} = -1$$

$$\tan \theta = \frac{y}{x} = \frac{0}{-1} = 0 \qquad \cot \theta = \frac{x}{y} \text{ is undefined}$$

Note that if you had picked any other $(x, 0)$, $x < 0$, then $r = |x|$ and the computations would give the same values (i.e., the trigonometric functions are well defined).

You should also observe that if θ is any other angle whose terminal side is the negative x-axis, the calculations are exactly the same. In other words, in Example 1 we found the trigonometric functions of θ for any angle θ whose terminal side is the negative x-axis.

The values of the trigonometric functions of any other quadrantal (between quadrant) angle can be found similarly. They are listed in Table 1 (where a dash means undefined).

TABLE 1

TERMINAL SIDE	$\sin \theta$	$\cos \theta$	$\tan \theta$	$\csc \theta$	$\sec \theta$	$\cot \theta$
Pos. x-axis (e.g., $0 = 0°$)	0	1	0	—	1	—
Pos. y-axis (e.g., $\pi/2 = 90°$)	1	0	—	1	—	0
Neg. x-axis (e.g., $\pi = 180°$)	0	-1	0	—	-1	—
Neg. y-axis (e.g., $3\pi/2 = 270°$)	-1	0	—	-1	—	0

To compute the values in Table 1, the easiest points to use are $(1, 0)$, $(0, 1)$, $(-1, 0)$, and $(0, -1)$, respectively. In fact, using these points it is so easy to derive the values in Table 1 that the table need not be memorized.

Unfortunately, there are few other angles for which it is easy to find the values of the trigonometric functions directly from the definitions. In the past, extensive tables were used to find the values of the trigonometric functions. However, now you can just use a calculator; the procedure for other angles is exactly the same as for acute angles. Remember, *make sure your calculator is in the proper mode*.

EXAMPLE 2 Find **(a)** $\sin 192°$, **(b)** $\cos 2$, and **(c)** $\tan 90°$.

SOLUTION
(a) With the calculator in degree mode, press [192][sin], obtaining $-.207912$.
(b) With the calculator in radian mode, press [2][cos], obtaining $-.416147$.
(c) With the calculator in degree mode, press [90][tan]. You should obtain the error indication for your calculator since $\tan 90°$ is undefined.

From Definition (1), it is still true that

$$\csc \theta = \frac{1}{\sin \theta}, \quad \sec \theta = \frac{1}{\cos \theta}, \quad \text{and} \quad \cot \theta = \frac{1}{\tan \theta}$$

These identities are used when evaluating cosecant, secant, or cotangent on a calculator.

EXAMPLE 3 Find $\csc(-218.2°)$.

SOLUTION With the calculator in degree mode, use $\csc(-218.2°) = 1/\sin(-218.2°)$, and press [218.2][+/−][sin][1/x], obtaining 1.61705.

Warning: Some calculators compute trigonometric functions only for a highly restricted set of values. For such calculators, the properties of the trigonometric functions described in the next two sections can be used when an angle is outside the restrictions.

We have mentioned that we could use any point on the terminal side of an angle θ to compute the trigonometric functions of θ from Definition (1). Conversely, when we are given an angle θ, trigonometric functions of θ computed on a calculator can be used to find the coordinates of any point on the terminal side of θ.

EXAMPLE 4 Find the coordinates of the points on the terminal side of $\theta = 327°$ of distances **(a)** 1 and **(b)** 5.8 units from the origin.

SOLUTION If (x,y) is a point on the terminal side of θ of distance r from the origin, we know from Definition (1) that $\cos\theta = x/r$ and $\sin\theta = y/r$. Multiplying these through by r yields

$$x = r\cos\theta \quad \text{and} \quad y = r\sin\theta$$

With $\theta = 327°$ and $r = 1$ and 5.8, respectively, we easily compute x and y:
(a) $x = \cos 327° \approx .838671$; $y = \sin 327° \approx -.544639$.
(b) $x = 5.8 \cos 327° \approx 4.86429$; $y = 5.8 \sin 327° \approx -3.15891$.

Remark: The angle θ may of course be measured either in degrees or radians. However, we should observe that in many mathematical situations, it is often useful to think of the trigonometric functions as functions of real numbers, where the real numbers do not represent angles. If you recall that radians may be viewed as real numbers, then it is natural to **evaluate a trigonometric function at a real number exactly as if the number were an angle measured in radians**. Because of this, your calculator does not have a separate mode for evaluating trigonometric functions of real numbers; you just use radian mode. For example, suppose f is a function whose rule is given by

$$f(t) = \sin t$$

where t is any real number. If you wish to evaluate this function at the real number 2, write $f(2) = \sin 2$, and compute the sine of 2 by putting your calculator in radian mode and pressing [2][sin] to obtain

$$f(2) = \sin 2 \approx .909297$$

EXERCISES

In Exercises 1–6, find the values of the six trigonometric functions of θ.

1. $\theta = 146.35°$
2. $\theta = 661.2°$
3. $\theta = -489°$
4. $\theta = 2.6$
5. $\theta = -2.6$
6. $\theta = 9.21685$

In Exercises 7–14, find the coordinates of the point P on the terminal side of the given θ satisfying the given condition.

7. $\theta = 167.08°, r = 16.3$
8. $\theta = -29°, x = 14$
9. $\theta = 4.895, y = -5.0631$
10. $\theta = 10.2, r = 1176$
11. $\theta = -1.89, x = -.184$
12. $\theta = -11.3, y = 683.9$
13. $\theta = 684.3°, y = -1.6802$
14. $\theta = -219.43°, x = -10$

In Exercises 15–18, use the definitions to compute the six trigonometric functions of θ directly.

15. The terminal side of θ is the positive x-axis.
16. The terminal side of θ is the negative x-axis.
17. The terminal side of θ is the negative y-axis.
18. The terminal side of θ is the positive y-axis.

In Exercises 19–24, evaluate the given expression without a calculator.

19. $\sin 90° \csc 630° + \tan 720°$
20. $\cos \dfrac{5\pi}{2} + \cos 3\pi + \cos \dfrac{7\pi}{2} + \cos 4\pi$
21. $\sin 540° + \tan 990°$
22. $\sec(-900°) \sec 900° + \tan 900° \sin 405°$
23. $\sin \dfrac{11\pi}{2} + \sin\left(-\dfrac{11\pi}{2}\right)$
24. $6 + 3\sin^2 \dfrac{35\pi}{2} + \cos^2\left(-\dfrac{13\pi}{2}\right)$

In Exercises 25–30, evaluate the given expression.

25. $8.23 \sin 142° + 4.65 \tan 206.2°$
26. $\dfrac{1}{\cos\left(3.2 + \frac{1}{6}\right)}$
27. $\tan 4.625 + (1.97 \sin 7.4)^{1/2}$
28. $\sin^2(-42°) + \cos^2(-42°)$
29. $\sqrt{\csc(-507.2°) + 2.0328}$
30. $6.2^{4.7} + \cos^3 9.25$

In Exercises 31–40, evaluate the given function at the given real number.

31. $f(t) = 2 \tan t, \ t = 3$
32. $f(x) = 5 \cot x, \ x = 5.8$
33. $f(u) = \frac{1}{6}\cot^2 u, \ u = 4.98$
34. $f(t) = 5 \sin t \cos t, \ t = \sqrt{3}$
35. $g(x) = 6 \sin^4 x - \tan^2 x, \ x = 2.19$
36. $g(u) = \dfrac{1}{1 + \cos u} + 4 \sec u, \ u = 1.89$
37. $f(t) = \tan(\sin t), \ t = 5.3$
38. $g(t) = \sin[\cos(t - 1.93)], \ t = 2.65$
39. $g(t) = t^3 + 3t \sin t, \ t = e^{2.1}$
40. $g(x) = 5x^{1/3} - 6 \cot(x^3), \ x = -2^{\sqrt{\pi}}$

336 GENERAL TRIGONOMETRY

SECTION 8.3. PROPERTIES OF THE TRIGONOMETRIC FUNCTIONS

In this section, we shall describe some of the properties of the trigonometric functions. One of the most fundamental properties is that the elementary identities still hold (whenever both sides are defined).

The Elementary Trigonometric Identities

(2)
$$\csc\theta = \frac{1}{\sin\theta}, \quad \sec\theta = \frac{1}{\cos\theta}, \quad \cot\theta = \frac{1}{\tan\theta}$$

(3)
$$\tan\theta = \frac{\sin\theta}{\cos\theta}, \quad \cot\theta = \frac{\cos\theta}{\sin\theta}$$

(4)
$$\sin^2\theta + \cos^2\theta = 1$$

It is immediate from Definition (1) that (2) and (3) are still true. We can verify (4) as follows:

$$\sin^2\theta + \cos^2\theta = \left(\frac{y}{r}\right)^2 + \left(\frac{x}{r}\right)^2 = \frac{y^2}{r^2} + \frac{x^2}{r^2} = \frac{y^2 + x^2}{r^2} = \frac{r^2}{r^2} = 1$$

since $r = \sqrt{x^2 + y^2}$.

There are two identities similar to (4) that can be obtained from (2)–(4). If (4) is divided by $\cos^2\theta$, we obtain

$$\frac{\sin^2\theta}{\cos^2\theta} + \frac{\cos^2\theta}{\cos^2\theta} = \frac{1}{\cos^2\theta}$$

or

$$\left(\frac{\sin\theta}{\cos\theta}\right)^2 + 1 = \left(\frac{1}{\cos\theta}\right)^2$$

By (2) and (3),

(5)
$$\tan^2\theta + 1 = \sec^2\theta$$

If instead we divide (4) by $\sin^2\theta$, we obtain

(6)
$$1 + \cot^2\theta = \csc^2\theta$$

There are two other identities that follow immediately from Definition (1) and from the fact that the terminal side of $-\theta$ is the reflection of the terminal side of θ in the x-axis [i.e., if (x,y) is on the terminal side of θ, then $(x, -y)$ is on the

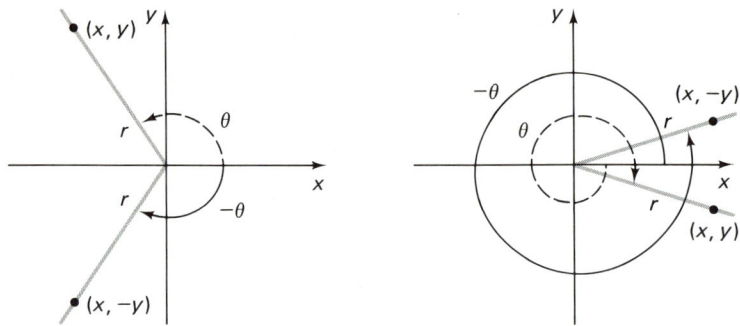

FIG. 9

terminal side of $-\theta$]. See Fig. 9. Since $\sqrt{x^2 + y^2} = r = \sqrt{x^2 + (-y)^2}$,

$$\sin(-\theta) = \frac{-y}{r} = -\left(\frac{y}{r}\right) = -\sin\theta$$

and

$$\cos(-\theta) = \frac{x}{r} = \cos\theta$$

Therefore we have shown the following:

(7)
$$\sin(-\theta) = -\sin\theta \quad \text{and} \quad \cos(-\theta) = \cos\theta$$

In Sec. 1 of this chapter, we saw that if we are given the value of one trigonometric function of θ, then we can find the values of the others directly from Definition (1). However, this type of problem can also be done using the elementary identities (2)–(4), just as we did in the previous chapter. If $\tan\theta$ or $\cot\theta$ is given, it is sometimes useful to use (5) or (6) as well.

EXAMPLE 1 Suppose $\tan \theta = -\frac{1}{2}$ and $\sin \theta$ is positive. Find the remaining trigonometric functions of θ.

SOLUTION Since $\tan \theta$ is negative and $\sin \theta$ is positive, the terminal side of θ must be in Q_2. Hence, $\cos \theta$ and $\sec \theta$ are negative. From (5), $\tan^2 \theta + 1 = \sec^2 \theta$, we have $(-\frac{1}{2})^2 + 1 = \sec^2 \theta$, or $\sec^2 \theta = \frac{5}{4}$. Thus, $\sec \theta = -\sqrt{5}/2$, and by (2), $\cos \theta = -2/\sqrt{5}$. From (3),

$$\tan \theta = \frac{\sin \theta}{\cos \theta}, \text{ or } \sin \theta = \cos \theta \tan \theta = \left(\frac{-2}{\sqrt{5}}\right)\left(\frac{-1}{2}\right) = \frac{1}{\sqrt{5}}$$

Thus by (2), $\csc \theta = \sqrt{5}$. Also by (2), since $\tan \theta = -1/2$, $\cot \theta = -2$.

There is a further observation that can be made from the definitions of the trigonometric functions of an angle θ, where θ is an angle in standard position. Since all you need to know are the coordinates of a point P on the terminal side of θ, the *trigonometric functions of θ depend only on the location of the terminal side, not on the direction of rotation or on how many times the initial side is rotated completely around before stopping at the terminal side.*

EXAMPLE 2 In Fig. 10, each of $\theta_1, \theta_2, \theta_3$ have the same terminal side. Hence $\sin \theta_1 = \sin \theta_2 = \sin \theta_3$, $\cos \theta_1 = \cos \theta_2 = \cos \theta_3$, etc.

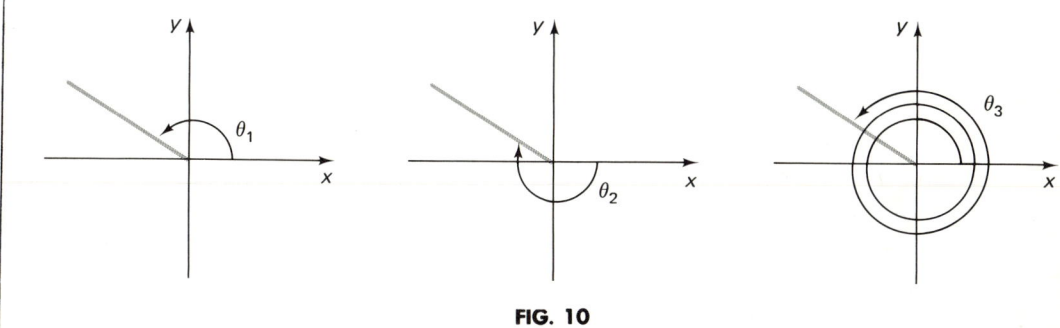

FIG. 10

Now two angles θ and φ in standard position have the same terminal side precisely when they differ by an integral number of compete rotations; i.e., in radians $\varphi = \theta + n \cdot 2\pi$, or in degrees $\varphi = \theta + n \cdot 360°$, where n is an integer (positive, negative, or zero). (In Example 2, using radians, $\theta_2 = \theta_1 - 2\pi$, $\theta_3 = \theta_1 + 2 \cdot 2\pi = \theta_2 + 3 \cdot 2\pi$.) Since $\theta + n \cdot 2\pi$ and θ have the same terminal side, for any integer n,

(8)
$$\sin(\theta + n \cdot 2\pi) = \sin \theta \qquad \csc(\theta + n \cdot 2\pi) = \csc \theta$$
$$\cos(\theta + n \cdot 2\pi) = \cos \theta \qquad \sec(\theta + n \cdot 2\pi) = \sec \theta$$
$$\tan(\theta + n \cdot 2\pi) = \tan \theta \qquad \cot(\theta + n \cdot 2\pi) = \cot \theta$$

A nonconstant function f is called **periodic** if there is a real number $p \neq 0$ such that $f(x + p) = f(x)$ for all x for which $f(x)$ is defined. If p_1 is the smallest positive real number such that $f(x + p_1) = f(x)$ for all x for which $f(x)$ is defined, then p_1 is called the **period** of f.

From equations (8) with $n = 1$, $\sin(\theta + 2\pi) = \sin\theta$, $\cos(\theta + 2\pi) = \cos\theta$, and similarly for the other trigonometric functions. Consequently, *each of the trigonometric functions is periodic of period at most* 2π. We shall see in Sec. 5 that $\sin\theta$, $\cos\theta$, $\csc\theta$, and $\sec\theta$ have period 2π, while $\tan\theta$ and $\cot\theta$ have period π. However, in this section, we shall just use the fact that the functions are periodic and equations (8) hold.

We have mentioned that some calculators have restrictions on the values of θ for which the trigonometric functions of θ can be calculated. As long as the restricted values include $0 \leq \theta < 2\pi$, we can use this together with equations (8) to evaluate a trigonometric function at any value of θ.

EXAMPLE 3 Suppose your calculator computes the trigonometric functions only for $0 \leq \theta \leq 2\pi$. Find **(a)** $\sin 478.2$ and **(b)** $\tan(-8712.41°)$.

SOLUTION **(a)** Divide 478.2 by 2π, obtaining 76.107893. Thus
$$478.2 \approx 2\pi(76.107893)$$
$$\approx 2\pi(76 + .107893)$$
$$\approx 2\pi(76) + 2\pi(.107893)$$
$$\approx 2\pi(76) + .6779168$$

Consequently,
$$\sin 478.2 = \sin[478.2 - 76(2\pi)]$$
$$\approx \sin .6779168 \approx .627172$$

(b) Divide $-8712.41°$ by $360°$, obtaining -24.201138. Thus
$$-8712.41° \approx 360°(-24.201138)$$
$$\approx -24(360°) - .201138(360°)$$
$$= -24(360°) - 72.41°$$

Consequently,
$$\tan(-8712.41°) = \tan[-8712.41° + 24(360°)]$$
$$= \tan(-72.41°)$$

However, $-74.21°$ is still not between $0°$ and $360°$. But
$$\tan(-72.41°) = \tan(-72.41° + 360°)$$
$$= \tan 287.59° \approx -3.15431$$

It may be wise to use the periodicity for large values of θ even if the calculator can handle them, because machine error begins to play a significant role for the larger values. For instance, one calculator computes $\sin(10^{10}\pi + \frac{1}{6}\pi)$ as 0 instead of $\sin\frac{1}{6}\pi = .5$. Moreover, if θ is measured in radians and $|\theta|$ is larger than say 10^7 or 10^8, then even using periodicity can lead to massive round-off error since the calculator uses only an 8–12 digit approximation of π. If θ is in either degrees or radians and is large, roughly $|\theta| \geq 10^{10}$, then there is almost no chance that your calculator will compute the trigonometric functions of that angle correctly.

EXERCISES

In Exercises 1–20, use the identities (2)–(7) to find the values of the trigonometric functions of θ.

1. $\sin \theta = \frac{5}{13}$
2. $\cos \theta = -\frac{7}{25}$
3. $\tan \theta = -\frac{4}{3}$
4. $\cot \theta = -\frac{12}{5}$
5. $\csc \theta = -\frac{25}{24}$
6. $\sec \theta = \frac{13}{5}$
7. $\sin \theta = \frac{1}{3}$, θ in Q_2
8. $\cos \theta = -\frac{1}{5}$, θ in Q_3
9. $\sec \theta = -2$, θ in Q_2
10. $\cot \theta = \sqrt{3}$, θ in Q_3
11. $\tan \theta = -1$, θ in Q_4
12. $\csc \theta = 3$, θ in Q_1
13. $\sin \theta = \frac{2}{3}$, $\tan \theta$ negative
14. $\cos \theta = -\frac{1}{4}$, $\sin(-\theta)$ negative
15. $\tan \theta = -\sqrt{3}$, $\sin(-\theta)$ positive
16. $\sin \theta = \frac{3}{5}$, $\cos(-\theta)$ positive
17. $\cot \theta = \frac{1}{2}$, $\cos(-\theta)$ negative
18. $\sec \theta = -\sqrt{5}$, $\tan \theta$ positive
19. $\cos \theta = \frac{1}{3}$, $\cot \theta$ negative
20. $\csc \theta = 2$, $\sec \theta$ negative

In Exercises 21–35, compute the value of the given expression. Use identities if possible. If not, use your calculator, but do so under the assumption that it only computes the trigonometric functions for $0 \leq \theta \leq 2\pi$ or $0° \leq \theta \leq 360°$, and use periodicity.

21. $\cos 1436.7°$
22. $\tan 64{,}291°$
23. $\sin(-42{,}385.6°)$
24. $\cos(-81{,}277{,}033°)$
25. $\sin 48{,}963.4$
26. $\tan(-74{,}306)$
27. $\cos 9621.47$
28. $\sin^2 864 + \cos^2 864$
29. $\sin(-4382°) + \sin 4382°$
30. $\cos(-11{,}227°) + \cos 11{,}227°$
31. $\sin^2 47.23 - \cos^2 47.23$
32. $(\sin 1098° - 1)(\sin 1098° + 1) + (\cos 1098°)[\cos(-1098°)]$
33. $\tan^2 108 + \sec^2 108$
34. $\cot^2 432{,}156° - \csc^2 432{,}156°$
35. $(\cos 99 + 1)(\cos 99 - 1) + (\sin 99 + 1)(\sin 99 - 1)$

SECTION 8.4. REFERENCE ANGLES

We now turn our attention to reference angles. This is one of the most useful concepts in trigonometry, and it provides a vital connection between general trigonometry and right triangle trigonometry.

> **(9) DEFINITION**
>
> If θ is an angle in standard position, the **reference angle** of θ is the smallest positive angle φ between the terminal side of θ and the x-axis.

If θ is a quadrantal angle, then the reference angle of θ is $\varphi = 0 = 0°$ if the terminal side is on the x-axis or $\varphi = \pi/2 = 90°$ if the terminal side is on the y-axis. Otherwise Fig. 11 demonstrates where φ is, depending on the quadrant where the terminal side of θ lies. In each case, one of the many possible angles θ with the given terminal side is shown. For that terminal side, φ is unique.

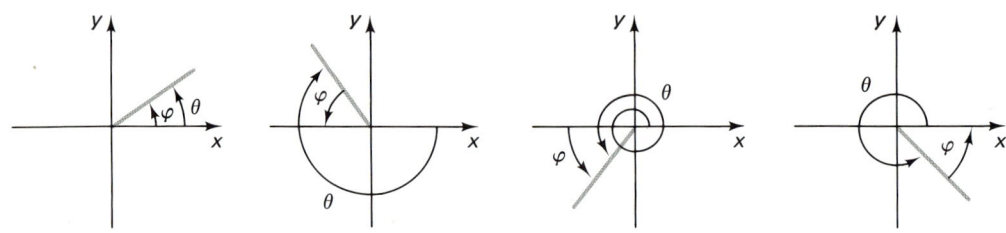

FIG. 11

EXAMPLE 1 Find the reference angle φ if θ is **(a)** 35°; **(b)** 114°; **(c)** 4.08; **(d)** 5.1963

SOLUTION
(a) $\varphi = \theta = 35°$. See Fig. 12(a).
(b) $\varphi = 180° - \theta = 180° - 114° = 66°$. See Fig. 12(b).
(c) $\varphi = \theta - \pi = 4.08 - \pi \approx .938407$. See Fig. 12(c).
(d) $\varphi = 2\pi - \theta = 2\pi - 5.1963 \approx 1.08689$. See Fig. 12(d).

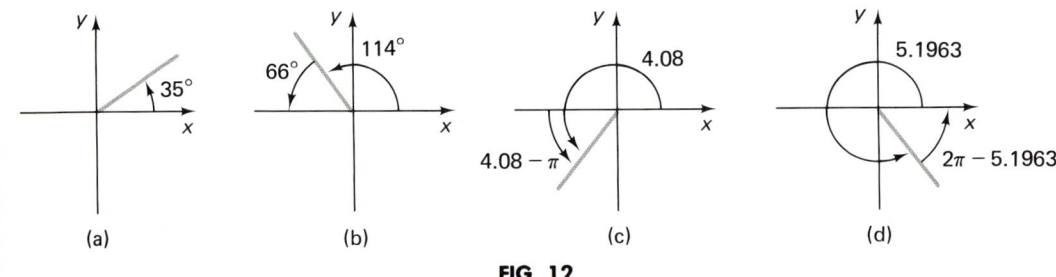

FIG. 12

In general when you are given an angle θ, you first determine the quadrant where the terminal side of θ lies and then proceed as in Example 1 to find the reference angle. In doing the latter, you can add or subtract multiples of $2\pi = 360°$ without changing the reference angle since complete rotations do not affect where the terminal side is located.

EXAMPLE 2 If $\theta = 7789°$, find the quadrant in which the terminal side lies, and then determine the reference angle φ.

SOLUTION

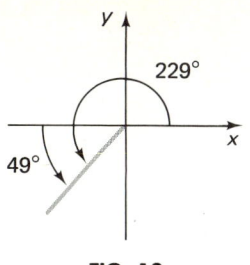

FIG. 13

We divide 7789° by 360°, obtaining 21.6361. Thus, 7789° is 21 complete rotations plus .6361 of a complete rotation. Since the last is between one-half and three-quarters of a complete rotation [it has measure 7789° − 21(360°) = 229°], the terminal side is in Q_3. The reference angle is $\varphi = 229° - 180° = 49°$. See Fig. 13.

Now let us see how reference angles relate general trigonometry and right triangle trigonometry. As we have seen, if θ is any angle in standard position, the trigonometric functions depend only on the terminal side, not on how many times or in what direction the initial side was rotated to get there. Assume for a moment that θ is not a quadrantal (between quadrant) angle. Then the terminal side of θ is in one of the quadrants. Now let P be any point (except the origin) on the terminal side, and drop a perpendicular from P to the x-axis. The possible cases are shown in Fig. 14. In each case, a natural right triangle is

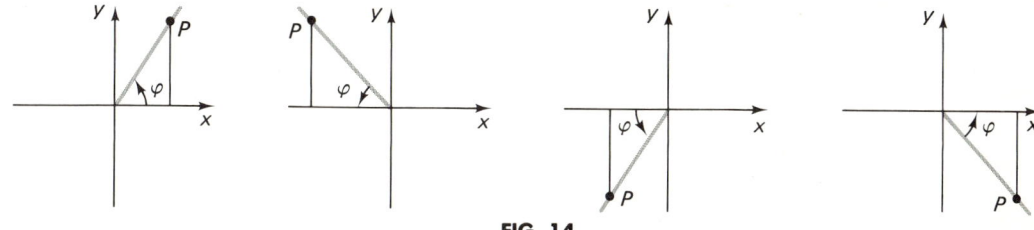

FIG. 14

formed, and we have used φ to denote the angle between the leg along the x-axis and the hypotenuse. Notice that φ is also the reference angle of θ in each case. If P has coordinates (x,y) and $r = \sqrt{x^2 + y^2}$, then all four right triangles have dimensions as shown in Fig. 15. Since $P = (x,y)$ is a point on the terminal side of θ, we can compute the trigonometric functions of θ and compare them with those of φ:

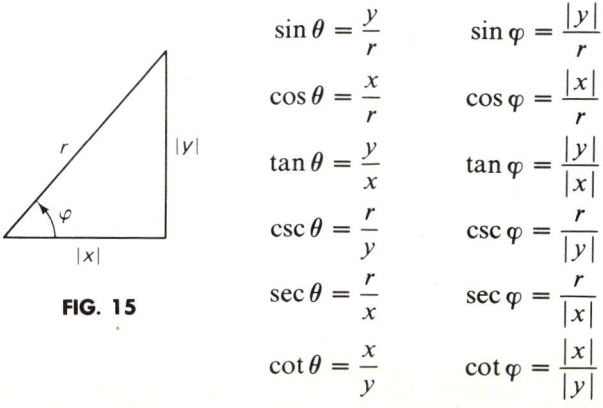

FIG. 15

$$\sin \theta = \frac{y}{r} \qquad \sin \varphi = \frac{|y|}{r}$$

$$\cos \theta = \frac{x}{r} \qquad \cos \varphi = \frac{|x|}{r}$$

$$\tan \theta = \frac{y}{x} \qquad \tan \varphi = \frac{|y|}{|x|}$$

$$\csc \theta = \frac{r}{y} \qquad \csc \varphi = \frac{r}{|y|}$$

$$\sec \theta = \frac{r}{x} \qquad \sec \varphi = \frac{r}{|x|}$$

$$\cot \theta = \frac{x}{y} \qquad \cot \varphi = \frac{|x|}{|y|}$$

The only possible differences are the signs.

> **(10)**
>
> If φ is the reference angle of an angle θ, then the value of a trigonometric function of θ and the value of that same function of φ agree, except possibly for the sign.

[Although quadrantal angles were excluded from the discussion because no right triangle is formed, result (10) holds for them as well.]

EXAMPLE 3 Suppose $(-3, 4)$ is on the terminal side of θ. Compare the trigonometric functions of θ and its reference angle φ.

SOLUTION Since $(-3, 4)$ is on the terminal side and $(-3, 4)$ is in Q_2, we have the graph in Fig. 16 and the corresponding right triangle. Thus

$\sin \theta = \frac{4}{5},$ $\sin \varphi = \frac{4}{5}$ $\csc \theta = \frac{5}{4},$ $\csc \varphi = \frac{5}{4}$

$\cos \theta = -\frac{3}{5},$ $\cos \varphi = \frac{3}{5}$ $\sec \theta = -\frac{5}{3},$ $\sec \varphi = \frac{5}{3}$

$\tan \theta = -\frac{4}{3},$ $\tan \varphi = \frac{4}{3}$ $\cot \theta = -\frac{3}{4},$ $\cot \varphi = \frac{3}{4}$

We see that corresponding functions agree, except for some of the signs.

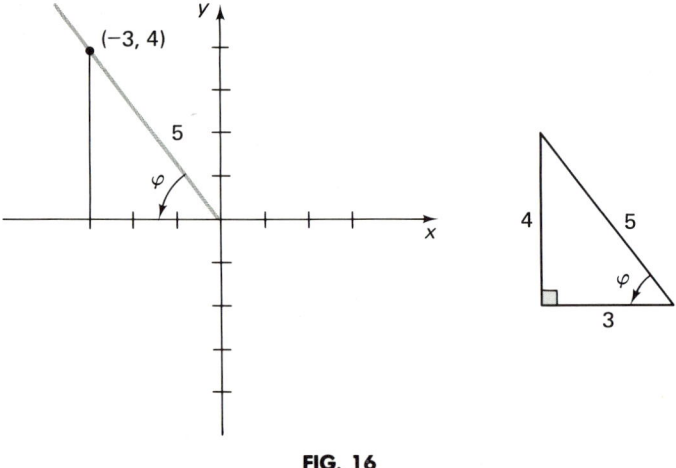

FIG. 16

We can exploit this agreement (except for signs) as follows: Suppose that you know in which quadrant the terminal side of an angle θ lies and that you know the values of the trigonometric functions of the reference angle φ of θ. Then by merely assigning signs from the sign charts, Fig. 4 or 5, you obtain the values of the trigonometric functions of θ. This is particularly useful if the reference angle is $\pi/6 = 30°$, $\pi/4 = 45°$, or $\pi/3 = 60°$, since we have already seen how to compute the trigonometric functions of these special angles without a calculator. Because of this, we now use the term **special angle** to denote any angle with one of these three angles as reference angle.

EXAMPLE 4 Find the trigonometric functions of 300° (without a calculator).

SOLUTION The terminal side of 300° is in Q_4, and the reference angle is 60°. We know that

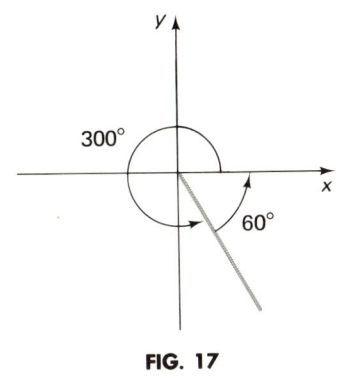

$$\sin 60° = \frac{\sqrt{3}}{2}, \quad \cos 60° = \frac{1}{2}, \quad \tan 60° = \sqrt{3}$$

$$\csc 60° = \frac{2}{\sqrt{3}}, \quad \sec 60° = 2, \quad \cot 60° = \frac{1}{\sqrt{3}}$$

In Q_4, cosine and secant are positive, and the rest negative. Hence

$$\sin 300° = -\frac{\sqrt{3}}{2}, \quad \cos 300° = \frac{1}{2}, \quad \tan 300° = -\sqrt{3}$$

$$\csc 300° = -\frac{2}{\sqrt{3}}, \quad \sec 300° = 2, \quad \cot 300° = -\frac{1}{\sqrt{3}}$$

FIG. 17

EXERCISES

In Exercises 1–16, the angle is in standard position. Find in which quadrant the terminal side lies, and determine the reference angle.

1. 237°
2. −38°
3. 103°
4. 409°
5. −716°
6. 903°
7. −331°15′
8. 497.23°
9. $\dfrac{7\pi}{9}$
10. $\dfrac{38\pi}{3}$
11. $-\dfrac{24\pi}{5}$
12. $\dfrac{11\pi}{6}$
13. 2
14. 5
15. 15
16. −11

17. Fill out the following table (without a calculator):

	$\dfrac{\pi}{6}$	$\dfrac{\pi}{4}$	$\dfrac{\pi}{3}$	$\dfrac{2\pi}{3}$	$\dfrac{3\pi}{4}$	$\dfrac{5\pi}{6}$	$\dfrac{7\pi}{6}$	$\dfrac{5\pi}{4}$	$\dfrac{4\pi}{3}$	$\dfrac{5\pi}{3}$	$\dfrac{7\pi}{4}$	$\dfrac{11\pi}{6}$
$\sin\theta$												
$\cos\theta$												
$\tan\theta$												

	$-\dfrac{\pi}{6}$	$-\dfrac{\pi}{4}$	$-\dfrac{\pi}{3}$	$-\dfrac{2\pi}{3}$	$-\dfrac{3\pi}{4}$	$-\dfrac{5\pi}{6}$	$-\dfrac{7\pi}{6}$	$-\dfrac{5\pi}{4}$	$-\dfrac{4\pi}{3}$	$-\dfrac{5\pi}{3}$	$-\dfrac{7\pi}{4}$	$-\dfrac{11\pi}{6}$
$\sin\theta$												
$\cos\theta$												
$\tan\theta$												

In Exercises 18–33, evaluate as follows: First, find the reference angle of the given angle, then evaluate the trigonometric function at that reference angle (using a calculator if necessary), and finally attach the appropriate sign. Check by evaluating the function directly.

18. $\cos 8$
19. $\sin 5$
20. $\tan 10.2$
21. $\sin(-2.87)$
22. $\tan(-6483)$
23. $\cos 479{,}362$
24. $\sec 16.435$
25. $\csc(-4)$
26. $\sin 491°$
27. $\cos(-903°)$
28. $\cos 315°$
29. $\sin 930°$
30. $\tan 812{,}714.5°$
31. $\sin(-60{,}235.46°)$
32. $\cot 8435°$
33. $\sec 154°$

In Exercises 34–41, verify the equations without a calculator.

34. $\sin^2 \frac{5\pi}{6} + \cos^2 \frac{5\pi}{6} = 1$
35. $\cos(\frac{3\pi}{2} + \frac{\pi}{6}) = \sin \frac{\pi}{6}$
36. $\sin \frac{\pi}{3} \cos \frac{\pi}{6} + \cos \frac{\pi}{3} \sin \frac{\pi}{6} = 1$
37. $\cos 1230° + \cos(-390°) = 0$
38. $\sin(\pi + \frac{\pi}{4}) = -\sin \frac{\pi}{4}$
39. $\sin(\frac{3\pi}{2} + \frac{\pi}{6}) = -\cos \frac{\pi}{6}$
40. $\cos \frac{\pi}{3} \cos \frac{\pi}{6} + \sin \frac{\pi}{3} \sin \frac{\pi}{6} = \cos \frac{\pi}{6}$
41. $\sin 270° \cos 990° + \sin 540° \cos 720° = 0$

SECTION 8.5. GRAPHS OF THE TRIGONOMETRIC FUNCTIONS

In this section, we shall determine the graphs of the trigonometric functions. In the following, you may consider x to be a real number or the measure of an angle in radians, since we have previously seen that these are essentially the same.

From our discussion of the periodicity of the trigonometric functions, we know that if f is any trigonometric function, then $f(x + 2\pi) = f(x)$. Hence we need only find the graphs of each function on an interval of length 2π, say $0 \leq x \leq 2\pi$ or $-\pi/2 \leq x \leq 3\pi/2$, and then use $f(x + 2\pi) = f(x)$ to get the graph everywhere.

We first consider the sine function, and we shall use the interval $0 \leq x \leq 2\pi$. In Fig. 18, let $P = (a, b)$ be a point one unit ($r = 1$) from the origin on the terminal side of an angle x (in radians) in standard position. Since $r = 1$, $\sin x = b/1 = b$; i.e., $\sin x$ is the second coordinate of P.

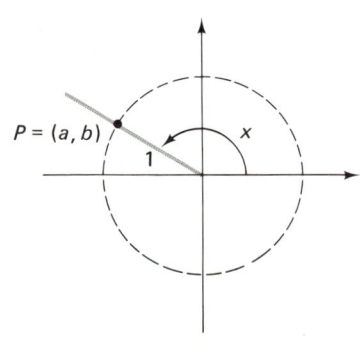

FIG. 18

Now consider what happens to $\sin x$ as x increases from 0 to 2π. When $x = 0$, $\sin x = 0$.

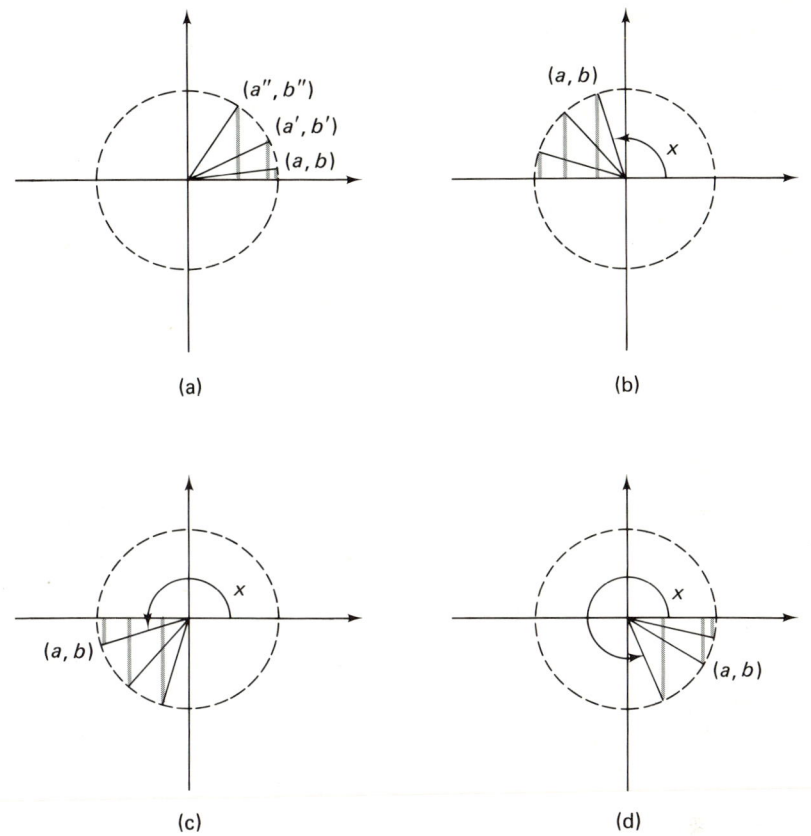

FIG. 19

As x increases from 0 to $\pi/2$, $b = \sin x$ increases from 0 to 1. See Fig. 19(a). As x increases from $\pi/2$ to π, $b = \sin x$ decreases from 1 to 0. See Fig. 19(b). As x increases from π to $3\pi/2$, $b = \sin x$ decreases from 0 to -1. See Fig. 19(c). As x increases from $3\pi/2$ to 2π, $b = \sin x$ increases from -1 to 0. See Fig. 19(d).

From this discussion, we now know the general nature of $y = \sin x$, $0 \le x \le 2\pi$. If we keep this in mind and plot several points, we can obtain a fairly accurate graph. We do this in Fig. 20(a), using the table of special angles (estimating $\sqrt{2}/2 \approx .71$ and $\sqrt{3}/2 \approx .87$) given as Table 2. Of course you could use a calculator to obtain points instead of using Table 2. From the fact that $\sin(x + 2\pi) = \sin x$ for all x (i.e., periodicity), if we repeat the graph in Fig. 20(a) every 2π units, we obtain the graph of $y = \sin x$ for all real x. See Fig. 20(b).

TABLE 2

x	0	$\frac{\pi}{6}$	$\frac{\pi}{4}$	$\frac{\pi}{3}$	$\frac{\pi}{2}$	$\frac{2\pi}{3}$	$\frac{3\pi}{4}$	$\frac{5\pi}{6}$	π	$\frac{7\pi}{6}$	$\frac{5\pi}{4}$	$\frac{4\pi}{3}$	$\frac{3\pi}{2}$	$\frac{5\pi}{3}$	$\frac{7\pi}{4}$	$\frac{11\pi}{6}$
$\sin x$	0	$\frac{1}{2}$	$\frac{\sqrt{2}}{2}$	$\frac{\sqrt{3}}{2}$	1	$\frac{\sqrt{3}}{2}$	$\frac{\sqrt{2}}{2}$	$\frac{1}{2}$	0	$-\frac{1}{2}$	$-\frac{\sqrt{2}}{2}$	$-\frac{\sqrt{3}}{2}$	-1	$-\frac{\sqrt{3}}{2}$	$-\frac{\sqrt{2}}{2}$	$-\frac{1}{2}$

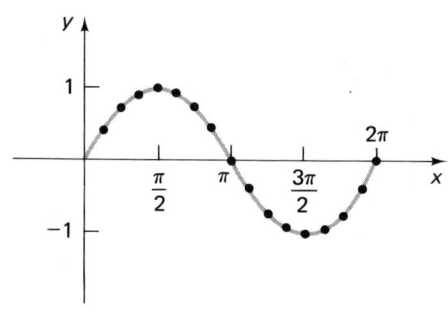

(a) $y = \sin x$, $0 \leq x \leq 2\pi$

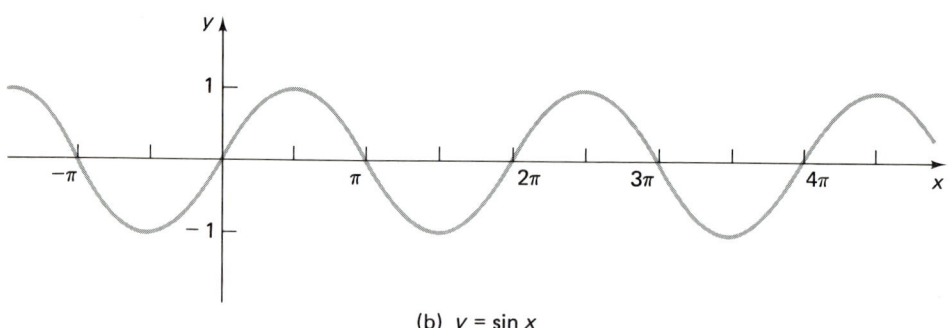

(b) $y = \sin x$

FIG. 20

We next consider the cosine function. We again use the interval $0 \leq x \leq 2\pi$ and Fig. 18. Recall that $P = (a, b)$ is one unit from the origin on the terminal side of x. Thus, $\cos x = a/1 = a$, so that $\cos x$ is the first coordinate of P. We now follow what happens to $\cos x$ as x increases from 0 to 2π. When $x = 0$, $\cos x = 1$.

As x increases from 0 to $\pi/2$, $a = \cos x$ decreases from 1 to 0. As x increases from $\pi/2$ to π, $a = \cos x$ decreases from 0 to -1. As x increases from π to $3\pi/2$, $a = \cos x$ increases from -1 to 0. As x increases from $3\pi/2$ to 2π, $a = \cos x$ increases from 0 to 1. See Fig. 21.

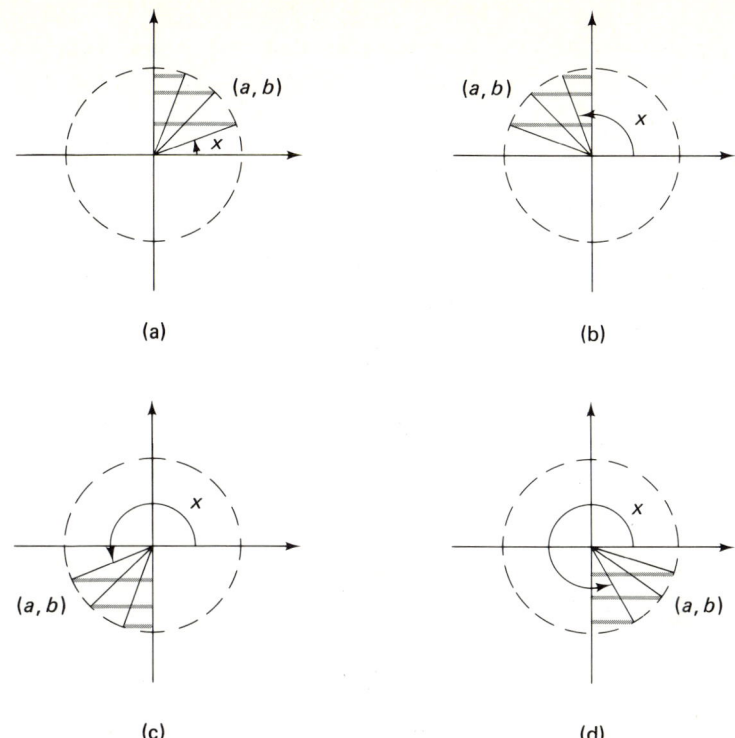

FIG. 21

From this discussion, we know the general nature of the graph of $y = \cos x$, $0 \leq x \leq 2\pi$. If we use this and plot some points from a table like Table 2, we obtain a fairly accurate graph, which is given in Fig. 22(a). From the fact that $\cos(x + 2\pi) = \cos x$ for all x (i.e., periodicity), if we repeat the graph in Fig. 22(a) every 2π units, we obtain the full graph of $y = \cos x$ for all real x. See Fig. 22(b).

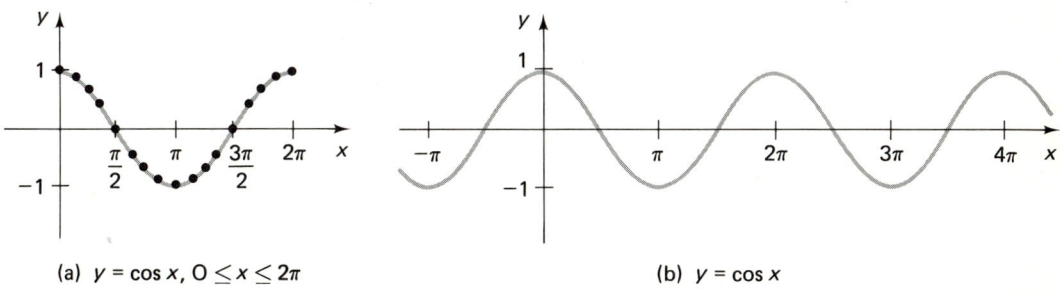

(a) $y = \cos x$, $0 \leq x \leq 2\pi$

(b) $y = \cos x$

FIG. 22

Section 8.5   Graphs of the Trigonometric Functions

From the graphs of $y = \sin x$ and $y = \cos x$, it is clear that there is no positive number $p < 2\pi$ for which $\sin(x + p) = \sin x$ for all x or $\cos(x + p) = \cos x$ for all x. Therefore, since we do know that $\sin(x + 2\pi) = \sin x$ and $\cos(x + 2\pi) = \cos x$ for all x, we have the following:

(11)

The functions sin x and cos x are periodic of period 2π.

By taking reciprocals, we obtain the following:

(12)

The functions csc x and sec x are periodic of period 2π.

Indeed, the graphs of $y = \csc x$ and $y = \sec x$ can also be obtained by taking reciprocals, either directly from the graphs of $y = \sin x$ and $y = \cos x$ as indicated in Fig. 23(a) and (b), or from a table of values, such as Table 2. Note the vertical asymptotes to the two graphs.

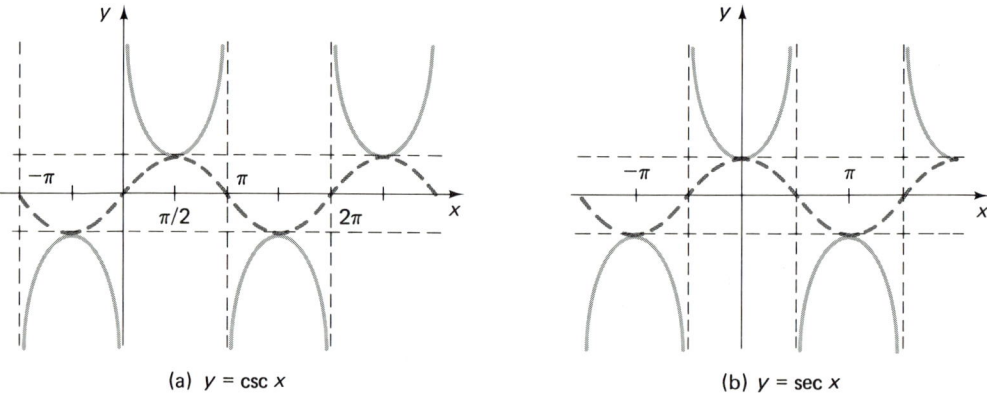

(a) $y = \csc x$ (b) $y = \sec x$

FIG. 23

We now consider the tangent function, and we shall use the interval $-\pi/2 \leq x \leq 3\pi/2$. When the terminal side of x is on the y-axis, i.e., for $x = -\pi/2, \pi/2$ or $3\pi/2$, tan x is undefined. When the terminal side of x is not on the y-axis, let P be the point where the terminal side of x meets the vertical line through $(1, 0)$ or through $(-1, 0)$. Thus, $P = (1, b)$ if $-\pi/2 < x < \pi/2$, and $P = (-1, b)$ if $\pi/2 < x < 3\pi/2$. By the definition of tan x,

$$\tan x = \frac{b}{1} = b \qquad \text{if} \qquad -\frac{\pi}{2} < x < \frac{\pi}{2}$$

$$\tan x = \frac{b}{-1} = -b \qquad \text{if} \qquad \frac{\pi}{2} < x < \frac{3\pi}{2}$$

Now consider what happens to $\tan x$ as x increases from near $-\pi/2$ to near $3\pi/2$. When x is near $-\pi/2$ (but $x > -\pi/2$), $\tan x = b$ is large and negative. As x increases from near $-\pi/2$, through 0, and then approaches $\pi/2$, the numbers $b = \tan x$ increase from very large (in absolute value) negative numbers, pass through zero, and become very large positive numbers. See Fig. 24(a). As $x > \pi/2$ increases from near $\pi/2$, through π, and then approaches $3\pi/2$, the numbers b decrease from very large positive numbers, pass through zero, and become very large (in absolute value) negative numbers. Hence the numbers $-b = \tan x$ increase from very large (in absolute value) negative numbers, go through zero, and become very large positive numbers. See Fig. 24(b).

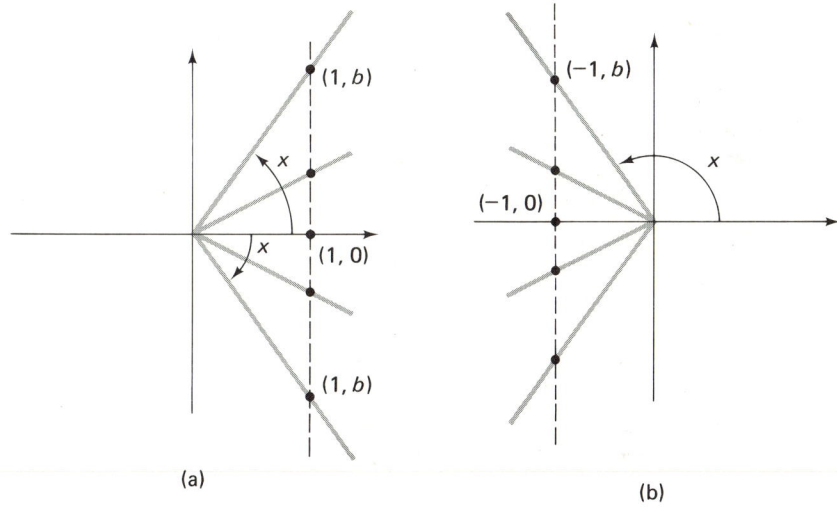

(a)   (b)

FIG. 24

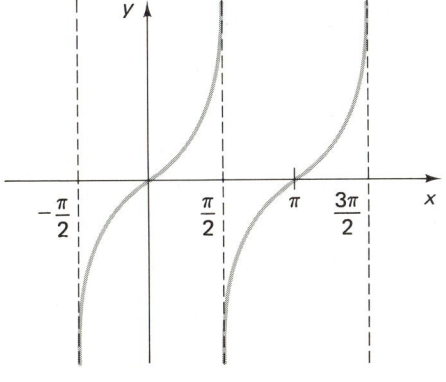

FIG. 25

From this discussion, we see that the general nature of the graph of $y = \tan x$, $-\pi/2 < x < 3\pi/2$, $x \ne \pi/2$, is like that sketched in Fig. 25. Of course, we could use a table of values or a calculator to plot points. Note in particular that the lines $x = -\pi/2, \pi/2, 3\pi/2$ are vertical asymptotes (and we may wish to use a calculator to obtain values close to these numbers).

From Fig. 25, it appears that $\tan x$ is periodic of period π. It can be shown that this is the case.

Section 8.5 Graphs of the Trigonometric Functions

> **(13)**
>
> The function tan x is periodic of period π.

From Fig. 25, we now use periodicity (observing that $\tan x$ is not defined for $x = \frac{1}{2}\pi + k\pi$, k an integer) and obtain the full graph of $y = \tan x$, all $x \neq \frac{1}{2}\pi + k\pi$; see Fig. 26(a).

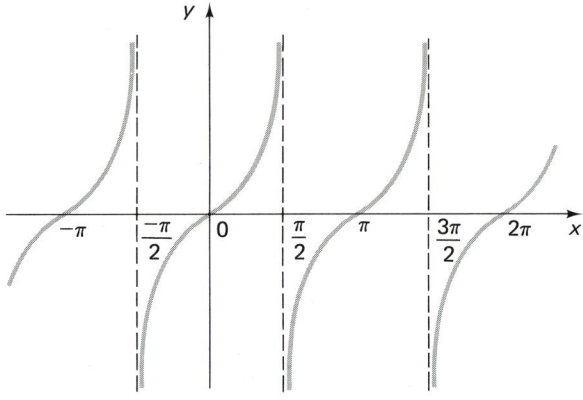

(a) $y = \tan x$

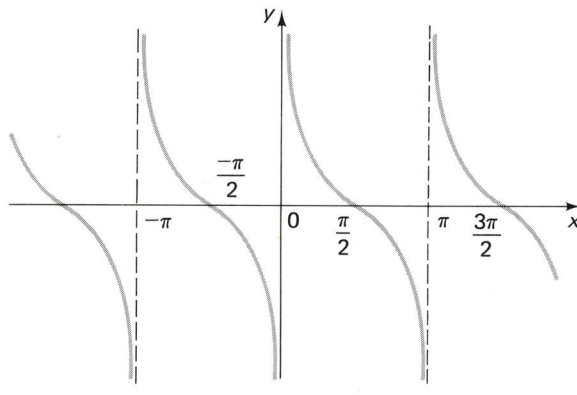

(b) $y = \cot x$

FIG. 26

The graph of $y = \cot x$ given in Fig. 26(b) is obtained in a manner very similar to that of $y = \tan x$. The important things to remember are that $\cot x$ is positive for $0 < x < \pi/2$, $\cot x$ is undefined when x is a multiple of π, and the lines $x = k\pi$ (k an integer) are vertical asymptotes.

Since $\cot x = 1/\tan x$, it follows from (13) that

> **(14)**
>
> The function cot x is periodic of period π.

EXAMPLE 1 Graph $y = \cos 2x$.

SOLUTION We could do this by plotting several points, either from a table such as Table 2 or from a calculator. However, it is easier to use graphing techniques. Recall that replacing x by $2x$ results in a contraction by a factor of 2 in the x direction. Thus the part of the graph that took 2π units (one period) is now contracted to π units and the rest of the graph accordingly. Problems of this nature are treated more thoroughly in the next section.

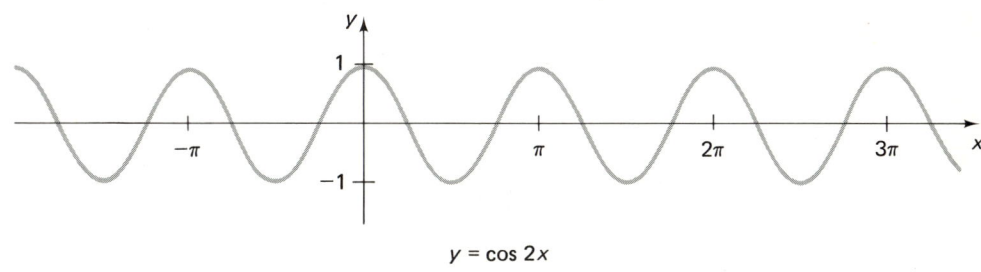

$y = \cos 2x$

EXERCISES

In Exercises 1–6, graph each function on the interval $0 \le x \le 2\pi$ by using your calculator to find the approximate values of the functions for a whole series of x's and then plotting the points.

1. $y = \sin x$
2. $y = \cos x$
3. $y = \tan x$
4. $y = \csc x$
5. $y = \sec x$
6. $y = \cot x$

In Exercises 7–18, sketch the graph.

7. $y = |\sin x|$
8. $y = |\cos x|$
9. $y = |\tan x|$
10. $y = |\csc x|$
11. $y = |\sec x|$
12. $y = |\cot x|$
13. $y = \sin|x|$
14. $y = \cos|x|$
15. $y = \tan|x|$
16. $y = \csc|x|$
17. $y = \sec|x|$
18. $y = \cot|x|$

In Exercises 19–30, sketch the graph, using graphing techniques.

19. $y = \sin(-x)$
20. $y = \cos(-x)$
21. $y = \tan(-x)$
22. $y = 2 \sin x$
23. $y = 3 \cos x$
24. $y = \tfrac{1}{2}\sin x$
25. $y = \sin(x + \tfrac{1}{2}\pi)$
26. $y = \cos(x - \tfrac{1}{4}\pi)$
27. $y = \tan(x - \tfrac{1}{2}\pi)$
28. $y = \sin 2x$
29. $y = \cos \tfrac{1}{2} x$
30. $y = 3 \sin x$

In Exercises 31–36, use the graphs in this section to determine how many numbers x in the interval $[-2, 6]$ satisfy the equation.

31. $\sin x = .2$
32. $\cos x = .2$
33. $\tan x = .2$
34. $\csc x = 5$
35. $\sec x = 5$
36. $\cot x = 5$

SECTION 8.6. THE GRAPHS OF $y = a\sin(bx - c)$ AND $y = a\cos(bx - c)$

In this section we see how to apply our knowledge of expansions, contractions, and translations of graphs to the graphs of $y = \sin x$ and $y = \cos x$ in order to obtain the graphs of $y = a\sin(bx - c)$ and $y = a\cos(bx - c)$. We shall do this in stages, first finding the significance of a, then b, and finally c.

From the graphs of $y = \sin x$ and $y = \cos x$, we know that both of these graphs regularly oscillate. (In fact, because of the shape of their graphs, these functions are sometimes called wave functions.) For any function whose graph is regularly oscillating, the **amplitude** is defined to be one-half of the difference between the maximum value and the minimum value of the function. For a function such as $\sin x$ or $\cos x$ which regularly oscillates an equal amount above and below the x-axis, the amplitude is just the maximum value of the function. Thus it is immediate that $\sin x$ and $\cos x$ have amplitude 1.

Let us now consider the function $y = a\sin x$ (the function $y = a\cos x$ is analogous). From our study of the graphing of functions we know the following: If $a > 0$, the graph of $y = a\sin x$ is obtained from the graph of $y = \sin x$ by expanding (if $a > 1$) or contracting (if $0 < a < 1$) in the y-direction, while if $a < 0$, we first expand or contract the graph of $y = \sin x$ in the y-direction and then reflect the result in the x-axis. That is, the graph of $y = a\sin x$ can be obtained from the graph of $y = \sin x$ by multiplying the y-coordinate of every point on the graph of $y = \sin x$ by a.

This can be seen by examining Example 1.

EXAMPLE 1 The graphs of $y = \sin x$, $y = 2\sin x$, $y = \frac{1}{2}\sin x$, and $y = -2\sin x$, all drawn on the same axes, are shown in Fig. 27.

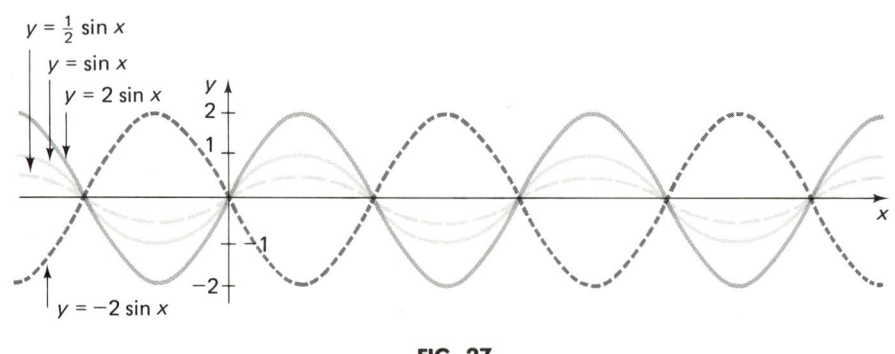

FIG. 27

From our discussion it should be clear that $y = a\sin x$ (and similarly $y = a\cos x$) has amplitude $|a|$. Moreover, by the previous section, $y = \sin x$ and $y = \cos x$ are periodic of period 2π. Since expansions or contractions in the y-direction do not affect the period (see, for example, Fig. 27), $y = a\sin x$ and $y = a\cos x$ have period 2π.

EXAMPLE 4 Graph $y = \cos(\frac{2}{3}x - \frac{1}{3}\pi) = \cos\frac{2}{3}(x - \frac{1}{2}\pi)$.

SOLUTION The function is of the form $y = \cos(bx - c) = \cos b(x - c/b)$ with $b = \frac{2}{3}$, $c = \frac{1}{3}\pi$. Thus the period is $2\pi/b = 2\pi/\frac{2}{3} = 3\pi$, and the phase shift is $c/b = \frac{1}{3}\pi/\frac{2}{3} = \frac{1}{2}\pi$. Some people have difficulty remembering the formula for these numbers. If so, just remember how we derived them: One period is $0 \le \frac{2}{3}(x - \frac{1}{2}\pi) \le 2\pi$, or $0 \le x - \frac{1}{2}\pi \le 3\pi$ (dividing by $\frac{2}{3}$), and finally $\frac{1}{2}\pi \le x \le \frac{1}{2}\pi + 3\pi$ (adding $\frac{1}{2}\pi$). Hence one period starts at $\frac{1}{2}\pi$ (the phase shift) and goes for 3π units (the period). This is the darkened part of the graph; the graph is completed using periodicity.

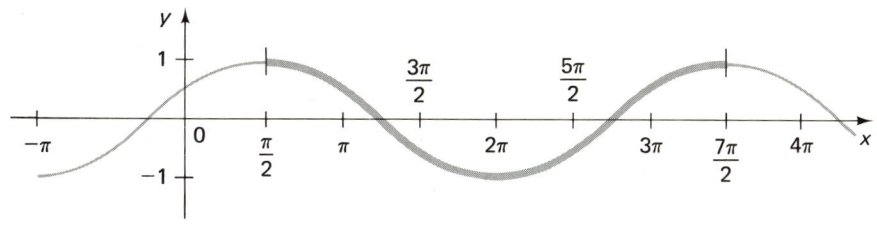

Finally we can now consider the functions $y = a\sin(bx - c)$ and $y = a\cos(bx - c)$. First, if b is negative, use $\sin(-\theta) = -\sin\theta$ or $\cos(-\theta) = \cos\theta$ to write it in the form where $b > 0$. For instance, if $y = 3\sin(-2x + 4)$, we write $y = 3\sin[-(2x - 4)] = 3[-\sin(2x - 4)]$, or $y = -3\sin(2x - 4)$, and graph this. Similarly, if $y = -7\cos(-\pi x - 5)$, we write $y = -7\cos[-(\pi x + 5)] = -7\cos(\pi x + 5)$ and graph this. Other than that, we know that the effect of multiplying by a is to change the amplitude to $|a|$, and if $a < 0$, to reflect the graph through the x-axis.

(18)

$$y = a\sin(bx - c) = a\sin b\left(x - \frac{c}{b}\right) \text{ and}$$

$$y = a\cos(bx - c) = a\cos b\left(x - \frac{c}{b}\right), b > 0,$$

have amplitude $|a|$, period $\dfrac{2\pi}{b}$, and phase shift $\dfrac{c}{b}$.

EXAMPLE 5 Graph $y = -3\sin(\frac{1}{2}x + 1)$.

SOLUTION Since $-3\sin(\frac{1}{2}x + 1) = -3\sin\frac{1}{2}[x - (-2)]$, $a = -3$, $b = \frac{1}{2}$, and $c = -1$. Thus the amplitude is 3, the period is $2\pi/\frac{1}{2} = 4\pi$, and the phase shift is

$-1/\frac{1}{2} = -2$. Since a is negative, we can draw the graph of $3\sin(\frac{1}{2}x + 1)$ and reflect it through the x-axis. See Fig. 29.

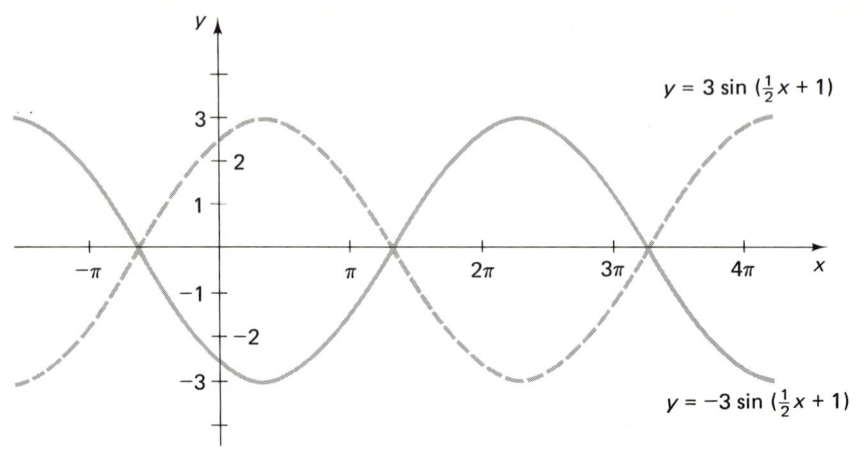

FIG. 29

Occasionally you may run across equations like $y = -3\sin(\frac{1}{2}x + 1) + 4$. To graph this, first graph $y = -3\sin(\frac{1}{2}x + 1)$ as in Fig. 29, and then translate that graph four units upward.

EXERCISES

In Exercises 1–25, state the amplitude, period, and phase shift (if nonzero), and then sketch the graph.

1. $y = 2\cos x$
2. $y = \frac{1}{2}\cos x$
3. $y = -\frac{1}{2}\cos x$
4. $y = -\frac{1}{2}\sin x$
5. $y = 3\sin x$
6. $y = \cos 2x$
7. $y = \cos\frac{1}{2}x$
8. $y = \sin 3x$
9. $y = \cos(-3x)$
10. $y = \sin(-\pi x)$
11. $y = \sin(x - \frac{1}{3}\pi)$
12. $y = \cos(x + \frac{5}{6}\pi)$
13. $y = \sin(x + 1)$
14. $y = \cos(x - 3)$
15. $y = \cos(\pi - x)$
16. $y = \sin(2x - \frac{1}{2}\pi)$
17. $y = \cos(3x - \pi)$
18. $y = \sin(\frac{1}{2}x + \frac{1}{3}\pi)$
19. $y = \sin(-3x + 6)$
20. $y = \cos(-\pi x + \pi)$
21. $y = 3\sin(\pi x + \frac{1}{2}\pi)$
22. $y = -2\cos(3x - 4\pi)$
23. $y = 4\cos(-3x + 3)$
24. $y = \frac{1}{2}\sin(-2x + \pi)$
25. $y = -\frac{1}{3}\cos(2\pi - x)$

In Exercises 26–31, sketch the graph.

26. $y = 2\cos x + 3$
27. $y = \sin 2x - 1$

28. $y = \cos \tfrac{1}{2} x - \tfrac{1}{2}$

29. $y = \sin(x - \tfrac{1}{3}\pi) + \tfrac{1}{3}\pi$

30. $y = 2\sin(\tfrac{1}{2} x + \pi) + 2$

31. $y = \tfrac{1}{2}\cos(\pi x - \pi) + 3$

In Exercises 32–40, sketch the graph, using the techniques demonstrated in this section.

32. $y = \tan(x - \tfrac{1}{2}\pi)$

33. $y = \tan \tfrac{1}{2} x$

34. $y = \csc(x + \tfrac{1}{2}\pi)$

35. $y = \csc 2x$

36. $y = \sec(\tfrac{3}{2} x - \tfrac{1}{2}\pi)$

37. $y = \sec(\pi x + \tfrac{1}{4}\pi)$

38. $y = \tan 3x - 1$

39. $y = \csc \pi x + \pi$

40. $y = 2\sec 3x - 4$

SECTION 8.7. INVERSE TRIGONOMETRIC FUNCTIONS

In the previous Chapter, we defined the six trigonometric functions of acute angles, and then we saw that each of these functions had an inverse function. Now that we have generalized the definitions of the six trigonometric functions, it is natural to ask if each of the functions still has an inverse. Recall that a function has an inverse function precisely when no horizontal line intersects the graph more than once, i.e., the function passes the **horizontal line test**. It is apparent from their graphs, Figs. 20(b), 22(b), 23, and 26, that each trigonometric function fails the horizontal line test, so it does not have an inverse function.

When a function does not have an inverse and it would be useful to have one, what we usually do is restrict the domain in such a way that the function, *with its restricted domain*, has an inverse. Accordingly, for each of the trigonometric functions, we shall restrict the domain subject to the following:

1. Each value in the range is taken on only once (i.e., the graph passes the horizontal line test).
2. The range of the function with the restricted domain is the same as the range of the original function.
3. The domain includes the most commonly used numbers (or angles), $0 < x < \pi/2$.
4. The graph is connected (if possible).

For the function $y = \sin x$, conditions 1–4 are satisfied if we restrict the domain to $-\pi/2 \le x \le \pi/2$:

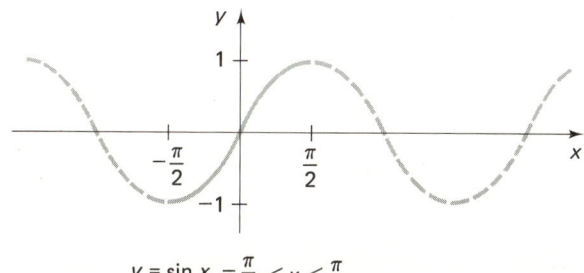

$y = \sin x$, $-\tfrac{\pi}{2} \le x \le \tfrac{\pi}{2}$

> **(19)** **DEFINITION**
>
> Suppose $-1 \leq x \leq 1$. Then $y = \sin^{-1} x$ (or $y = \arcsin x$) if and only if $x = \sin y$ and $-\pi/2 \leq y \leq \pi/2$.

In other words, if $-1 \leq x \leq 1$, then the equation $y = \sin^{-1} x$ is saying "y is the number (angle) between $-\pi/2$ and $\pi/2$ whose sine is x."

Recall that the graph of the inverse f^{-1} can be obtained by reflecting the graph of f through the line $y = x$. Accordingly, the graph of $y = \sin^{-1} x$, $-1 \leq x \leq 1$, is obtained by reflecting the graph of $y = \sin x$, $-\pi/2 \leq x \leq \pi/2$, through the line $y = x$. See Fig. 30.

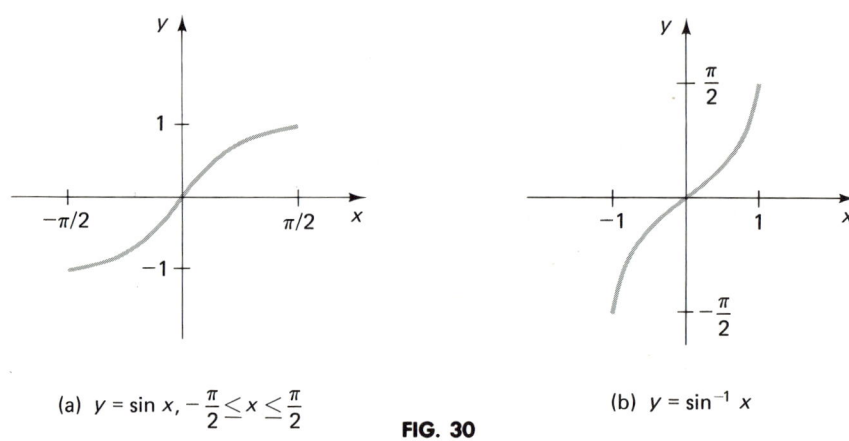

(a) $y = \sin x,\ -\dfrac{\pi}{2} \leq x \leq \dfrac{\pi}{2}$ (b) $y = \sin^{-1} x$

FIG. 30

For the function $y = \cos x$, conditions 1–4 are satisfied if we restrict the domain to $0 \leq x \leq \pi$. See Fig. 31(a).

> **(20)** **DEFINITION**
>
> Suppose $-1 \leq x \leq 1$. Then $y = \cos^{-1} x$ (or $y = \arccos x$) if and only if $x = \cos y$ and $0 \leq y \leq \pi$.

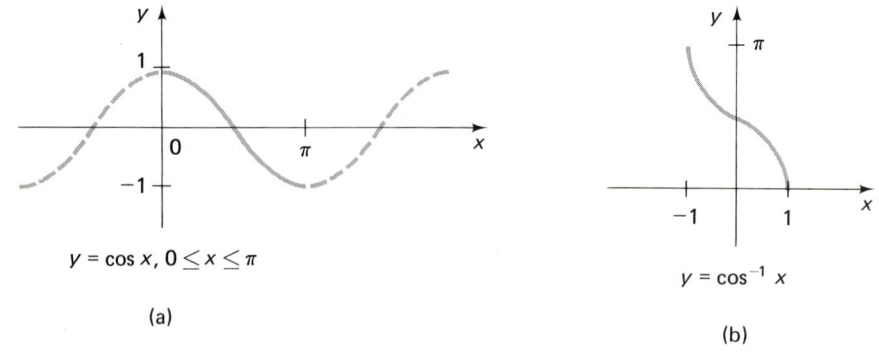

$y = \cos x,\ 0 \leq x \leq \pi$

(a)

$y = \cos^{-1} x$

(b)

FIG. 31

360 GENERAL TRIGONOMETRY

In other words, if $-1 \le x \le 1$, then the equation $y = \cos^{-1} x$ is saying "y is the number (angle) between 0 and π whose cosine is x." The graph of $y = \cos^{-1} x$, Fig. 31(b), is obtained from Fig. 31(a) by reflection through the line $y = x$.

For the function $y = \tan x$, conditions 1–4 are satisfied if we restrict the domain to $-\pi/2 < x < \pi/2$. See Fig. 32(a).

---

(21)   **DEFINITION**

Suppose x is any real number. Then $y = \tan^{-1} x$ (or $y = \arctan x$) if and only if $x = \tan y$ and $-\pi/2 < y < \pi/2$.

In other words, if x is any real number, the equation $y = \tan^{-1} x$ is saying "y is the number (angle) strictly between $-\pi/2$ and $\pi/2$ whose tangent is x." The graph of $y = \tan^{-1} x$ is shown in Fig. 32(b).

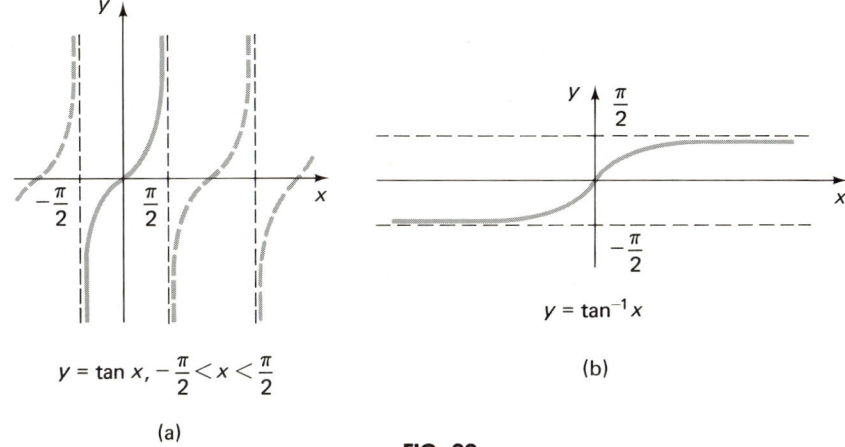

$y = \tan x, \ -\dfrac{\pi}{2} < x < \dfrac{\pi}{2}$

(a)

$y = \tan^{-1} x$

(b)

FIG. 32

The remaining trigonometric functions also have inverses if their domains are appropriately restricted.

---

(22)   **DEFINITION**

Suppose $|x| \ge 1$. Then $y = \csc^{-1} x$ (or $y = \text{arccsc } x$)   if and only if $x = \csc y$ and $-\pi/2 \le y \le \pi/2$, $y \ne 0$.

---

(23)   **DEFINITION**

Suppose $|x| \ge 1$. Then $y = \sec^{-1} x$ (or $y = \text{arcsec } x$)   if and only if $x = \sec y$ and $0 \le y \le \pi$, $y \ne \pi/2$.

---

(24)   **DEFINITION**

Suppose x is any real number. Then $y = \cot^{-1} x$ (or $y = \text{arccot } x$) if and only if $x = \cot y$ and $0 < y < \pi$.

In each of the definitions of the inverse trigonometric functions, y may be viewed as an angle measured in degrees. Just change the restriction on y to the corresponding inequality in degrees, such as $-90° \leq y \leq 90°$ in Definition (19).

The process of finding inverse sine, inverse cosine, or inverse tangent on a calculator is exactly the same as was described in the previous chapter. First, *make sure your calculator is in the proper mode,* and remember that if you are computing with trigonometric functions or their inverses and you want all values to be real numbers, the calculator must be in radian mode. Then on either an RPN or an algebraic calculator, press (1) the number; (2) the button [ARC], [INV], [F], or the equivalent, whichever is on your calculator; and finally (3) the appropriate button of [sin], [cos], or [tan].

EXAMPLE 1 Find $\sin^{-1}(-.5)$.

SOLUTION In radian mode, press [.5][+/−][INV][sin], obtaining $-.523599$. In degree mode pressing the same buttons yields $-30°$.

EXAMPLE 2 Find arctan 56.47.

SOLUTION Press [56.47][INV][tan]. In radian mode you would obtain 1.55309, while in degree mode you would obtain 88.9855°.

EXAMPLE 3 Find $\cos^{-1} 2.3$.

SOLUTION If you press [2.3][INV][cos], you will get an error indication (in either mode), because $\cos^{-1} x$ is only defined for $-1 \leq x \leq 1$.

Although it is easy to use the calculator for inverse trigonometric functions, there are many times where it is not necessary to do so, and you gain much more understanding of the functions if you do not use your calculator. For instance, Example 1 can be done without a calculator.

EXAMPLE 4 Find $\sin^{-1}(-.5)$ directly from the definition (without using a calculator).

SOLUTION Set $\theta = \sin^{-1}(-.5)$. Then by Definition (19), $\theta = \sin^{-1}(-.5)$ if and only if $\sin \theta = -.5$ and $-\pi/2 \leq \theta \leq \pi/2$. Since $\sin \theta$ is negative and $-\pi/2 \leq \theta \leq \pi/2$, we have $-\pi/2 \leq \theta \leq 0$ (i.e., θ is in Q_4). If φ is the reference angle of θ, then $\sin \varphi = .5$. Therefore from our knowledge of special angles, $\varphi = \pi/6$. But the only angle θ with $-\pi/2 \leq \theta \leq 0$ and reference angle $\pi/6$ is $\theta = -\pi/6$. Of course, in degrees, $\theta = -30°$.

The following problem cannot be done on a calculator since there are no numbers involved. However, it can be done by techniques similar to the preceding.

EXAMPLE 5 Find $\sin(\cos^{-1} u)$ as an algebraic expression in u.

SOLUTION Set $\theta = \cos^{-1} u$. We want $\sin \theta$ in terms of u. From Definition (20), $\theta = \cos^{-1} u$ if and only if $\cos \theta = u$ and $0 \leq \theta \leq \pi$. Since we have $\cos \theta = u$ and want $\sin \theta$ in terms of u, we use the elementary identity relating $\sin \theta$ and $\cos \theta$, namely, $\sin^2 \theta + \cos^2 \theta = 1$. We obtain

$$\sin^2 \theta = 1 - \cos^2 \theta = 1 - u^2 \quad \text{or} \quad \sin \theta = \pm \sqrt{1 - u^2}$$

The sign can be determined in this case. Since $0 \leq \theta \leq \pi$ and $\sin \theta$ is positive in Q_1 and Q_2, the sign must be $+$. Thus

$$\sin \theta = \sin(\cos^{-1} u) = \sqrt{1 - u^2}$$

From the definition of inverse functions, we know that $f(f^{-1}(x)) = x$ for all x in the domain of f^{-1} and that $f^{-1}(f(x)) = x$ for all x in the domain of f. Therefore these relations hold for the *restricted* trigonometric functions and their inverses.

EXAMPLE 6 (a) $\sin[\sin^{-1}(-.17)] = -.17$. (b) $\arctan(\tan \pi/5) = \pi/5$.
(c) $\cos^{-1}(\cos 171°) = 171°$.

If the inverse function is on the inside, this always works. However, if the trigonometric function is on the inside, this works only if the angle is within the restricted domain of that function.

EXAMPLE 7 Find $\arcsin(\sin 589°)$ without using your calculator.

SOLUTION

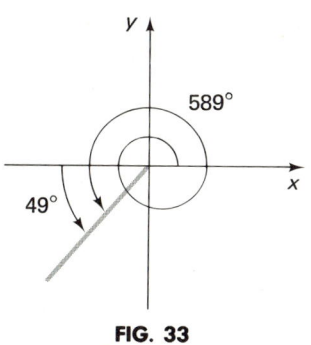

FIG. 33

The answer is not 589°, because 589° is not in the restricted domain of the sine function, $-90° \leq \theta \leq 90°$. If we set $\theta = \arcsin(\sin 589°)$, then from Definition (19),

$$\theta = \arcsin(\sin 589°) \text{ if and only if}$$

$$\sin \theta = \sin 589° \text{ and } -90° \leq \theta \leq 90°$$

The angle 589° is in Q_3 with reference angle 49°. See Fig. 33. Since sine is negative in Q_3, θ must satisfy the following: (1) $-90° \leq \theta \leq 90°$, (2) the reference angle of θ is 49°, and (3) $\sin \theta$ is negative. Clearly, $\theta = -49°$ is the desired angle. Thus $\arcsin(\sin 589°) = -49°$. Of course this problem could have been done on a calculator by pressing (in degree mode) [589] [sin] [INV] [sin].

Even though your calculator does not have buttons for cosecant, secant, or cotangent, their inverse functions can still be evaluated on a calculator. For instance, suppose you are given x, $|x| \geq 1$, and you wish to find $\csc^{-1} x$. Set $y = \csc^{-1} x$. Then, using the notation $\Leftrightarrow$ for "if and only if,"

$$y = \csc^{-1} x \Leftrightarrow x = \csc y \quad \text{and} \quad -\frac{\pi}{2} \leq y \leq \frac{\pi}{2}, y \neq 0 \quad \text{[by Definition (22)]}$$
$$\Leftrightarrow x = \frac{1}{\sin y} \quad \text{and} \quad -\frac{\pi}{2} \leq y \leq \frac{\pi}{2}, y \neq 0 \quad (\text{since } \csc y = 1/\sin y)$$
$$\Leftrightarrow \frac{1}{x} = \sin y \quad \text{and} \quad -\frac{\pi}{2} \leq y \leq \frac{\pi}{2}, y \neq 0 \quad (\text{taking reciprocals})$$
$$\Leftrightarrow y = \sin^{-1}\left(\frac{1}{x}\right) \quad \text{[by Definition (19)]}.$$

Consequently, we have shown the following:

(25)
$$\text{If } |x| \geq 1, \quad \csc^{-1} x = \sin^{-1}\left(\frac{1}{x}\right).$$

Using formula (25), $\csc^{-1} x$ can easily be found on a calculator.

EXAMPLE 8 Find arccsc 5.3.

SOLUTION By formula (25), arccsc $5.3 = \arcsin(1/5.3)$. Thus we press [5.3][1/x][INV][sin], obtaining .189817 (in radian mode) or 10.8757° (in degree mode).

There is a formula similar to (25) for $\sec^{-1} x$ (see Exercise 86). By different methods, a formula for $\cot^{-1} x$ is

$$\cot^{-1} x = \tfrac{1}{2}\pi - \tan^{-1} x, \text{ for all } x$$

EXERCISES

In Exercises 1–8, evaluate in three ways, assuming the expression is (a) a real number, (b) an angle measured in radians, and (c) an angle measured in degrees.

1. $\sin^{-1} .82$
2. $\arcsin(-.35)$
3. $\arccos(-.1267)$
4. $\cos^{-1}(.0013)$
5. $\tan^{-1}(-4.9)$
6. $\arctan 12.2$
7. $\operatorname{arccsc} 12.2$
8. $\csc^{-1}(-5.876)$

In Exercises 9–20, evaluate, assuming all numbers involved are real numbers.

9. $\cos 7^{-1}$
10. $\cos^{-1} 7^{-1}$
11. $(\cos^{-1} 7^{-1})^{-1}$
12. $[(\cos^{-1} 7^{-1})^{-1}]^{-1}$
13. $\sin(\cos^{-1} \tfrac{2}{3})$
14. $\cos(\arcsin \tfrac{2}{3})$

15. sec(arctan 60.85)

16. $4.2 \sin^{-1} .367 + 9.12 \cos 12.1$

17. $\cos[\frac{1}{28} \tan^{-1}(-.805)] - 36\pi$

18. $\dfrac{\cos^{-1}(-.75)}{3 + \sin 48.5}$

19. $\arcsin\left(\dfrac{4}{3.6 + 11.73} - \operatorname{arccsc} 4\right)$

20. $[\tan^{-1}(-128)]^5 + 16 \arccos 3^{-5}$

In Exercises 21–36, evaluate the inverse functions directly from the definition (without using a calculator).

21. $\sin^{-1}(\frac{1}{2})$
22. $\arcsin(-1/2)$
23. $\arccos(-1/2)$
24. $\cos^{-1}(\frac{1}{2})$
25. $\tan^{-1}(-1)$
26. $\arctan 1$
27. $\operatorname{arccot} 1$
28. $\cot^{-1}(-1)$
29. $\sin^{-1} \dfrac{1}{\sqrt{2}}$
30. $\operatorname{arccsc}(-\sqrt{2})$
31. $\operatorname{arcsec} \sqrt{2}$
32. $\cos^{-1}(-1/\sqrt{2})$
33. $\sec^{-1}(-2/\sqrt{3})$
34. $\operatorname{arccsc}(2/\sqrt{3})$
35. $\operatorname{arccot}(-\sqrt{3})$
36. $\tan^{-1} \sqrt{3}$

In Exercises 37–48, state which one of (A), (B), and (C) holds without evaluating the problem: (A) the expression is undefined, (B) the expression could be evaluated without using a calculator, or (C) a calculator should be used in the evaluation.

37. $\sin .5$
38. $\sin^{-1} .5$
39. $\cos 1.5$
40. $\arccos 1.5$
41. $\tan\left(-\dfrac{\pi}{2}\right)$
42. $\tan^{-1}\left(-\dfrac{\pi}{2}\right)$
43. $\csc 0$
44. $\csc^{-1} 0$
45. $\cot \sqrt{3}$
46. $\operatorname{arccot} \sqrt{3}$
47. $\sec(-.5)$
48. $\sec^{-1}(-.5)$

In Exercises 49–64, evaluate the expression first without using the trigonometric buttons on your calculator and then using these buttons to check your work.

49. $\sin(\arcsin .2)$
50. $\cos(\arccos .643)$
51. $\tan^{-1}\left(\tan \dfrac{11\pi}{3}\right)$
52. $\tan(\tan^{-1} 12.9)$
53. $\sin^{-1}(\sin 38°)$
54. $\arccos[\cos(-16°)]$
55. $\cos^{-1}\left(\cos \dfrac{5\pi}{8}\right)$
56. $\tan^{-1}\left(\tan \dfrac{\pi}{3}\right)$
57. $\sin^{-1}(\sin 642°)$
58. $\csc(\operatorname{arccsc} 6.89)$
59. $\arctan[\tan(-843°)]$
60. $\arcsin[\sin(-11.7352)]$
61. $\arccos(\cos 6514°)$
62. $\sin^{-1}[\sin(-11.7352°)]$
63. $\tan[\tan^{-1}(-479)]$
64. $\tan^{-1}[\tan(-479°)]$

65. Suppose $\pi/2 < x \le \pi$. Then $\sin^{-1}(\sin x) = x$ is not true. Using reference angles, find m and b such that, for $\pi/2 < x \le \pi$, $\sin^{-1}(\sin x) = mx + b$.

In Exercises 66–70, rewrite the given expression as an algebraic expression in u.

66. $\cos(\sin^{-1} u)$
67. $\sec(\sin^{-1} u)$

68. $\sin(\sec^{-1} u)$ **69.** $\tan(\sec^{-1} u)$
70. $\cot(\text{arccsc } u)$

In Exercises 71–85, sketch the graphs by applying your knowledge of graphing techniques (translations, contractions, and expansions) to the graphs of the inverse trigonometric functions.

71. $y = 2 \sin^{-1} x$ **72.** $y = \tfrac{1}{2}\arcsin x$

73. $y = \arccos 3x$ **74.** $y = \cos^{-1}(\tfrac{1}{3} x)$

75. $y = \tan^{-1}(x + \pi)$ **76.** $y = \arctan(x - 2)$

77. $y = 2 \arccos x$ **78.** $y = \arcsin(x - 5)$

79. $y = \dfrac{5}{\pi}\tan^{-1} x$ **80.** $y = 2 \sin^{-1}(3x - 4)$

81. $y = \tfrac{1}{3}\arccos(\tfrac{1}{2} x + 2)$ **82.** $y = \dfrac{1}{\pi}\arctan(x + 3)$

83. $y = 5 \tan^{-1}(2x - 3)$ **84.** $y = \dfrac{6}{\pi}\sin^{-1}[\tfrac{1}{3}(x + \pi)]$

85. $y = \tfrac{1}{6}\cos^{-1}(2x - \tfrac{5}{2})$

86. Develop a formula for $\sec^{-1} x$ by using similar steps to those used to develop formula (25).
87. Use the formula developed in Exercise 86 to evaluate $\sec^{-1}(-21.73)$ and $\text{arcsec } 1.489$.
88. Graph $y = \pi/2 - \tan^{-1} x$ (using graphing techniques) and compare it with the graph of $y = \cot^{-1} x$.

SECTION 8.8. APPLICATIONS

Anyone who has sat on a beach watching the waves roll in has at least an intuitive idea as to what wave motion is. The surface of the water is shaped a little like a sine curve (if you take a cross section in the right direction), and the crests and troughs of the waves move along through the water. If you get into the water and float for a while, you find yourself bobbing up and down and not moving with the waves. Hence you can realize that the water itself is not flowing along in the waves; it is energy moving through the water that causes the waves. (For example, a boat moving through the water temporarily pushes the water aside, and the energy from this push travels through the water in waves.) The water molecules themselves just move around in one region, transmitting this energy along.

Water waves provide one example of many in which energy moves through our environment in *waves*. Others include sound, light, and even earthquake waves. Trigonometry is the basic mathematical tool that is used to analyze wave motion. Our purpose in this section is to discuss briefly a few familiar physical phenomena and give an indication as to how trigonometry is used to describe them. Unfortunately, we are forced to keep the mathematics fairly simple, for otherwise the mathematics and physics become too deep for this brief survey.

Wave motion can be quite complicated, so we shall consider only the simplest wave motion.

---

(26)   **DEFINITION**

A phenomenon is said to be **simple harmonic** or have **simple harmonic motion** if it can be described by

$$y = a \sin(bx - c)$$

where $b > 0$ and x is distance. If so, it has amplitude $|a|$, wavelength $\lambda = 2\pi/b$, and phase shift c/b. See Fig. 34.

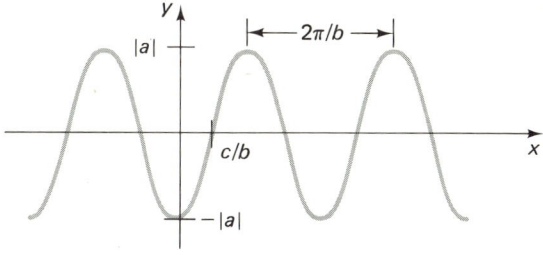

FIG. 34

In the situations we describe, the waves will be moving as time changes. Different kinds of waves move at different speeds. The time it takes for one complete wave to pass a fixed point is called the **period** of the motion; the number of waves that pass a fixed point in one unit of time is called the **frequency** of the motion. These two quantities are related by

---

(27)
$$\text{Frequency} = \frac{1}{\text{period}}$$

In addition, it is not hard to see that there is a relationship between frequency, wavelength, and the speed that the waves are traveling:

---

(28)
$$\text{Frequency} \times \text{wavelength} = \text{speed}$$

We now discuss a few natural phenomena.

Sound

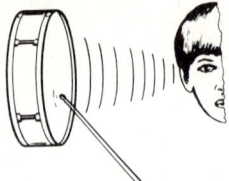

The term *sound* has two distinct uses. A physiologist uses the word sound in connection with the sense of hearing and the effects on a human ear produced by certain vibrations in the air. However, we shall follow the usage of the physicists, who consider sound to be those air vibrations themselves.* The air vibrations are caused by something (such as a drum head or a violin string) moving back and forth. The vibrating object pushes the air aside, causing sound waves in somewhat the same way that a boat pushes water aside, causing water waves.

When made to vibrate, a tuning fork will produce a **pure tone**; this means that only one frequency is produced. A tuning fork that produces a middle C vibrates at approximately 260 cycles per second (cps) or 260 Hz (1 Hz is 1 hertz, which is 1 cps). Now sound moves through air at approximately 1100 feet per second. From relationship (28), frequency × wavelength = speed, we see that the wavelength λ of middle C is approximately

$$\lambda \approx \frac{1100}{260} \approx 4.23 \text{ ft}$$

This is of great interest to anyone who wants to build an open organ pipe to produce a middle C. An open pipe must be exactly one-half wavelength long to produce that frequency of sound (as a fundamental frequency). A closed pipe (for example, a cola bottle) need only be one-fourth wavelength (because the closed end reflects the wave and hence doubles the effect of the pipe). Try blowing across a bottle filled to different levels with fluid, and hear what happens.

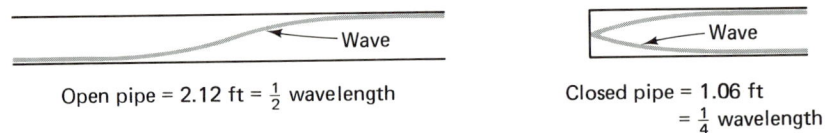

Open pipe = 2.12 ft = $\frac{1}{2}$ wavelength   Closed pipe = 1.06 ft = $\frac{1}{4}$ wavelength

Length of pipe to produce middle C

If you have ever tuned a stringed instrument, you know that the tighter the string, the higher the sound which it produces. This occurs because a tighter string vibrates more quickly, and the higher the frequency (and correspondingly the shorter the wavelength), the higher the pitch of the sound.

If the stringed instrument is like a violin or a guitar, you also know that by pressing the string against the neck, a higher sound will be produced when the string is vibrated. Again, a shorter string (of the same tension) vibrates more quickly and hence produces a higher sound.

*If a tree falls on a deserted island, does it produce a sound? It depends on which sense of the word *sound* you use.

Usually when an object such as a drum, a bell, or a string is vibrating, more than one frequency is involved. The lowest frequency is called the **fundamental frequency**. The other frequencies that are present are multiples of the fundamental frequency and are called **overtones**. If the multiples are integers, the overtones are called **harmonics**.

Harmonics can be seen quite vividly on a stringed instrument, such as a violin. Suppose a violin string is tuned to produce middle C, so that it is vibrating at approximately 260 cps. See Fig. 35(a). But this is just the fundamental frequency; at the same time the string is vibrating twice as quickly, 520 cps, as indicated in Fig. 35(b), producing the C above middle C. This is not all, for at the same time the string is also vibrating at three times the fundamental frequency (i.e., 780 cps), etc. In fact, the string is vibrating at all integral multiples of the fundamental frequency, producing all harmonics. Indeed, you can physically hear the harmonics. If you lightly touch a vibrating string in the middle, this will cancel all odd multiples of the fundamental frequency (including the fundamental frequency itself), and the sound will immediately rise an octave (to the first harmonic). Any reasonably experienced violinist knows how to produce a few harmonics this way (though he or she may not know why it works).

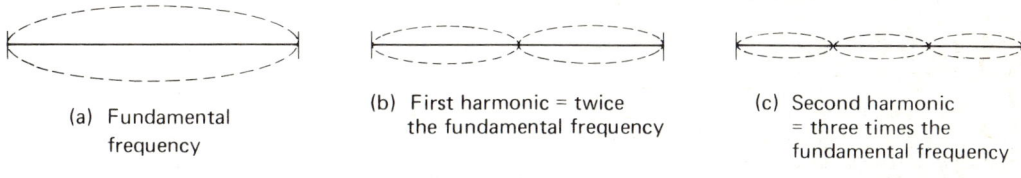

(a) Fundamental frequency

(b) First harmonic = twice the fundamental frequency

(c) Second harmonic = three times the fundamental frequency

FIG. 35

Radio Waves and Other Electromagnetic Waves

The signal sent from the broadcast tower of a radio station to the antenna of your radio is an electromagnetic wave. There is a main wave present which is called a carrier wave, and this wave is basically a sine wave. If the station is an AM station, then the amplitude of the carrier wave is changed to "carry" sound waves, as indicated in Fig. 36(a). AM stands for **amplitude modulation**. If the station is an FM station, then the frequency of the carrier wave is changed to "carry" sound waves, as indicated in Fig. 36(b). FM stands for **frequency modulation**.

To obtain a rough idea as to dimensions, the middle of the AM dial is approximately 1200 kilohertz (kHz), which stands for 1200 kilocycles per second or $1,200,000 = 1.2 \times 10^6$ cps. Radio waves travel at the speed of light, which is approximately 3×10^8 meters per second. From relation (28), frequency $\times$ wavelength = speed, we see that the wavelength of this radio band is approximately $(3 \times 10^8)/(1.2 \times 10^6)$ meters ≈ 200 meters long. (Using 1 meter $\approx$ 1 yard, we see that this is about two football fields long.) For a comparison, CB radios broadcast at wavelengths of about 11 meters.

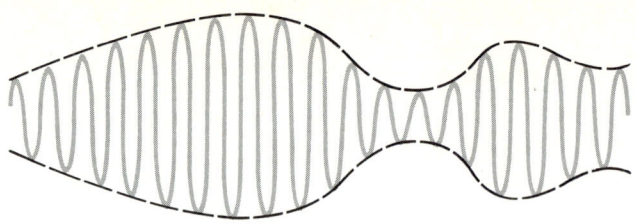

(a) AM: Carrier wave *amplitude* varied by sound source. Frequency is constant.

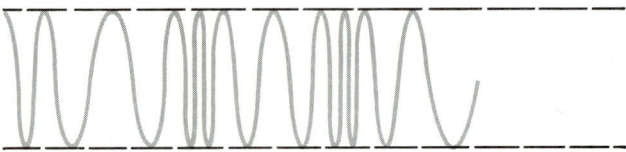

(b) FM: Carrier wave *frequency* varied by sound source. Amplitude is constant.

FIG. 36

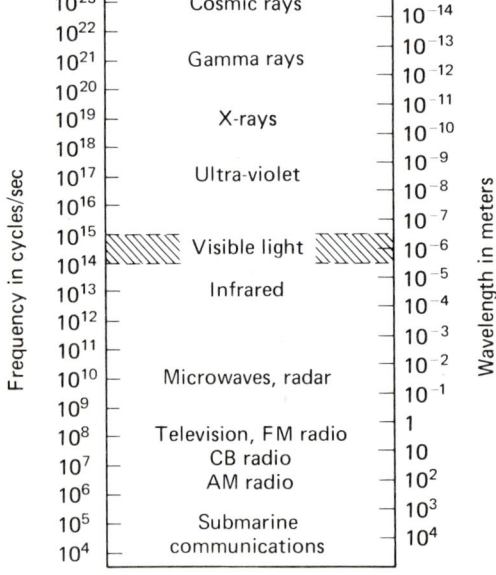

The Electromagnetic Spectrum
(Freq. × Wavelength ≈ 3×10^8 m/sec)

FIG. 37

There are many other electromagnetic waves besides radio waves. These include television waves, microwaves, cosmic rays, and many others. In a famous work, Maxwell* showed that light could be explained as electromagnetic waves. There is a whole spectrum of electromagnetic waves as Fig. 37 illustrates.

*James Clerk Maxwell (1831–1879), a Scottish theoretical physicist who became famous for his mathematical theory of electricity, magnetism, and light.

EXERCISES

In Exercises 1–8, assume that middle C has a frequency of approximately 260 Hz and that the A above middle C is 440 Hz (1 Hz is 1 cps). Assume all data are correct to three significant figures.

1. What is the wavelength of C above middle C?
2. What is the wavelength of the A above middle C?
3. What is the frequency of A above the C above middle C?
4. What is the frequency of the C below middle C?
5. If the speed of sound in water is about 4800 feet per second, what is the wavelength of middle C in water?
6. The speed of sound in steel is about 16,000 feet per second. Suppose that an A above middle C tuning fork is vibrating at one end of a 20-foot steel beam and that the sound is coming out of the other end. How many wavelengths are there from one end of the beam to the other?
7. How long should you make an open organ pipe to produce an A above middle C? How long should a closed pipe be to produce that A?
8. Suppose Neptune wanted to build a water organ to play under the sea. How long should he make an open pipe to produce an A above middle C? A closed pipe? (See Exercise 5.)
9. Red light has a wavelength of about 7×10^{-7} meter. What is its frequency?
10. A certain radar set, used by police to check for speeders, uses microwaves of length .1 meter. What is the frequency?
11. The ideal length of a radio transmitting antenna is any multiple of one-half wavelength. What is the shortest antenna a radio station should use if it is broadcasting at 1400 kHz on the AM dial? Give the answer to the nearest meter.
12. What is the shortest antenna a radio station should use if it is broadcasting at 103 megahertz (MHz) on the FM dial? (See Exercise 11.)

REVIEW EXERCISES

In Exercises 1–5, use the information given to find the values of the six trigonometric functions of θ.

1. $(-1, 4)$ is on the terminal side of θ.
2. $(-3, -2)$ is on the terminal side of θ.
3. For some point on the terminal side of θ, $r = 5$, $x = 2$.
4. For some point on the terminal side of θ, $r = 5$, $y = -1$.
5. For every point on the terminal side of θ, $y = -3x$.

In Exercises 6–13, find the values of the trigonometric functions of θ.

6. $\cos \theta = -\frac{5}{13}$
7. $\tan \theta = -\frac{12}{5}$
8. $\sin \theta = 0$
9. $\cos \theta = \frac{3}{5}$, $\sin(-\theta)$ negative
10. $\sin \theta = \frac{3}{5}$, $\cos(-\theta)$ negative
11. $\tan \theta = -5$, $\sin \theta$ positive
12. $\csc \theta = 3$, $\sec \theta$ positive
13. $\sec \theta = -4$, $\sin(-\theta)$ positive

In Exercises 14 and 15, find the coordinates of the point P on the terminal side of the given θ satisfying the given condition.

14. $\theta = -432.65°$, $r = 25.8$
15. $\theta = 8.63$, $x = -1.36$

In Exercises 16–26, evaluate the given expression, using a calculator only when necessary.

16. $\sin^2 1465° + \cos^2 1465°$

17. $\tan 540° \sec 493° + \cos 450° \sin(-493°)$

18. $\sqrt{6.84 \tan 3.63 - 1.895}$

19. $\dfrac{\sin 4.619}{\cos 4.619} - \tan 4.619 + \cot 4.619$

20. $\cos^4(-63.2°) - \sin^4(-63.2°) + \sin^2(-63.2°)$

21. $\sin \dfrac{11\pi}{6} \cos 3\pi - \tan \dfrac{13\pi}{4} \sec \dfrac{4\pi}{3}$

22. $\tan 1950° + \sin 1950° + \cos 1950°$

23. $\tan 1127° \cos 1127° + \sin(-1127°)$

24. $\arcsin(-.308)$

25. $3\sin^2 2.3 + 4\cos^{-1} .6294$

26. $\sin(1 + \arctan 7.065)$

In Exercises 27–30, find the reference angle, and then evaluate trigonometric functions of the reference angle and attach the appropriate sign.

27. $\sin 11.47$

28. $\tan(-4.198)$

29. $\cos 1265°$

30. $\sec(-15{,}688°)$

In Exercises 31–34, state the amplitude, period, and phase shift (if nonzero), and then sketch the graph.

31. $y = 3\cos\left(x - \dfrac{\pi}{3}\right)$

32. $y = -\tfrac{1}{2}\sin\left(2x + \dfrac{\pi}{2}\right)$

33. $y = 2\sin\left(\dfrac{1}{2}x - \dfrac{\pi}{4}\right) - 2$

34. $y = \cos(\pi - 2x)$

In Exercises 35 and 36, sketch the graph.

35. $y = 3\sin^{-1} x$

36. $y = \cos^{-1}\left(\tfrac{1}{2}x\right)$

In Exercises 37–40, evaluate without a calculator.

37. $\sin^{-1}\left(-\dfrac{1}{2}\right)$

38. $\tan^{-1}\sqrt{3}$

39. $\sin^{-1}(\sin 658°)$

40. $\cos^{-1}\left[\cos\left(\dfrac{-17\pi}{7}\right)\right]$

Analytic and Geometric Trigonometry

CHAPTER NINE

Many of the identities and formulas presented in this chapter were known to the Greeks 2000 years ago (in different but equivalent forms). The addition and half-angle formulas as well as the Law of Sines were probably known to Hipparchus (about 150 B.C.) and definitely to Ptolemy (about 150 A.D.). The Law of Cosines was known to Euclid (about 300 B.C.).

The Greeks worked with the chord function (where $\text{crd } 2\theta = 2 \sin \theta$), but the identity $\sin^2 \theta + \cos^2 \theta = 1$ became obvious soon after the Hindus invented the sine function (around 300–400 A.D.).

The Arabs discovered many identities. For example, identities like

$$2 \cos x \cos y = \cos(x+y) + \cos(x-y)$$

were discovered by ibn-Yunus (around 1000 A.D.). Such identities were used by astronomers in the sixteenth century to convert large products into more manageable sums, before logarithms were invented by Napier for this purpose.

Vectors were invented much later in Europe by physicists and did not become a topic of serious mathematical study until the nineteenth century.

SECTION 9.1. TRIGONOMETRIC IDENTITIES

An equation involving a variable is called an **identity** if equality holds for every value of the variable for which all terms in the equation are defined. For instance, $x + 2 = (x + 2)(x - 1)/(x - 1)$ is an identity; the right-hand side is undefined at $x = 1$, but for every other value of x, equality holds. On the other hand, $x^2 - 4 = 0$ is not an identity; the terms are defined for all real values of x, but equality holds only when $x = 2$ or $x = -2$. An equation of this type is called a **conditional equation**. In this section, we shall study **trigonometric identities**, that is, identities which involve trigonometric functions, while in the next section, we shall study conditional trigonometric equations.

We have already seen a number of trigonometric identities. The most important are the identities we called the elementary trigonometric identities:

(1) $$\csc \theta = \frac{1}{\sin \theta}, \qquad \sec \theta = \frac{1}{\cos \theta}, \qquad \cot \theta = \frac{1}{\tan \theta}$$

(2) $$\tan \theta = \frac{\sin \theta}{\cos \theta}, \qquad \cot \theta = \frac{\cos \theta}{\sin \theta}$$

(3) $$\sin^2 \theta + \cos^2 \theta = 1$$

Some other trigonometric identities we have seen are

$$\tan^2 \theta + 1 = \sec^2 \theta, \qquad 1 + \cot^2 \theta = \csc^2 \theta,$$
$$\sin(-\theta) = -\sin \theta, \quad \text{and} \quad \cos(-\theta) = \cos \theta$$

In this section, we shall be learning how to prove that certain trigonometric equations are identities. The preferred method of carrying out such a proof is to start with one side of the equation and by using *known trigonometric identities*, reduce that side to the other side. The known trigonometric identities used are mainly the elementary identities, but they may be any identity you have previously proved.

EXAMPLE 1 Prove $\sin\theta \tan\theta + \cos\theta = \sec\theta$ is an identity.

SOLUTION We usually start with the more complicated side, in this case $\sin\theta \tan\theta + \cos\theta$, and try to reduce it to the other side. We proceed as follows:

$$\sin\theta \tan\theta + \cos\theta = \sin\theta \frac{\sin\theta}{\cos\theta} + \cos\theta \quad \left[\text{by (2), } \tan\theta = \frac{\sin\theta}{\cos\theta}\right]$$

$$= \frac{\sin^2\theta}{\cos\theta} + \frac{\cos^2\theta}{\cos\theta} \quad (\cos\theta \text{ is the least common denominator})$$

$$= \frac{\sin^2\theta + \cos^2\theta}{\cos\theta} \quad \text{(adding)}$$

$$= \frac{1}{\cos\theta} \quad [\text{by (3), } \sin^2\theta + \cos^2\theta = 1]$$

$$= \sec\theta \quad \left[\text{by (1), } \sec\theta = \frac{1}{\cos\theta}\right]$$

Unfortunately, there is no general method of proof of identities which works for all identities, but the following are a few hints as to what to try.

1. Work with the more complicated side (if there is one).
2. If you do not see directly how to reduce one side to the other, use elementary identities (1) and (2) to change everything into sines and cosines.
3. Look for ways to use identity (3), $\sin^2\theta + \cos^2\theta = 1$, or one of its other forms: $\cos^2\theta = 1 - \sin^2\theta = (1 - \sin\theta)(1 + \sin\theta)$, $\sin^2\theta = 1 - \cos^2\theta = (1 - \cos\theta)(1 + \cos\theta)$.
4. Above all, keep in mind the expression that you want in the end—this usually dictates what you should do.

The following are a series of examples demonstrating the verification of identities.

EXAMPLE 2 Verify the identity $\cot\theta + \tan\theta = \csc\theta \sec\theta$

SOLUTION

$$\cot\theta + \tan\theta = \frac{\cos\theta}{\sin\theta} + \frac{\sin\theta}{\cos\theta}$$

$$= \frac{\cos^2\theta + \sin^2\theta}{\sin\theta \cos\theta}$$

$$= \frac{1}{\sin\theta \cos\theta}$$

$$= \frac{1}{\sin\theta} \frac{1}{\cos\theta}$$

$$= \csc\theta \sec\theta$$

EXAMPLE 3 Verify the identity
$$\frac{1 + \sin x}{\cos x} = \frac{\cos x}{1 - \sin x}$$

SOLUTION There is no "more complicated" side, so we arbitrarily choose the left-hand side. Looking at the right-hand side, we see $1 - \sin x$ in the denominator. One way to get a term like that in the denominator of the left-hand side is to multiply the left-hand side by $1 = (1 - \sin x)/(1 - \sin x)$:

$$\frac{1 + \sin x}{\cos x} = \frac{1 + \sin x}{\cos x} \cdot \frac{1 - \sin x}{1 - \sin x}$$
$$= \frac{1 - \sin^2 x}{\cos x (1 - \sin x)}$$
$$= \frac{\cos^2 x}{\cos x (1 - \sin x)}$$
$$= \frac{\cos x}{1 - \sin x}$$

EXAMPLE 4 Verify the identity $\sin^4 t - \cos^4 t = 2\sin^2 t - 1$.

SOLUTION Sometimes it pays to factor:
$$\sin^4 t - \cos^4 t = (\sin^2 t - \cos^2 t)(\sin^2 t + \cos^2 t)$$
$$= (\sin^2 t - \cos^2 t) \cdot 1$$
$$= \sin^2 t - (1 - \sin^2 t)$$
$$= 2\sin^2 t - 1$$

EXAMPLE 5 Verify the identity $\sin 2t + \cos 2t \cot 2t = \csc 2t$.

SOLUTION Do not be confused by the $2t$. That is just the angle. You can think of it as θ if you like.

$$\sin 2t + \cos 2t \cot 2t = \sin 2t + \cos 2t \frac{\cos 2t}{\sin 2t}$$
$$= \frac{\sin^2 2t + \cos^2 2t}{\sin 2t}$$
$$= \frac{1}{\sin 2t}$$
$$= \csc 2t$$

If an identity involves trigonometric functions of different angles, we have to be very careful. For instance, $\sin^2 2t + \cos^2 t = 1$ is *not* an identity. We shall discuss equations like this in later sections.

Verifying identities can be difficult if the expressions on both sides of the equation are complicated. It is permissible to work on both sides *separately* as long as the steps are reversible.

EXAMPLE 6 Verify the identity

$$\frac{\cos^2\theta + \cot\theta}{\cos^2\theta - \cot\theta} = \frac{\cos^2\theta \tan\theta + 1}{\cos^2\theta \tan\theta - 1}$$

SOLUTION Start with the left-hand side:

$$\frac{\cos^2\theta + \cot\theta}{\cos^2\theta - \cot\theta} = \frac{\cos^2\theta + \dfrac{\cos\theta}{\sin\theta}}{\cos^2\theta - \dfrac{\cos\theta}{\sin\theta}}$$

$$= \frac{\cos\theta\left(\cos\theta + \dfrac{1}{\sin\theta}\right)}{\cos\theta\left(\cos\theta - \dfrac{1}{\sin\theta}\right)}$$

$$= \frac{\cos\theta + \dfrac{1}{\sin\theta}}{\cos\theta - \dfrac{1}{\sin\theta}}$$

$$= \frac{\dfrac{\cos\theta \sin\theta + 1}{\sin\theta}}{\dfrac{\cos\theta \sin\theta - 1}{\sin\theta}}$$

$$= \frac{\cos\theta \sin\theta + 1}{\cos\theta \sin\theta - 1}$$

If you do not see how to proceed from here, try the right-hand side:

$$\frac{\cos^2\theta \tan\theta + 1}{\cos^2\theta \tan\theta - 1} = \frac{\cos^2\theta \dfrac{\sin\theta}{\cos\theta} + 1}{\cos^2\theta \dfrac{\sin\theta}{\cos\theta} - 1}$$

$$= \frac{\cos\theta \sin\theta + 1}{\cos\theta \sin\theta - 1}$$

Since the last term is the same term we obtained when working on the left-hand side and the steps used on the right-hand side are reversible, the identity is verified.

EXERCISES

In Exercise 1–40, verify the identities.

1. $\sin\theta \cot\theta = \cos\theta$

2. $\sin\theta \csc\theta = 1$

3. $\sec\theta/\csc\theta = \tan\theta$
4. $\cot\theta \tan\theta = 1$
5. $\sin x + \cot x \cos x = \csc x$
6. $\sin\theta(\csc\theta - \sin\theta) = \cos^2\theta$
7. $1 - 2\sin^2 t = 2\cos^2 t - 1$
8. $\dfrac{\sin x}{1 + \cos x} + \dfrac{1 + \cos x}{\sin x} = 2\csc x$
9. $\csc^2 x/(1 + \tan^2 x) = \cot^2 x$
10. $\dfrac{1 + \tan u}{1 - \tan u} + \dfrac{1 + \cot u}{1 - \cot u} = 0$
11. $\dfrac{\sin\theta}{\sin\theta - \cos\theta} = \dfrac{1}{1 - \cot\theta}$
12. $(\csc t - \cot t)(\sec t + 1) = \tan t$
13. $(\sin 2y + \cos 2y)(\tan 2y + \cot 2y) = \sec 2y + \csc 2y$
14. $\dfrac{\sin u + \cos u}{\sec u + \csc u} = \dfrac{\sin u}{\sec u}$
15. $\dfrac{\sec^2 3\theta}{\sec 3\theta + 1} = \dfrac{\sec 3\theta - 1}{\sin^2 3\theta}$
16. $\sec^2\theta \cot^2\theta - \cos^2\theta \csc^2\theta = 1$
17. $\dfrac{\sin^3\theta + \cos^3\theta}{\sin\theta + \cos\theta} = 1 - \sin\theta\cos\theta$
18. $\csc^4\theta - \cot^4\theta = \csc^2\theta + \cot^2\theta$
19. $\dfrac{\csc^2\theta - 1}{\cos\theta} = \cot\theta \csc\theta$
20. $\dfrac{\sin u}{1 - \cos u} = \dfrac{\tan u}{\sec u - 1}$
21. $\cos^4\theta - \sin^4\theta = 2\cos^2\theta - 1$
22. $\dfrac{\tan^2 u}{\sec u + 1} = \dfrac{1 - \cos u}{\cos u}$
23. $(\sec^2\theta - \tan^2\theta)^3 = 1$
24. $(\cot\theta - \csc\theta)^3(\cot\theta + \csc\theta)^3 = -1$
25. $\dfrac{\sin^2\theta + 2}{\sin^2\theta - 2} = \dfrac{1 - 3\sec^2\theta}{1 + \sec^2\theta}$
26. $\dfrac{\cot x}{\csc x - 1} = \dfrac{\csc x + 1}{\cot x}$
27. $\dfrac{\tan\theta - \cot\theta}{\cos\theta + \sin\theta} = \sec\theta - \csc\theta$
28. $\dfrac{\sin x + \cos x}{\tan^2 x - 1} = \dfrac{\cos^2 x}{\sin x - \cos x}$
29. $\dfrac{1 + \sin\theta}{1 - \sin\theta} = \dfrac{\csc\theta + 1}{\csc\theta - 1}$
30. $\tan^2\theta - \sin^2\theta = \tan^2\theta \sin^2\theta$
31. $\dfrac{1 + \cos^2\varphi}{\sin^2\varphi} = 2\csc^2\varphi - 1$
32. $\dfrac{\sin\theta}{\sin\theta + \cos\theta} = \dfrac{\tan\theta}{1 + \tan\theta}$
33. $\sec\theta + \tan\theta = \dfrac{\cos\theta}{1 - \sin\theta}$
34. $\sin^4\theta + \cos^2\theta = \cos^4\theta + \sin^2\theta$
35. $\dfrac{\sin^2\theta}{\sin^2\theta + \cos\theta} = \dfrac{\tan\theta}{\tan\theta + \csc\theta}$
36. $\dfrac{\cot\varphi - 1}{1 - \tan\varphi} = \dfrac{1}{\tan\varphi}$
37. $\left(\dfrac{\sec^3 x}{\tan^6 x}\right)^2 \left(\dfrac{\cos^2 x}{\cot^4 x}\right)^3 = 1$
38. $\sec^4\theta(1 - \sin^4\theta) = \sec^2\theta + \dfrac{1}{\cot^2\theta}$
39. $\dfrac{\cos\alpha - \sin\alpha \tan\beta}{\sin\alpha + \cos\alpha \tan\beta} = \dfrac{\cot\alpha - \tan\beta}{1 + \tan\beta \cot\alpha}$
40. $\dfrac{\sin\alpha \cos\beta - \cos\alpha \sin\beta}{\cos\alpha \cos\beta + \sin\alpha \sin\beta} = \dfrac{\tan\alpha - \tan\beta}{1 + \tan\alpha \tan\beta}$

In Exercises 41–48, either verify the identity or find a value of θ for which equality does not hold (but make sure that this is not just due to calculator error).

41. $\sin^2\theta(1 + \tan^2\theta) = \tan\theta$
42. $\dfrac{\sin\theta \cot\theta + \cos\theta}{\cot\theta} = 2\cos\theta$
43. $\dfrac{\sin\theta}{1 + \cos\theta} = \dfrac{1 - \cos\theta}{\sin\theta}$
44. $\dfrac{\sec\theta}{\sin\theta} - \dfrac{\sin\theta}{\cos\theta} = \cot\theta$
45. $\dfrac{\sin^4\theta - \cos^4\theta}{\tan^4\theta - 1} = \cos^2\theta$
46. $\csc^2\theta \tan^2\theta - \sin^2\theta \sec^2\theta = \tan^4\theta$
47. $\sec^2\theta + \csc^2\theta = 1$
48. $\tan\theta + \cot\theta = \tan\theta \csc^2\theta$

SECTION 9.2. CONDITIONAL TRIGONOMETRIC EQUATIONS

A conditional trigonometric equation is an equation involving trigonometric functions which is not an identity. In other words, there are values of the variables for which the terms in the equation are all defined but equality does not hold. A simple example is $\sin x = 1$. The terms are defined for all x, but equality only holds when $x = \frac{1}{2}\pi + n \cdot 2\pi$, where n is an integer.

When we are given a conditional trigonometric equation, we usually are interested in finding its solutions. To do so, we may use any method normally employed in finding the solutions to an ordinary (conditional) equation, such as factoring, the quadratic formula, etc. In addition, we may have to use trigonometric identities and our knowledge of the trigonometric functions. We shall demonstrate some of these methods in the following examples.

EXAMPLE 1 Find all solutions to $2 \cos x + \sqrt{2} = 0$.

SOLUTION It is clearly easy to solve the equation for $\cos x$, obtaining $\cos x = -\sqrt{2}/2$. One solution to this equation is $x = \cos^{-1}(-\sqrt{2}/2) = 3\pi/4$. However, there are many more solutions. For any such solution x, the reference angle must be $\varphi = \cos^{-1}(\sqrt{2}/2) = \pi/4$. Since $\cos x$ is negative ($\cos x = -\sqrt{2}/2$) the terminal side of x must be in the second or third quadrant. The solutions are all those angles with terminal side in the second or third quadrant and reference angle $\pi/4$. That is, the solutions are all the numbers $x = \frac{3}{4}\pi + n \cdot 2\pi$ and $x = \frac{5}{4}\pi + n \cdot 2\pi$, where n is an integer. See accompanying figure.

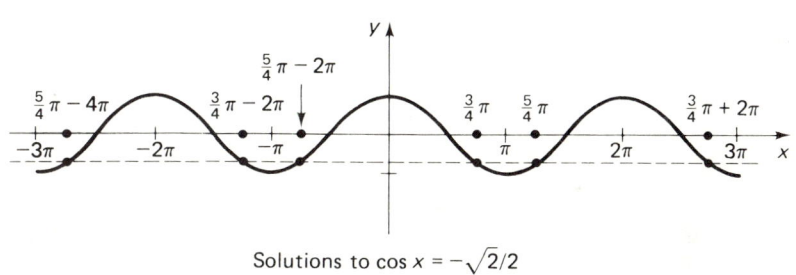

Solutions to $\cos x = -\sqrt{2}/2$

Remark: As in Example 1, the solutions to most conditional trigonometric equations are written in terms of n, where n is any integer. Hereafter, we shall leave the phrase "where n is any integer" as understood. Also, hereafter we shall write $n \cdot 2\pi$ as $2n\pi$.

Most conditional equations are more complicated than that in Example 1, but often the final steps are like the preceding. One of the techniques frequently used in solving conditional equations is factoring.

EXAMPLE 2 Find all solutions to $\cos x \tan x = \cos x$.

SOLUTION It would be wrong to divide by $\cos x$, as you would miss all solutions where $\cos x = 0$. Instead, subtract $\cos x$ from both sides:

$$\cos x \tan x - \cos x = 0$$

Then factor out $\cos x$:
$$\cos x(\tan x - 1) = 0$$

The only way the product of two factors can be zero is if one of the factors is zero:
$$\cos x = 0 \quad \text{or} \quad \tan x - 1 = 0$$

Now $\cos x = 0$ has solutions $x = \frac{1}{2}\pi + n\pi$, while $\tan x = 1$ has solutions $x = \frac{1}{4}\pi + n\pi$. Thus, $\cos x \tan x = \cos x$ has solutions $x = \frac{1}{2}\pi + n\pi$ and $x = \frac{1}{4}\pi + n\pi$.

It may be necessary to use a trigonometric identity before factoring.

EXAMPLE 3 Find all solutions to $-3\cos^2 x - 7\sin x + 5 = 0$.

SOLUTION If we use the identity $\cos^2 x = 1 - \sin^2 x$, the equation becomes a quadratic equation in $\sin x$:
$$-3(1 - \sin^2 x) - 7\sin x + 5 = 0$$
$$3\sin^2 x - 7\sin x + 2 = 0$$

Thinking $u = \sin x$, this quadratic factors as
$$(3\sin x - 1)(\sin x - 2) = 0$$

This equation is true if either
$$3\sin x - 1 = 0 \quad \text{or} \quad \sin x - 2 = 0$$

If $\sin x - 2 = 0$, then $\sin x = 2$, but this has no solutions.

If $3\sin x - 1 = 0$, then $\sin x = \frac{1}{3}$, so x is in Q_1 or Q_2 with reference angle $\sin^{-1}(\frac{1}{3}) \approx .339837$. Therefore $x \approx .339837 + 2n\pi$ or $x \approx (\pi - .339837) + 2n\pi \approx .280176 + 2n\pi$.

A technique often employed in solving equations involving radicals is squaring both sides. This may be employed here, but we must remember that extraneous solutions may be introduced, so all solutions should be checked in the original equation.

EXAMPLE 4 Find all solutions to $\sin x = \cos x - 1$.

SOLUTION Since $\sin^2 x = 1 - \cos^2 x$, $\sin x = \pm\sqrt{1 - \cos^2 x}$. Thus the equation we wish to solve is
$$\pm\sqrt{1 - \cos^2 x} = \cos x - 1$$

Squaring, we obtain
$$1 - \cos^2 x = \cos^2 x - 2\cos x + 1$$

Thus
$$-\cos^2 x = \cos^2 x - 2\cos x$$
$$0 = 2\cos^2 x - 2\cos x$$
$$0 = 2\cos x(\cos x - 1)$$

Consequently $\cos x = 0$, in which case $x = \frac{1}{2}\pi + n\pi$, or $\cos x = 1$, in which case $x = 2n\pi$. We now check these in our original equation, $\sin x = \cos x - 1$. When $x = 2n\pi$, both sides are zero, so the solutions check. However, when $x = \frac{1}{2}\pi + n\pi$, the right-hand side is -1, but the left-hand side is -1 only when n is odd (the left-hand side is $+1$ when n is even). Thus the solutions are $x = \frac{1}{2}\pi + n\pi$, n odd, and $x = 2n\pi$. We sometimes write $x = \frac{1}{2}\pi + n\pi$, n odd, as $x = \frac{1}{2}\pi + (2k+1)\pi$.

When an equation can be put into quadratic form but it is not easy to factor, we usually use the quadratic formula.

EXAMPLE 5 Find all solutions to $\sec^2 x - 3 \sec x + 1 = 0$.

SOLUTION By the quadratic formula,

$$\sec x = \frac{3 \pm \sqrt{3^2 - 4 \cdot 1 \cdot 1}}{2 \cdot 1} = \frac{3 \pm \sqrt{5}}{2} \approx 2.6180339, \quad .381966$$

$\sec x \approx .381966$ has no solution since $|\sec x| \geq 1$. If $\sec x \approx 2.6180339$, then the reference angle is $\sec^{-1}(2.6180339) \approx 1.17887$. Since $\sec x > 0$, x is in Q_1 or Q_4 with this reference angle. Thus the solutions are $x \approx 1.17887 + 2n\pi$ and $x \approx -1.17887 + 2n\pi$.

EXERCISES

Find all solutions for each of the following equations:

1. $\sqrt{2} \sin x - 1 = 0$
2. $\sqrt{3} \cot x - 1 = 0$
3. $\cot x + \sqrt{3} = 0$
4. $\sqrt{3} \tan x + 1 = 0$
5. $2 \cos x - 1 = 0$
6. $2 \cos x + \sqrt{3} = 0$
7. $\sec x - 2 = 0$
8. $4 \sin x + 3 = 0$
9. $2 \cos x - 3 = 0$
10. $3 \cos x - 2 = 0$
11. $\sin x \cos x = 2 \sin x$
12. $2 \sin x \cos x = \sin x$
13. $\sec^2 x - \sec x - 2 = 0$
14. $4 \sin^2 x - 3 = 0$
15. $6 \sin^2 x - \sin x = 1$
16. $\tan^2 x + 6 \tan x + 5 = 0$
17. $-\sin^2 x - 4 \cos x + 4 = 0$
18. $\cos^4 x - \sin^4 x = 0$
19. $11 + \cos^2 x + \sin x = 0$
20. $\sec^2 x + 3 \tan x - 1 = 0$
21. $\sin x = -\cos x$
22. $\sin x = 1 - \cos x$
23. $\sin x = 1 + \cos x$
24. $\csc x + \cot x = 1$
25. $\tan^5 x - 9 \tan x = 0$
26. $4 \sin x \cos x + 2 \sin x + 2\cos x + 1 = 0$
27. $4 \sin^2 x + \sin x - 1 = 0$
28. $3 \cos^2 x + 5 \cos x + 1 = 0$

SECTION 9.3. THE ADDITION FORMULAS

In this section, we shall derive the following identities, which are called **the addition formulas**:

(4)
$$\sin(u+v) = \sin u \cos v + \cos u \sin v$$

(5)
$$\sin(u-v) = \sin u \cos v - \cos u \sin v$$

(6)
$$\cos(u+v) = \cos u \cos v - \sin u \sin v$$

(7)
$$\cos(u-v) = \cos u \cos v + \sin u \sin v$$

(8)
$$\tan(u+v) = \frac{\tan u + \tan v}{1 - \tan u \tan v}$$

(9)
$$\tan(u-v) = \frac{\tan u - \tan v}{1 + \tan u \tan v}$$

As we can see from these, an addition formula for a trigonometric function is an identity which relates the trigonometric function of the sum or difference of two angles to trigonometric functions of the individual angles.

Now let us derive the addition formulas (4)–(9). We begin by deriving the addition formula for $\cos(u - v)$. Let u and v be any two angles in standard position. Let $P_1 = (x_1, y_1)$ and $P_2 = (x_2, y_2)$ be points on the terminal side of u and v, respectively, each of distance 1 from the origin ($r = 1$). See Fig. 1(a). Then $\sin u = y_1/r = y_1$, $\cos u = x_1/r = x_1$, $\sin v = y_2/r = y_2$, and $\cos v = x_2/r = x_2$. Now rotate the whole angle $u - v$ until it is in standard position (the side that was the terminal side of v is now the initial side). Let $P_3 = (x_3, y_3)$ be the point on the terminal side of $u - v$ of distance 1 from the origin. See Fig. 1(b). Then $\sin(u - v) = y_3$ and $\cos(u - v) = x_3$. Now clearly under the rotation, P_2 has been rotated to the point $(1, 0)$, and P_1 has been rotated to P_3. Therefore *the distance between P_1 and P_2 is the same as the distance between P_3 and $(1, 0)$.* By

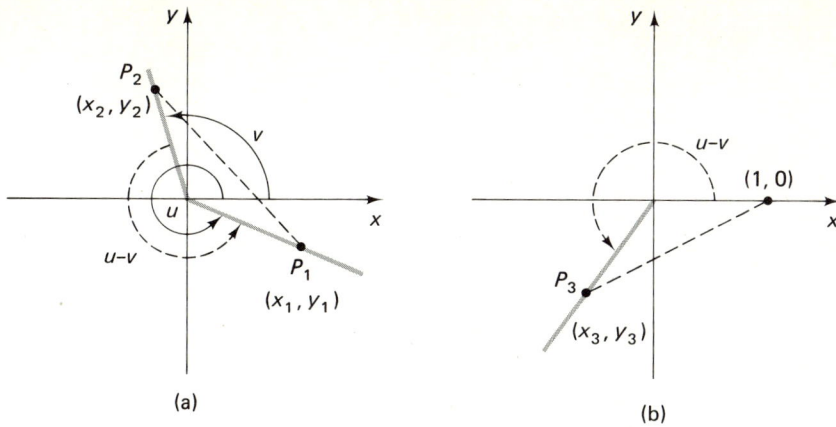

FIG. 1

using the distance formula, this statement becomes

$$\sqrt{(x_2 - x_1)^2 + (y_2 - y_1)^2} = \sqrt{(x_3 - 1)^2 + y_3^2}$$

By using $x_1 = \cos u$, $y_1 = \sin u$, $x_2 = \cos v$, $y_2 = \sin v$, $x_3 = \cos(u - v)$, and $y_3 = \sin(u - v)$, the equation becomes

$$\sqrt{(\cos v - \cos u)^2 + (\sin v - \sin u)^2} = \sqrt{[\cos(u - v) - 1]^2 + \sin^2(u - v)}$$

Let us simplify this. First square both sides and then multiply out:

$$(\cos v - \cos u)^2 + (\sin v - \sin u)^2 = [\cos(u - v) - 1]^2 + \sin^2(u - v)$$
$$\cos^2 v - 2\cos u \cos v + \cos^2 u + \sin^2 v - 2\sin u \sin v + \sin^2 u$$
$$= \cos^2(u - v) - 2\cos(u - v) + 1 + \sin^2(u - v)$$

If we reorder the last equation, we see that we can use the identity $\sin^2 \theta + \cos^2 \theta = 1$ three times:

$$(\sin^2 v + \cos^2 v) + (\sin^2 u + \cos^2 u) - 2\cos u \cos v - 2\sin u \sin v$$
$$= [\sin^2(u - v) + \cos^2(u - v)] + 1 - 2\cos(u - v)$$
$$2 - 2\cos u \cos v - 2\sin u \sin v = 2 - 2\cos(u - v)$$

Subtract 2 and then divide by -2 to obtain equation (7):

$$\cos(u - v) = \cos u \cos v + \sin u \sin v$$

Thus we have derived our first addition formula. Fortunately, we do not have to go to so much work to derive the others, for they may be obtained from this one. Before we do that, let us look at an example. The main reason for doing problems like the following is to become familiar with the formulas and how they work so that you can readily use them when they arise in more complicated situations.

EXAMPLE 1 Use an addition formula to find the exact value of cos 15°.

SOLUTION We use $u = 45°$ and $v = 30°$ since we know the trigonometric functions of these angles and that their difference is 15°:

$$\cos 15° = \cos(45° - 30°) = \cos 45° \cos 30° + \sin 45° \sin 30°$$
$$= \frac{\sqrt{2}}{2} \frac{\sqrt{3}}{2} + \frac{\sqrt{2}}{2} \frac{1}{2} = \frac{\sqrt{2}}{4}(\sqrt{3} + 1)$$

You might check on your calculator that the decimal approximation to this is the same as what your calculator gives for cos 15°.

To simplify our derivation of the remaining addition formulas, we first verify the following identities:

(10) $$\cos\left(\frac{\pi}{2} - x\right) = \sin x$$

(11) $$\sin\left(\frac{\pi}{2} - x\right) = \cos x$$

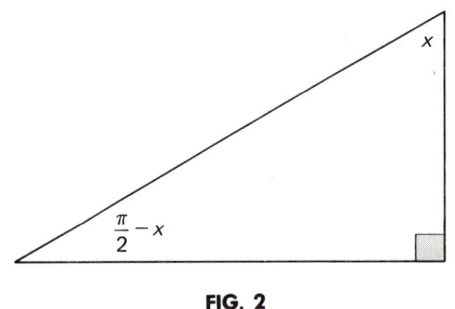

FIG. 2

These identities are obviously true if x is an acute angle in a right triangle. For the other acute angle is $\frac{1}{2}\pi - x$, and the side adjacent to $\frac{1}{2}\pi - x$ is the side opposite to x, and vice versa. See Fig. 2. Thus

$$\cos\left(\frac{\pi}{2} - x\right) = \frac{\text{side adj. to }\left(\frac{1}{2}\pi - x\right)}{\text{hypotenuse}}$$

$$= \frac{\text{side opp. to } x}{\text{hypotenuse}} = \sin x$$

and similarly, $\sin(\pi/2 - x) = \cos x$. We now show that these identities are true for any angle x.

Verification of (10): $\cos(\frac{1}{2}\pi - x) = \sin x$:

$$\cos\left(\frac{\pi}{2} - x\right) = \cos\frac{\pi}{2}\cos x + \sin\frac{\pi}{2}\sin x \quad [\text{by (7)}]$$
$$= 0 \cdot \cos x + 1 \cdot \sin x$$
$$= \sin x$$

Verification of (11): $\sin(\tfrac{1}{2}\pi - x) = \cos x$.

Let $y = \tfrac{1}{2}\pi - x$. By (10), $\sin y = \cos(\tfrac{1}{2}\pi - y)$. Replace y by $\tfrac{1}{2}\pi - x$:

$$\sin(\tfrac{1}{2}\pi - x) = \cos\left[\tfrac{1}{2}\pi - \left(\tfrac{1}{2}\pi - x\right)\right] = \cos x$$

We also need the following two identities, which we derived in the previous chapter using the definitions of the trigonometric functions:

(12) $\qquad\qquad\qquad\qquad \cos(-x) = \cos x$
(13) $\qquad\qquad\qquad\qquad \sin(-x) = -\sin x$

Since $\tan(-x) = \sin(-x)/\cos(-x) = (-\sin x)/\cos x = -\tan x$, we also have

(14) $\qquad\qquad\qquad\qquad \tan(-x) = -\tan x$

Finally we are in position to derive the remaining addition formulas.

Verification of (6): $\cos(u + v) = \cos u \cos v - \sin u \sin v$:

$$\begin{aligned}
\cos(u + v) &= \cos[u - (-v)] \\
&= \cos u \cos(-v) + \sin u \sin(-v) &&\text{[by (7)]} \\
&= \cos u \cos v + \sin u(-\sin v) &&\text{[by (12) and (13)]} \\
&= \cos u \cos v - \sin u \sin v
\end{aligned}$$

Verification of (4): $\sin(u + v) = \sin u \cos v + \cos u \sin v$:

$$\begin{aligned}
\sin(u + v) &= \cos\left[\frac{\pi}{2} - (u + v)\right] &&\text{[by (10)]} \\
&= \cos\left[\left(\frac{\pi}{2} - u\right) - v\right] \\
&= \cos\left(\frac{\pi}{2} - u\right)\cos v + \sin\left(\frac{\pi}{2} - u\right)\sin v &&\text{[by (7)]} \\
&= \sin u \cos v + \cos u \sin v &&\text{[by (10) and (11)]}
\end{aligned}$$

Verification of (5): $\sin(u - v) = \sin u \cos v - \cos u \sin v$:

$$\begin{aligned}
\sin(u - v) &= \sin[u + (-v)] \\
&= \sin u \cos(-v) + \cos u \sin(-v) &&\text{[by (4)]} \\
&= \sin u \cos v + \cos u(-\sin v) &&\text{[by (12) and (13)]} \\
&= \sin u \cos v - \cos u \sin v
\end{aligned}$$

Verification of (8): $\tan(u + v) = \dfrac{\tan u + \tan v}{1 - \tan u \tan v}$:

$$\begin{aligned}
\tan(u + v) &= \frac{\sin(u + v)}{\cos(u + v)} \\
&= \frac{\sin u \cos v + \cos u \sin v}{\cos u \cos v - \sin u \sin v} &&\text{[by (4) and (6)]}
\end{aligned}$$

$$= \frac{\dfrac{\sin u \cos v}{\cos u \cos v} + \dfrac{\cos u \sin v}{\cos u \cos v}}{\dfrac{\cos u \cos v}{\cos u \cos v} - \dfrac{\sin u \sin v}{\cos u \cos v}} \qquad \text{[divide top and bottom by } \cos u \cos v \text{ (if nonzero)]}$$

$$= \frac{\dfrac{\sin u}{\cos u} + \dfrac{\sin v}{\cos v}}{1 - \dfrac{\sin u}{\cos u}\dfrac{\sin v}{\cos v}}$$

$$= \frac{\tan u + \tan v}{1 - \tan u \tan v}$$

Note: In the third step we used $\cos u \cos v \neq 0$. If $\cos u \cos v = 0$, then either $\cos u = 0$ or $\cos v = 0$, which in turn means either $\tan u$ or $\tan v$ is not defined. Since an equation is an identity if equality holds whenever both sides are defined, equation (8) is still an identity.

Verification of (9): $\tan(u - v) = \dfrac{\tan u - \tan v}{1 + \tan u \tan v}$:

$$\tan(u - v) = \tan[u + (-v)]$$
$$= \frac{\tan u + \tan(-v)}{1 - \tan u \tan(-v)} \qquad \text{[by (8)]}$$
$$= \frac{\tan u - \tan v}{1 + \tan u \tan v} \qquad \text{[by (14)]}$$

Now that we have derived the addition formulas, let us look at some examples.

EXAMPLE 2: Suppose $\sin u = \frac{1}{3}$, $\cos v = -\frac{3}{4}$, u is in Q_2, and v is in Q_2. Find $\sin(u + v)$, $\cos(u + v)$, and $\tan(u + v)$, and determine in which quadrant $u + v$ lies.

SOLUTION From the given information, $\cos u = -\sqrt{1 - \sin^2 u} = -\frac{2}{3}\sqrt{2}$, $\tan u = \frac{1}{3}/(-\frac{2}{3}\sqrt{2}) = -\frac{1}{4}\sqrt{2}$, $\sin v = \sqrt{1 - \cos^2 v} = \frac{1}{4}\sqrt{7}$, and $\tan v = \frac{1}{4}\sqrt{7}/(-\frac{3}{4}) = -\frac{1}{3}\sqrt{7}$. Thus

$$\sin(u + v) = \sin u \cos v + \cos u \sin v = \frac{1}{3}\left(\frac{-3}{4}\right) + \left(\frac{-2\sqrt{2}}{3}\right)\frac{\sqrt{7}}{4}$$
$$= -\frac{3 + 2\sqrt{14}}{12}$$

$$\cos(u + v) = \cos u \cos v - \sin u \sin v = \left(\frac{-2\sqrt{2}}{3}\right)\left(\frac{-3}{4}\right) - \frac{1}{3}\frac{\sqrt{7}}{4}$$
$$= \frac{6\sqrt{2} - \sqrt{7}}{12}$$

$$\tan(u+v) = \frac{\tan u + \tan v}{1 - \tan u \tan v} = -\frac{\frac{1}{4}\sqrt{2} - \frac{1}{3}\sqrt{7}}{1 - \left(-\frac{1}{4}\sqrt{2}\right)\left(-\frac{1}{3}\sqrt{7}\right)} = -\frac{3\sqrt{2} + 4\sqrt{7}}{12 - \sqrt{14}}$$

Since $\sin(u+v)$ is negative and $\cos(u+v)$ is positive, $u+v$ must be in Q_4.

EXAMPLE 3 Verify the identity $\cos(x + \frac{1}{2}\pi) = -\sin x$.

SOLUTION
$$\cos\left(x + \frac{\pi}{2}\right) = \cos x \cos \frac{\pi}{2} - \sin x \sin \frac{\pi}{2}$$
$$= (\cos x) \cdot 0 - (\sin x) \cdot 1$$
$$= -\sin x$$

EXERCISES

In Exercises 1–8, use $15° = 45° - 30°$, $5\pi/12 = \pi/4 + \pi/6$, etc., and the addition formulas to find the exact value of the functions.

1. $\sin 15°$
2. $\tan 15°$
3. $\cos \frac{5\pi}{12}$
4. $\sin \frac{5\pi}{12}$
5. $\tan \frac{7\pi}{12}$
6. $\cos 195°$
7. $\csc \frac{11\pi}{12}$
8. $\cos 345°$

In Exercises 9–14, use an addition formula to find the exact value.

9. $\sin 12° \cos 33° + \cos 12° \sin 33°$
10. $\sin 14° \cos 44° - \cos 14° \sin 44°$
11. $\cos 15° \cos 75° + \sin 15° \sin 75°$
12. $\cos 15° \cos 75° - \sin 15° \sin 75°$
13. $\cos \frac{\pi}{12} \sin \frac{5\pi}{12} + \sin \frac{\pi}{12} \cos \frac{5\pi}{12}$
14. $\frac{\tan 15° - \tan 75°}{1 + \tan 15° \tan 75°}$

In Exercises 15–18, find $\sin(u+v)$, $\sin(u-v)$, $\cos(u+v)$, $\cos(u-v)$, $\tan(u+v)$, and $\tan(u-v)$ from the given information.

15. $\sin u = \frac{3}{5}$, $\cos v = -\frac{5}{13}$, u is in Q_2, v is in Q_3
16. $\sec u = 3$, $\sec v = \frac{3}{2}$, $\csc u$ is positive, $\csc v$ is negative
17. $\tan u = -\sqrt{3}$, $\cot v = -\sqrt{2}$, u is in Q_2, v is not in Q_2
18. $\csc u = -\frac{5}{4}$, $\tan u = \frac{4}{3}$, $\sec v = -\frac{25}{7}$, $\sin v = -\frac{24}{25}$

In Exercises 19–36, verify the identities.

19. $\sin(x + \pi) = -\sin x$
20. $\cos(x + \pi) = -\cos x$
21. $\tan(x + \pi) = \tan x$
22. $\sin\left(x + \frac{\pi}{2}\right) = \cos x$
23. $\cos\left(\frac{3\pi}{2} - x\right) = -\sin x$
24. $\csc\left(\frac{3\pi}{2} - x\right) = -\sec x$
25. $\sec(x + 2\pi) = \sec x$
26. $\sec\left(x - \frac{\pi}{2}\right) = \csc x$

27. $\sin\left(x + \dfrac{\pi}{6}\right) = \dfrac{\sqrt{3}}{2}\sin x + \dfrac{1}{2}\cos x$

28. $\cos\left(x + \dfrac{\pi}{4}\right) = \dfrac{\sqrt{2}}{2}(\cos x - \sin x)$

29. $\cos\left(\dfrac{\pi}{3} - x\right) = \dfrac{1}{2}\cos x + \dfrac{\sqrt{3}}{2}\sin x$

30. $\sin\left(\dfrac{\pi}{4} - x\right) = \dfrac{\sqrt{2}}{2}(\cos x - \sin x)$

31. $\tan\left(x - \dfrac{\pi}{4}\right) = \dfrac{\tan x - 1}{\tan x + 1}$

32. $\tan\left(x + \dfrac{\pi}{4}\right) = \dfrac{\tan x + 1}{\tan x - 1}$

33. $\sin\left(x + \dfrac{5\pi}{4}\right) = -\dfrac{\sqrt{2}}{2}(\sin x + \cos x)$

34. $\cos\left(x - \dfrac{3\pi}{4}\right) = -\dfrac{\sqrt{2}}{2}(\cos x - \sin x)$

35. $\sin(u + v)\sin(u - v) = \sin^2 u - \sin^2 v$

36. $\cos(u + v)\cos(u - v) = \cos^2 u - \sin^2 v$

In Exercises 37–40, verify the identities. (The identities are called the **product formulas**. They allow you to replace the product on the right with the sum on the left.)

37. $\sin(u + v) + \sin(u - v) = 2\sin u \cos v$

38. $\sin(u + v) - \sin(u - v) = 2\cos u \sin v$

39. $\cos(u + v) + \cos(u - v) = 2\cos u \cos v$

40. $\cos(u - v) - \cos(u + v) = 2\sin u \sin v$

In Exercises 41–44, verify the identities. Use Exercises 37–40 with $a = u + v$, $b = u - v$. (The identities are called the **sum formulas**.)

41. $\sin a + \sin b = 2\sin\dfrac{a+b}{2}\cos\dfrac{a-b}{2}$

42. $\sin a - \sin b = 2\cos\dfrac{a+b}{2}\sin\dfrac{a-b}{2}$

43. $\cos a + \cos b = 2\cos\dfrac{a+b}{2}\cos\dfrac{a-b}{2}$

44. $\cos b - \cos a = 2\sin\dfrac{a+b}{2}\sin\dfrac{a-b}{2}$

In Exercises 45–50, use Exercises 37–40 to express the product as a sum (or difference).

45. $2\sin 3\theta \cos\theta$

46. $6\cos 4\theta \cos 7\theta$

47. $-3\sin 5\theta \cos 2\theta$

48. $\cos(-5t)\cos 3t$

49. $\sin 12t \sin 6t$

50. $3\cos 5t \sin 8t$

In Exercises 51–56, use Exercises 41–44 to express the sum (or difference) as a product.

51. $\sin 5\theta + \sin 3\theta$

52. $\cos 2\theta + \cos 6\theta$

53. $5\cos 3\theta + 5\cos 4\theta$

54. $\sin(-3t) - \sin t$

55. $\sin 2x - \sin(-x)$

56. $\cos 7t + \cos 3t$

SECTION 9.4. THE MULTIPLE-ANGLE FORMULAS

In this section, we shall obtain formulas for the trigonometric functions of $2u$ (the **double-angle formulas**) and $\frac{1}{2}u$ (the **half-angle formulas**) in terms of the trigonometric functions of u. There are many aspects of integral calculus where these formulas play a key role.

If we start with the addition formula

$$\sin(u + v) = \sin u \cos v + \cos u \sin v$$

and let $u = v$, the formula becomes

$$\sin 2u = \sin(u + u) = \sin u \cos u + \cos u \sin u$$

Thus

(15)
$$\sin 2u = 2 \sin u \cos u$$

Similarly, if we start with $\cos(u + v) = \cos u \cos v - \sin u \sin v$ and let $u = v$, we obtain
$$\cos 2u = \cos(u + u) = \cos u \cos u - \sin u \sin u$$

Thus

(16)
$$\cos 2u = \cos^2 u - \sin^2 u$$

There are two other forms of the double-angle formulas for cosine. Using $\cos^2 u = 1 - \sin^2 u$, we obtain $\cos 2u = (1 - \sin^2 u) - \sin^2 u$, or

(17)
$$\cos 2u = 1 - 2 \sin^2 u$$

Using $\sin^2 u = 1 - \cos^2 u$, we obtain $\cos 2u = \cos^2 u - (1 - \cos^2 u)$, or

(18)
$$\cos 2u = 2 \cos^2 u - 1$$

In the same manner as for $\sin 2u$ and $\cos 2u$, one can obtain the formula for $\tan 2u$:

(19)
$$\tan 2u = \frac{2 \tan u}{1 - \tan^2 u}$$

EXAMPLE 1 Verify the identity $\sin 2u = \tan u(1 + \cos 2u)$.

SOLUTION If we start with the right-hand side, we must choose which of (16), (17), or (18) to use for $\cos 2u$. The best choice is (18) since the 1's then drop out:

$$\begin{aligned}
\tan u(1 + \cos 2u) &= \tan u\left[1 + (2\cos^2 u - 1)\right] \\
&= \tan u(2 \cos^2 u) \\
&= 2\frac{\sin u}{\cos u} \cos^2 u \\
&= 2 \sin u \cos u \\
&= \sin 2u
\end{aligned}$$

Identities (17) and (18) can easily be transformed into two other identities called the *half-angle formulas*:

(20)
$$\sin \frac{u}{2} = \pm \sqrt{\frac{1 - \cos u}{2}}$$

(21)
$$\cos \frac{u}{2} = \pm \sqrt{\frac{1 + \cos u}{2}}$$

where the sign is determined by the quadrant in which $u/2$ lies.

Let us see how to obtain (20). From (17),
$$\cos 2v = 1 - 2\sin^2 v$$
If we solve this for $\sin v$, we obtain
$$2\sin^2 v = 1 - \cos 2v$$
$$\sin^2 v = \frac{1 - \cos 2v}{2}$$
$$\sin v = \pm \sqrt{\frac{1 - \cos 2v}{2}}$$
Now if we let $u = 2v$, then $v = u/2$, and the formula becomes
$$\sin \frac{u}{2} = \pm \sqrt{\frac{1 - \cos u}{2}}$$

The derivation of (21) is similar.

EXAMPLE 2 Find $\sin 22.5°$ and $\cos 22.5°$ by the half-angle formulas.

SOLUTION
$$\sin 22.5° = \sqrt{\frac{1 - \cos 45°}{2}} = \sqrt{\frac{1 - \sqrt{2}/2}{2}} = \frac{\sqrt{2 - \sqrt{2}}}{2}$$
and
$$\cos 22.5° = \sqrt{\frac{1 + \cos 45°}{2}} = \sqrt{\frac{1 + \sqrt{2}/2}{2}} = \frac{\sqrt{2 + \sqrt{2}}}{2}$$

The signs are both $+$ since $22.5°$ is in Q_1.

EXERCISES

In Exercises 1–6, use the given information and the double-angle formulas to find $\sin 2u$, $\cos 2u$, and $\tan 2u$.

1. $u = \dfrac{\pi}{3}$
2. $\sin u = \dfrac{4}{5}$, u in Q_2
3. $\cos u = -\dfrac{5}{13}$, $\tan u < 0$
4. $\csc u = 3$, u acute
5. $\cot u = -\dfrac{4}{3}$, $90° < u < 180°$
6. $\sec u = \dfrac{25}{7}$, $\csc u = -\dfrac{25}{24}$

In Exercises 7–14, use the half-angle formulas to find the exact value of the function.

7. $\sin 15°$
8. $\cos 67.5°$
9. $\cos \dfrac{5\pi}{8}$
10. $\sin\left(\dfrac{-5\pi}{8}\right)$
11. $\cos \dfrac{u}{2}$ if $\cos u = \dfrac{1}{8}$, $-90° < u < 0$
12. $\sin \dfrac{u}{2}$ if $\sin u = \dfrac{1}{8}$, $450° < u < 540°$
13. $\sin \dfrac{\theta}{2}$ if $\sec \theta = -7$, $\dfrac{\pi}{2} < \theta < \pi$
14. $\cos \dfrac{\theta}{2}$ if $\tan \theta = \dfrac{3}{4}$, $\pi < \theta < \dfrac{3\pi}{2}$

In Exercises 15–26, verify the identities.

15. $\tan x = \dfrac{\sin 2x}{1 + \cos 2x}$
16. $\tan x = \dfrac{1 - \cos 2x}{\sin 2x}$
17. $\sin 4x = 2 \sin 2x \cos 2x$
18. $\cos 3x = 4 \cos^3 x - 3 \cos x$
19. $\sin 3x = 3 \sin x - 4 \sin^3 x$
20. $\tan 3x = \dfrac{3 \tan x - \tan^3 x}{1 - 3 \tan^2 x}$
21. $\sin 4x = \cos x (4 \sin x - 8 \sin^3 x)$
22. $\cos 4x = 8 \cos^4 x - 8 \cos^2 x + 1$
23. $\cos^4 x - \sin^4 x = \cos 2x$
24. $\dfrac{1 + \sin 2x}{\cos 2x} = \dfrac{\cos x + \sin x}{\cos x - \sin x}$
25. $\sec 2u = \dfrac{\sec^2 u}{2 - \sec^2 u}$
26. $\dfrac{\sin 3x + \cos 3x}{\cos x - \sin x} = 1 + 2 \sin x$

In Exercises 27–36, find all solutions to the conditional equations in the interval $[0, 2\pi]$.

27. $\sin 2x = -\dfrac{1}{2}$
28. $\cos 2x = \dfrac{1}{2}\sqrt{3}$
29. $\sin 2x + \cos x = 0$
30. $\sin 2x - \sin x = 0$
31. $\cos 2x - \sin x = 0$
32. $\cos 2x + \sin x = 1$
33. $\cos 2x + \cos x = 0$
34. $\cos x + \sin 2x \sin x = 0$
35. $\tan 2x = \tan x$
36. $\cos 2x - 1 - \tan x = 0$

SECTION 9.5. OBLIQUE TRIANGLES; THE LAW OF SINES

An oblique triangle is a triangle which is not a right triangle. If at least three of the sides and/or angles are given, including at least one side, one can use trigonometry to find the remaining parts of the triangle. This is called **solving the triangle**. When doing this, the following law is often helpful.

(22)   **THE LAW OF SINES**

In any triangle with sides a, b, c and opposite angles α, β, γ, respectively,

$$\frac{a}{\sin \alpha} = \frac{b}{\sin \beta} = \frac{c}{\sin \gamma}$$

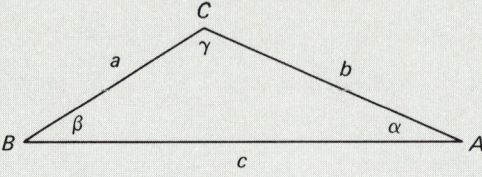

Let us first see why this is true; then we shall use it to solve some triangles. First we draw the triangle but in addition drop an altitude from C and from B to points D and E, respectively. See Fig. 3. Suppose CD is of length r and BE is of

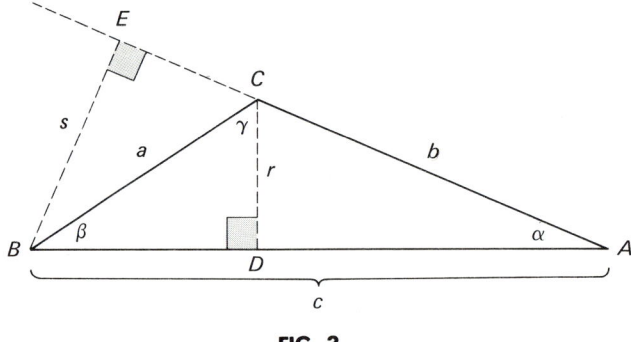

FIG. 3

length s. Then in triangle ACD, $\sin \alpha = r/b$, or $r = b \sin \alpha$, while in triangle BCD, $\sin \beta = r/a$, or $r = a \sin \beta$. Therefore, $b \sin \alpha = a \sin \beta$. Dividing both sides by $\sin \alpha \sin \beta$ yields

$$\frac{b \sin \alpha}{\sin \alpha \sin \beta} = \frac{a \sin \beta}{\sin \alpha \sin \beta} \quad \text{or} \quad \frac{b}{\sin \beta} = \frac{a}{\sin \alpha}$$

Similarly in triangle ABE, $\sin \alpha = s/c$, or $s = c \sin \alpha$, while in triangle CBE, $s/a = \sin(\angle BCE) = \sin(180° - \gamma) = \sin 180° \cos \gamma - \cos 180° \sin \gamma = (0) \cos \gamma - (-1) \sin \gamma = \sin \gamma$. Thus $s = a \sin \gamma$, and so $a \sin \gamma = c \sin \alpha$. Dividing both sides by $\sin \alpha \sin \gamma$ yields $a/\sin \alpha = c/\sin \gamma$. But we have already shown that $a/\sin \alpha = b/\sin \beta$. Therefore

$$\frac{a}{\sin \alpha} = \frac{b}{\sin \beta} = \frac{c}{\sin \gamma}$$

Now that we have verified the Law of Sines, let us see how to use it to solve triangles.

Angle-Side-Angle (ASA) or Angle-Angle-Side (AAS)

Suppose we are given two angles and one side. If α and β are the given angles, then since $\alpha + \beta + \gamma = 180°$, the third angle $\gamma = 180° - \alpha - \beta$. Since we know all three angles and we know one side, say side a, we can use the Law of Sines to find the other sides: $b/\sin\beta = a/\sin\alpha$, or $b = a\sin\beta/\sin\alpha$, while $c/\sin\gamma = a/\sin\alpha$, or $c = a\sin\gamma/\sin\alpha$.

EXAMPLE 1 Solve the triangle ABC if $\alpha = 32°$, $\gamma = 103.2°$, and $b = 12.93$.

SOLUTION The third angle is $\beta = 180° - 32° - 103.2° = 44.8°$. By the Law of Sines, $\frac{a}{\sin\alpha} = \frac{b}{\sin\beta}$, or $a = \frac{b\sin\alpha}{\sin\beta}$. Thus, $a = \frac{12.93 \sin 32°}{\sin 44.8°} \approx 9.724$. Similarly, $\frac{c}{\sin\gamma} = \frac{b}{\sin\beta}$, or $c = \frac{b\sin\gamma}{\sin\beta} = \frac{12.93 \sin 103.2°}{\sin 44.8°} \approx 17.8651$.

Side-Side-Angle (SSA)

Suppose two sides, say a and b, and the angle opposite one of them, say angle α, are given. Suppose angle α is obtuse. If $a > b$, there is a unique triangle [see Fig. 4(a)], while if $a < b$, no such triangle exists [see Fig. 4(b)].

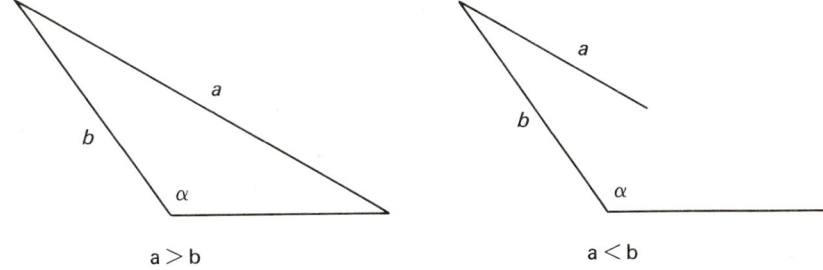

FIG. 4

Suppose α is acute. Again if $a > b$, there is a unique triangle, while the case $a < b$ is ambiguous: There may be no triangle, a unique triangle, or two triangles as Fig. 5 indicates.

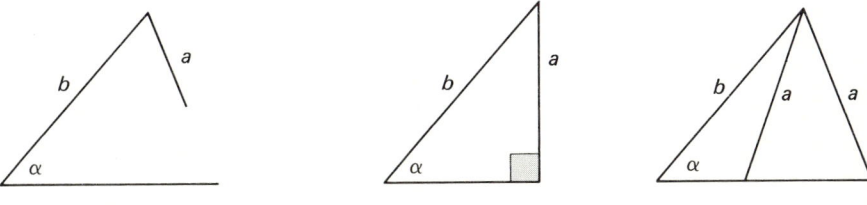

FIG. 5

Fortunately, you do not have to worry about which particular situation you are in, as it will become apparent during the computations. The usual method of attack is to use the Law of Sines to find the unknown angle opposite the known side (this is angle β if a, b, and α are given). Then find the third angle (using the three angles sum to 180°), and then use the Law of Sines to find the remaining side.

EXAMPLE 2 Solve the triangle ABC if $a = 6.3$, $b = 11.2$, and $\alpha = 71.7°$.

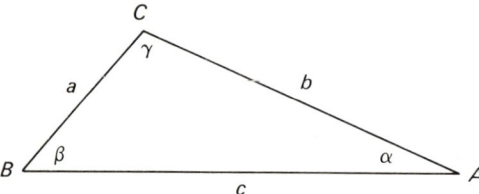

SOLUTION By the Law of Sines, $6.3/\sin 71.7° = 11.2/\sin \beta$. Thus $\sin \beta = 11.2 \sin 71.7°/6.3 \approx 1.68787$. But there is no angle β with $\sin \beta$ greater than 1. Hence there is no such triangle.

EXAMPLE 3 Solve the triangle ABC if $a = 12.4$, $b = 10.1$, and $\beta = 51°$.

SOLUTION By the Law of Sines,

$$\frac{12.4}{\sin \alpha} = \frac{10.1}{\sin 51°} \quad \text{or} \quad \sin \alpha = \frac{12.4 \sin 51°}{10.1} \approx .954198$$

Therefore, either $\alpha = \alpha_1 \approx \sin^{-1}(.954198) \approx 72.5769°$ or $\alpha = \alpha_2 \approx 180° - 72.5769° \approx 107.423°$. If $\alpha = \alpha_1$, then $\gamma_1 = 180° - 51° - \alpha_1 \approx 56.4231°$ and

$$c_1 = \frac{b \sin \gamma_1}{\sin \beta} \approx \frac{10.1 \sin 56.4231°}{\sin 51°} \approx 10.8278$$

On the other hand, if $\alpha = \alpha_2$, then $\gamma_2 = 180° - 51° - \alpha_2 \approx 21.5769°$ and

$$c_2 \approx \frac{b \sin \gamma_2}{\sin \beta} \approx \frac{10.1 \sin 21.5769°}{\sin 51°} \approx 4.77938$$

Thus we have found two possible triangles.

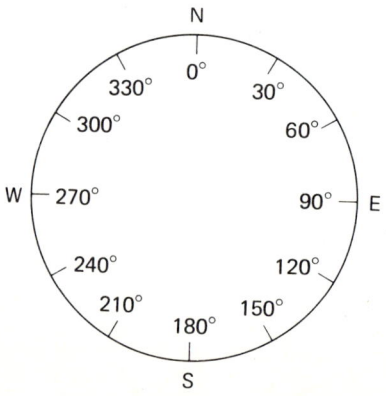

Angles are used in navigation to designate the course of a plane, boat, etc. Thus it is important to understand what is meant by the phrase "The object is on a course of 90°." In what direction is the object headed? North? East? South? West? Somewhere in between? The answer in this case is due east. This is because a navigator's compass is marked off in degrees in a clockwise direction, with 0° at due north (see diagram).

The following are examples of courses of 315° and 60°, respectively:

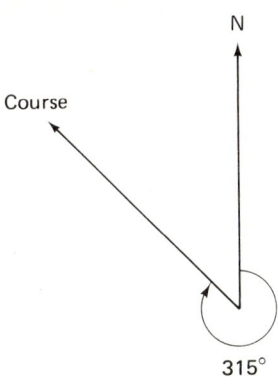
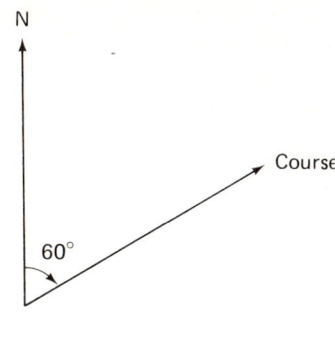

EXAMPLE 4 From a ship, port A is 43.6 nautical miles away on a course of 306.2°, while port B is on a course of 163.3°. If the ports are 87.5 nautical miles apart, how far is the ship from port B?

SOLUTION

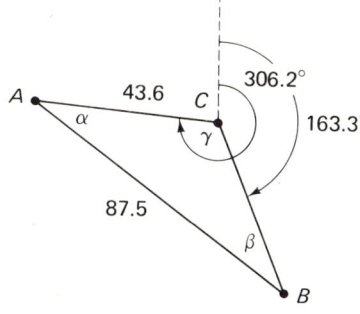

First we draw a picture. We let C denote the ship and label the angles as shown. Then $\gamma = 306.2° - 163.3° = 142.9°$, and we have side-side-angle. By the Law of Sines,

$$\frac{87.5}{\sin 142.9°} = \frac{43.6}{\sin \beta} \quad \text{or} \quad \sin \beta = \frac{43.6 \sin 142.9°}{87.5} \approx .30056992$$

Thus $\beta \approx 17.491837°$, and $\alpha = 180° - \beta - \gamma \approx 19.608163°$. Therefore

$$\frac{87.5}{\sin 142.9°} = \frac{|BC|}{\sin \alpha} \quad \text{or} \quad |BC| = \frac{87.5 \sin \alpha}{\sin 142.9°} \approx 48.6793 \text{ nautical mi}$$

EXERCISES

In Exercises 1–20, solve the triangle ABC having the given parts.

1. $\alpha = 36.12°$, $\beta = 61.27°$, $c = 407.1$
2. $\beta = 101°$, $\gamma = 26.34°$, $c = 31.04$
3. $\beta = 1.62$ rad, $\gamma = .53$ rad, $a = 96.5$
4. $\alpha = 1.21$ rad, $\beta = .97$ rad, $a = 625.5$
5. $\gamma = .34$ rad, $\alpha = .83$ rad, $c = .0145$
6. $\alpha = 46°$, $\gamma = 79.123°$, $b = 1.107$
7. $\gamma = 122.2°$, $\beta = 13.19°$, $b = 1492$
8. $\alpha = 1.4812$ rad, $\gamma = .8667$ rad, $a = 271.063$

9. $\beta = 88.42°$, $\alpha = 16.21°$, $b = 24.68$
10. $\alpha = 110°$, $a = 27.12$, $c = 11.78$
11. $\beta = 2.12$ rad, $a = 27.12$, $b = 11.78$
12. $\alpha = 43°$, $a = 943.16$, $b = 476.5$
13. $\gamma = 56°27'$, $b = 21.23$, $c = 32.21$
14. $\beta = .634$ rad, $b = 637$, $c = 1239$
15. $\alpha = .4957$ rad, $c = 17$, $a = 9$
16. $\gamma = 2.46$ rad, $c = 1096$, $a = 548$
17. $\alpha = 132.64°$, $a = 6.02$, $b = 23.51$
18. $\beta = 68.142°$, $b = 113.715$, $c = 84.957$
19. $\gamma = .3172$ rad, $b = 67.624$, $c = 61.1$
20. $\gamma = 36°13'25''$, $a = 191.4$, $c = 134$

21. A tree standing (vertically) on a slope inclined at an angle of 11° to the horizontal casts a shadow of length 51 feet up the slope. If the angle of elevation of the sun is 72.5°, how high is the tree?

22. A stretch of a river has straight banks. A man on one bank observes a tree on the other bank. His line of sight makes an angle of 51° with his bank. After walking 40 feet along his bank, he reaches a tree on his side of the river. The angle which his line of sight to the original tree makes with his bank has increased to 64°. He wants to stretch a rope between the two trees. How far apart are the trees?

23. A pole is slanting 13° away from vertical. To hold the pole up, a guy wire is attached to the top of the pole and anchored in the ground directly opposite the slant. If the pole is 96 feet long and the guy wire makes an angle of 47° with the (level) ground, how long is the guy wire?

24. Towns A, B, and C are connected by straight roads. It is 13.3 miles from A to B and 22.7 miles from B to C. At A, the roads to B and C meet at an angle of 63°24'. How far is A from C?

25. Points A and B are on opposite sides of a pond. A woman walks 67 feet from A straight to a point C (off to the side of the pond) and then turns and walks 91 feet straight to B. If she then observes that the angle between BA and BC is 38°, how far is it from A to B? (Can you determine the answer uniquely from the given information?)

26. A ramp 7.603 meters long is inclined from level ground to a loading dock. It makes an angle of 34.92° with the ground. If it is replaced with a ramp which makes an angle of 12.7° with the ground, how long is the new ramp?

27. Airport B is 200 miles due east of airport A. To reach airport C, a plane flies from A on a course of 346.35° or from B on a course of 302.2°. Find the distance from A to C and from B to C.

28. Airport B is 200 miles from airport A on a course of 52°. To reach airport C, a plane flies from A on a course of 346° or from B on a course of 302°. Find the distance from A to C and from B to C.

29. A ship leaves a small island on a course of 106°. The ship is traveling at 25 knots (nautical miles per hour). After 2 hours, the course is changed to 221°. Sometime later, the ship receives a message to return to the island. To do so, the ship turns to a heading of 33°30'. What is the total length of the trip (in nautical miles)?

30. A plane has crashed and activated an emergency transmitter. The signal is being received by two rescue units, A and B. A is 8.63 kilometers due north of B. From the signal, the rescuers determine that they must take a course of 127.25° from A or 43.08° from B to reach the plane. How far is each rescue unit from the plane?

31. Two planes take off from the same airport. The first plane flies on a course of 220.1°. The second plane flies on a course of 154.4°. After the first plane flies 362.4 kilometers, the course from it to the second plane is 83.5°. How far is the second plane from the airport?

SECTION 9.6. THE LAW OF COSINES

The Law of Sines is not sufficient to solve a triangle if we are given two sides and the included angle. For this we need another relationship between the sides and the angles. We derive such a relationship by placing the triangle ABC in the coordinate plane in such a way that A is at the origin and AB is along the positive x-axis. The two diagrams in Fig. 6 indicate this positioning of the

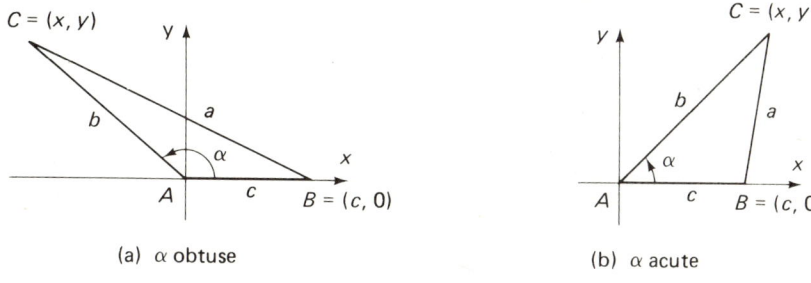

(a) α obtuse (b) α acute

FIG. 6

triangle when α is obtuse and when α is acute. In either case $C = (x,y)$ is a point on the terminal side of α, an angle in standard position, and C is $r = b$ units from the origin. By the definition of the trigonometric functions, $\sin \alpha = y/r = y/b$, and $\cos \alpha = x/r = x/b$. Multiplying each equation by b yields $y = b \sin \alpha$ and $x = b \cos \alpha$. Since these are the coordinates of C, $C = (b \cos \alpha, b \sin \alpha)$. Since side AB has length c, B has coordinates $B = (c, 0)$. We obtain a relationship between a, b, c, and α by observing that the length of the side BC is not only a but is also the distance between B and C. Therefore

$$a = \sqrt{(b \cos \alpha - c)^2 + (b \sin \alpha)^2}$$

Squaring both sides yields

$$a^2 = (b \cos \alpha - c)^2 + (b \sin \alpha)^2$$

Multiplying out the right-hand side yields

$$\begin{aligned} a^2 &= (b \cos \alpha)^2 - 2bc \cos \alpha + c^2 + (b \sin \alpha)^2 \\ &= b^2 \cos^2 \alpha - 2bc \cos \alpha + c^2 + b^2 \sin^2 \alpha \\ &= b^2(\cos^2 \alpha + \sin^2 \alpha) + c^2 - 2bc \cos \alpha \end{aligned}$$

Since $\cos^2 \alpha + \sin^2 \alpha = 1$,

$$a^2 = b^2 + c^2 - 2bc \cos \alpha$$

The last equation is the relationship we were seeking. If we had placed B or C at the origin, we would have obtained

$$b^2 = a^2 + c^2 - 2ac \cos \beta \quad \text{or} \quad c^2 = a^2 + b^2 - 2ab \cos \gamma$$

Any one of these three equations is an instance of the Law of Cosines:

THE LAW OF COSINES

In a triangle, the square of the length of one side is equal to the sum of the squares of the lengths of the other two sides minus twice their product times the cosine of the angle between them. That is,

(23) $$a^2 = b^2 + c^2 - 2bc \cos \alpha$$
(24) $$b^2 = a^2 + c^2 - 2ac \cos \beta$$
(25) $$c^2 = a^2 + b^2 - 2ab \cos \gamma$$

The Law of Cosines is useful in solving triangles where you are given two sides and the included angle or three sides.

Side-Angle-Side (SAS)

If you are given two sides and the included angle (SAS), the method of solution is as follows:

First use the Law of Cosines in order to find the remaining side. Once you know three sides and one angle, you can use either the Law of Cosines or the Law of Sines to find a second angle. [The third is 180° (or π) minus the sum of the other two.] The Law of Sines is easier to use, but there is a possible problem. If, say, a, b, c, and α are known and you try to find β, you get $a/\sin \alpha = b/\sin \beta$, or $\sin \beta = (b/a) \sin \alpha$. If β is acute, $\beta = \sin^{-1}[(b/a) \sin \alpha]$, but if β is obtuse, $\sin^{-1}[(b/a) \sin \alpha]$ is just the reference angle of β. Thus if you do not know ahead of time whether β is acute or obtuse, then you do not know whether $\sin^{-1}[(b/a) \sin \alpha]$ is β or the reference angle of β, and consequently the Law of Sines cannot be used. However, you can avoid this difficulty by making sure that you are looking for an acute angle. Since in a triangle the smaller angles are opposite the smaller sides and the smallest two angles are always acute, if you calculate the smaller of the two remaining angles, you will be sure it is acute. Summarizing,

> If three sides and one angle in a triangle are known and the Law of Sines is being used to find a second angle, make sure that angle is the smaller of the two remaining angles (i.e., opposite the smaller of the two remaining sides).

EXAMPLE 1 Solve the triangle ABC if $a = 6.31$, $c = 4.74$, and $\beta = 42°$.

SOLUTION The remaining side is found by the Law of Cosines. We use equation (24) since a, c, and β are known, and we wish to find b:

$$b^2 = a^2 + c^2 - 2ac \cos \beta$$
$$= (6.31)^2 + (4.74)^2 - 2(6.31)(4.74) \cos 42° \approx 17.829657$$

Therefore, $b \approx 4.22252$. Now since $c < a$, we know that $\gamma < \alpha$ and that γ is acute. We compute γ by the Law of Sines:

$$\frac{c}{\sin \gamma} = \frac{b}{\sin \beta} \quad \text{or} \quad \sin \gamma = \frac{c \sin \beta}{b} \approx \frac{4.74 \sin 42°}{4.2225179} \approx .7511349$$

Since γ is acute, $\gamma \approx \sin^{-1}(.7511349) \approx 48.6888°$. (*Note*: In doing these calculations on your calculator, intermediate results should be kept to as many digits as is possible on your calculator so that the final answer is as accurate as possible.) Finally, the angle α is $\alpha = 180° - \beta - \gamma \approx 180° - 42° - 48.6888° \approx 89.3112°$.

EXAMPLE 2 Plane I leaves an airport at 1 p.m. and flies a straight course at 400 miles per hour. Plane II leaves the same airport at 1:30 p.m. and flies at 300 miles per hour on a straight course which makes an angle of 78.3° with that of plane I. How far apart are the planes at 3 p.m.?

SOLUTION

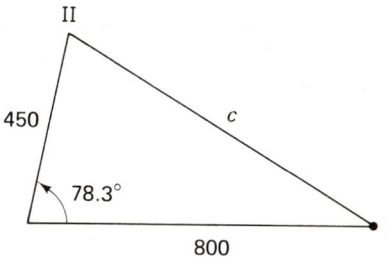

At 3 p.m., plane I will have flown 800 miles, while plane II will have flown 450 miles. The distance between the planes at 3 p.m. is c in the triangle shown, where $a = 450$, $b = 800$, and $\gamma = 78.3°$. Thus by the Law of Cosines, equation (25),

$$c^2 = a^2 + b^2 - 2ab \cos \gamma$$
$$= (450)^2 + (800)^2 - 2(450)(800) \cos 78.3° \approx 696,493.36$$

Thus $c \approx \sqrt{696,493.36} \approx 834.562$ miles.

Side-Side-Side (SSS)

If three sides of a triangle are given, we can use any of the three formulas for the Law of Cosines to find one angle. Then proceed as in the case SAS to find the remaining two angles.

EXAMPLE 3 Solve the triangle ABC if $a = 12.6$, $b = 7.42$, and $c = 11.03$.

SOLUTION Let us find angle α first. We use the formula for the Law of Cosines in which α appears, namely (23), and solve the equation for $\cos \alpha$:

$$a^2 = b^2 + c^2 - 2bc \cos \alpha$$
$$2bc \cos \alpha = b^2 + c^2 - a^2$$
$$\cos \alpha = \frac{b^2 + c^2 - a^2}{2bc}$$
$$= \frac{(7.42)^2 + (11.03)^2 - (12.6)^2}{2(7.42)(11.03)}$$
$$\approx .1097063$$

Therefore, $\alpha \approx \cos^{-1}(.1097063) \approx 83.7016°$. Next we find the smaller angle β by the Law of Sines:

$$\sin \beta = \frac{b \sin \alpha}{a} \approx .5853343$$

$$\beta \approx \sin^{-1}(.5853343) \approx 35.8266°$$

Therefore, $\gamma = 180° - \beta - \alpha \approx 60.4718°$.

EXERCISES

In Exercises 1–15, solve the triangle ABC having the given parts.

1. $a = 10.7, b = 4.75, \gamma = 60°$
2. $a = 413, c = 981, \beta = 102°$
3. $b = .051, c = .034, \alpha = 122°$
4. $b = .051, c = .034, \alpha = 81.47°$
5. $a = 1024, c = 685, \beta = .385$
6. $a = 63.714, b = 84.105, \gamma = 2.12$
7. $b = 141, c = 298, \alpha = 14.41°$
8. $a = 735, b = 379, \gamma = 141°$
9. $a = 6.39, c = 12.78, \beta = 40°$
10. $a = 14.92, b = 7.49, c = 12.51$
11. $a = .12, b = .67, c = 1.03$
12. $a = 5, b = 8, c = 11$
13. $a = 221, b = 284, c = 148$
14. $a = .0035, b = .0024, c = .0017$
15. $a = 2743, b = 967, c = 2102$

16. A triangular field has sides of length 147, 206, and 182 meters. Find the angles of the triangle.

17. A ship intends to sail from port A to port B, a distance of 350 miles. It sails at a constant speed of 35 miles per hour. After 2 hours, the captain discovers that they have been sailing off course by 18°. He then corrects the course. What is the total time of the trip?

18. A, B, and C are three airports. The distances from A to B and A to C are 348 and 296 miles, respectively. The angle between the flight paths from A to B and A to C is 71.4333°. How far is it from B to C?

19. Some pipe is to be laid directly through a small hill. From a point off to the side, a straight rope is run to the entry point, and another straight rope is run to the exit point. The first rope is 42 feet long, while the second is 33.6 feet long, and the two ropes meet at an angle of 56.245°. How long is the pipe, and at what angle with the first rope should it be driven so that it exits at the correct place?

20. A plane leaves an airport and flies at 360 miles per hour on a course of 103.25° for 1 hour and 33 minutes. Then it changes course to 236.75° and flies for 48 minutes. How far is it from the airport? If it wished to return, what would be its course?

21. From one house you can go 1.68 kilometers along a straight road to a corner and then turn and go 2.35 kilometers along another straight road to a second house. The angle between the roads is 111°. There is a footpath that runs straight between the two houses. How much shorter is it to go on the footpath instead of the roads?

22. A guy wire runs 15.3 meters from the top of a pole to the (level) ground. It is replaced by a 21.16-meter wire which runs from the top of the pole to a spot on the ground 8.47 meters farther from the pole than the old wire. What angle does the new wire make with the ground?

23. A small electronic component is in the shape of a triangle with sides 6.23, 8.146, and 11.392 millimeters. Find the largest angle.

24. Two boats leave a dock at the same time on straight courses. The angle between their courses is 51.2°. One boat is traveling at 14.7 kilometers per hour, while the other is traveling at 18.3 kilometers per hour. After 2 hours, how far apart are the boats?

25. In Exercise 24, how far apart are the boats after 2 hours and 16 minutes?

SECTION 9.7. VECTORS

Many familiar concepts have numbers associated with them, such as distance, area, volume, etc. However, other concepts not only have numbers associated with them but direction as well. For instance, the velocity of an object is its speed together with the direction it is heading (e.g., the velocity of a car may be 50 miles per hour (mph) due north). Other concepts that involve both a magnitude and a direction include acceleration and force.

---

(26)   **DEFINITION**

A quantity which has both a magnitude and a direction is called a **vector**.

Note: It is sometimes convenient to use ordered pairs in order to denote vectors, where the first entry represents magnitude and the second entry represents direction. For example, the velocity of the airplane is (250 mph, due west).

When working with vectors, it often helps to represent them geometrically. Now when we want to represent a number geometrically, we usually draw a line segment. Similarly, we usually represent a vector by drawing an *arrow* or *directed line segment* whose length has the given magnitude *and* which is pointed in the given direction. If the arrow goes from a point A (called the **initial point**) to the point B (called the **terminal point**), we denote the vector by $\overline{AB}$. See Fig. 7(a). If the initial and terminal points are not specified, we may denote vectors by small Roman boldface letters, such as **v**. See Fig. 7(b). The length of the arrow is the **length** or **magnitude** of the vector and is denoted by putting vertical lines on either side of the vector (like absolute values). Thus the lengths of the vectors in Fig. 7 are $|\overline{AB}|$ and $|\mathbf{v}|$.

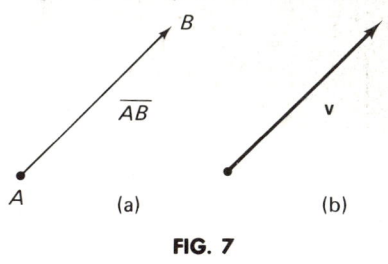

FIG. 7

There is one vector which is special, namely, the vector of zero length. This is called the **zero vector**. It is denoted by **0**, and you may assume it has any direction that you wish.

We say that two vectors are **equal**, or that two arrows represent the same vector, if they have the same magnitude *and* the same direction. For example, in Fig. 7, $\overline{AB} = \mathbf{v}$. A consequence of this definition of equality is that when you want to represent a vector geometrically, you may put the initial point anywhere that is convenient (and then draw an arrow of the appropriate length and direction).

Now consider what happens when two forces act on the same thing. For example, suppose you are rowing a boat toward the north at 4 mph across a river which is flowing toward the east at 3 mph. See Fig. 8. Then the boat actually moves as though it were being propelled at a speed and in a direction which is some combination of the speed and direction of the water and the speed and direction of the rowing. This combination is called the **vector sum** of the two vectors.

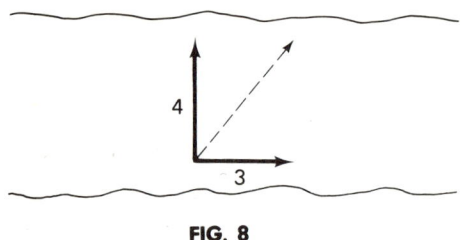

FIG. 8

If two vectors **v** and **w** are not in the same or opposite directions, the rule for their sum is as follows: We first draw the arrows corresponding to **v** and **w** with the same initial point O. See Fig. 9(a). Thus, $\mathbf{v} = \overline{OA}$ and $\mathbf{w} = \overline{OB}$. Then we complete this to a parallelogram $OACB$ and draw in the diagonal $\overline{OC}$. See Fig. 9(b).

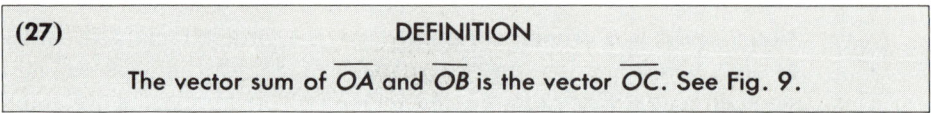

(27)   **DEFINITION**
The vector sum of $\overline{OA}$ and $\overline{OB}$ is the vector $\overline{OC}$. See Fig. 9.

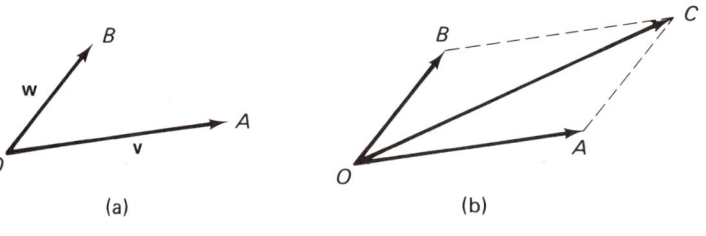

FIG. 9

Note that since opposite sides of a parallelogram are equal and parallel, the vectors $\overline{OA}$ and $\overline{BC}$ are equal, as are $\overline{OB}$ and $\overline{AC}$. Thus we could have defined the vector sum of **v** and **w** by saying the following: First draw any vector $\overline{OA}$ equal to **v**. Then using A as the initial point, draw the vector $\overline{AC} = \mathbf{w}$. Then $\overline{OC}$ is the sum of **v** and **w**. This is what we use in the special cases when **v** and **w** are in the same or opposite directions.

If two vectors have the same direction, then their sum has that same direction, and the magnitude of the sum is the sum of the two magnitudes.

If two vectors have opposite directions, then either

1. Their lengths are equal, and their sum is the zero vector, or
2. Their lengths are unequal, and their sum has the direction of the vector with larger magnitude and a magnitude which is the larger minus the smaller of the magnitudes of the given vectors.

From this it follows that the zero vector **0** acts as an additive identity:
$$\mathbf{0} + \mathbf{v} = \mathbf{v} + \mathbf{0} = \mathbf{v}$$
For, we can think of **0** as having zero length and the direction of **v** (recall that we may think of **0** as having any direction which is convenient).

The general case of adding two vectors was defined using a parallelogram, and we shall need the following properties:

(28)   In a parallelogram,

i. Opposite sides are equal.

ii. Opposite angles are equal.

iii. The sum of two adjacent angles is 180°.

Suppose we want to compute the vector sum of two arbitrary vectors. First, we shall assume that the vectors are described by giving their lengths and their directions on a navigator's compass. In this context, Example 1 illustrates how to find the sum using the Laws of Sines and Cosines.

EXAMPLE 1 Suppose a small plane is flying on a heading of 253° (the heading is the direction the plane is pointed), and suppose there is a wind of 25 mph in the direction of 150°. If the airspeed of the plane is 115 mph (the airspeed is the speed relative to the air), find the ground speed and actual course of the plane.

SOLUTION First we draw a diagram indicating the given vectors, and to keep the diagram uncluttered, we also draw a separate parallelogram. See Fig. 10. There

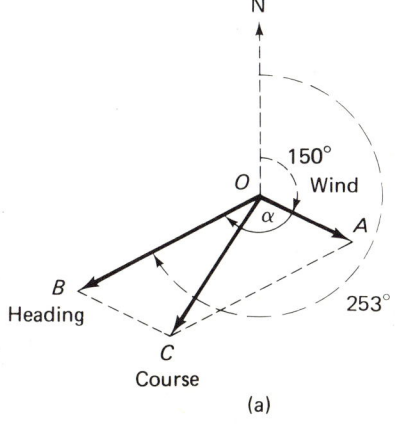
(a)

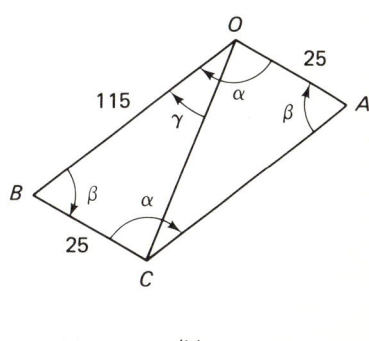
(b)

FIG. 10

Section 9.7   Vectors   403

are two things to find: the length $|\overline{OC}|$ and the angle $\angle NOC$. We find the length $|\overline{OC}|$ first. Referring to Fig. 10(b), since $OBCA$ is a parallelogram, $|\overline{BC}| = |\overline{OA}| = 25$. If we knew angle β, we could easily use the Law of Cosines to compute $|\overline{OC}|$. We do know that $150° + \alpha = 253°$ [see Fig. 10(a)] so that $\alpha = 253° - 150° = 103°$. Now by (28iii), $\alpha + \beta = 180°$, so $\beta = 180° - 103° = 77°$. By the Law of Cosines,

$$|\overline{OC}|^2 = 115^2 + 25^2 - 2(115)(25)\cos 77° \approx 12{,}556.531$$

Thus, $|\overline{OC}| \approx 112.056$ is the ground speed (in mph) of the plane.

If we knew angle γ, then we could easily find the heading since $\gamma + \angle NOC = 253°$. By the Law of Sines,

$$\frac{\sin \gamma}{|\overline{BC}|} = \frac{\sin \beta}{|\overline{OC}|}$$

so

$$\sin \gamma = \frac{|\overline{BC}| \sin \beta}{|\overline{OC}|} \approx \frac{25 \sin 77°}{112.0559} \approx .02173848$$

Hence

$$\gamma \approx \sin^{-1}(.02173848) \approx 12.5555°$$

and the actual course is

$$253° - 12.5555° \approx 240.445°$$

Some problems require a little *vector algebra*, so we should point out that vector addition can be seen to satisfy the same properties of addition as the real numbers. These properties will be useful in Example 2.

EXAMPLE 2 Port B is 200 miles from port A on a course of 60°. The current is due east (i.e., in a direction of 90°) at 6 mph. Determine the heading and water speed of a ship in order for that ship to sail directly from A to B in 8 hours.

SOLUTION In this problem, we are given one vector **v** (6 mph, 90°), and we want to find a second vector **x**, so that the vector sum is **w** (200/8 = 25 mph, 60°). Thus we want to solve

$$\mathbf{v} + \mathbf{x} = \mathbf{w}$$

for **x**. See Fig. 11(a). Since vector addition behaves like addition of numbers, we can solve this for **x**, obtaining

$$\mathbf{x} = \mathbf{w} + (-\mathbf{v})$$

where $-\mathbf{v}$ is (6 mph, 270°). See Fig. 11(b). Writing this as an addition problem allows us to follow the pattern of Example 1. We first see that $\alpha = 150°$ ($= 90° + 60°$). Since $\alpha + \beta = 180°$, $\beta = 180° - 150° = 30°$. By the Law of Cosines,

$$|\overline{OC}|^2 = 25^2 + 6^2 - 2(25)(6)\cos 30° \approx 401.1925$$

404 ANALYTIC AND GEOMETRIC TRIGONOMETRY

so that

$$|\overline{OC}| \approx 20.02979$$

is the water speed in mph.

To find γ, by the Law of Sines,

$$\sin \gamma = \frac{|\overline{CB}| \sin \beta}{|\overline{OC}|} \approx \frac{6 \sin 30°}{20.02979} \approx .1497768$$

so that

$$\gamma \approx \sin^{-1}(.1497768) \approx 8.614°$$

By Fig. 11(b), $\angle NOC + \gamma = 60°$, so that the heading is $\angle NOC \approx 60° - 8.614° \approx 51.386°$.

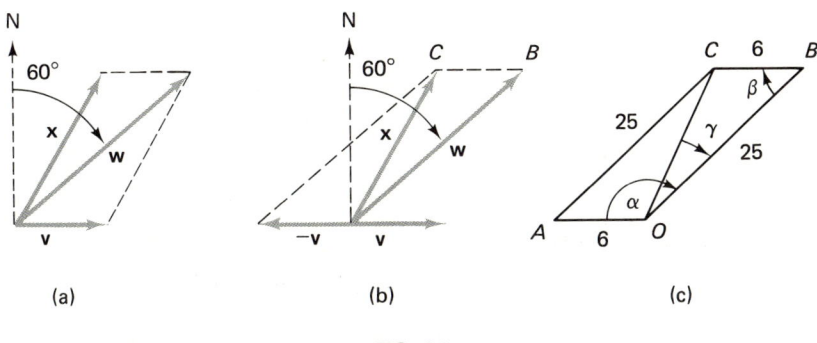

FIG. 11

If **u**, **v**, and **w** are vectors and **w** = **u** + **v**, then **u** and **v** are called **components** of **w** (in the sense that they "make up" **w**). Thus each vector **w** has many pairs of components. There are many problems in which you are given one vector **w**, and you are asked to find one or both of its components with certain properties. This process is called **resolving** the vector.

EXAMPLE 3 A 50-pound block of ice is on a ramp inclined at 20°. **(a)** What force pushing up the ramp is necessary to keep the block from sliding? **(b)** What is the actual force caused by the weight of the ice against the ramp? (We assume friction is negligible.)

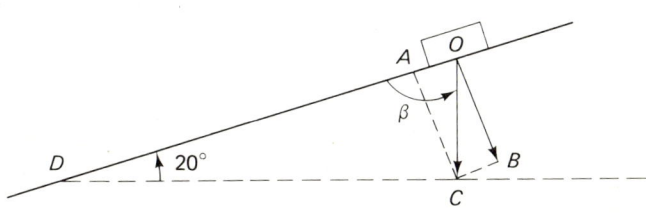

FIG. 12

SOLUTION The weight of the block is a vector $\overline{OC}$ which points vertically downward. The vector $\overline{OC}$ can be resolved into two vectors: $\overline{OB}$, which is perpendicular to the ramp and represents the actual force of the block against the ramp, and $\overline{OA}$, which is parallel to the ramp pointing downwards and represents the force of the block down the ramp. Thus we want to compute $|\overline{OA}|$ and $|\overline{OB}|$. See Fig. 12. Note that $\overline{OA}$ and $\overline{OB}$ are perpendicular. Since ODC is a right triangle, $\beta = 90° - 20° = 70°$. Since OAC is a right triangle,

$$|\overline{OA}| = |\overline{OC}| \cos \beta = 50 \cos 70° \approx 17.101 \text{ lb}$$

Since $OACB$ is a rectangle,

$$|\overline{OB}| = |\overline{AC}| = |\overline{OC}| \sin \beta = 50 \sin 70° \approx 46.9847 \text{ lb}$$

EXERCISES

In Exercises 1–8, find the sum.

	FIRST VECTOR		**SECOND VECTOR**	
	Magnitude	**Direction**	**Magnitude**	**Direction**
1.	50	90°	25	60°
2.	1046	45°	837	135°
3.	6.28	215°	4.65	334°
4.	96.256	196.3°	12.31	83.05°
5.	11.14	33.9°	25.91	351°
6.	31.47	171°	42.81	259.3°
7.	819	106.51°	819	286.51°
8.	64.8	303°	100.2	303°

9. A man can row his boat at 3 mph. He is in a stream flowing from north to south at 4 mph. How fast and in what direction will he travel if he rows
  a. Directly downstream?
  b. Directly upstream?
  c. Due east (across the stream)?
  d. Northeast (45° on a navigator's compass)?
  e. Southeast (135°)?

10. A duck is flying directly north at 25 km/hr (kilometers per hour). How fast and in what direction will the bird actually travel if the wind is blowing at 30 km/hr
  a. Directly north?
  b. Directly south?
  c. From the west?
  d. Toward the southeast (135°)?
  e. Toward the northwest (315°)?

11. Assume that a portion of the Mississippi River is 1 mile wide and that the current is 6 mph. Points A and B are directly opposite each other on the banks. If a man starts from A rowing at 4 mph and keeps his boat parallel to $\overline{AB}$,
   a. What is his actual velocity?
   b. How far downriver from B will he land?
   c. How long will it take him to land?
   d. How long would it take him to land if there were no current?

12. An airplane flies over the northwest corner of Colorado heading due south at 200 mph with a crosswind of 25 mph blowing directly east. If it is 275 miles between the northern and southern borders of Colorado,
   a. What is the actual velocity of the plane?
   b. How far east of the southwestern corner of Colorado will the plane cross the border?
   c. How long will it take the plane to cross the southern border of Colorado?
   d. How long would it take the plane to reach the southern border if there were no wind?

13. Referring to Exercise 11, suppose the man puts a motor on the boat which propels it at 9 mph.
   a. In what direction should he head in order to actually land at B?
   b. How long would it take him to arrive at B?
   c. Would it be possible for him to arrive at B by rowing at 4 mph?

14. Referring to Exercise 12, suppose the pilot wanted to fly directly to the southwestern corner of Colorado.
   a. In what direction should he head?
   b. How long will it take him to reach the corner?

15. A plane is flying at an airspeed of 400 mph with a heading of 328°. There is a wind of 40 mph in the direction of 243°. Find the course and ground speed of the plane.

16. A ship is sailing at 21 km/hr on a heading of 107° through a crosscurrent of 3 mph in the direction of 21°. Find the actual speed and direction of the ship.

17. There is a 50-km/hr wind in the direction of 36°. A pilot wishes to fly on a course of 121° with a ground speed of 400 km/hr. Determine her heading and airspeed.

18. Two people want to take their motorboat to an island 20 miles away on a heading of 349°. If there is a 2 mph crosscurrent in the direction of 40°, in what direction and at what speed should they head if they want to take 2 hours for the trip?

19. A patrol boat leaves its base on a heading of 17° traveling at 17 km/hr. After 2 hours it turns to a new heading of 310° and slows to 12 km/hr. After 45 minutes on this new heading, how far and in what direction is the boat from its base?

20. An airplane flies due north at 200 mph. After 3 hours, it turns and flies southeast (135°) at 210 mph. After an hour on this new heading, how far is it from its starting point?

21. A 38-kilogram weight is on a ramp inclined at 27°. If there is no friction, what force, parallel to the ramp, is necessary to keep the weight at rest?

22. A 2000-pound car is on a hill inclined at 15°. How much force, pushing up the hill, is necessary to keep the car from rolling?

23. A 50-pound weight on an incline is being held at rest by a force of 21 pounds pushing up the incline. What is the angle of inclination?

24. A force of 48 pounds is required to keep a weight of 70 pounds from sliding down an inclined plane. What is the angle of inclination?

25. A manual lawnmower requires a force of 10 pounds parallel to the ground in order to move it through the grass. If the handle is inclined at 41° to the ground, how much force is necessary along the handle? See Fig. 13(a).

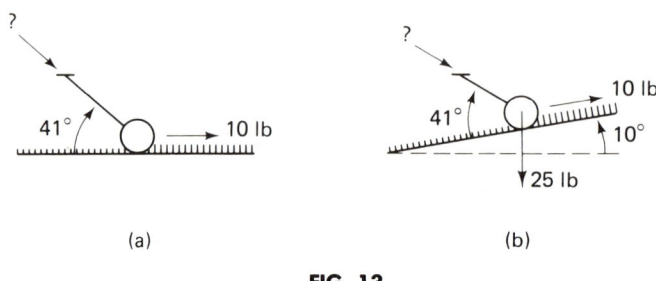

(a) (b)

FIG. 13

26. If the lawnmower of Exercise 25 weighs 25 pounds and is pushed up a hill of 10°, how much force exerted along the handle is necessary to keep it moving through the grass? The handle is still inclined 41° to the slope, and you must take into account both the 10 pounds from Exercise 25 and the weight of the lawnmower. See Fig. 13(b).

SECTION 9.8. POLAR COORDINATES

You are all familiar with the way we coordinatize the plane with a Cartesian (rectangular) coordinate system. We draw a horizontal number line (the x-axis) and a vertical line (the y-axis) as shown in Fig. 14. Associated with each point P, there is a unique ordered pair of numbers (x,y), called the coordinates of P. Conversely, with each pair (x,y) we associate the unique point P which has (x,y) as coordinates.

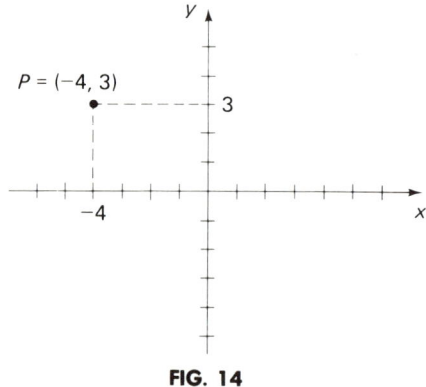

FIG. 14

In this section we introduce another system of coordinates, the system of **polar coordinates**. We start with a point O, called the **origin**, or **pole**, and a horizontal half line with endpoint O, called the **polar axis**. We think of the polar axis as being the nonnegative half of a number line. This gives us a scale so that we can measure distance in the plane.

---

(29)   **DEFINITION**

We say that a point P has polar coordinates (r, θ), $r \geq 0$, if θ has the polar axis as initial side and

   i. P is on the terminal side of θ.

   ii. $|OP| = r$ (i.e., the distance from the origin to P is r).

See Fig. 15(a).

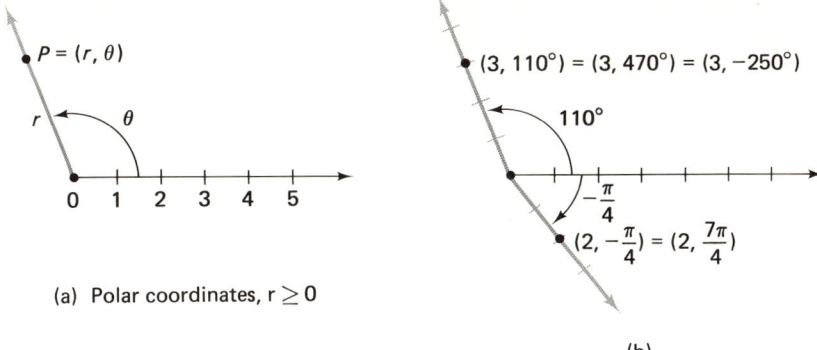

(a) Polar coordinates, $r \geq 0$

(b)

FIG. 15

The angle θ is considered to be an angle in standard position; it is positive for counterclockwise rotations and negative for clockwise rotations. Since we know that any angles which differ by complete rotations have the same terminal side, (r, θ) and $(r, \theta + 360°n)$ are different polar coordinates for the same point. [Of course θ could be measured in radians, in which case we write (r, θ) and $(r, \theta + 2n\pi)$.] Some examples of points in polar coordinates and a few of the many different ways these coordinates may be written are illustrated in Fig. 15(b). The origin O has coordinates $(0, \theta)$ for any angle θ.

It is convenient for applications to allow r to be negative.

---

(30) **DEFINITION**

We say that a point P has polar coordinates (r, θ), $r < 0$, if we draw the terminal side of θ and then extend that line $|r|$ units on the other side of the origin to P. See Fig. 16(a).

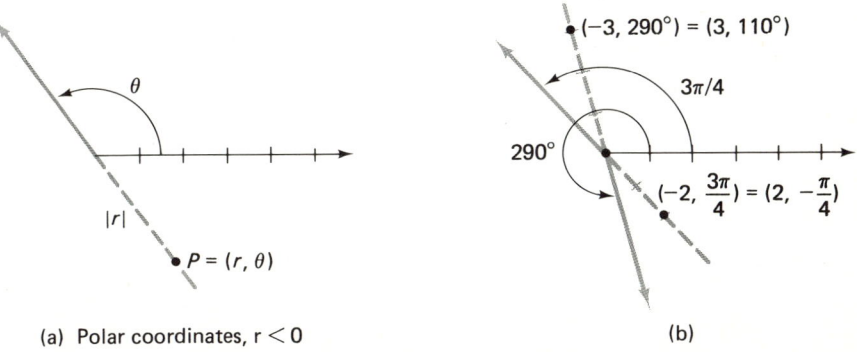

(a) Polar coordinates, $r < 0$

(b)

FIG. 16

Section 9.8 Polar Coordinates

In Fig. 16(b), we illustrate how polar coordinates of the points in Fig. 15(b) can be written with r negative. Notice that if P has polar coordinates (r, θ), $r < 0$, then $(|r|, \theta + 180°)$ are also polar coordinates for P.

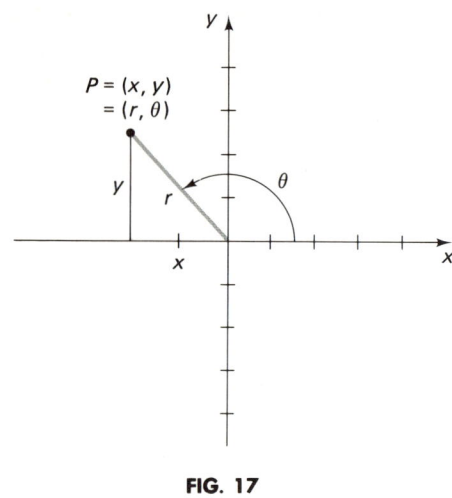

There are many times when it is very useful to be able to relate polar coordinates and Cartesian coordinates. To do this, we let the origins coincide and let the polar axis be the positive x-axis. See Fig. 17.

Given the Cartesian coordinates (x, y) of a point P, we can obtain polar coordinates (r, θ) by setting $r = \sqrt{x^2 + y^2}$ and finding θ from any of the following equations (which you will recognize as the defining equations of the three trigonometric functions):

FIG. 17

(31)
$$\left. \begin{array}{l} \sin \theta = \dfrac{y}{r} \\[4pt] \cos \theta = \dfrac{x}{r} \\[4pt] \tan \theta = \dfrac{y}{x} \end{array} \right\} \quad \text{where } r = \sqrt{x^2 + y^2}$$

You just have to keep in mind what quadrant θ is in.

EXAMPLE 1 Find polar coordinates (r, θ) of the points with Cartesian coordinates **(a)** $(-2, 3)$ and **(b)** $(-2, -3)$.

SOLUTION **(a)** Set $r = \sqrt{(-2)^2 + 3^2} = \sqrt{13}$. Since arccosine gives angles in Q_1 or Q_2 and $(-2, 3)$ is in Q_2, we can use $\cos \theta = x/r = -2/\sqrt{13}$ and solve $\theta = \cos^{-1}(-2/\sqrt{13}) \approx 123.69°$. Thus, $(r, \theta) \approx (\sqrt{13}, 123.69°)$. **(b)** Set $r = \sqrt{(-2)^2 + (-3)^2} = \sqrt{13}$. Since no inverse trigonometric function gives an angle in Q_3, we use our knowledge of reference angles to find θ. If φ is the reference angle, $\tan \varphi = y/x = -3/(-2) = 1.5$. Thus, $\varphi = \tan^{-1}(1.5) \approx 56.3099°$. Since θ is in Q_3, $\theta = \varphi + 180° \approx 236.310°$.

Given the polar coordinates (r, θ) of a point P, we can easily obtain the Cartesian coordinates (x, y). If $r \geq 0$, we just solve the first two equations in (31) for x and y, obtaining

(32)
$$x = r\cos\theta \quad \text{and} \quad y = r\sin\theta \quad \text{if } r \geq 0$$

If $r < 0$, then $(|r|, \theta + \pi)$ are also polar coordinates for P, so we can use these in (32) to obtain
$$x = |r|\cos(\theta + \pi) \quad \text{and} \quad y = |r|\sin(\theta + \pi) \quad \text{if } r < 0$$
However, these can be simplified, since if $r < 0$,
$$x = |r|\cos(\theta + \pi) = |r|(\cos\theta\cos\pi - \sin\theta\sin\pi)$$
$$= -|r|\cos\theta = r\cos\theta$$
and
$$y = |r|\sin(\theta + \pi) = |r|(\sin\theta\cos\pi + \cos\theta\sin\pi)$$
$$= -|r|\sin\theta = r\sin\theta$$
In other words, (32) holds for all r.

(33)
$$x = r\cos\theta \quad \text{and} \quad y = r\sin\theta$$

EXAMPLE 2 Find the Cartesian coordinates of the point with polar coordinates $(-2, 4.819 \text{ rad})$.

SOLUTION Observe that the angle is in radians:
$$x = -2\cos(4.819) \approx -.212818$$
$$y = -2\sin(4.819) \approx 1.98864$$

We now consider the graphs of some equations in polar coordinates.

EXAMPLE 3 Sketch the graph of $r = 5$.

SOLUTION

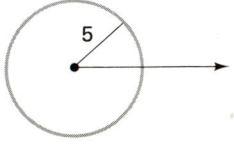

The circle $r = 5$

FIG. 18

The set of all points of distance 5 from the origin (and no restriction on θ) is a circle of radius 5 centered at the origin. See Fig. 18. This circle also has equation $r = -5$, since $(-5, \theta)$ and $(5, \theta + \pi)$ are different polar coordinates for the same point.

EXAMPLE 4 Sketch the graph of $\theta = 130°$.

SOLUTION

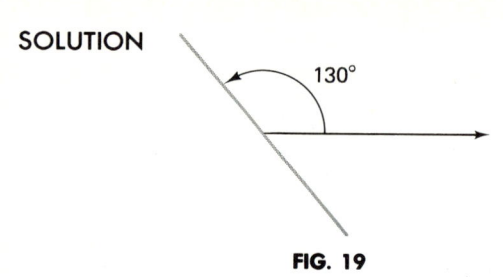

FIG. 19

The graph of $\theta = 130°$ is all points on the terminal side of θ ($r \geq 0$) together with all points opposite the origin from these ($r < 0$). Thus it is the straight line in Fig. 19.

EXAMPLE 5 Convert $r = \cos\theta + 2\sin\theta$ to Cartesian coordinates.

SOLUTION The easiest way is to multiply both sides by r, obtaining $r^2 = r\cos\theta + 2r\sin\theta$. Since $r^2 = x^2 + y^2$ and $x = r\cos\theta$, $y = r\sin\theta$, this is

$$x^2 + y^2 = x + 2y$$

EXAMPLE 6 Sketch the graph of $r = 2(1 - \cos\theta)$.

SOLUTION We first observe that since $\cos(-\theta) = \cos\theta$, the point (r, θ) is on the graph precisely when the point $(r, -\theta)$ is on the graph. In other words, the graph is symmetric about the x-axis. Thus if we sketch the graph when $0 \leq \theta \leq \pi$ and reflect the result in the x-axis, we obtain the whole graph. We plot the points from the following table, although we could use a calculator:

θ	0	$\dfrac{\pi}{6}$	$\dfrac{\pi}{4}$	$\dfrac{\pi}{3}$	$\dfrac{\pi}{2}$	$\dfrac{2\pi}{3}$	$\dfrac{3\pi}{4}$	$\dfrac{5\pi}{6}$	π
$r = 2(1 - \cos\theta)$	0	$2 - \sqrt{3}$	$2 - \sqrt{2}$	1	2	3	$2 + \sqrt{2}$	$2 + \sqrt{3}$	4

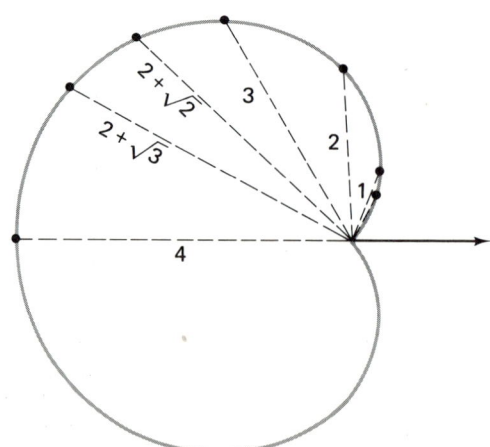

This curve is called a cardioid.

EXERCISES

In Exercises 1–10, plot the points with the given polar coordinates.

1. $(2, 40°)$
2. $(-3, 120°)$
3. $(-1, 18°)$
4. $(1, 198°)$
5. $(2.3, -400°)$
6. $\left(3, \dfrac{\pi}{2} \text{ rad}\right)$
7. $\left(-1, \dfrac{7\pi}{6} \text{ rad}\right)$
8. $(-3.5, 12\pi \text{ rad})$
9. $(5, 5 \text{ rad})$
10. $(2, 15 \text{ rad})$

In Exercises 11–18, find a set of polar coordinates for the points with the given Cartesian coordinates.

11. $(-4, 4)$
12. $(-7, -2)$
13. $(1.3, -6.8)$
14. $(3\sqrt{3}, 3)$
15. $(-6, 0)$
16. $(-3, -4)$
17. $(1, -3)$
18. $(-1, 3)$

In Exercises 19–26, find Cartesian coordinates for the points with the given polar coordinates.

19. $(2, 120°)$
20. $(-3, 45°)$
21. $(-6, -12°)$
22. $(4.65, 349.2°)$
23. $(0, 1.63 \text{ rad})$
24. $\left(12.3, \dfrac{5\pi}{6} \text{ rad}\right)$
25. $(-2, 2 \text{ rad})$
26. $(-3, 27.415 \text{ rad})$

In Exercises 27–32, convert the equations to polar coordinates.

27. $x = 7$
28. $x^2 + y^2 = 9$
29. $y = 2x$
30. $(x + 2)^2 + y^2 = 4$
31. $\dfrac{x^2}{4} + y^2 = 1$
32. $x^2 - y^2 = 1$

In Exercises 33–40, convert the equations to Cartesian coordinates.

33. $r = 3$
34. $r = -4$
35. $\theta = \dfrac{3\pi}{4}$
36. $r \cos \theta = -4$
37. $r \sin \theta = 7$
38. $r = \cos \theta$
39. $r^2 \cos \theta = 3 \sin \theta - \cos \theta$
40. $r = 3 \sin \theta + 1$

In Exercises 41–52, sketch the graph of the given equation.

41. $r = 2$
42. $r = -3$
43. $\theta = 100°$
44. $\theta = \dfrac{\pi}{4}$
45. $r \cos \theta = 3$
46. $r \sin \theta = 2$
47. $r = 1 - \cos \theta$
48. $r = 1 + \sin \theta$
49. $r = 2 \sin \theta - 1$
50. $r = 1 - 2 \cos \theta$
51. $r = \sin 2\theta$
52. $r = \sin 3\theta$

REVIEW EXERCISES

In Exercises 1–10, verify the identities.

1. $\cos\theta + \tan\theta \sin\theta = \sec\theta$

2. $\dfrac{\sin x - \cos x}{\csc x - \sec x} = -\dfrac{1}{\csc x \sec x}$

3. $\sec^4\theta - \tan^4\theta = \dfrac{1 + \sin^2\theta}{\cos^2\theta}$

4. $\dfrac{\tan u + \cot u}{\tan u - \cot u} = \dfrac{-1}{\cos 2u}$

5. $\dfrac{1}{1 - \sin x} - \dfrac{1}{1 + \sin x} = 2\tan x \sec x$

6. $\sin\left(\theta - \dfrac{\pi}{2}\right) = -\cos\theta$

7. $\tan\left(x + \dfrac{5\pi}{4}\right) = \dfrac{1 + \tan x}{1 - \tan x}$

8. $\cos\left(\dfrac{2\pi}{3} - x\right) = \tfrac{1}{2}(\sqrt{3}\sin x - \cos x)$

9. $\sin 4x = 4\sin x \cos^3 x - 4\cos x \sin^3 x$

10. $\dfrac{\sin^2 2x}{1 + \cos 2x} = 2\sin^2 x$

In Exercises 11–16, find all solutions.

11. $2\sin x + 1 = 0$
12. $2\cos x \csc x = \cos x$
13. $3\sin x \cos x = \sin x$
14. $\tan^2 x + 5\tan x + 6 = 0$
15. $\cos 2x - \cos x = 0$
16. $\sin 2x - \tan x = 0$

17. If $\sin u = -\tfrac{5}{13}$ and $\cos v = \tfrac{4}{5}$, with u in Q_3 and v in Q_4, find $\sin(u + v)$, $\cos(u + v)$, and $\tan(u + v)$.

18. Use an addition formula to find the exact value of $\sin(7\pi/12)$.

In Exercises 19–23, solve the triangle ABC having the given parts.

19. $\alpha = 24.3°$, $\beta = 85.2°$, $c = 108.1$
20. $\gamma = \dfrac{\pi}{5}$, $a = 27.95$, $c = 15.98$
21. $\gamma = \dfrac{\pi}{5}$, $a = 27.95$, $c = 18.11$
22. $a = 4065$, $b = 4849$, $c = 6507$
23. $\alpha = .98$, $b = .926$, $c = .305$

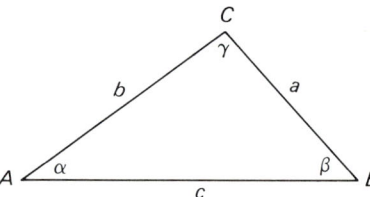

24. From one observer, the angle of elevation of a balloon is 38°. From an observer 1431 meters away on the opposite side, the angle of elevation of the balloon is 55°. How far is the balloon from each observer?

25. A plane leaves an airport and flies 83.65 miles on one course and then turns and flies 98.15 miles on a course of 26.2°. At that point, the plane then turns to a course of 167.9° in order to head back to the airport. How far is it from the airport?

26. A triangular wooden object is being built. Two sides of lengths 157.3 and 126.9 centimeters are to meet in an angle of 84°. How long should the third side be?

27. A golf ball is 186 yards from the hole. The ball is then hit (straight) 172.4 yards and ends up 31.3 yards from the hole. What is the angle between the line of flight of the ball and a direct line to the hole?

In Exercises 28 and 29, find the sum $\mathbf{u} + \mathbf{v}$.

28. $\mathbf{u}$ has magnitude 60, direction 150°; $\mathbf{v}$ has magnitude 40, direction 225°.
29. $\mathbf{u}$ has magnitude 117.51, direction 321.2°; $\mathbf{v}$ has magnitude 139.08, direction 84.25°.

30. A river is .5 kilometers wide and the current is 8 km/hr. Points A and B are directly opposite each other on the banks. If a boat propelled at 6 km/hr leaves A and is always parallel to $\overline{AB}$,
  a. What is its actual velocity?
  b. How far downriver from B will it land?
  c. How far will it travel?
  d. How long will it take it to land?

31. A small plane is flying at an airspeed of 150 mph with a heading of 247°. There is a wind of 35 mph in the direction of 132°. Find the course and ground speed of the plane.

32. There is a 50-mph wind in the direction of 346°. A pilot wishes to fly on a course of 321° with a ground speed of 250 mph. Determine the heading and airspeed.

Chapter Ten: Complex Numbers

Complex numbers first arose in trying to understand the Cardan-Tartaglia formulas for solving cubics. For example, Cardan (1501–1576) knew (for other reasons) that there was only one positive solution to the equation $x^3 = 15x + 4$ and that 4 worked by direct substitution. However, the formulas, which worked well for many equations, gave $x = \sqrt[3]{2 + \sqrt{-121}} + \sqrt[3]{2 - \sqrt{-121}}$ as that only positive solution. Historically, the ancient Greeks had denied the existence of negative numbers. By Cardan's time, negative numbers were on shaky ground, though made plausible by the number line. Cardan used negative numbers but called them "numeri ficti." Roots of negative numbers were "unimaginable." Several years later, an Italian algebraist Rafael Bombelli (1526–1573) had a "wild idea" and wrote the answer as $\sqrt[3]{2 + 11\sqrt{-1}} + \sqrt[3]{2 - 11\sqrt{-1}}$, and then he was able to show that this is $2 + \sqrt{-1} + 2 - \sqrt{-1} = 4$. However, $\sqrt{-1}$ was still regarded as imaginary for the next two centuries. In 1797, Caspar Wessel (1745–1818) invented the complex plane. This went virtually unnoticed until 30 years later when it was used extensively by Carl Friedrich Gauss (1777–1855). This geometric interpretation brought about the acceptance of roots of negative numbers, but the use of the word "imaginary" persists.

SECTION 10.1. THE COMPLEX NUMBERS

In most practical problems we encounter, we are interested in those solutions which are real numbers. However, there are situations where it is useful to have solutions even though they are not real numbers. For these situations, we define a number system larger than the reals which contains these solutions.

For example, consider the equation $x^2 = -1$. There is no real number whose square is -1, so this equation has no real solutions. Let us invent a solution and call it i, so that $i^2 = -1$. Thus, i is a solution to the equation $x^2 = -1$, and we would like i to be in some larger number system. This new number system must contain the reals $\mathbb{R}$ (in the same sense that the reals contain the rationals, and the rationals contain the integers) and the number i. Being a number system (containing the reals), it should possess the operations of addition and multiplication which must obey all the fundamental laws, such as the associative, commutative, and distributive laws. For example, since $i^2 = -1$,

$$(-i)^2 = (-i)(-i) = (-1)i(-1)i = (-1)(-1)ii = 1 \cdot i^2 = -1$$

so $-i$ is also a solution to $x^2 = -1$ in the new number system.

This new number system is called the system of **complex numbers**, and the set of all complex numbers is denoted by $\mathbb{C}$. Let us see what complex numbers look like. Suppose b is in $\mathbb{R}$; then it is also in $\mathbb{C}$ along with i so that $b \cdot i = bi$ is in $\mathbb{C}$. If a is another real number, a is also in $\mathbb{C}$ so $a + bi$ is in $\mathbb{C}$. Thus, $\mathbb{C}$ must contain all numbers of the form $a + bi$, a and b real. In fact, $\mathbb{C}$ is the set of all numbers of this form.

DEFINITION

The set of **complex numbers** is $\mathbb{C} = \{a + bi \mid a \text{ and } b \text{ are real numbers}\}$

The form $a + bi$ is called the **standard form** of a complex number.

Note that the set of all complex numbers $a + bi$ with $b = 0$ is the set of real numbers. Because of this, in a complex number $z = a + bi$, a is called the **real part** of z. In $z = a + bi$, b is called the **imaginary part** of z. (Note that b, *not* bi, is the imaginary part of $a + bi$.) If $a = 0$, z is called **pure imaginary**.

Before we can define operations like addition and multiplication of complex numbers, we need to understand what it means for two complex numbers to be equal.

DEFINITION OF EQUALITY OF COMPLEX NUMBERS

$a + bi = c + di$, with a, b, c, d real, if and only if $a = c$ and $b = d$.

This definition can be used to solve certain types of equations.

EXAMPLE 1 Solve for x and y real:
$$3x + 4i = -6 + 3yi$$

SOLUTION We recognize that this is an equation of the form
$$a + bi = c + di$$
where $a = 3x$, $b = 4$, $c = -6$, $d = 3y$. Since x and y are *real* numbers, a, b, c, and d are all real numbers. Using the definition of equality of complex numbers, we have
$$a = c \quad \text{and} \quad b = d$$
or
$$3x = -6 \quad \text{and} \quad 4 = 3y$$
Solving, we obtain $x = -2$ and $y = 4/3$.

We now turn to the problem of defining the sum and product of any two complex numbers. If two complex numbers are in standard form, we would like their sum and product also to be in standard form. We start with $z = a + bi$ and $w = c + di$, a, b, c, and d real, and try to add and multiply them as though they were real numbers but keeping in mind that $i^2 = -1$:

$$z + w = (a + bi) + (c + di) = a + c + bi + di = (a + c) + (b + d)i$$

$$\begin{aligned} z \cdot w = (a + bi) \cdot (c + di) &= ac + adi + bci + bdi^2 \\ &= ac + (ad + bc)i + bd(-1) \\ &= (ac - bd) + (ad + bc)i \end{aligned}$$

Thus we see that addition and multiplication can be defined in $\mathbb{C}$ as follows:

DEFINITION

If $z = a + bi$ and $w = c + di$, a, b, c, and d real, then
$$z + w = (a + c) + (b + d)i$$
$$z \cdot w = (ac - bd) + (ad + bc)i$$

However, this definition need not be memorized to do computations. In practice, to compute $z + w$ or $z \cdot w$, just proceed in the "natural" way, replacing i^2 by -1 when it occurs.

EXAMPLE 2 Find $(5 + 3i) + (-7 + 6i)$ and $(5 + 3i)(-7 + 6i)$.

SOLUTION $(5 + 3i) + (-7 + 6i) = 5 - 7 + 3i + 6i = -2 + 9i$ and $(5 + 3i)(-7 + 6i) = -35 + 30i - 21i + 18i^2 = -35 + 9i - 18 = -53 + 9i$.

We can also subtract in the "natural" way:
$$(5 + 3i) - (-7 + 6i) = 5 + 3i + 7 - 6i = 12 - 3i$$

One type of multiplication is special:
$$(a + bi)(a - bi) = a^2 - (bi)^2 = a^2 - b^2 i^2 = a^2 + b^2$$

The result $a^2 + b^2$ is zero if $a = b = 0$, but otherwise it is a *positive real* number. Thus the complex numbers $a + bi$ and $a - bi$ have a special relationship.

DEFINITION

If $z = a + bi$, we define the **complex conjugate** $\bar{z}$ of z to be the complex number $\bar{z} = a - bi$.

EXAMPLE 3 $\overline{5 + 3i} = 5 - 3i$; $\overline{5 - 3i} = 5 + 3i$; $\overline{-7 + 6i} = -7 - 6i$; $\overline{2 + 0i} = 2 - 0i = 2$.

Example 3 leads us to the observation that for $z = a + bi$, we have $z = \bar{z}$ (that is, $a + bi = a - bi$) if and only if $a = a$ and $b = -b$; i.e., $b = 0$. Thus

---

$z = \bar{z}$ if and only if z is a real number.

From the preceding, we see that if $z = a + bi$, then

$$z\bar{z} = (a + bi)(a - bi) = a^2 + b^2$$

Let us determine some further properties of conjugation. Suppose $z = a + bi$ and $w = c + di$. Then

$$\overline{z + w} = \overline{(a + bi) + (c + di)} = \overline{(a + c) + (b + d)i} = (a + c) - (b + d)i$$
$$= a + c - bi - di = (a - bi) + (c - di) = \bar{z} + \bar{w}$$

Similarly,

$$\overline{zw} = \overline{(a + bi)(c + di)} = \overline{(ac - bd) + (ad + bc)i} = (ac - bd) - (ad + bc)i$$
$$= [ac - (-b)(-d)] + [a(-d) + (-b)c]i = (a - bi)(c - di) = \bar{z}\bar{w}$$

Thus we have seen that two properties of complex conjugation are

$$\overline{z+w}=\bar{z}+\bar{w} \quad \text{and} \quad \overline{z\cdot w}=\bar{z}\cdot\bar{w}$$

EXAMPLE 4 $\overline{(5+3i)+(-7+6i)} = \overline{-2+9i} = -2-9i$, while
$\overline{(5+3i)}+\overline{(-7+6i)} = (5-3i)+(-7-6i) = -2-9i$.

EXAMPLE 5

$\overline{(5+3i)(-7+6i)} = \overline{(-35+30i-21i-18)} = \overline{-53+9i} = -53-9i$, while
$\overline{(5+3i)}\,\overline{(-7+6i)} = (5-3i)(-7-6i) = -35-30i+21i-18$
$= -53-9i$.

The complex conjugate is particularly useful when doing division of complex numbers. To divide $z = a + bi$ by a nonzero complex number $w = c + di$ and then put the answer in standard form, we merely multiply numerator and denominator of z/w by the complex conjugate $\bar{w}$ of the denominator:

$$\frac{z}{w} = \frac{z}{w}\cdot\frac{\bar{w}}{\bar{w}} = \frac{a+bi}{c+di}\cdot\frac{c-di}{c-di} = \frac{(a+bi)(c-di)}{c^2+d^2} = \frac{ac+bd}{c^2+d^2} + \frac{-ad+bc}{c^2+d^2}i$$

EXAMPLE 6 $\dfrac{5+3i}{-7+6i} = \dfrac{(5+3i)(-7-6i)}{(-7+6i)(-7-6i)} = \dfrac{-35-30i-21i+18}{(-7)^2+6^2}$

$= -\dfrac{17}{85} - \dfrac{51}{85}i = -\dfrac{1}{5} - \dfrac{3}{5}i$

Computing powers of i is an interesting problem. Starting with the first four powers of i, we see that

$$i^1 = i \qquad\qquad i^2 = -1$$
$$i^3 = i^2\cdot i = (-1)i = -i \qquad i^4 = i^2\cdot i^2 = (-1)(-1) = 1$$

Then the cycle starts again for the next four powers:

$$i^5 = i^4\cdot i = 1\cdot i = i, \qquad\qquad i^6 = i^4\cdot i^2 = 1\cdot(-1) = -1,$$
$$i^7 = i^4\cdot i^3 = 1\cdot(-i) = -i, \qquad i^8 = i^4\cdot i^4 = 1\cdot 1 = 1$$

This cycle repeats every four powers, so to compute i to any integral power, we simply split off all the powers of 4 in the exponent and compute what remains.

EXAMPLE 7 Find **(a)** i^{49}; **(b)** i^{91}; **(c)** $i^{487,123}$.

SOLUTION (a) $i^{49} = i^{48+1} = (i^4)^{12} \cdot i = 1^{12} \cdot i = i$.
(b) $i^{91} = i^{88+3} = (i^4)^{22} \cdot i^3 = 1^{22} \cdot (-i) = -i$.
(c) Dividing 487,123 by 4, we get 121,780.75. This means that 487,123 divided by 4 is 121,780 with a remainder of 3. That is, $487,123 = 4 \cdot 121,780 + 3$. Thus

$$i^{487,123} = i^{4 \cdot 121,780 + 3} = (i^4)^{121,780} i^3 = 1^{121,780} \cdot (-i) = -i$$

EXERCISES

In Exercises 1–48, simplify. Write the answer in the form $a + bi$.

1. $\overline{3 + i}$
2. $\overline{4 - 2i}$
3. $\overline{5}$
4. $\overline{2i}$
5. $(3 + 2i) + (4 + 5i)$
6. $(7 + 9i) + (3 - 8i)$
7. $(11 - 4i) + (4 - 3i)$
8. $(-2 - 7i) + (-5 + 15i)$
9. $(4 + 3i) + \overline{(-4 + 3i)}$
10. $(-9 - 2i) + \overline{(9 - 3i)}$
11. $(2 - 3i) - (-4 - 2i)$
12. $(18 - 2i) - (13 - 8i)$
13. $-(-3 + 8i) - \overline{(-7 + 6i)}$
14. $-\overline{(4 - 5i)} - (-7 - 9i)$
15. $5 - (2 - 3i)$
16. $-9 - (-7 + 2i)$
17. $7i - (3 + 9i)$
18. $(15 + 2i) - 8i$
19. $(-5 - \sqrt{-4}) - (3 + \sqrt{-9})$
20. $(14 + \sqrt{-25}) + (3 - \sqrt{-64})$
21. $i - (4 + \sqrt{-144})$
22. $(-11 - \sqrt{-121}) - (4 - 2i)$
23. $(2 - 3i)(4 + i)$
24. $(7 - i)(6 + 4i)$
25. $\overline{(1 + 2i)(4 - i)}$
26. $\overline{(7 + 4i)(5 - i)}$
27. $i(3 - i)$
28. $\overline{(3 - 2i)4i}$
29. $2(4 + 7i) + i(1 - i)$
30. $(3i + 1)(2i) + (2i)(3 - i)$
31. $(2 - 3i)^2$
32. $(7 + 2i)^2$
33. i^{17}
34. i^{30}
35. i^{3600}
36. $i^{51,271}$
37. $i^{29} + i^{21} + i$
38. $i^{4219} - i^{8121} + i^{3802}$
39. $\dfrac{2 + i}{3 + 2i}$
40. $\dfrac{4 - i}{2 - 5i}$
41. $\dfrac{1 + 3i}{4 - i}$
42. $\dfrac{1 - i}{1 + i}$
43. $\dfrac{1}{1 + 2i}$
44. $\dfrac{1}{3 - i}$
45. $\dfrac{4 + 3i}{i}$
46. $\dfrac{2 - i}{3i}$
47. $\dfrac{1}{i}$
48. $-\dfrac{3}{2i}$

In Exercises 49–58, solve for x and y real.

49. $4x + 7i = -8 + 3yi$
50. $7 - 2yi = 18x + 3i$
51. $(x + yi)i + 2 = 3x + y - 4i$
52. $(x - y) + (x + y)i = 4 - y - i$
53. $2^x + \frac{1}{9}i = 8 + (\log_3 y)i$
54. $\log 2 + 3e^{2x}i = 10^y + 6i$

55. $8^y + 27i = 2 + 9^x i$
56. $5 + (\log_4 y)i = \log_5 x + 2i$
57. $\sin x + \tfrac{1}{2}i = 1 + (\cos y)i$
58. $-1 + (\tan^{-1} x)i = \tan y + \sqrt{3}\, i$

SECTION 10.2. COMPLEX ROOTS OF EQUATIONS

In this section, we shall study equations in which some of the solutions are complex numbers. We begin with linear equations. Linear equations cannot have nonreal solutions unless they have nonreal coefficients. We shall see that this is not true for higher-degree equations.

Recall that the general strategy for solving a linear equation in x is to do the following:

1. Get all the x's on one side and everything else on the other.
2. Combine similar terms.
3. Divide by the coefficient of x, and simplify.

This same general strategy works for solving a linear equation involving complex numbers.

EXAMPLE 1 Solve for x complex, putting the answer in the form $a + bi$:

$$(2 - 3i)x + 4 - i = (6 + 3i)x + 7 + 4i$$

SOLUTION First, get all the x's on one side and everything else on the other:

$$(2 - 3i)x - (6 + 3i)x = -4 + i + 7 + 4i$$

Next, combine similar terms:

$$(-4 - 6i)x = 3 + 5i$$

Now, divide by the coefficient of x:

$$x = \frac{3 + 5i}{-4 - 6i}$$

This is the answer, but we want it expressed in the form $a + bi$. So we carry out the division as in the previous section, by multiplying top and bottom by the conjugate of the denominator:

$$x = \frac{3 + 5i}{-4 - 6i} \cdot \frac{-4 + 6i}{-4 + 6i} = \frac{-12 - 2i + 30i^2}{16 + 36}$$

$$= \frac{-42 - 2i}{52} = -\frac{21}{26} - \frac{1}{26}i$$

We now turn to quadratic equations and begin with the equation $x^2 = d$. Recall that if d is a positive real number, the equation $x^2 = d$ has two solutions, $x = +\sqrt{d}$ and $x = -\sqrt{d}$. The symbol $\sqrt{d}$ represents $+\sqrt{d}$ and is called the **principal square root** of the positive real number d.

Now consider the equation $x^2 = -d$, where d is again a positive real number. This equation has no real solutions, but it does have two complex solutions, $x = \sqrt{d}\,i$ and $x = -\sqrt{d}\,i$. To see this, first observe that

$$(x - \sqrt{d}\,i)(x + \sqrt{d}\,i) = x^2 - (\sqrt{d})^2(i)^2 = x^2 - d(-1) = x^2 + d$$

Thus we can solve $x^2 = -d$ by factoring:

$$x^2 + d = 0, \quad (x + \sqrt{d}\,i)(x - \sqrt{d}\,i) = 0$$
$$x = \sqrt{d}\,i \quad \text{or} \quad x = -\sqrt{d}\,i$$

Therefore, in the same way that we solve $x^2 = d$, $d > 0$, by writing $x = \pm\sqrt{d}$, we can now solve $x^2 = -d$, $d > 0$, by writing $x = \pm\sqrt{d}\,i$.

EXAMPLE 2 Solve **(a)** $x^2 = -16$ and **(b)** $x^2 = -17$.

SOLUTION **(a)** $x^2 = -16$, $x = \pm\sqrt{16}\,i$, $x = \pm 4i$.
(b) $x^2 = -17$, $x = \pm\sqrt{17}\,i$, $x \approx \pm 4.12311i$.

We also extend the definition of principal root, following the pattern in the real case:

DEFINITION

If d is a positive real number, then the **principal square root** of $-d$ is denoted by $\sqrt{-d}$ and is defined to be the complex number $\sqrt{d}\,i$.

Warning: If a and b are positive real numbers, $\sqrt{-a}\,\sqrt{-b}$ is *not* the same as $\sqrt{(-a)(-b)}$.

EXAMPLE 3 Simplify **(a)** $\sqrt{-4}\,\sqrt{-9}$ and **(b)** $\sqrt{(-4)(-9)}$.

SOLUTION **(a)** $\sqrt{-4}\,\sqrt{-9} = (2i)(3i) = 6i^2 = -6$. **(b)** $\sqrt{(-4)(-9)} = \sqrt{36} = +6$. Hence $\sqrt{-4}\,\sqrt{-9} \neq \sqrt{(-4)(-9)}$. To avoid errors, always convert expressions involving radicals to the form $a + bi$ before any algebraic manipulations.

EXAMPLE 4 Simplify $(2 - \sqrt{-75})(-3 + \sqrt{-3})$.

SOLUTION $(2 - \sqrt{-75})(-3 + \sqrt{-3}) = (2 - 5\sqrt{3}\,i)(-3 + \sqrt{3}\,i)$

$$= -6 + 17\sqrt{3}\,i - 5(\sqrt{3})^2 i^2$$

$$= -6 + 17\sqrt{3}\,i - 15(-1) = 9 + 17\sqrt{3}\,i$$

We are now ready to solve the general quadratic equation $ax^2 + bx + c = 0$, $a \neq 0$, a, b, c real numbers. The solutions are given by the quadratic formula:

$$x = \frac{-b + \sqrt{b^2 - 4ac}}{2a} \quad \text{and} \quad x = \frac{-b - \sqrt{b^2 - 4ac}}{2a}$$

When $b^2 - 4ac$ is less than zero, the equation $ax^2 + bx + c = 0$ has no real roots. But since we now know square roots of negative real numbers are complex numbers, we see that the quadratic formula gives us two complex solutions. The formula can be evaluated by the methods already given in this section.

EXAMPLE 5 Solve $3x^2 + 2x + 1 = 0$.

SOLUTION By the quadratic formula we have

$$x = \frac{-2 \pm \sqrt{2^2 - 4(3)(1)}}{2(3)} = \frac{-2 \pm \sqrt{-8}}{6}$$

$$= \frac{-2 \pm 2\sqrt{2}\, i}{6} = -\frac{1}{3} \pm \frac{\sqrt{2}}{3} i$$

Remark. These answers can be verified by direct substitution. See Exercise 22.

Note that the two roots in Example 5 are conjugates of each other. This illustrates a general principle concerning complex roots. Suppose you are solving an equation of the form

$$a_n x^n + a_{n-1} x^{n-1} + \cdots + a_1 x + a_0 = 0, \quad a_n \neq 0$$

It follows from the Fundamental Theorem of Algebra that this equation has exactly n roots. If all the coefficients are *real* numbers, then whenever one complex number is a root, its conjugate is also a root. Thus there are always an even number of roots which are nonreal, and they can be grouped in pairs, each with its conjugate. This general principle will be illustrated in some of the following examples.

Sometimes higher-degree equations can be solved by factoring the expressions into linear and quadratic factors and handling these individually.

EXAMPLE 6 Solve $x^3 + 1 = 0$.

SOLUTION $x^3 + 1 = 0$. Factoring, we obtain $(x + 1)(x^2 - x + 1) = 0$. Thus, $x + 1 = 0$ or $x^2 - x + 1 = 0$. Solving, we obtain

$$x = -1 \quad \text{or} \quad x = \frac{-(-1) \pm \sqrt{(-1)^2 - 4(1)(1)}}{2} = \frac{1}{2} \pm \frac{\sqrt{3}}{2} i$$

Therefore the solutions to $x^3 + 1 = 0$ are

$$x = -1, \quad \frac{1}{2} + \frac{\sqrt{3}}{2}i, \quad \frac{1}{2} - \frac{\sqrt{3}}{2}i$$

Note that the two nonreal roots are conjugates.

The quadratic formula can also be used to solve quadratic equations with complex coefficients.

EXAMPLE 7 Solve $2x^2 + 3ix + 2 = 0$.

SOLUTION By the quadratic formula we have

$$x = \frac{-3i \pm \sqrt{(3i)^2 - 4(2)(2)}}{2(2)} = \frac{-3i \pm \sqrt{-25}}{4}$$

$$= \frac{-3i \pm 5i}{4} = -2i, \frac{1}{2}i$$

Note that this example shows that when some of the coefficients are not real, it is not necessary for conjugates of roots to be roots.

Equations in which $b^2 - 4ac$ is a complex number can also be solved. However, we must put off this discussion until the next two sections, where we develop techniques for finding square roots of complex numbers.

If you start with several complex numbers, it is straightforward to construct an equation with these numbers as roots. You simply multiply the appropriate linear factors together.

EXAMPLE 8 Find an equation with roots $1, -i, 2i$.

SOLUTION First form $(x - 1)(x + i)(x - 2i) = 0$ and then multiply the factors together:

$$(x - 1)(x^2 - ix + 2) = 0$$

$$x^3 + (-1 - i)x^2 + (2 + i)x - 2 = 0$$

The interesting observation to make is that if all the complex roots appear in conjugate pairs, then the resulting equation has all real coefficients. In such situations, when multiplying the factors together, first multiply together the factors with conjugate roots, and do this by regrouping as illustrated in Example 9.

EXAMPLE 9 Find an equation with roots $-2, 1+i, 1-i$.

SOLUTION
$$(x+2)[x-(1+i)][x-(1-i)] = 0$$
$$(x+2)[(x-1)-i][(x-1)+i] = 0$$
$$(x+2)[(x-1)^2 - i^2] = 0$$
$$(x+2)(x^2 - 2x + 1 + 1) = 0$$
$$(x+2)(x^2 - 2x + 2) = 0$$
$$x^3 - 2x + 4 = 0$$

EXERCISES

In Exercises 1–20, simplify. Write the answer in the form $a + bi$.

1. $\sqrt{-4}$
2. $-\sqrt{-16}$
3. $-\sqrt{-20}$
4. $\sqrt{-1200}$
5. $\sqrt{-25} + \sqrt{-36}$
6. $\sqrt{-16} + \sqrt{-4}$
7. $\sqrt{-100}\sqrt{-81}$
8. $\sqrt{-144}\sqrt{-9}$
9. $-\sqrt{-12}\sqrt{-75}$
10. $-\sqrt{-20}\sqrt{-45}$
11. $\dfrac{\sqrt{-169}}{\sqrt{-4}}$
12. $\dfrac{-\sqrt{-36}}{\sqrt{-4}}$
13. $(3 + \sqrt{-4})(2 - \sqrt{-9})$
14. $(-7 - \sqrt{-9})(8 + \sqrt{-25})$
15. $(2 - \sqrt{-12})(5 - \sqrt{-75})$
16. $(-4 + \sqrt{-45})(-3 + \sqrt{-80})$
17. $\sqrt{-4}(5 - \sqrt{-9})$
18. $\sqrt{-8}(\sqrt{-48} + \sqrt{40})$
19. $(\sqrt{-8})^3$
20. $(\sqrt{-12})^3 + \sqrt{-12})^5$

In Exercises 21–24, show by direct substitution that the answers satisfy the given equation.

21. $x^2 + x + 1 = 0,\ x = -\frac{1}{2} \pm \frac{1}{2}\sqrt{3}\ i$
22. $3x^2 + 2x + 1 = 0,\ x = -\frac{1}{3} \pm \frac{1}{3}\sqrt{2}\ i$
23. $2x^2 + 3ix + 2 = 0,\ x = -2i, \frac{1}{2}i$
24. $x^3 + 1 = 0,\ x = -1, \frac{1}{2} \pm \frac{1}{2}\sqrt{3}\ i$

In Exercises 25–64, solve the given equation, putting all answers in the form $a + bi$.

25. $3ix + 4 = 5$
26. $48x - 1 = 6ix$
27. $(2 + i)x + 2i = 7i$
28. $(2 - 2i)x + 4 = -6$
29. $4ix - 5 = 0$
30. $(7 + i)x - 8 = (7 + 3i)x$
31. $(-1 - 2i)x - 4i = 5 + i$
32. $(4 - i)x - 6 = 8 + 3i$
33. $6ix + 2 - i = (3 + 2i)x - 4i$
34. $(2 - i)x + 3 = 4 - i - (3 + 2i)x$
35. $2x - 3i = ix - 4$
36. $ix - 4 = 3x + 2i$
37. $x^2 - 2x + 2 = 0$
38. $x^2 + 2x + 5 = 0$
39. $2x^2 - 3x + 2 = 0$
40. $3x^2 - x + 2 = 0$
41. $3x^2 + 4x + 2 = 0$
42. $-2x^2 + x - 4 = 0$
43. $x^3 - 1 = 0$
44. $x^3 + 8 = 0$
45. $x^3 - 125 = 0$
46. $x^3 + 64 = 0$
47. $x^4 - 16 = 0$
48. $x^4 - \frac{1}{81} = 0$

49. $x^6 - 64 = 0$
50. $x^3 + \frac{1}{8} = 0$
51. $2x^4 + 3x^2 + 1 = 0$
52. $4x^4 + 21x^2 + 27 = 0$
53. $x^4 + x^3 + x^2 = 0$
54. $9x^4 + 13x^2 + 4 = 0$
55. $x^2 + ix + 2 = 0$
56. $2x^2 - ix + 3 = 0$
57. $x^2 + (1 - 2i)x - i = 0$
58. $3x^2 + (2 + 3i)x + i = 0$
59. $2x^2 + (2 - 2i)x - i = 0$
60. $9x^2 + (3 + 3i)x + \frac{1}{2}i = 0$
61. $.21x^2 - .35x + .49 = 0$
62. $813.2x^2 + 141.7x + 731.3 = 0$
63. $4.191x^2 - 3.198x + 1.092 = 0$
64. $x^3 + 5.128 = 0$

In Exercises 65–76, find an equation having the given numbers as roots.

65. $2i, -2i$
66. $3 + i, 3 - i$
67. $-2 + 3i, -2 - 3i$
68. $\frac{1}{3} - \frac{1}{2}i, \frac{1}{3} + \frac{1}{2}i$
69. $\sqrt{3}\,i, -\sqrt{3}\,i$
70. $1 + \sqrt{2}\,i, 1 - \sqrt{2}\,i$
71. $1, i, -3i$
72. $4, -2i, -i$
73. $-1, 1 - i, 1 + i$
74. $3, -3 + i, -3 - i$
75. $i, 2 + i, 2 - i$
76. $-i, 1 - 2i, 1 + 2i$

SECTION 10.3. TRIGONOMETRIC FORM OF COMPLEX NUMBERS

We have previously said that two complex numbers $a + bi$ and $c + di$ were equal if and only if $a = c$ and $b = d$. Similarly, two ordered pairs (a, b) and (c, d) are equal if and only if $a = c$ and $b = d$. Thus it is natural to associate the complex number $a + bi$ with the ordered pair (a, b). This correspondence is *one-to-one*. That is, each complex number corresponds to exactly one ordered pair of real numbers and vice versa.

Now given an ordered pair (a, b) of real numbers, we normally think of that ordered pair as representing the point in the plane which has coordinates $x = a$ and $y = b$. Thus using the preceding correspondence, we may view each complex number $a + bi$ geometrically as the point (a, b) in the Cartesian plane. See Fig. 1.

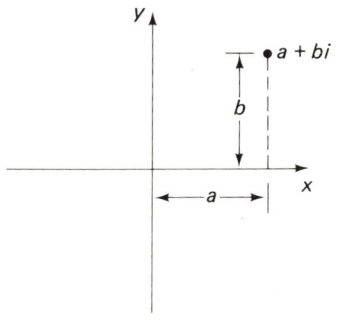

FIG. 1

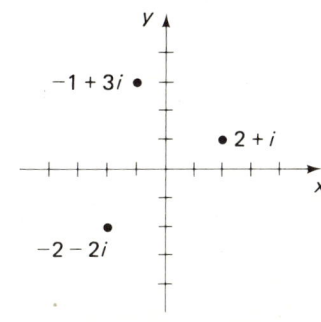

FIG. 2

EXAMPLE 1 Represent $2 + i$, $-1 + 3i$, and $-2 - 2i$ geometrically.

SOLUTION See Fig. 2.

When the plane is viewed this way, it is usually referred to as the **complex plane**. Since the numbers $a + 0i$, i.e., the real numbers, are the points on the x-axis, the x-axis is called the **real axis**. Similarly, the y-axis is called the **imaginary axis** since points there correspond to pure imaginary numbers.

In the complex plane, it is sometimes useful to view a complex number $z = a + bi$ as a vector from the origin to (a, b). For instance, if $z = -4 + i$ and $w = 3 + 2i$, then $z + w = -1 + 3i$. Now if we consider z and w as vectors **z** and **w** and add them by vector addition (using parallelograms), we see that adding z and w as vectors coincides with adding them as complex numbers. See Fig. 3. This illustrates the following general principle:

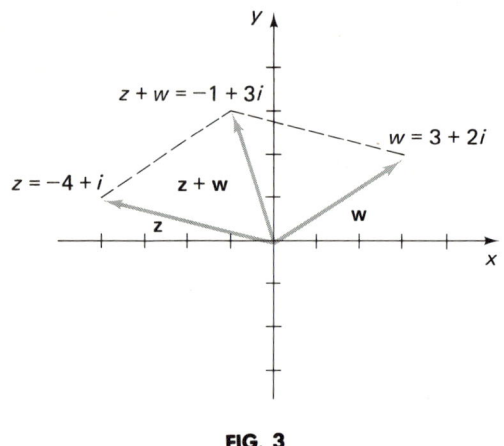

FIG. 3

Addition of complex numbers may be described geometrically as addition of vectors.

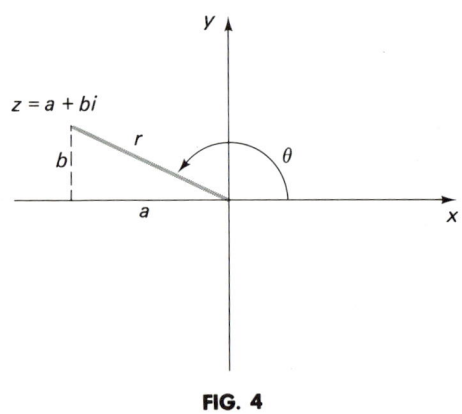

FIG. 4

To get a nice geometric description of multiplication of complex numbers, we look at a different form of complex numbers, obtained as follows. Suppose $z = a + bi$ is plotted in the complex plane. Draw a line from the point (a, b) to the origin. See Fig. 4. The length of this line is $r = \sqrt{a^2 + b^2}$, the distance from the point to the origin. Let θ be any angle in standard position for which (a, b) is on the terminal side. Then by the definition of the trigonometric functions, $\cos \theta = a/r$ and $\sin \theta = b/r$. It follows that $a = r \cos \theta$ and $b = r \sin \theta$. Therefore, $z = a + bi = r \cos \theta + (r \sin \theta)i$, or

$$z = r(\cos \theta + i \sin \theta)$$

Note: We write $i \sin \theta$ instead of $(\sin \theta)i$ to eliminate parentheses and because writing $\sin \theta i$ makes it appear that the angle is θi.

The expression $r(\cos\theta + i\sin\theta)$ is called the **trigonometric form** or **polar form** of z. [Those students who have had polar coordinates will recognize r and θ as the polar coordinates of (a,b).] It is *not* unique, since θ is not unique. For example,

$$1 + i = \sqrt{2}\left(\cos\tfrac{1}{4}\pi + i\sin\tfrac{1}{4}\pi\right) = \sqrt{2}\left(\cos\tfrac{9}{4}\pi + i\sin\tfrac{9}{4}\pi\right) = \cdots$$

In addition, if $z = 0$, then $r = 0$, and θ can be *any* angle. If $z = r(\cos\theta + i\sin\theta)$, the angle θ is called the **argument** of z, and r is called the **absolute value** (or **modulus**) of z. Just as in the reals, the absolute value of z is the distance between z and the origin and is denoted by $|z|$. Since $r = \sqrt{a^2 + b^2}$ and $z\bar{z} = a^2 + b^2$,

$$|z| = \sqrt{z\bar{z}}$$

Since the expression $\cos\theta + i\sin\theta$ is somewhat cumbersome to write, we usually shorten this to cis θ, the abbreviation *cis* being obtained from $\underline{c}os\,\theta + \underline{i}\,\underline{s}in\,\theta$.

NOTATION:

$$\text{cis }\theta = \cos\theta + i\sin\theta.$$

Thus the trigonometric form is abbreviated by

$$z = r(\cos\theta + i\sin\theta) = r\,\text{cis}\,\theta$$

In the first few examples, we shall write complex numbers in the trigonometric form both as $r(\cos\theta + i\sin\theta)$ and as the shortened version $r\,\text{cis}\,\theta$. After that, we shall use the longer form only where necessary for clarity.

EXAMPLE 2 Put the numbers $z_1 = 1 + i$, $z_2 = \sqrt{3} - i$, and $z_3 = -3 + 4i$ in trigonometric form.

SOLUTION

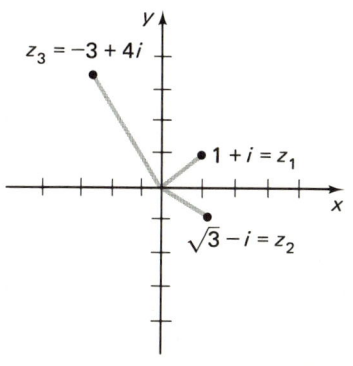

FIG. 5

See Fig. 5. For z_1: $r = \sqrt{1 + 1} = \sqrt{2}$; $\cos\theta = 1/\sqrt{2}$, and $\sin\theta = 1/\sqrt{2}$. One such angle θ is $\theta = 45°$. Thus, $z_1 = \sqrt{2}(\cos 45° + i\sin 45°) = \sqrt{2}\,\text{cis}\,45°$.

For z_2: $r = \sqrt{(\sqrt{3})^2 + (-1)^2} = \sqrt{4} = 2$; $\cos\theta = \sqrt{3}/2$, and $\sin\theta = -\tfrac{1}{2}$. Choose $\theta = -\pi/6$. Then $z_2 = 2[\cos(-\pi/6) + i\sin(-\pi/6)] = 2\,\text{cis}(-\pi/6)$.

For z_3: $r = \sqrt{(-3)^2 + 4^2} = \sqrt{25} = 5$; $\cos\theta = -3/5$, and $\sin\theta = \frac{4}{5}$. One such angle is $\theta = \cos^{-1}(-3/5) \approx 126.869°$. Thus, $z_3 \approx 5(\cos 126.869° + i\sin 126.869°) = 5\operatorname{cis} 126.869°$.

The virtue of the trigonometric form of complex numbers is that it allows us to simplify multiplication and division. Suppose z and w are two complex numbers in trigonometric form, $z = r\operatorname{cis}\theta = r(\cos\theta + i\sin\theta)$ and $w = s\operatorname{cis}\varphi = s(\cos\varphi + i\sin\varphi)$. Then, using the rules of multiplication of complex numbers, we obtain

$$zw = rs(\cos\theta + i\sin\theta)(\cos\varphi + i\sin\varphi)$$
$$= rs\big[(\cos\theta\cos\varphi - \sin\theta\sin\varphi) + (\sin\theta\cos\varphi + \cos\theta\sin\varphi)i\big]$$

The quantities in parentheses can be simplified using the addition formulas:

$$\cos\theta\cos\varphi - \sin\theta\sin\varphi = \cos(\theta + \varphi)$$
$$\sin\theta\cos\varphi + \cos\theta\sin\varphi = \sin(\theta + \varphi)$$

Therefore, $zw = rs[\cos(\theta + \varphi) + i\sin(\theta + \varphi)] = rs\operatorname{cis}(\theta + \varphi)$, a number again in trigonometric form.

> **Multiplication of complex numbers in trigonometric form:**
> $$(r\operatorname{cis}\theta)(s\operatorname{cis}\varphi) = rs\operatorname{cis}(\theta + \varphi)$$

What this says is *when multiplying complex numbers, multiply the absolute values and add the arguments.*

EXAMPLE 3 $(3\operatorname{cis} 41°)(2\operatorname{cis} 197°) = 6\operatorname{cis} 238°$. See Fig. 6(a).

EXAMPLE 4 $(3\operatorname{cis} 108°)^2 = 9\operatorname{cis} 216°$. See Fig. 6(b).

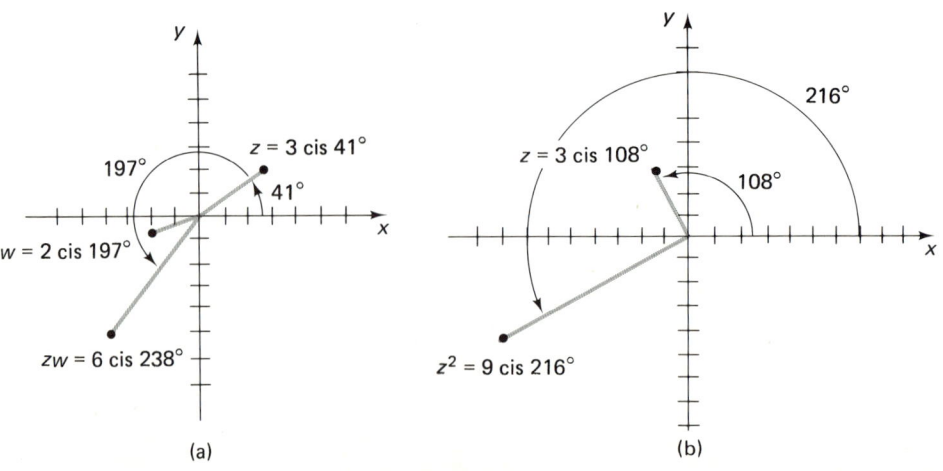

FIG. 6

The rule for inverses is quite interesting. If $z = r \operatorname{cis} \theta = r(\cos \theta + i \sin \theta)$,

$$\frac{1}{z} = \frac{1}{r(\cos\theta + i\sin\theta)}$$
$$= \frac{1}{r} \frac{1}{\cos\theta + i\sin\theta} \frac{\cos\theta - i\sin\theta}{\cos\theta - i\sin\theta}$$
$$= \frac{1}{r} \frac{\cos\theta - i\sin\theta}{\cos^2\theta + \sin^2\theta}$$
$$= r^{-1}[\cos(-\theta) + i\sin(-\theta)] = r^{-1}\operatorname{cis}(-\theta)$$

The last line follows from the identities $\sin^2\theta + \cos^2\theta = 1$, $\cos(-\theta) = \cos\theta$, and $\sin(-\theta) = -\sin\theta$. Hence we have shown the following:

The reciprocal of complex numbers in trigonometric form:
$$(r\operatorname{cis}\theta)^{-1} = r^{-1}\operatorname{cis}(-\theta)$$

What this says is *to find the reciprocal of a complex number, you find the reciprocal of its absolute value and the negative of its argument.*

EXAMPLE 5 $(.5 \operatorname{cis} 40°)^{-1} = 2 \operatorname{cis}(-40°)$. See Fig. 7(a).

EXAMPLE 6 $[3\operatorname{cis}(-170°)]^{-1} = \frac{1}{3}\operatorname{cis} 170°$. See Fig. 7(b).

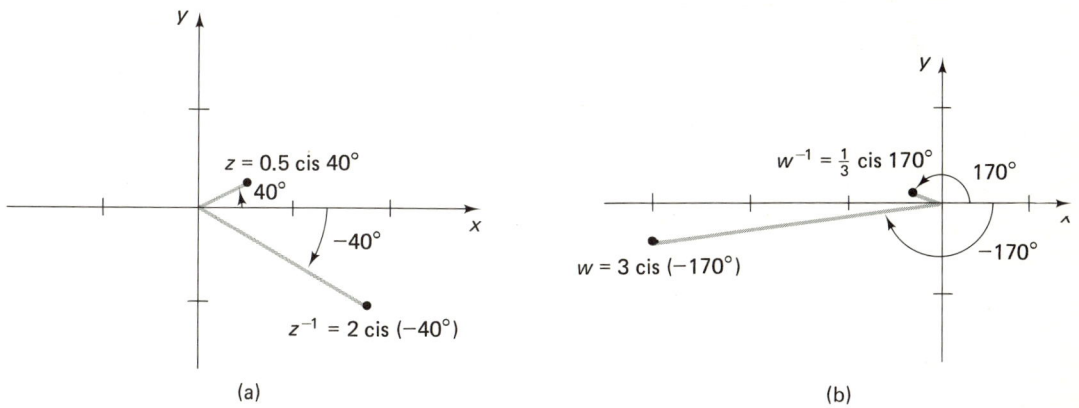

FIG. 7

Now using the multiplication and inverse rules, we see that if $z = r\operatorname{cis}\theta$, $w = s\operatorname{cis}\varphi$,

$$\frac{z}{w} = zw^{-1} = (r\operatorname{cis}\theta)\left[\frac{1}{s}\operatorname{cis}(-\varphi)\right]$$
$$= \frac{r}{s}\operatorname{cis}(\theta - \varphi)$$

Therefore,

> **Division of complex numbers in trigonometric form:**
>
> $$\frac{r \operatorname{cis} \theta}{s \operatorname{cis} \varphi} = \frac{r}{s} \operatorname{cis}(\theta - \varphi)$$

What this says is *to divide complex numbers, you divide the absolute values and subtract the arguments.*

EXAMPLE 7

$$\frac{8 \operatorname{cis} 41°}{2 \operatorname{cis} 147°} = \frac{8}{2} \operatorname{cis}(41° - 147°) = 4 \operatorname{cis}(-106°)$$

See Fig. 8.

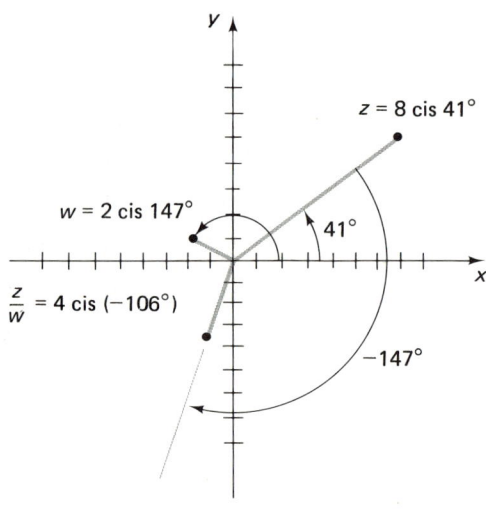

FIG. 8

EXAMPLE 8

$$\frac{4 \operatorname{cis} 3.17}{8 \operatorname{cis} 1.49} = .5 \operatorname{cis} 1.68$$

EXERCISES

In Exercises 1–5, plot each complex number, and then find its absolute value and put it into trigonometric form (long and short version).

1. $z_1 = 1 + i$, $z_2 = -1 + i$, $z_3 = -1 - i$, $z_4 = 1 - i$
2. $z_1 = 1 + \sqrt{3}\,i$, $z_2 = -1 + \sqrt{3}\,i$, $z_3 = -1 - \sqrt{3}\,i$, $z_4 = 1 - \sqrt{3}\,i$
3. $z_1 = \sqrt{3} + i$, $z_2 = -\sqrt{3} + i$, $z_3 = -\sqrt{3} - i$, $z_4 = \sqrt{3} - i$
4. $z_1 = 2 - 2i$, $z_2 = 4 - 4i$

5. $z_1 = 6 + 2i$, $z_2 = -3 + i$, $z_3 = -9 + 3i$, $z_4 = 6 - 2i$

In Exercises 6–10, change each complex number from trigonometric form to the form $a + bi$.

6. 3 cis 120°
7. 6 cis 210°
8. 2 cis 27°
9. 2.48 cis 1.73
10. $8 \text{ cis } \dfrac{7\pi}{4}$

In Exercises 11–14, multiply. Sketch the problem graphically.

11. (3 cis 98°)(6.4 cis 321°)
12. $\left(81.7 \text{ cis } \dfrac{\pi}{5}\right)\left(.031 \text{ cis } \dfrac{\pi}{3}\right)$
13. (11.516 cis 4.32)(64.3 cis 5.8)
14. (3.914 cis 21°42′)(13.47 cis 261.3°)

In Exercises 15–18, find the inverse of the given number. Sketch.

15. 2 cis 70°
16. 1.5 cis(5π/6)
17. .6 cis(−π/5)
18. 1.2 cis(−160°)

In Exercises 19–22, divide. Sketch.

19. $\dfrac{6.42 \text{ cis } 284°}{3 \text{ cis } 107°}$
20. $\dfrac{3 \text{ cis } 107°}{6.42 \text{ cis } 284°}$
21. $\dfrac{61.28 \text{ cis}(2\pi/3)}{46.5 \text{ cis}(4\pi/5)}$
22. $\dfrac{4916 \text{ cis } .71}{739 \text{ cis } 2.03}$

23. Compute **(a)** $(1 + i)(\sqrt{3} - i)(-4 + 4i)(-1 - \sqrt{3}i)$ and **(b)** $(1 + i)(-1 - \sqrt{3}i)/(\sqrt{3} - i)(-4 + 4i)$ directly. Then change each of the four complex numbers to trigonometric form, and recompute the answers.

24. Show that cis θ = cis($\theta + 2n\pi$), n an integer.

SECTION 10.4. POWERS AND ROOTS OF COMPLEX NUMBERS

Having a complex number in trigonometric form is particularly useful if you want to raise the number to an integer power. Let us look at some small powers of $r \text{ cis } \theta$:

$$(r \text{ cis } \theta)^2 = (r \text{ cis } \theta)(r \text{ cis } \theta)$$
$$= r \cdot r \text{ cis}(\theta + \theta)$$
$$= r^2 \text{ cis } 2\theta$$
$$(r \text{ cis } \theta)^3 = (r \text{ cis } \theta)^2 (r \text{ cis } \theta)$$
$$= (r^2 \text{ cis } 2\theta)(r \text{ cis } \theta)$$
$$= r^2 \cdot r \text{ cis}(2\theta + \theta)$$
$$= r^3 \text{ cis } 3\theta$$
$$(r \text{ cis } \theta)^4 = (r \text{ cis } \theta)^3 (r \text{ cis } \theta)$$
$$= (r^3 \text{ cis } 3\theta)(r \text{ cis } \theta)$$
$$= r^4 \text{ cis } 4\theta$$

These examples are following a very simple pattern:

---
(1)   If n is a positive integer,
$$(r \operatorname{cis} \theta)^n = r^n \operatorname{cis} n\theta$$

This formula can be shown to be true for all positive integers n by applying the principle of mathematical induction. We now extend this to all integers.

To be consistent with the rules for powers of real numbers, we define any nonzero complex number raised to the zero power to be 1:

DEFINITION
$$(a+bi)^0 = 1 \quad \text{if } a+bi \neq 0.$$

Since the trigonometric form of 1 is $1(\cos 0 + i \sin 0)$, formula (1) holds with $n = 0$ whenever $r(\cos \theta + i \sin \theta) \neq 0$, i.e., whenever $r \neq 0$:

$$(r \operatorname{cis} \theta)^0 = r^0 \operatorname{cis} 0\theta$$
$$= r^0(\cos 0\theta + i \sin 0\theta)$$
$$= 1(\cos 0 + i \sin 0) = 1$$

Let us look at negative integer powers. From the previous section, we know that

$$(r \operatorname{cis} \theta)^{-1} = r^{-1} \operatorname{cis}(-\theta) = r^{-1} \operatorname{cis}[(-1)\theta]$$

More generally, if n is any *positive* integer and (1) holds,

$$(r \operatorname{cis} \theta)^{-n} = \left[(r \operatorname{cis} \theta)^{-1}\right]^n$$
$$= \left[r^{-1} \operatorname{cis}(-\theta)\right]^n$$
$$= (r^{-1})^n \operatorname{cis}[n(-\theta)] \quad [\text{by (1)}]$$
$$= r^{-n} \operatorname{cis}[(-n)\theta]$$

Thus we have the following formula, called **De Moivre's Theorem**:

DE MOIVRE'S THEOREM

For any integer n, $(r \operatorname{cis} \theta)^n = r^n \operatorname{cis} n\theta$

EXAMPLE 1

$$(2 \operatorname{cis} 10°)^5 = 2^5 \operatorname{cis} 50°$$
$$(2 \operatorname{cis} 10°)^{-7} = 2^{-7} \operatorname{cis}(-70°)$$
$$\left(3.2 \operatorname{cis} \frac{3\pi}{13}\right)^{10} = (3.2)^{10} \operatorname{cis} \frac{30\pi}{13}$$

The advantage of De Moivre's Theorem becomes apparent if one tries to calculate powers of complex numbers directly. It would take a lot of work to compute $(-\sqrt{3} + i)^{53}$ directly. However, $-\sqrt{3} + i$ can be put in trigonometric form, $-\sqrt{3} + i = 2 \text{ cis } 150°$. Thus

$$\begin{aligned}(-\sqrt{3} + i)^{53} &= (2 \text{ cis } 150°)^{53} \\ &= 2^{53} \text{ cis}[53(150°)] \\ &= 2^{53} \text{ cis } 7950° \\ &= 2^{53} \text{ cis } 30°, \quad \text{since } 7950° = 30° + 22(360°) \\ &= 2^{53}(\cos 30° + i \sin 30°) \\ &= 2^{53}\left(\frac{\sqrt{3}}{2} + \frac{1}{2}i\right) = 2^{52}\sqrt{3} + 2^{52}i\end{aligned}$$

In the preceding example, we could go from

$$\cos 30° + i \sin 30° \quad \text{to} \quad \frac{\sqrt{3}}{2} + \frac{1}{2}i$$

without using a calculator. It is usually better to avoid using a calculator in such situations, because the answer is in a more compact form that can be handled more easily in later computations. However, there are situations which require the use of a calculator.

EXAMPLE 2 Find $(1.102 - \sqrt[3]{2}\, i)^{12}$.

SOLUTION First put $z = 1.102 - \sqrt[3]{2}\, i$ in trigonometric form:

$$|z| = r = \sqrt{(1.102)^2 + (\sqrt[3]{2})^2} \approx 1.6738585$$

Since $\cos \theta = 1.102/r$ and $\sin \theta = -\sqrt[3]{2}/r$, θ is in Q_4 with a reference angle of $\cos^{-1}(1.102/r) \approx .852160$ rad; i.e., $\theta \approx -.852160$. Therefore

$$\begin{aligned}z^{12} &\approx [1.6738585 \text{ cis}(-.852160)]^{12} \\ &\approx (1.6738585)^{12} \text{ cis}[12(-.852160)] \\ &\approx 483.754[\cos(-10.2259) + i \sin(-10.2259)] \\ &\approx 336.638 + 347.409i\end{aligned}$$

We started our discussion of the complex numbers by stating that we wanted to be able to solve the equation $x^2 + 1 = 0$. Not only can this equation be solved in the complex numbers, but so can any polynomial equation with complex coefficients. This is not obvious, and the proof requires some rather deep mathematics. Moreover, it follows from the Fundamental Theorem of Algebra that any polynomial equation of degree n has precisely n roots. One such equation is $x^n - a = 0$, $a \neq 0$, or, equivalently,

(2) $$x^n = s \text{ cis } \varphi, \quad s \neq 0$$

where $a = s \operatorname{cis} \varphi$ is expressed in polar form. Now we shall see how De Moivre's Theorem can be employed to find the n solutions to equation (2), i.e., to find the n distinct nth roots of the complex number $a = s \operatorname{cis} \varphi$.

Suppose $r \operatorname{cis} \theta$ is one solution to (2). Then

$$(r \operatorname{cis} \theta)^n = s \operatorname{cis} \varphi \quad \text{or} \quad r^n \operatorname{cis} n\theta = s \operatorname{cis} \varphi$$

Written out in the nonabbreviated trigonometric form, this is

$$r^n(\cos n\theta + i \sin n\theta) = s(\cos \varphi + i \sin \varphi)$$

For this equality to hold, we must have $r^n = s$, $\cos n\theta = \cos \varphi$, and $\sin n\theta = \sin \varphi$. Now if r and s are positive real numbers and $r^n = s$, then $r = s^{1/n}$. Moreover, if $\cos n\theta = \cos \varphi$ and $\sin n\theta = \sin \varphi$, then $n\theta$ and φ must have the same terminal side. In other words, $n\theta = \varphi + 360°k$ for some integer k. Thus, $\theta = \varphi/n + 360°k/n$. Therefore we have seen that for any integer k, $z_k = s^{1/n} \operatorname{cis}(\varphi/n + 360°k/n)$ is a solution to equation (2): $x^n = s \operatorname{cis} \varphi$. Of course, if the angles are in radians, then $z_k = s^{1/n} \operatorname{cis}(\varphi/n + 2k\pi/n)$.

EXAMPLE 3 Solve $x^5 = 3 \operatorname{cis} 40°$.

SOLUTION By the preceding, the solutions are

$$z_k = 3^{1/5} \operatorname{cis}\left(\frac{40°}{5} + \frac{k}{5} \cdot 360°\right) = 3^{1/5} \operatorname{cis}(8° + 72°k)$$

where k is an integer. Let us look at some of them:

$z_0 = 3^{1/5} \operatorname{cis} 8° \approx 1.23361 + .173372i$

$z_1 = 3^{1/5} \operatorname{cis}(8° + 72°) = 3^{1/5} \operatorname{cis} 80° \approx .216319 + 1.22681i$

$z_2 = 3^{1/5} \operatorname{cis}(8° + 144°) = 3^{1/5} \operatorname{cis} 152° \approx -1.09992 + .584835i$

$z_3 = 3^{1/5} \operatorname{cis}(8° + 216°) = 3^{1/5} \operatorname{cis} 224° \approx -.896104 - .865357i$

$z_4 = 3^{1/5} \operatorname{cis}(8° + 288°) = 3^{1/5} \operatorname{cis} 296° \approx .54609 - 1.11966i$

$z_5 = 3^{1/5} \operatorname{cis}(8° + 360°) = 3^{1/5} \operatorname{cis} 8° = z_0$

$z_6 = 3^{1/5} \operatorname{cis}(8° + 360° + 72°) = 3^{1/5} \operatorname{cis} 80° = z_1, \quad$ etc.

The first five solutions, z_0, z_1, z_2, z_3, z_4, are distinct. All others are just one of these. Note that this agrees with our statement that a polynomial equation of degree 5 should have five zeros.

In general we can find the n distinct solutions to the equation in the same way:

(3)

The n distinct solutions to $x^n = s \operatorname{cis} \varphi$ are

$$z_k = s^{1/n} \operatorname{cis}\left(\frac{\varphi}{n} + \frac{360°k}{n}\right) \quad \text{if } \varphi \text{ is in degrees}$$

$$z_k = s^{1/n} \operatorname{cis}\left(\frac{\varphi}{n} + \frac{2k\pi}{n}\right) \quad \text{if } \varphi \text{ is in radians}$$

for $k = 0, 1, 2, \ldots, n-1$.

EXAMPLE 4 Find all solutions to $x^6 = \sqrt{3} - i$.

SOLUTION Since $\sqrt{3} - i = 2\operatorname{cis}(-\pi/6)$, the solutions are

$$z_k = 2^{1/6} \operatorname{cis}\left(\frac{-\pi/6}{6} + \frac{2k\pi}{6}\right) \quad \text{for } k = 0, 1, 2, 3, 4, 5$$

They are

$$z_0 = 2^{1/6} \operatorname{cis}\left(-\frac{\pi}{36}\right) \approx 1.11822 - .0978290i$$

$$z_1 = 2^{1/6} \operatorname{cis}\left(-\frac{\pi}{36} + \frac{\pi}{3}\right) = 2^{1/6} \operatorname{cis}\left(\frac{11\pi}{36}\right) \approx .643818 + .919467i$$

$$z_2 = 2^{1/6} \operatorname{cis}\left(-\frac{\pi}{36} + \frac{2\pi}{3}\right) = 2^{1/6} \operatorname{cis}\left(\frac{23\pi}{36}\right) \approx -.474373 + 1.01730i$$

$$z_3 = 2^{1/6} \operatorname{cis}\left(-\frac{\pi}{36} + \pi\right) = 2^{1/6} \operatorname{cis}\left(\frac{35\pi}{36}\right) \approx -1.11822 + .0978290i$$

$$z_4 = 2^{1/6} \operatorname{cis}\left(-\frac{\pi}{36} + \frac{4\pi}{3}\right) = 2^{1/6} \operatorname{cis}\left(\frac{47\pi}{36}\right) \approx -.643818 - .919467i$$

$$z_5 = 2^{1/6} \operatorname{cis}\left(-\frac{\pi}{36} + \frac{5\pi}{3}\right) = 2^{1/6} \operatorname{cis}\left(\frac{59\pi}{36}\right) \approx .474373 - 1.01730i$$

Note that the last three answers are the negatives of the first three.

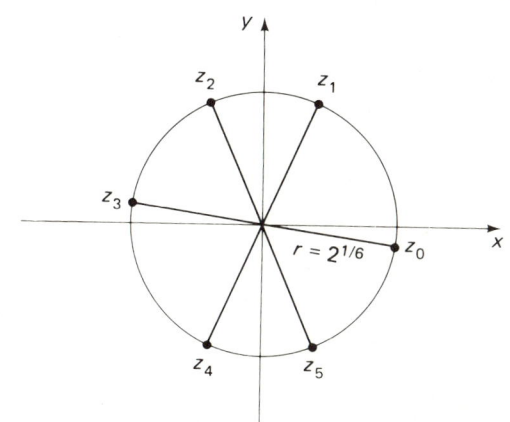

Let us plot the points $z_0, z_1, z_2, z_3, z_4, z_5$ from Example 4 in the complex plane. They all have the same absolute value, $2^{1/6}$. Thus they all lie on a circle of radius $2^{1/6}$ with center at the origin. First plot z_0; it has argument $\varphi/n = -\pi/36 = -5°$. Then the points $z_0, z_1, \ldots, z_5$ divide the circumference of the circle into $n = 6$ equal arcs, each subtending an angle of $2\pi/n = \pi/3 = 60°$.

Such a geometric picture is always possible. The n roots of $x^n = s(\cos \varphi + i \sin \varphi)$ all lie on a circle about the origin of radius $s^{1/n}$. They divide the circumference into n equal arcs, each subtending an angle of $360°/n$ (or $2\pi/n$ rad) starting from z_0, which has an argument of φ/n.

Now that we know how to find roots of complex numbers, we can solve any quadratic equation $ax^2 + bx + c = 0$, $a \neq 0$, with complex coefficients. Previously when we used the quadratic formula

$$x = \frac{-b \pm \sqrt{b^2 - 4ac}}{2a}$$

we considered examples where $b^2 - 4ac$ was a real number. But we could not handle the case where $b^2 - 4ac$ was complex but not real, because we did not know how to take square roots of such numbers. Now we use the techniques we have just learned.

EXAMPLE 5 Solve $2ix^2 - (3 + 6i)x + 7 = 0$.

SOLUTION By the quadratic formula,

$$x = \frac{3 + 6i \pm \sqrt{[-(3 + 6i)]^2 - 4(2i)7}}{4i}$$

$$= \frac{3 + 6i \pm \sqrt{(3 + 6i)^2 - 56i}}{4i}$$

$$= \frac{3 + 6i \pm \sqrt{9 + 36i - 36 - 56i}}{4i}$$

$$= \frac{3 + 6i \pm \sqrt{-27 - 20i}}{4i}$$

Letting $z = -27 - 20i \approx 33.600595 \text{ cis } 216.5289°$, we see that the square roots of z are

$$z_0 \approx 33.600595^{1/2} \text{ cis } \frac{216.5289°}{2} \approx 5.79660 \text{ cis } 108.264°$$

$$\approx -1.81663 + 5.50459i$$

$$z_1 \approx 33.600595^{1/2} \text{ cis}\left(\frac{216.5289°}{2} + 180°\right) \approx 5.79660 \text{ cis } 288.264°$$

$$\approx 1.81663 - 5.50459i$$

(Note that z_1 is the negative of z_0.) Consequently,

$$x \approx \frac{3 + 6i \pm (-1.81663 + 5.50459i)}{4i}$$

$$x \approx \frac{1.18337 + 11.50459i}{4i} \quad \text{or} \quad \frac{4.81663 + .49541i}{4i}$$

Multiplying top and bottom by $-4i$ yields

$$x \approx 2.87615 - .295843i \quad \text{or} \quad .12385 - 1.20416i$$

EXERCISES

In Exercises 1–6, simplify using De Moivre's theorem.

1. $(3 \text{ cis } 102°)^6$
2. $(8 \text{ cis } 37°)^{100}$
3. $(2 \text{ cis } 213°)^{-8}$
4. $\left(6.4 \text{ cis } \frac{\pi}{5}\right)^{11}$

5. $\left[3.7\left(\cos\frac{\pi}{3} - i\sin\frac{\pi}{3}\right)\right]^5$
6. $(1.31 \text{ cis } 8.92)^{20}$

In Exercises 7–12, simplify to the form $a + bi$.

7. $\left(-\frac{\sqrt{2}}{2} + \frac{\sqrt{2}}{2}i\right)^{93}$
8. $\left(-\frac{\sqrt{3}}{2} + \frac{1}{2}i\right)^{207}$
9. $(1.3 - 1.82i)^8$
10. $(3 + 4i)^7$
11. $(2.3 + 1.037i)^9$
12. $(-.604 - 1.425i)^5$

In Exercises 13–18, find all n distinct roots. Put the answer in both polar form and in rectangular form, $a + bi$.

13. $x^3 = 4 - 4\sqrt{3}\,i$
14. $x^5 = -1 - i$
15. $x^6 = 64i$
16. $x^7 = 5 + i$
17. $x^5 = -3 + 2i$
18. $x^2 = -7.301 - 13.86i$
19. Construct the fifth roots of $1 + i$ and of 32 geometrically.
20. Find all solutions to $(3 - i)z^7 - 2 + 4i = 0$.

In Exercises 21–28, solve for x.

21. $x^2 + (5 - 3i)x + 4 + 2i = 0$
22. $3x^2 - (3 - i)x + \frac{5}{6} = 0$
23. $ix^2 + (1 + i)x - 2 = 0$
24. $4ix^2 - 5x - 1 + i = 0$
25. $2x^2 + (1 + 2i)x - 3 - 4i = 0$
26. $(1 + i)x^2 + 4x + 4 = 0$

REVIEW EXERCISES

In Exercises 1–16, write in the form $a + bi$.

1. $(3 - 2i) + \overline{(4 - 7i)}$
2. $(-2 + 3i)\overline{(5 + 6i)}$
3. $(-5 + \sqrt{-9})(2 - \sqrt{-4})$
4. $(3 - 4i)^2$
5. $i^{3219} - i^{427} + i^{18}$
6. $(3 - i)/(4 + 2i)$
7. $(1 - i)/(2i)$
8. $\sqrt{-81}\,\sqrt{-27}$
9. $(\sqrt{-27})^3$
10. $(2 + \sqrt{-4})/(3 - \sqrt{-8})$
11. $4\text{ cis } 45°$
12. $3\text{ cis}(-\pi/2)$
13. $(2\text{ cis } 3)(3\text{ cis } 2)$
14. $(4\text{ cis } 30°)^{-1}$
15. $(3\text{ cis } 150°)^{10}$
16. $[16\text{ cis}(-\pi/2)]^{-4}$

In Exercises 17 and 18, solve for x and y real.

17. $3 - 2yi = 9^x - 7i$
18. $\cos x - \pi i = .5 + i\tan y$

In Exercises 19–25, solve.

19. $4ix - 5 = 6x - i$
20. $3ix - 2 = 0$
21. $x^3 - 8 = 0$
22. $2x^4 - 3x^2 + 1 = 0$
23. $3x^2 + (2 - 3i)x - i = 0$
24. $ix^2 + (2 - 3i)x - 2 - i = 0$
25. $x^5 = 4 - i$

Polynomials

CHAPTER ELEVEN

Finding solutions to polynomial equations has been a problem of interest throughout mathematical history. General solutions to quadratic equations can be found in Old Babylonian texts over 4000 years old. However, a general solution for cubic and higher-degree polynomial equations remained an outstanding problem until the early sixteenth century.

The first person to make significant progress toward solving cubic equations was an Italian, Scipione del Ferro (1465–1526), who discovered how to solve certain special cases of the cubic equation. However, his work was not published.

Perhaps inspired by rumors from Italy, Niccolo Tartaglia (ca. 1500–1557) found a complete solution to the general cubic by 1541, but he did not publish it either. He wanted to wait to publish it as a crowning jewel of all his achievements. Geronimo Cardan (1501–1576), hearing rumors of Tartaglia's achievement, invited Tartaglia to his home. Cardan, after taking a solumn oath of secrecy, enticed Tartaglia to tell him the secret. Later, Cardan's technical secretary Ludovico Ferrari (1522–1565) figured out how to modify Tartaglia's method to solve fourth-degree polynomial equations. Cardan, having such fantastic gems in his possession, could not restrain himself, broke his oath, and in 1545 published Ars Magna ("Great Art"), which contained both solutions (giving full credit to both Tartaglia and Ferrari but completely embittering Tartaglia).

No further progress was made on solving higher-degree polynomials until 1824 when the young Norwegian Henrik Abel (1802–1829) proved there is no general solution to fifth- and higher-degree polynomials. The story was completed in 1832 by the French mathematician Evariste Galois (1811–1832), who gave necessary and sufficient conditions for the solvability of any polynomial in terms of rational operations and nth roots on the coefficients. Galois sketched out much of his brilliant work in a letter to a friend the night before he was killed in a political duel at the age of 20.

SECTION 11.1. POLYNOMIALS; REMAINDER AND FACTOR THEOREMS

We began the discussion of polynomials in Sec. 4.2. There we defined a polynomial (in x) to be an algebraic expression of the form

(1) $$a_n x^n + a_{n-1} x^{n-1} + \cdots + a_1 x + a_0$$

where n is a nonnegative integer and the coefficients $a_0, a_1, \ldots, a_n$ are numbers (real or complex). In this section, we discuss various algebraic properties of polynomials.

First, we say that two polynomials are **equal** if and only if the coefficients of the corresponding powers are equal. Thus, $ax^3 + bx^2 + cx + d = -3x^3 + 2x^2 - 4$ if and only if $a = -3$, $b = 2$, $c = 0$, and $d = -4$.

Addition, subtraction, and multiplication of polynomials follow from the rules of elementary algebra.

EXAMPLE 1 If $f(x) = 2x^3 - 3x^2 + 4$ and $g(x) = -2x^3 + 5x - 2$, find (a) $f(x) + g(x)$, (b) $f(x) - g(x)$, and (c) $f(x) \cdot g(x)$.

SOLUTION

(a) $f(x) + g(x) = (2x^3 - 3x^2 + 4) + (-2x^3 + 5x - 2)$
$= [2 + (-2)]x^3 - 3x^2 + 5x + 4 + (-2)$
$= -3x^2 + 5x + 2$

Note that the sum has degree 2.

(b) $f(x) - g(x) = (2x^3 - 3x^2 + 4) - (-2x^3 + 5x - 2)$
$= [2 - (-2)]x^3 - 3x^2 - (5x) + 4 - (-2)$
$= 4x^3 - 3x^2 - 5x + 6$

Note that the difference has degree 3.

(c) $f(x) \cdot g(x) = (2x^3 - 3x^2 + 4)(-2x^3 + 5x - 2)$
$= 2x^3(-2x^3 + 5x - 2) - 3x^2(-2x^3 + 5x - 2)$
$\quad + 4(-2x^3 + 5x - 2)$
$= -4x^6 + 10x^4 - 4x^3 + 6x^5 - 15x^3$
$\quad + 6x^2 - 8x^3 + 20x - 8$
$= -4x^6 + 6x^5 + 10x^4 - 27x^3 + 6x^2 + 20x - 8$

Note that the product has degree 6.

The general principles relating the degree of sums, differences, and products are as follows:

---

(2)

If $f(x)$ and $g(x)$ are polynomials of degrees m and n, respectively, then

i. $f(x) + g(x)$ and $f(x) - g(x)$ are polynomials of degree less than or equal to the larger of m and n.
ii. $f(x) \cdot g(x)$ is a polynomial of degree $m + n$.

These principles apply whether the coefficients are real or complex.

EXAMPLE 2 If $f(x) = (2 - i)x^3 - 3x + i$ and $g(x) = 3ix^3 + (1 - i)x - 4$, then the product $f(x)g(x) = (3 + 6i)x^6 + (1 - 12i)x^4 - (11 - 4i)x^3 - (3 - 3i)x^2 + (13 + i)x - 4i$ has degree 6.

Division of polynomials is analogous to division of integers, so it is helpful to compare these two. First, recall that if k, m, and n are integers and $k = m \cdot n$, then we say that m and n are **factors** of k and that k is **divisible** by m and n. For example, 12 has positive integer factors 1, 2, 3, 4, 6, and 12, and 12 is divisible by each of these numbers.

This same language is used for polynomials. For example, by Example 1, if $h(x) = -4x^6 + 6x^5 + 10x^4 - 27x^3 + 6x^2 + 20x - 8$, then $h(x) = (2x^3 - 3x^2 + 4)(-2x^3 + 5x - 2)$. We say that $2x^3 - 3x^2 + 4$ and $-2x^3 + 5x - 2$ are **factors** of $h(x)$ and that $h(x)$ is **divisible** by them.

Returning to integers, if we divide 9617 by a number such as 23, then we go through a process called long division and get a quotient and remainder. For these numbers, the long division can be arranged as follows:

$$\begin{array}{r} 418 \\ 23 \overline{)9617} \\ \underline{92} \\ 41 \\ \underline{23} \\ 187 \\ \underline{184} \\ 3 \end{array}$$

Here, the number 418 is called the **quotient** and the number 3 is called the **remainder**. (Also, 23 is called the **divisor** and 9617 is called the **dividend**.) Of course, the remainder is always less than the divisor.

The preceding computation can be rewritten as

$$\frac{9617}{23} = 418 + \frac{3}{23}$$

Multiplying by 23 gives

$$9617 = 23(418) + 3$$

or

$$\text{Dividend} = (\text{divisor})(\text{quotient}) + \text{remainder}$$

This formula relating the dividend, divisor, quotient, and remainder of a division process is very useful for theoretical purposes and is formalized as follows:

---

(3)   **DIVISION ALGORITHM FOR INTEGERS**

If a and b are integers with $b > 0$, then there are unique integers q and r such that

$$a = bq + r \qquad \text{where } 0 \leq r < b.$$

It is also possible to define long division for polynomials. For example, if we want to divide $4x^4 + 3x^3 + \frac{1}{4}x - \frac{1}{4}$ by $2x^2 - 3x + 1$, the long division is arranged as follows:

$$
\begin{array}{r}
2x^2 + \frac{9}{2}x + \frac{23}{4} \\
2x^2 - 3x + 1 \overline{\smash{)}4x^4 + 3x^3 + 0x^2 + \tfrac{1}{4}x - \tfrac{1}{4}} \\
\underline{4x^4 - 6x^3 + 2x^2} \\
9x^3 - 2x^2 + \tfrac{1}{4}x \\
\underline{9x^3 - \tfrac{27}{2}x^2 + \tfrac{9}{2}x} \\
\tfrac{23}{2}x^2 - \tfrac{17}{4}x - \tfrac{1}{4} \\
\underline{\tfrac{23}{2}x^2 - \tfrac{69}{4}x + \tfrac{23}{4}} \\
\tfrac{52}{4}x - \tfrac{24}{4} = 13x - 6
\end{array}
$$

Just as with integers, the polynomial $2x^2 + \frac{9}{2}x + \frac{23}{4}$ is called the *quotient*, $13x - 6$ is called the *remainder*, $2x^2 - 3x + 1$ is called the *divisor*, and $4x^4 + 3x^3 + \frac{1}{4}x - \frac{1}{4}$ is the *dividend*.

There are two things to remember about the process of long division:

1. You must put the terms in order of descending powers, and either put in any missing terms with coefficient 0 (as we put in $0x^2$) or else leave a corresponding amount of space, so that like terms can be vertically aligned.

2. You keep the division process going until either you get 0 (in which case the divisor is a factor of the dividend) or the degree of the remainder is less than the degree of the divisor.

The preceding computation can be rewritten as

$$\frac{4x^4 + 3x^3 + \frac{1}{4}x - \frac{1}{4}}{2x^2 - 3x + 1} = \left(2x^2 + \frac{9}{2}x + \frac{23}{4}\right) + \frac{13x - 6}{2x^2 - 3x + 1}$$

Multiplying by $2x^2 - 3x + 1$,

$$4x^4 + 3x^3 + \tfrac{1}{4}x - \tfrac{1}{4} = (2x^2 - 3x + 1)\left(2x^2 + \tfrac{9}{2}x + \tfrac{23}{4}\right) + (13x - 6)$$

Thus, just as with integers,

$$\text{Dividend} = (\text{divisor})(\text{quotient}) + \text{remainder}$$

which is formalized as follows:

---

(4)   **DIVISION ALGORITHM FOR POLYNOMIALS**

If $f(x)$ and $g(x)$ are polynomials and $g(x) \neq 0$, then there exist unique polynomials $q(x)$ and $r(x)$ such that

$$f(x) = g(x)q(x) + r(x)$$

where either $r(x) = 0$ or $\deg r(x) < \deg g(x)$.

The special case where $g(x) = x - c$, c a number, is very important. In this case, either $r(x) = 0$ or $\deg r(x) < \deg(x - c) = 1$ [i.e., $\deg r(x) = 0$]. If $\deg r(x) = 0$, then $r(x)$ is just a nonzero number. Therefore, $r(x)$ is just a number, either zero or nonzero. Thus when $g(x) = x - c$, we can simply set $r(x) = r$. Then the division algorithm can be written as

---

(5)

$$f(x) = (x - c)q(x) + r$$

This equation holds for all values of x. In particular, it holds for the value c:

$$f(c) = (c - c)q(c) + r$$
$$f(c) = 0 \cdot q(c) + r$$
$$f(c) = r$$

Hence the remainder r is $f(c)$, which proves the following theorem:

---

(6)   **REMAINDER THEOREM**

If c is a number and the polynomial $f(x)$ is divided by $x - c$, then the remainder is $f(c)$.

EXAMPLE 3 Let $f(x) = 3x^3 - 2x^2 + 5x - 6$ and $c = 2$. Find $f(2)$ directly and by the remainder theorem.

SOLUTION We easily obtain $f(2)$ directly by substitution:
$$f(2) = 3(2^3) - 2(2^2) + 5(2) - 6 = 24 - 8 + 10 - 6 = 20$$
Now we use long division of $f(x)$ by $x - 2$:

$$\begin{array}{r}
3x^2 + 4x + 13 \\
x - 2 \overline{) 3x^3 - 2x^2 + 5x - 6} \\
\underline{3x^3 - 6x^2 } \\
4x^2 + 5x \\
\underline{4x^2 - 8x } \\
13x - 6 \\
\underline{13x - 26} \\
20
\end{array}$$

Thus the remainder is 20, which is $f(2)$. In the next section, we shall show how to do this long division much more quickly.

A second important theorem is a consequence of the remainder theorem.

---

(7)   **FACTOR THEOREM**

A polynomial $f(x)$ has a factor $x - c$ if and only if $f(c) = 0$.

Proof: If $x - c$ is a factor of $f(x)$, then the division of $f(x)$ by $x - c$ is exact, so that the remainder is 0. By the remainder theorem, $f(c) = 0$. On the other hand, if $f(c) = 0$, then $f(x) = (x - c)q(x) + 0$; i.e., $x - c$ is a factor of $f(x)$.

EXAMPLE 4 Show that $x + 2$ is a factor of $x^5 + 32$.

SOLUTION Set $f(x) = x^5 + 32$. Since $x + 2 = x - (-2)$, $c = -2$. Then $f(-2) = (-2)^5 + 32 = -32 + 32 = 0$. Thus by the factor theorem, $x + 2$ is a factor of $x^5 + 32$. We could have obtained the same result by long division, but that would obviously have been more work.

EXERCISES

In Exercises 1–6, write the given expression in the form
$$a_n x^n + a_{n-1} x^{n-1} + \cdots + a_1 x + a_0$$

1. $(x - 2)(x + 6) - x(x^3 - 1)$
2. $(x^2 - 4)(x + 3) - 2(x - 5)$
3. $(x^{325} - x^{112})(x^4 - 1) - 3x(x^{447} - 1)$
4. $(x^{10} + 3)(x^{10} - 3)(x^{10} + 1)$

5. $(x - i)(2x^2 - 3ix + 1) - i(x^2 - 4 - i)$
6. $(2i - x^2)(x - ix^2 - i) - i(x - ix^2 + 1)$

In Exercises 7–10, $f(x)$ and $g(x)$ are polynomials. Find the degrees of **(a)** $f(x)g(x)$, **(b)** $f(x) + g(x)$, **(c)** $f^3(x)$, **(d)** $q(x)$, and **(e)** $r(x)$, where $f(x) = q(x)g(x) + r(x)$ is obtained by the division algorithm.

7. $\deg f(x) = 7$, $\deg g(x) = 4$
8. $\deg f(x) = 1983$, $\deg g(x) = 1729$
9. $\deg f(x) = 8$, $\deg g(x) = 8$
10. $\deg f(x) = 57$, $\deg g(x) = 92$

In Exercises 11–16, find the quotient $q(x)$ and remainder $r(x)$ if $f(x)$ is divided by $g(x)$.

11. $f(x) = 2x^4 - 3x^3 + x^2 - 5x + 3$, $g(x) = x^2 - 2x + 5$
12. $f(x) = 4x^4 - x^2 - x + 1$, $g(x) = x^2 + 3x$
13. $f(x) = 6x^5 - x^2 + 1$, $g(x) = 2x^3 + 3x^2 - 1$
14. $f(x) = 4x^3 - 3x^2 - 2x$, $g(x) = 2x^2 - x + 1$
15. $f(x) = x^2 - x + 1$, $g(x) = x^3 + x^2 - 1$
16. $f(x) = x^8 + 1$, $g(x) = x^2 + 1$

In Exercises 17–22, use the remainder theorem to find $f(c)$.

17. $f(x) = 3x^3 - x^2 - x + 4$, $c = -2$
18. $f(x) = 4x^3 - 3x^2 + 2$, $c = 4$
19. $f(x) = -3x^4 + x - 4$, $c = 3$
20. $f(x) = 2x^4 - 3x^3 - 5$, $c = -3$
21. $f(x) = x^6 - 2x^4 + 3x^2 - 1$, $c = \sqrt{2}$
22. $f(x) = x^5 + x^4 - x^3 - x^2 + x + 1$, $c = i$

23. Explain why we get the same remainder when a given polynomial $f(x)$ is divided by $x - \frac{1}{3}$ or by $3x - 1$. What, if anything, is different?
24. What is the remainder when $3x^{50} + 2$ is divided by $x - 1$?
25. What is the remainder when $2x^{100} - 3x^{75} + 4x^{40} - 3$ is divided by $x + 1$?
26. What is the remainder when $4x^{300} - 3x^{101} + 2x^{43} + 2$ is divided by $x - i$?

In Exercises 27–36, use the factor theorem.

27. Show that $x - 2$ is a factor of $f(x) = x^5 - 3x^4 + x^3 - x^2 + 4x + 4$.
28. Show that $x + 1$ is a factor of $f(x) = x^6 + 2x^5 + x^3 + x + 3$.
29. Show that $x + 2$ is a factor of $f(x) = x^7 + 128$.
30. Show that $x - a$ is a factor of $f(x) = x^{1351} - a^{1351}$.
31. Show that $x - a$ is a factor of $f(x) = x^n - a^n$, n a positive integer.
32. Show that $x + a$ is a factor of $f(x) = x^n + a^n$, n an odd positive integer.
33. Determine k such that $x - 1$ is a factor of $3x^3 + x^2 + 2kx + 4$.
34. Determine k such that $x + 3$ is a factor of $x^3 + kx^2 - kx + 9$.
35. Determine k such that $x - 2$ is a factor of $x^3 + k^2x^2 - 2kx - 10$.
36. Determine k such that $x - k$ is a factor of $2x^2 + 5x - 3$.

SECTION 11.2. SYNTHETIC DIVISION

If a polynomial $f(x)$ is divided by $x - c$, we know from the remainder theorem in the last section that the remainder is $f(c)$. In this section, we shall develop a quick and efficient method for performing the division, called *synthetic division*.

This, in turn, is related to the *nested* form of a polynomial which yields the most efficient method of evaluating polynomials and is used on most computers and programmable calculators.

Synthetic division is derived by condensing the ordinary process as much as possible. We illustrate the derivation with the following example.

Suppose we wish to divide $3x^4 - 7x^3 + 5x + 1$ by $x - 2$. Using long division, we obtain

$$
\begin{array}{r}
3x^3 - x^2 - 2x + 1 \\
x - 2 \overline{\smash{)}3x^4 - 7x^3 + 0x^2 + 5x + 1} \\
\underline{3x^4 - 6x^3} \\
-x^3 + 0x^2 \\
\underline{-x^3 + 2x^2} \\
-2x^2 + 5x \\
\underline{-2x^2 + 4x} \\
x + 1 \\
\underline{x - 2} \\
3
\end{array}
$$

where the term $0x^2$ has been added to help align the powers.

First, observe that in this division process the x's (i.e., x^4, x^3, x^2, x) are used mainly as place holders. We could omit the x's and just write the coefficients as long as we realize the x's should be there. (Do not forget to write the coefficients 1 of x and -1 of x^3.)

$$
\begin{array}{r}
3\ \ -1\ \ -2\ \ \ 1 \\
1\ -2\ \overline{\smash{)}3\ \ -7\ \ \ \ 0\ \ \ \ 5\ \ \ \ 1} \\
\boxed{3}\ -6 \\
\overline{-1\ \ \boxed{0}} \\
\boxed{-1}\ \ \ 2 \\
\overline{-2\ \ \boxed{5}} \\
\boxed{-2}\ \ \ 4 \\
\overline{1\ \ \boxed{1}} \\
\boxed{1}\ -2 \\
\overline{3}
\end{array}
$$

Section 11.2 Synthetic Division

Next, we observe that each number in a circled position is always identical to the number directly above it. We can eliminate these repetitions and then compress what remains:

$$
\begin{array}{r|rrrrr}
 & 3 & -1 & -2 & 1 & \\
1 \;\; -2 & 3 & -7 & 0 & 5 & 1 \\
 & & -6 & 2 & 4 & -2 \\
\hline
 & & -1 & -2 & 1 & 3
\end{array}
$$

If we "bring down" the first coefficient to the bottom line, then the first four numbers in the bottom line are precisely the numbers on the top line (which are the coefficients of the quotient). Hence the top line can be omitted. Also the leading 1 of the divisor can be omitted. We now have

$$
\begin{array}{r|rrrrr}
-2 & 3 & -7 & 0 & 5 & 1 \\
 & & -6 & 2 & 4 & -2 \\
\hline
 & 3 & -1 & -2 & 1 & 3
\end{array}
$$

Each number in the third row, except the first, is obtained by subtracting the number in the second row from the number above it. Each number in the second row is obtained by multiplying the preceding entry in the third row by the number in the partial box, -2 in this case. By changing the sign of the -2 in the box, we could use addition instead of subtraction to obtain line 3. This helps to cut down on arithmetic errors.

We now have the following scheme:

---

(8)   **SYNTHETIC DIVISION**

Scheme for dividing a polynomial $f(x)$ by $x - c$ using *synthetic division*; for example, to divide $3x^4 - 7x^3 + 5x + 1$ by $x - 2$:

①   Place c here.   ②   Place coefficients of $f(x)$ here. Make sure powers are in descending order and insert 0's for missing terms.

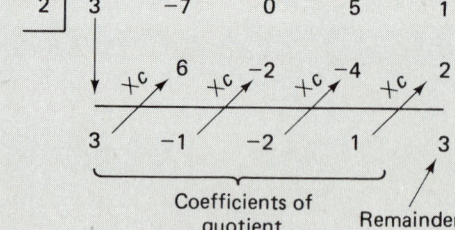

Coefficients of quotient   Remainder

③   Bring down leading coefficient.

④   Multiply the entry in the bottom row by c, put the result in the next column, and add as indicated.

The quotient is $3x^3 - x^2 - 2x + 1$, and the remainder is 3.

We consider several examples.

EXAMPLE 1 Divide $5x^5 - 4x^3 + x + 2x^2$ by $x + 3$.

SOLUTION First, since we are to divide by $x + 3$, we must be careful to recognize that the c in the quantity $x - c$ is -3, since $x + 3 = x - (-3)$. Next, we must write the powers in *descending* order, $5x^5 - 4x^3 + 2x^2 + x$, and put in the *two* missing terms, $0x^4$ and 0. Then we follow the scheme in (8) for synthetic division:

$$\begin{array}{r|rrrrrr} -3 & 5 & 0 & -4 & 2 & 1 & 0 \\ & & -15 & 45 & -123 & 363 & -1092 \\ \hline & 5 & -15 & 41 & -121 & 364 & -1092 \end{array}$$

The remainder is -1092, and the quotient is $5x^4 - 15x^3 + 41x^2 - 121x + 364$.

EXAMPLE 2 If $f(x) = x^5 - 2x^4 + 2x^2 - 1$, use synthetic division to find $f(3)$.

SOLUTION By the remainder theorem, $f(3)$ is the remainder when $f(x)$ is divided by $x - 3$. Using synthetic division, we obtain

$$\begin{array}{r|rrrrrr} 3 & 1 & -2 & 0 & 2 & 0 & -1 \\ & & 3 & 3 & 9 & 33 & 99 \\ \hline & 1 & 1 & 3 & 11 & 33 & 98 \end{array}$$

Since the remainder is 98, $f(3) = 98$.

EXAMPLE 3 Use synthetic division to show that $x + i$ is a factor of $f(x) = x^4 + 2x^3 + 4x^2 + 2x + 3$.

SOLUTION By the factor theorem, $x + i = x - (-i)$ is a factor of $f(x)$ if and only if $f(-i)$ is 0. We use synthetic division to compute $f(-i)$ as in Example 2:

$$\begin{array}{r|rrrrr} -i & 1 & 2 & 4 & 2 & 3 \\ & & -i & -1-2i & -2-3i & -3 \\ \hline & 1 & 2-i & 3-2i & -3i & 0 \end{array}$$

Since the remainder is $0 = f(-i)$, $x - (-i)$ is a factor.

Synthetic division as used in Example 2 is the most efficient way to evaluate almost all polynomials. This is an important consideration not only if we are evaluating a polynomial by hand but also if we are using a calculator or computer. Even though computers are very fast, the number of steps taken becomes important both in terms of time and money, especially when a particular job requires many evaluations. When writing a program to evaluate a polynomial on a calculator or computer, another form of a polynomial, called the nested form, is usually used. Evaluating a polynomial in nested form requires very few steps ($2n$ steps for a polynomial of degree n). We shall see that evaluating a polynomial in nested form is equivalent to using synthetic division.

The nested form of a polynomial is obtained from the standard form of a polynomial,

$$f(x) = a_n x^n + a_{n-1} x^{n-1} + a_{n-2} x^{n-2} + \cdots + a_1 x + a_0$$

as follows:

First factor the term $a_n x^n$ as $(a_n)x^n$. Then add the next term, $a_{n-1}x^{n-1}$, and factor out x^{n-1}:

$$(a_n)x^n + a_{n-1}x^{n-1} = ((a_n)x + a_{n-1})x^{n-1}$$

Then add the next term, $a_{n-2}x^{n-2}$, and factor out x^{n-2}:

$$((a_n)x + a_{n-1})x^{n-1} + a_{n-2}x^{n-2} = (((a_n)x + a_{n-1})x + a_{n-2})x^{n-2}$$

Continue doing this until you have added in all the terms. When you are finished, you have a whole collection of terms in parentheses, with one set of parentheses "nested" inside the next. Consequently, this final form

(9) $\quad f(x) = (\cdots(((a_n)x + a_{n-1})x + a_{n-2})x + \cdots + a_1)x + a_0$

is called the **nested** form of a polynomial.

EXAMPLE 4 Put $f(x) = 3x^5 - 4x^4 + x^3 - 4x^2 - 2x + 5$ into nested form. Then evaluate $f(2)$, and compare the results with synthetic division.

SOLUTION In nested form,

$$f(x) = (((((3)x - 4)x + 1)x - 4)x - 2)x + 5$$

Now evaluating at $x = 2$, we obtain

$$\begin{aligned}
f(2) &= (((((3)2 - 4)2 + 1)2 - 4)2 - 2)2 + 5 \\
&= ((((2)2 + 1)2 - 4)2 - 2)2 + 5 \\
&= (((5)2 - 4)2 - 2)2 + 5 \\
&= ((6)2 - 2)2 + 5 \\
&= (10)2 + 5 \\
&= 25
\end{aligned}$$

Using synthetic division, we obtain

$$\underline{2\,|}\ \ \begin{array}{rrrrrr} 3 & -4 & 1 & -4 & -2 & 5 \\ & 6 & 4 & 10 & 12 & 20 \\ \hline 3 & 2 & 5 & 6 & 10 & 25 \end{array}$$

Note in Example 4 that the numbers in the innermost parentheses at each step are exactly the numbers, in order, on the bottom line of synthetic division. This is no accident. The two processes are essentially the same. To evaluate $f(c)$, at each stage you simply multiply what you have by c and then add in the next lower coefficient. For the example $f(x) = 3x^5 - 4x^4 + x^3 - 4x^2 - 2x + 5$, to find $f(2)$, you simply

> Take the leading coefficient 3, multiply by 2
> add -4, multiply by 2
> add 1, multiply by 2
> add -4, multiply by 2
> add -2, multiply by 2
> add 5.

Note that once you understand the process, you can do the whole computation in your head or on a calculator without writing anything down. To do this computation on your scientific calculator, you would press the following:

ALGEBRAIC	R.P.N.
[3] [×] [2]	[3] [ENT] [2] [×]
[+] [−4] [=] [×] [2]	[−4] [+] [2] [×]
[+] [1] [=] [×] [2]	[1] [+] [2] [×]
[+] [−4] [=] [×] [2]	[−4] [+] [2] [×]
[+] [−2] [=] [×] [2]	[−2] [+] [2] [×]
[+] [5] [=]	[5] [+]

Note that the algebraic program is written for calculators with algebraic hierarchy. If you have such a calculator, you should make sure you understand why all the [=]'s are necessary. If you have an algebraic calculator without hierarchy, you can eliminate all the [=]'s except for the last.

Remark: If the polynomial in Example 4 were $f(x) = 5x^5 + 1 + 2x^2 - 4x^3$, it would be crucial that your first step be to rewrite the terms in order of descending powers of x and to include the terms with zero coefficients:

$$f(x) = 5x^5 + 0x^4 - 4x^3 + 2x^2 + 0x + 1$$

EXERCISES

In Exercises 1–15, use synthetic division to find the quotient $q(x)$ and remainder r in each division.

1. $(3x^4 - 2x^3 + x^2 + x - 4) \div (x - 2)$
2. $(7x^8 + 3x^5 - x^2 + 10) \div (x - 1)$
3. $(3x^2 - 2.3x + 4.5) \div (x - 2.1)$
4. $(x^5 + x^4 - 4x^3 + 2) \div (x + 3)$
5. $(x^3 - 6x^4 + 1 - x^2) \div (x + 8)$
6. $(3 - x + x^2 - x^3) \div (x - 10)$
7. $(-11t^4 + 7t^2 + 2t) \div (t + 4)$
8. $(14u^3 - 8 + u) \div (u + 6)$
9. $(21t^4 - 38t^3 + 6t^2 - 25t + 11) \div (t - 2)$
10. $(-s^7 + 4.3s^4 - 7.2s^2) \div (s + 2)$

11. $(x^3 + 3.1x^2 - 8.7x - 11.31) \div (x - .9)$
12. $(4x - 3x^3 - .384) \div (x + 1.2)$
13. $(x^3 + 3cx^2 + 6c^2x + 42c^3) \div (x + c)$
14. $(2x^4 + 5a^2x^2 - 8a^4) \div (x - 2a)$
15. $(-3b^3x + x^4 - b^2x^2 - 81b^4) \div (x + 3b)$

In Exercises 16–21, use synthetic division to find the required function values. Note any zeros you find.

16. Find $f(1), f(2), f(-3)$ if $f(x) = 5x^6 + 8x^5 + 11x^4 + 13x^2 + 5x$.
17. Find $g(-1), g(4), g(3)$ if $g(x) = -x^4 + 15x^3 + 6x^2 - 12x + 3$.
18. Find $f(6), f(2), f(-2)$ if $f(x) = x^3 - 6x^2 - 4x + 24$.
19. Find $f(4), f(-5), f(3.2)$ if $f(x) = 6.2x^2 - 4x^3 + 3.81x + 4.28$.
20. Find $f(a), f(2a), f(-a)$ if $f(x) = 6x^3 - ax^2 + 4a^2x + a^3$.
21. Find $f(.1), f(.2), f(.3)$ if $f(x) = x^6 + 3x^4 - 24x^2 + 3.075$.

In Exercises 22–25, write the polynomials in nested form.

22. $f(x) = 2x^3 - 4x^2 + 11x + 15$
23. $f(x) = -6x^4 + 18x^3 + x^2 - 47$
24. $f(x) = x^5 + 97x^3 + 64x^4 + 32x^2 + 69x - 82$
25. $f(t) = t^8 + 6t^5 - 11t^4 - 14t + 1$

In Exercises 26–30, use your calculator to find the required function values. Use the method of nested polynomials without actually writing down the polynomials in nested form.

26. Find $f(2), f(8), f(-5)$ if $f(x) = x^5 - 6x^4 + 8x^3 + 17x^2 + 3x - 48$.
27. Find $f(10), f(-7), f(4.1)$ if $f(x) = 12x^6 - 19x^4 + 41x^3 + 16x - 21$.
28. Find $f(3), f(5), f(-5)$ if $f(x) = 3x^4 - 69x^2 - 4x^3 + 100x - 150$.
29. Find $f(.6), f(1.4), f(-.1)$ if $f(x) = 4.3x^3 - 1.742x + 12.384$.
30. Find $f(2), f(-3), f(-5)$ if $f(x) = x^8 + x^7 + x^6 + x^5 + x^4 + x^3 + x^2 + x + 1$.

SECTION 11.3. ZEROS OF POLYNOMIALS

In this section, we shall learn some facts about zeros of polynomials of any degree n with complex coefficients. (Polynomials with complex coefficients include polynomials with real coefficients, the case which interests us the most, since the complex numbers contain the real numbers.) The main result is so important that it is called the *Fundamental Theorem of Algebra*. Its proof requires some deep mathematics that is beyond the scope of this book.

---

(10)   THE FUNDAMENTAL THEOREM OF ALGEBRA

Every polynomial of positive degree which has complex coefficients has at least one complex zero.

Let us see how we can use this theorem. Let

$$f(x) = a_0 + a_1x + \cdots + a_nx^n, \quad a_n \neq 0$$

be a polynomial of degree n, $n \geq 1$, with complex coefficients. By the Fundamental Theorem of Algebra, $f(x)$ has a zero r_1. By the factor theorem, $x - r_1$ is a factor of $f(x)$, say $f(x) = (x - r_1)f_1(x)$. Note that since $f(x)$ has degree n and $x - r_1$ has degree 1, $f_1(x)$ has degree $n - 1$. If this is still at least 1, then we can apply the fundamental theorem again to obtain a zero r_2 of $f_1(x)$. Then by the factor theorem, $f_1(x) = (x - r_2)f_2(x)$. This means that

$$f(x) = (x - r_1)(x - r_2)f_2(x)$$

If degree $f_2(x) = n - 2$ is still at least 1, we can continue the process. We continue until we reach a polynomial of degree 0, which is simply a constant polynomial c. Since each $f_i(x)$ is of degree one less than the previous one, there will be precisely n of the factors $x - r$. That is,

$$f(x) = (x - r_1)(x - r_2) \cdots (x - r_n)c$$

If you multiply these out, you see that the coefficient of x^n is c. But the coefficient of x^n in $f(x)$ is a_n, so $c = a_n$. The constant a_n is usually written in front of the factors. Thus

(11)
$$f(x) = a_n(x - r_1)(x - r_2) \cdots (x - r_n)$$

EXAMPLE 1 $f(x) = 5x^4 + 5x^3 - 25x^2 + 5x - 30$ factors as $f(x) = 5(x - i)(x + i)(x - 2)(x + 3)$. (Verify this.)

Note in Example 1 that the degree of $f(x)$ is 4, so there are four factors of the form $x - r$. The constant in front of these factors is 5, the same as the coefficient of x^4.

It may be that some of the factors are the same.

EXAMPLE 2 $x^4 + 5x^3 + 9x^2 + 7x + 2 = (x + 1)^3(x + 2)$. (Verify this.)

When a factor $x - r$ appears k times, we say that r is a zero of **multiplicity k**. In Example 2, -1 is a zero of multiplicity 3, and -2 is a zero of multiplicity 1, or a **simple zero**.

The following result is now clear:

(12) **THEOREM**
If we count each zero according to its multiplicity, every polynomial of degree $n \geq 1$ with complex coefficients has precisely n (complex) zeros.

EXAMPLE 3 Write a polynomial in factored form which has the following zeros: i with multiplicity 2, 6 with multiplicity 4, $3 - \sqrt{2}i$ with multiplicity 5, and 7 with multiplicity 1.

SOLUTION $f(x) = (x - i)^2(x - 6)^4[x - (3 - \sqrt{2}i)]^5(x - 7)$ is such a polynomial. Note that $f(x)$ has degree 12, the sum of the multiplicities.

When we were studying quadratic polynomials, we saw that if a quadratic polynomial has real coefficients and one complex zero $r = a + bi$, then the complex conjugate of that zero, $\bar{r} = a - bi$, is also a zero. This can be generalized.

---

(13)   **THEOREM**

If $f(x) = a_0 + a_1 x + \cdots + a_n x^n$ has real coefficients (i.e., $a_0, a_1, \ldots, a_n$ are all real numbers) and $f(x)$ has a complex zero r, then $\bar{r}$ is also a zero of $f(x)$.

Proof: Since r is a zero, $a_0 + a_1 r + \cdots + a_n r^n = 0$. Take the conjugate of both sides: $\overline{a_0 + a_1 r + \cdots + a_n r^n} = \bar{0}$. On the right-hand side, $\bar{0} = 0$. On the left, we use the fact that the conjugate of a sum is the sum of the conjugates, obtaining $\overline{a_0} + \overline{a_1 r} + \cdots + \overline{a_n r^n} = 0$. But the conjugate of a product is also the product of the conjugates. Thus, $\bar{a}_0 + \bar{a}_1 \bar{r} + \cdots + \bar{a}_n (\bar{r})^n = 0$. Now each coefficient a_i is real, and the conjugate of a real number is itself; i.e., $\bar{a}_i = a_i$. Consequently, $a_0 + a_1 \bar{r} + \cdots + a_n (\bar{r})^n = 0$. But this is just the equation we need to say that $f(\bar{r}) = 0$, which means that $\bar{r}$ is a zero of $f(x)$.

EXAMPLE 4 $1 - 2i$ is one zero of $f(x) = x^4 - 4x^3 + 11x^2 - 14x + 10$. Find the remaining zeros.

SOLUTION   Since $f(x)$ has real coefficients and $1 - 2i$ is a zero, its conjugate, $1 + 2i$, is also a zero. Hence $f(x) = [x - (1 - 2i)][x - (1 + 2i)] f_1(x)$, where $f_1(x)$ is quadratic. We can find $f_1(x)$ either by multiplying out the two linear factors and dividing the result into $f(x)$ or by using synthetic division twice. We do the latter:

$1 - 2i$	1	-4	11	-14	10
		$1 - 2i$	$-7 + 4i$	$12 - 4i$	-10
	1	$-3 - 2i$	$4 + 4i$	$-2 - 4i$	0

$1 + 2i$	1	$-3 - 2i$	$4 + 4i$	$-2 - 4i$
		$1 + 2i$	$-2 - 4i$	$2 + 4i$
	1	-2	2	0

Thus, $f_1(x) = x^2 - 2x + 2$, which we can solve by the quadratic formula to obtain $x = (2 \pm \sqrt{4 - 8})/2 = 1 \pm i$. Therefore the zeros of $f(x)$ are $x = 1 \pm 2i, 1 \pm i$.

EXAMPLE 5 Find a polynomial $f(x)$ of lowest degree with zeros $2 + 3i$ of multiplicity 1, $-2i$ of multiplicity 2, and 5 of multiplicity 1 if **(a)** the coefficients of $f(x)$ are allowed to be complex and **(b)** the coefficients of $f(x)$ are real.

SOLUTION **(a)** A polynomial with complex coefficients and the given zeros is

$$f(x) = [x - (2 + 3i)](x + 2i)^2(x - 5)$$
$$= x^4 - (7 - i)x^3 + (18 - 13i)x^2 - (32 - 28i)x - 40 + 60i$$

(b) If $f(x)$ is to have real coefficients, then $f(x)$ must also have the zeros $2 - 3i$ with multiplicity 1 and $2i$ of multiplicity 2. (Note that the conjugate of 5 is 5. In this case the theorem says "If 5 is a zero, then 5 is a zero." Thus no new zeros arise.) Thus we may choose

$$f(x) = [x - (2 + 3i)][x - (2 - 3i)](x + 2i)^2(x - 2i)^2(x - 5)$$
$$= (x^2 - 4x + 13)(x^2 + 4)^2(x - 5)$$
$$= x^7 - 9x^6 + 41x^5 - 137x^4 + 280x^3 - 664x^2 + 528x - 1040$$

EXERCISES

In Exercises 1–8, leave the answer in factored form.

1. Find a polynomial of degree 3 with zeros $1, -2, 5$.
2. Find a polynomial of degree 4 with zeros $-6, 4, 7, -2$.
3. Find a polynomial of degree 3 with zeros $i, 3, -2 + 5i$.
4. Find a polynomial of degree 4 with zeros $2 + i, 2 - i, 1 - 3i, 1 + 3i$.
5. Find a polynomial of degree 5 with zeros -1 of multiplicity 3 and -3 of multiplicity 2.
6. Find a polynomial of degree 5 with zeros 2 of multiplicity 2, -2 with multiplicity 2, and 7 with multiplicity 1.
7. Find a polynomial of degree 5 with zeros i of multiplicity 3, 6 with multiplicity 1, and -6 with multiplicity 1.
8. Find a polynomial of degree 7 with zeros $1 + i$ of multiplicity 2, $1 - i$ of multiplicity 2, 1 with multiplicity 2, and -10 with multiplicity 1.

In Exercises 9–12, find a polynomial of lowest degree with the given zeros if **(a)** the coefficients may be nonreal and **(b)** the coefficients *must* be real.

9. $2i, 3, 1 - i$
10. $3 + 2i, 3 - 2i, 2 + 3i$
11. $3i$ with multiplicity 3
12. $-1 + 2i$ with multiplicity 2, 1 with multiplicity 2

In Exercises 13–18, find the zeros of the given polynomials together with their multiplicities.

13. $(x^2 - 4x + 3)(x^2 + 3x - 4)$
14. $(x^2 - 25)^3(x^2 - 3)$
15. $(x^2 + 2x + 4)^4(x - \sqrt{2})^3$
16. $(x^4 - 1)^5$
17. $(x - 1 + i)^3(x^3 + 8)$
18. $(x^3 - 8)^3(x^2 + 5)^2(x + 3 - i)^4$

In Exercises 19–22, use the given zeros to find the remaining zeros of the polynomials.

19. $3x^3 + 20x^2 + 46x + 35$ has $-\frac{5}{3}$ as a zero.

20. $x^4 - (2 + 6i)x^3 - (14 - 12i)x^2 + (18 + 30i)x + 45$ has $3i$ as a zero of multiplicity 2.
21. $x^4 + 5x^3 + 13x^2 + 12x + 8$ has $-2 - 2i$ as a zero.
22. $x^7 - 3x^6 + 3x^5 - x^4 - 16x^3 + 48x + 16$ has 1 as a zero of multiplicity 3.
23. Find a polynomial with zeros $6, -4, -7, 2$, and compare with your answer to Exercise 2. (Multiply both out.) In general, if $f(x) = a_n x^n + a_{n-1} x^{n-1} + a_{n-2} x^{n-2} + \cdots + a_0$ has zeros $r_1, r_2, \ldots, r_n$, find a polynomial which has zeros $-r_1, -r_2, \ldots, -r_n$.
24. Find the polynomial with integer coefficients with zeros $1, -\frac{1}{2}, \frac{1}{3}$, and compare with your answer to Exercise 1. In general, if $f(x) = a_n x^n + a_{n-1} x^{n-1} + \cdots + a_1 x + a_0$ has all its zeros $r_1, \ldots, r_n$ nonzero, find a polynomial which has zeros $1/r_1, 1/r_2, \ldots, 1/r_n$.
25. Show that if $f(x) = x^n + a_{n-1} x^{n-1} + \cdots + a_1 x + a_0$ has all integral coefficients, then any rational zero of $f(x)$ is an integer.

SECTION 11.4. THE RATIONAL ROOT THEOREM

The problem of finding the zeros of a polynomial, or equivalently the roots of a polynomial equation, has occupied the attention of mathematicians for ages. Of course, it is trivial to find the zero of a linear polynomial, and the quadratic formula may be used to find the zeros of a quadratic polynomial. We have mentioned that there are also formulas to find the zeros of third- and fourth-degree polynomials but that Abel and Galois proved that there *cannot* be a general formula to find the zeros of a fifth-degree polynomial (or of a polynomial of larger degree).

Although there is no general formula for finding all the zeros of a polynomial of degree 5 or more, the following theorem enables us to list all *possible rational* zeros of any polynomial whose *coefficients are all integers*:

(14)   THE RATIONAL ROOT THEOREM

Let $f(x) = a_n x^n + a_{n-1} x^{n-1} + \cdots + a_1 x + a_0$, $a_n \neq 0$ be a polynomial with integer coefficients. If $f(x) = 0$ has a rational root s/t and s/t is in lowest terms, then s is a factor of a_0 and t is a factor of a_n.

EXAMPLE 1 It can be shown that the only rational zeros of $f(x) = 10x^4 - 69x^3 + 10x^2 - 107x - 84$ are $-\frac{3}{5}$ and $7 = \frac{7}{1}$. Note that the numerators, -3 and 7, are factors of -84, the constant coefficient in $f(x)$, and that the denominators, 5 and 1, are factors of 10, the coefficient of the highest power of x in $f(x)$.

Proof of the Theorem: We know that s/t being in lowest terms means that the integers s and t have no common factors other than ± 1. Since s/t is a zero of $f(x)$,

$$a_n \left(\frac{s}{t}\right)^n + a_{n-1} \left(\frac{s}{t}\right)^{n-1} + \cdots + a_1 \left(\frac{s}{t}\right)^1 + a_0 = 0$$

Multiply this equation by t^n to obtain

(15) $a_n s^n + a_{n-1} s^{n-1} t + \cdots + a_1 s t^{n-1} + a_0 t^n = 0$

Subtract $a_0 t^n$ from both sides:

$$a_n s^n + a_{n-1} s^{n-1} t + \cdots + a_1 s t^{n-1} = -a_0 t^n$$

Factor s out of the left:

$$s(a_n s^{n-1} + a_{n-1} s^{n-2} t + \cdots + a_1 t^{n-1}) = -a_0 t^n$$

Since s is a factor of the left, it is a factor of the right (both sides are integers). But s and t have no common factor other than ± 1, so s and t^n have no common factors other than ± 1. Since s is a factor of $-a_0 t^n$, it must be that s is a factor of a_0. This proves the first part of the theorem. The proof of the second part is similar: We start with equation (15) and subtract $a_n s^n$ from both sides:

$$a_{n-1} s^{n-1} t + \cdots + a_0 t^n = -a_n s^n$$

Then factor t from the left, obtaining

$$t(a_{n-1} s^{n-1} + \cdots + a_0 t^{n-1}) = -a_n s^n$$

Then since t must be a factor of $-a_n s^n$, we can conclude, by the same type of reasoning, that t is a factor of a_n.

Let us see how we can use the rational root theorem to find all the rational zeros of a polynomial with integer coefficients.

EXAMPLE 2 Find all the rational zeros of $f(x) = 6x^4 + 11x^3 + 8x^2 - 6x - 4$

SOLUTION If s/t is a rational zero of $f(x)$ in lowest terms, then s is a factor of the constant term, -4, and t is a factor of the coefficient, 6, of the highest power of x. Thus the possibilities for s and t are $s = \pm 1, \pm 2, \pm 4$ and $t = \pm 1, \pm 2, \pm 3, \pm 6$. This means that the possible rational zeros are

$$\pm 1, \pm 2, \pm 4, \pm \tfrac{1}{2}, \pm \tfrac{1}{3}, \pm \tfrac{2}{3}, \pm \tfrac{4}{3}, \pm \tfrac{1}{6}$$

The only way to determine which of these (if any) are actually zeros of $f(x)$ is to try them. The usual way of doing this is to use synthetic division. With a calculator, you may wish to just evaluate the polynomial at each of these numbers. If so, you should use nested polynomials, which we have seen to be equivalent to synthetic division.

In this problem, if we proceed down the list, we find that $\pm 1, \pm 2, \pm 4$, and $+\tfrac{1}{2}$ are not zeros but that $-\tfrac{1}{2}$ is:

$$\begin{array}{r|rrrrr} -\tfrac{1}{2} & 6 & 11 & 8 & -6 & -4 \\ & & -3 & -4 & -2 & 4 \\ \hline & 6 & 8 & 4 & -8 & 0 \end{array}$$

Thus, $f(x) = (x + \tfrac{1}{2})(6x^3 + 8x^2 + 4x - 8)$. We factor out a 2 from the second factor to make the following work easier:

$$f(x) = (x + \tfrac{1}{2}) 2 (3x^3 + 4x^2 + 2x - 4)$$

Any remaining rational zeros of $f(x)$ must be zeros of $3x^3 + 4x^2 + 2x - 4$. It usually pays to list again the possible rational zeros of this polynomial, for there may be some on our original list which are no longer possible. We obtain $s = \pm 1, \pm 2, \pm 4$ and $t = \pm 1, \pm 3$, so $s/t = \pm 1, \pm 2, \pm 4, \pm \frac{1}{3}, \pm \frac{2}{3}, \pm \frac{4}{3}$. Note that if we had not made this new list, we would have had to try $-\frac{1}{2}$ again (it could have been a multiple zero), and we would have had to try $\pm \frac{1}{6}$. We have already tried $\pm 1, \pm 2, \pm 4$. Continuing, we find that $\pm \frac{1}{3}$ are not zeros, while $+\frac{2}{3}$ is a zero:

$$\begin{array}{r|rrrr} \frac{2}{3} & 3 & 4 & 2 & -4 \\ & & 2 & 4 & 4 \\ \hline & 3 & 6 & 6 & 0 \end{array}$$

(*Note*: If you are using a calculator, you will probably not get zero, because you will have approximated 2/3 by, say, .6666667. Whenever you do so and get a very small number, such as -8×10^{-12}, you are probably not getting zero because of round-off error. Check the exact solution by hand.)

We now have

$$f(x) = \left(x + \tfrac{1}{2}\right)2\left(x - \tfrac{2}{3}\right)(3x^2 + 6x + 6)$$
$$= \left(x + \tfrac{1}{2}\right)2\left(x - \tfrac{2}{3}\right)3(x^2 + 2x + 2)$$

The zeros of the last factor can be found by the quadratic formula. However, since the rational root theorem tells us that the only possible rational zeros of this factor are ± 1 and ± 2 and we have already tried these, we know there are no further rational zeros. Thus the only rational zeros of $f(x)$ are $-\frac{1}{2}$ and $\frac{2}{3}$.

If the coefficients of a polynomial $g(x)$ are all rational numbers, it is still possible to find all its rational zeros. We simply multiply through both sides of the equation $g(x) = 0$ by the least common multiple of the denominators of the coefficients. This gives a polynomial equation with *integers* for all coefficients, so we can find its rational roots by the rational root theorem. Clearly, the new equation has the same roots as $g(x) = 0$.

EXAMPLE 3 Find the rational zeros of $g(x) = x^4 + \tfrac{11}{6}x^3 + \tfrac{4}{3}x^2 - x - \tfrac{2}{3}$.

SOLUTION If we multiply the equation $g(x) = 0$ by 6, the least common multiple of the denominators 6, 3, and 3 of the coefficients, we obtain

$$6x^4 + 11x^3 + 8x^2 - 6x - 4 = 0$$

We now use the rational root theorem to determine the rational roots of this equation and hence of $g(x) = 0$. This was done in Example 2, so the rational zeros of $g(x)$ are $-\frac{1}{2}, \frac{2}{3}$.

There are some observations you can make that help reduce the number of the possible zeros you must try.

EXAMPLE 4 If $f(x) = 5x^6 + 3x^4 + 7x + 6$, then the possible rational zeros of $f(x)$ given by the rational root theorem are ± 1, ± 2, ± 3, ± 6, $\pm \frac{1}{5}$, $\pm \frac{2}{5}$, $\pm \frac{3}{5}$, $\pm \frac{6}{5}$. However, the coefficients of $f(x)$ are all nonnegative. So if c is any positive number, $f(c) = 5c^6 + 3c^4 + 7c + 6$ is a sum of four positive numbers and hence is positive (and not 0). Thus in trying to find the rational zeros of $f(x)$, we need only consider the negative possibilities.

EXAMPLE 5 If $g(x) = 3x^5 - x^4 + 6x^3 - 2x^2 + 8x - 5$, then the possible rational zeros given by the rational root theorem are ± 1, ± 5, $\pm \frac{1}{3}$, $\pm \frac{5}{3}$. However, the coefficients are *alternating* in sign, that is, the signs are $+, -, +, -, \ldots$. So if c is any negative number, for example, if $c = -5$, then

$$g(c) = g(-5) = 3(-5)^5 - (-5)^4 + 6(-5)^3 - 2(-5)^2 + 8(-5) - 5$$
$$= -3(5)^3 - (5)^4 - 6(5)^3 - 2(5)^2 - 8(5) - 5$$

is a sum of negative numbers and hence is negative (and not zero). Thus in trying to find the rational zeros of $g(x)$, we need only consider the positive possibilities. [*Note*: The coefficients of $h(x) = 2x^3 - x + 1$ are *not* alternating in sign because the coefficients are 2, 0, -1, 1 and not just 2, -1, 1. In particular, observe that -1 is a zero of $h(x)$.]

EXERCISES

In Exercises 1–10, use the rational root theorem to determine the possible rational zeros of the given polynomial, but do not determine which of the possibilities actually are zeros.

1. $2x^5 - x^4 - 5x - 3$
2. $5x^7 - x^5 - 12x^2 - 2$
3. $x^6 - x^4 - 5x^3 + 3x^2 - 1$
4. $11x^{12} - 2x^5 - x^3 + 6$
5. $12x^6 + 3x^5 + 5x^4 + 4x^3 + 2x^2 + 4x + 4$
6. $-8x^4 - 3x^3 - 4x^2 - 2x - 6$
7. $3x^4 - \frac{1}{2}x^3 + \frac{1}{3}x^2 - 2x + 1$
8. $2x^5 - x^4 + \frac{3}{5}x^3 - x^2 + \frac{4}{5}x - 2$
9. $2x^7 - \frac{1}{4}x^5 + \frac{1}{2}x^3 - 5x$
10. $3x^7 - \frac{1}{5}x^5 + \frac{2}{3}x^3$

11. Explain why $f(x) = 2x^6 + x^4 + 4x^2 + 3$ can have no real zeros and hence no rational zeros.

12. Explain why $f(x) = -3x^5 - 4x^3 - 5x$ can have no real zeros other than zero and hence has no rational zeros other than zero.

In Exercises 13 – 32, find all the rational zeros of the given polynomial. If possible, find all the zeros.

13. $x^3 + 5x^2 - x - 5$
14. $x^6 - 1$
15. $2x^4 + x^3 + 2x^2 - x - 1$
16. $5x^3 - x^2 - 15x + 3$
17. $3x^3 - 2x^2 + 9x - 6$
18. $30x^3 - 53x^2 + 6x + 9$
19. $6x^4 + 3x^3 - 4x^2 - 5x - 10$
20. $6x^5 + 11x^4 + 51x^3 + 43x^2 + 63x + 18$
21. $2x^5 + 5x^4 - 2x^3 - 7x^2 - 4x - 12$
22. $9x^6 + 45x^5 - 25x^4 - 105x^3 + 96x^2 - 20x$
23. $14x^5 + 89x^4 + 78x^3 + 78x^2 + 64x - 11$
24. $6x^4 + 5x^3 + 14x^2 - 6x - 9$
25. $210x^4 - 667x^3 + 385x^2 - 77x + 5$
26. $120x^4 - 196x^3 + 114x^2 - 27x + 2$

27. $9x^6 + 60x^5 + 154x^4 + 192x^3 + 121x^2 + 36x + 4$
28. $x^3 + \frac{29}{6}x^2 + 7x + 3$
29. $\frac{1}{2}x^4 + \frac{9}{4}x^3 + 4x^2 + \frac{13}{4}x + \frac{5}{2}$
30. $x^{11} + 3x^2 - 2$
31. $64x^6 + 2x^2 - x - 1$
32. $120x^3 - 284x^2 + 172x - 24$

33. The height of a rectangular solid is 1 foot less than the width and the length is 1 foot more than twice the width. Find the dimensions if the volume is 3 cubic feet.

34. If one side of a cube is doubled, the second side is increased by 3 meters, and the third side is left alone, then the volume of the cube is increased by 4/3 cubic meters. What is the length of one side of the cube?

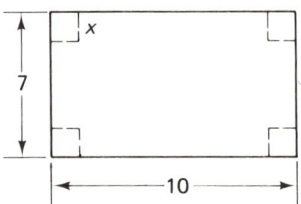

35. Identical squares are cut from each corner of a rectangular piece of tin 7 by 10 centimeters. The sides are then folded up to make a box (with an open top). If the volume of the box is 42 square centimeters, how long is the side of each square?

REVIEW EXERCISES

1. Use the remainder theorem to find $f(-3)$ if $f(x) = 2x^4 - x^3 - 6x^2 + 4x - 5$.
2. What is the remainder when $7x^{163} + 5x^{96} - 11x^{41} + 6$ is divided by $x + 1$?
3. Show that $x - 2$ is a factor of $f(x) = x^8 - 256$.
4. Determine k such that $x - 3$ is a factor of $x^4 - 4kx^2 + 5x + 12k$.

In Exercises 5–7, use synthetic division to find the quotient $q(x)$ and the remainder r.

5. $(5x^6 - 7x^4 - 16x^3 - 12x^2 - 5) \div (x + 2)$
6. $(x^3 - 4.5x^2 + 2.48x - 5.035) \div (x - 1.2)$
7. $[2x^4 + (3 - 2i)x^3 - (1 - i)x^2 + (5 + 3i)x + 1 + i] \div (x - i)$
8. Use synthetic division to find $f(-1), f(2)$, and $f(4)$ if $f(x) = -3x^5 + x^4 + 6x^3 - x^2 - 5x + 7$.
9. Write $f(x) = -2x^5 + 6x^4 + x^3 - 5x^2 - 11x + 8$ in nested form.
10. Use the method of nested polynomials without writing $f(x)$ in that form to find $f(.5), f(-4)$, and $f(2.7)$ if $f(x) = x^6 - 13.5x^5 + 11x^4 + 17x^2 + 86.42$.
11. Find a polynomial of degree 4 with zeros $2, 3, -3, 4$.
12. Find a polynomial of degree 7 with zeros 5 of multiplicity 2, -8 of multiplicity 4, and 16 of multiplicity 1.
13. Find a polynomial of lowest degree with zeros $1 - i, 1 + i, 2, 1 + 2i$ if **(a)** the coefficients may be nonreal or **(b)** the coefficients must be real.
14. $f(x) = 8x^5 + 4x^4 - 2x^3 + 7x^2 - 10x + 3$ has $-i$ as one zero. Find the remaining zeros.
15. Find all the rational zeros of $f(x) = 18x^3 + 9x^2 - 23x + 6$.
16. Find all the zeros of $f(x) = 6x^4 - 23x^3 + 20x^2 - 14x - 3$.

Sequences and Enumeration

CHAPTER TWELVE

Sequences and series appear in some of the earliest recorded mathematical works. Geometric progressions appear in the *Rhind Papyrus*, which was written in Egypt around 1650 B.C. However, the problems on this papyrus had been around for several centuries. A formula giving the sum of a finite geometric progression appears in Book IX of Euclid's *Elements*, which was written around 300 B.C. Archimedes (287–212 B.C.), who is noted for his explanation of buoyancy and leverage, knew how to sum infinite geometric series.

Although rudimentary forms of mathematical induction appeared in the middle sixteenth century, the first mathematicians to make well-formulated use of this method of proof were Fermat (1601–1665) and Pascal (1623–1662). Fermat used induction, formulated in a way that he called *infinite descent*, to prove number theoretic results. Pascal, who was the first person to give a clear and easily understandable explanation of how and why induction works, used induction to prove the relationships he discovered among binomial coefficients.

Pascal became very involved with binomial coefficients during his analysis of probability. In fact, the numerical triangle shown, which is made up of the binomial coefficients from the triangle at the beginning of Sec. 6, is often called a *Pascal triangle*, because Pascal was the first person to study relationships among the coefficients. However, it was known five centuries earlier both in China and in Arabia.

```
              1
            1   1
          1   2   1
        1   3   3   1
      1   4   6   4   1
    1   5  10  10   5   1
  1   6  15  20  15   6   1
```
Pascal triangle

SECTION 12.1. SEQUENCES AND ARITHMETIC PROGRESSIONS

An **infinite sequence** of numbers is an ordered collection of numbers

$$a_1, a_2, a_3, \ldots, a_n, \ldots$$

in which there is one number for each positive integer. The first number a_1 is called the **first term**, the second number a_2 is called the **second term**, and, in general, the nth number a_n is called the **nth term**. If there is a formula given for a_n in terms of n, then this is called the **general term**. Order is important, so that the two sequences

$$1, 2, 3, 4, \ldots \quad \text{and} \quad 1, 3, 2, 4, \ldots$$

are different sequences even if all the remaining terms are equal.

One way to describe a sequence is to list the first several terms of a sequence *and* to indicate the general term. A second way is just to give the general term.

The following are some examples of sequences:

a. $1, \frac{1}{2}, \frac{1}{3}, \frac{1}{4}, \ldots, 1/n, \ldots$ or $a_n = 1/n$
b. $2, \frac{3}{2}, \frac{4}{3}, \frac{5}{4}, \ldots, (n+1)/n, \ldots$ or $a_n = (n+1)/n$
c. $0, 2, 0, 2, 0, 2, \ldots, 1 + (-1)^n, \ldots$ or $a_n = 1 + (-1)^n$
d. $2, 3, 5, 7, 11, \ldots, a_n, \ldots$, where a_n is the nth prime number

A third method of defining the terms in a sequence is called a **recursive** definition. The most common type we run across is when a_1 is given, and there is a formula giving a_n in terms of a_{n-1}. Thus there is not a formula for a_n where you can just plug in a number, say 23, and find the value of a_{23}. Example 1 illustrates how to use a recursive definition.

EXAMPLE 1 Suppose $a_1 = 1$ and $a_n = na_{n-1}$ for $n > 1$. Find the first six terms of the sequence.

SOLUTION We already know $a_1 = 1$. We use this and the formula $a_n = na_{n-1}$ with $n = 2$ to find a_2, $a_2 = 2a_{2-1} = 2a_1 = 2 \cdot 1 = 2$. Since we now know a_2, we can use the formula $a_n = na_{n-1}$ with $n = 3$ to find a_3, $a_3 = 3a_{3-1} = 3a_2 = 3 \cdot 2 = 6$. Similarly, $a_4 = 4a_3 = 24$, $a_5 = 5a_4 = 120$, $a_6 = 6a_5 = 720$.

Since given the value of one term, you can use the (recursive) formula $a_n = na_{n-1}$ to find the value of the next term, you can eventually find the value of any term you desire. Of course in this case you may soon realize that $a_n = n(n-1)(n-2) \cdots 3 \cdot 2 \cdot 1$ and compute any term directly. But in most cases where a sequence is given recursively, a formula for a_n is not so obvious.

Sometimes a recursive formula may give a_n not just in terms of a_{n-1}, but in terms of a_{n-2}, or additional previous terms as well. If so, you need more initial values to get started.

EXAMPLE 2 Suppose $a_1 = 1$, $a_2 = 1$ and $a_n = a_{n-1} + a_{n-2}$ for $n > 2$. Find the first six terms of the sequence.

SOLUTION We use $a_1 = a_2 = 1$ and the formula $a_n = a_{n-1} + a_{n-2}$ with $n = 3$ to compute a_3, $a_3 = a_2 + a_1 = 1 + 1 = 2$. We next use the formula $a_n = a_{n-1} + a_{n-2}$ with $n = 4$ to compute a_4, $a_4 = a_3 + a_2 = 2 + 1 = 3$. Similarly, $a_5 = a_4 + a_3 = 3 + 2 = 5$, and $a_6 = a_5 + a_4 = 5 + 3 = 8$.

The sequence given in Example 2 is an example of a **Fibonacci sequence**, named after the Italian mathematician Leonardo Fibonacci (1170?–1250?). It arises naturally in biology from certain types of growth situations. See Exercises 19 and 20.

An **arithmetic progression** (or **arithmetic sequence**) is a sequence in which the difference between any two successive terms is a constant d. If a sequence is defined by giving a_1 and then saying the sequence satisfies the recursive equation $a_n = a_{n-1} + d$, where d is a constant, then the sequence is an arithmetic progression. This is because the recursive equation $a_n = a_{n-1} + d$ can be rewritten as $a_n - a_{n-1} = d$; i.e., the difference between any two successive terms is d. (The number d is called the **common difference**.) The first few terms in the sequence are

$$(1) \qquad a_1, a_1 + d, a_1 + 2d, a_1 + 3d, a_1 + 4d, \ldots$$

EXAMPLE 3 Write out the first five terms in the arithmetic progressions in which $a_1 = 3$ and for $n > 1$, a_n is defined by **(a)** $a_n = a_{n-1} + 2$; **(b)** $a_n = a_{n-1} - 2$; **(c)** $a_n = a_{n-1} + \pi$.

SOLUTION **(a)** $a_1 = 3$, $a_2 = 5$, $a_3 = 7$, $a_4 = 9$, $a_5 = 11$.
(b) $a_1 = 3$, $a_2 = 1$, $a_3 = -1$, $a_4 = -3$, $a_5 = -5$.
(c) $a_1 = 3$, $a_2 = 3 + \pi$, $a_3 = 3 + 2\pi$, $a_4 = 3 + 3\pi$, $a_5 = 3 + 4\pi$.

From (1), it appears that a formula for a_n is

(2)
$$a_n = a_1 + (n-1)d$$

This is in fact the case; it can be proved by mathematical induction. Formula (2) can be used to find any term in an arithmetic progression once a_1 and d are known.

EXAMPLE 4 Suppose the first five terms in an arithmetic progression are $-1, 3, 7, 11, 15$. Find the 81st term.

SOLUTION We may subtract any term from the next term to find d. Thus, $d = 3 - (-1) = 7 - 3 = 11 - 7 = 15 - 11$, or $d = 4$. Then we know that $a_1 = -1$ and $n = 81$. We can use (2) to find the 81st term: $a_{81} = (-1) + (81 - 1)4 = 319$.

If we are told that we have an arithmetic progression but are only told some nonsuccessive terms, we may have to solve a system of two equations in two unknowns to find a_1 and d.

EXAMPLE 5 Find the 23rd term in an arithmetic progression if the fourth term is 8 and the tenth term is 29.

SOLUTION We know that $a_4 = 8$ and $a_{10} = 29$. But by (2) we also know that $a_4 = a_1 + 3d$ and $a_{10} = a_1 + 9d$. Thus we have the system of equations

$$a_1 + 3d = 8$$
$$a_1 + 9d = 29$$

If we solve this system, we obtain $d = 3.5$ and $a_1 = -2.5$. Thus by (2), $a_{23} = -2.5 + 22(3.5) = 74.5$.

It is sometimes of interest to find the sum of the first n terms of an arithmetic progression, which we call the nth **partial sum** S_n of the progression. With a calculator, this is not very hard (unless n is large), but it may be very time-consuming. However, it is not necessary to add up all n terms on a calculator, for there is a very simple formula.

If an arithmetic progression has first term a_1 and nth term a_n, then the nth partial sum S_n is given by

(3)
$$S_n = \frac{n(a_1 + a_n)}{2}$$

This can be seen by first writing S_n in ascending and descending order:
$$S_n = a_1 + a_2 + a_3 + \cdots + a_n$$
$$S_n = a_n + a_{n-1} + a_{n-2} + \cdots + a_1$$

Now rewrite these two equations using the first term given and the difference d:
$$S_n = a_1 + (a_1 + d) + (a_1 + 2d) + \cdots + [a_1 + (n-1)d]$$
$$S_n = a_n + (a_n - d) + (a_n - 2d) + \cdots + [a_n - (n-1)d]$$

Adding the last two equations yields
$$2S_n = (a_1 + a_n) + (a_1 + a_n) + (a_1 + a_n) + \cdots + (a_1 + a_n)$$

Since there are n terms of the form $a_1 + a_n$,

(4)
$$2S_n = n(a_1 + a_n) \quad \text{or} \quad S_n = \frac{n(a_1 + a_n)}{2}$$

EXAMPLE 6 Find the sum of the first 85 terms in the arithmetic progression whose first three terms are 2, 5, 8.

SOLUTION $a_1 = 2$, and $d = 5 - 2 = 8 - 5 = 3$. Therefore, $a_{85} = 2 + 84(3) = 254$ by (2). Consequently, $S_{85} = 85(2 + 254)/2 = 10{,}880$.

EXERCISES

In Exercises 1–10, find the first five terms and the eleventh term in the sequence which has the given nth term.

1. $a_n = n$
2. $a_n = n^3$
3. $a_n = 2n + 5$
4. $a_n = \dfrac{n+2}{4n-3}$
5. $a_n = \sqrt{n} + \dfrac{1}{n}$
6. $a_n = 2^{n-1}$
7. $a_n = 8$
8. $a_n = n^2 + 2n - 3$
9. $a_n = \sin(n\pi/6)$
10. $a_n = \cos\dfrac{n\pi}{3}$

In problems 11–18, find the first five terms of the sequences defined recursively.

11. $a_1 = 3$, $a_n = -a_{n-1} + 4$ for $n > 1$
12. $a_1 = -4$, $a_n = 2a_{n-1} + 10$ for $n > 1$
13. $a_1 = -4$, $a_n = a_{n-1}/2$ for $n > 1$
14. $a_1 = -4$, $a_n = (a_{n-1})^{-1}$ for $n > 1$
15. $a_1 = 2$, $a_n = (-a_{n-1})^n$ for $n > 1$
16. $a_1 = 3$, $a_2 = 1$, $a_n = a_{n-1} + 2a_{n-2}$ for $n > 2$
17. $a_1 = 2$, $a_2 = -3$, $a_n = a_{n-1}a_{n-2}$ for $n > 2$
18. $a_1 = 1$, $a_2 = 2$, $a_n = a_{n-1}^2 + a_{n-2}^2$ for $n > 2$

19. A certain species of virus multiplies by each mature cell splitting off one new cell at the end of each hour. Each new cell must mature for one hour and then begin splitting off a new cell at the end of the second hour. Suppose you start off with one new cell.

    (a) Write down the first six terms of the sequence whose nth term is how many cells you have at the beginning of the nth hour. (Compare with Example 2.)

    (b) How many cells do you have after a day?

20. Each couple, one male and one female, of a certain species of rabbit produces another couple at the end of each month. The new couple must mature for a month and then start producing its own offspring at the end of the second month. Suppose you start with one couple of baby rabbits.

    (a) Write down the first six terms of the sequence whose nth term is the number of rabbits you have at the beginning of the nth month.

    (b) How many rabbits do you have after a year?

In Exercises 21–28, write the first four terms and the 100th term of an arithmetic progression with the given a_1 and d.

21. $a_1 = 4, d = 2$
22. $a_1 = 9, d = -4$
23. $a_1 = 6.3, d = .2$
24. $a_1 = 98, d = -2.5$
25. $a_1 = 11, d = 3$
26. $a_1 = \log 8, d = \log 2$
27. $a_1 = 3 - 2i, d = 2 + i$, where $i^2 = -1$
28. $a_1 = -\frac{3}{8}, d = \frac{5}{8}$

In Exercises 29–36, the first three terms of an arithmetic progression are given. Find the indicated term and partial sum.

29. $1, 4, 7$; find a_{47} and S_{47}.
30. $20, 19.5, 19$; find a_{64} and S_{64}.
31. $-86, -80, -74$; find a_{183} and S_{183}.
32. $\frac{35}{3}, \frac{33}{3}, \frac{31}{3}$; find a_{206} and S_{206}.
33. $48, 147, 246$; find a_{34} and S_{34}.
34. $.62, .69, .76$; find a_{94} and S_{94}.
35. $5x + 3y, 6x + y, 7x - y$; find a_{27} and S_{27}.
36. $4a - 6b, 2a - 3b, 0$; find a_{37} and S_{37}.

In Exercises 37–42, the terms given are from an arithmetic progression. Find a_1, d, a_n, and S_n for the given n.

37. $a_{17} = 4, a_{18} = 7; n = 31$
38. $a_{93} = 102, a_{94} = 63; n = 6$
39. $a_6 = 27, a_{13} = 55; n = 48$
40. $a_7 = 6, a_{18} = -16; n = 108$
41. $a_{31} = 17.3, a_{41} = 22.7; n = 50$
42. $a_1 = 0, a_{13} = \pi; n = 45$

43. Find the sum of all the even positive integers less than 100.
44. Find the sum of all the positive multiples of 3 which are less than 100.
45. How many integers between 100 and 1000 are divisible by 7? What is their sum?
46. How many integers between 50 and 500 are divisible by 9? What is their sum?
47. A stack of bricks has 61 bricks in the bottom layer, 58 bricks in the second layer, 55 bricks in the third layer, and so on until there are 10 bricks in the last layer.

    a. How many layers are there?

    b. How many bricks are in the eleventh layer?

    c. How many bricks are there altogether?

48. A bomb drops from a plane and falls 16 feet during the first second, 48 feet during the second second, and so on, forming an arithmetic progression. Suppose the bomb hits the ground

at the beginning of the sixteenth second.
   a. How far did the bomb fall during the fifteenth second?
   b. How high was the airplane when the bomb was dropped?

49. A girl on a bicycle coasts downhill, covering 4 feet the first second, 12 feet the second second, and, in general, 8 feet more each second than the previous second. If she reaches the bottom at the end of 14 seconds, how far did she coast? How far did she travel the tenth second?

50. The seats in a theater are arranged so that there are 70 seats in the first row, 72 seats in the second row, and so on for 30 rows altogether. How many seats are in the theater? How many seats are in the next to last row?

51. Once a month a man put some money into a cookie jar. Each month he put 50 cents more into the jar than the month before. After 12 years he counted his money; he had $5436. How much money did he put in the jar the first month and the last month?

52. A woman accepts a position at a salary of $16,000 for the first year with an increase of $2000 for each year thereafter. What will be her earnings after 12 years? How many years will she have to work for her total earnings to equal $1 million?

SECTION 12.2. GEOMETRIC PROGRESSIONS AND GEOMETRIC SERIES

In the previous section we studied arithmetic progressions. These are sequences in which any two successive terms have a common difference. A second kind of sequence which is quite useful is a sequence in which any two successive terms have a **common ratio** r, i.e.,

$$\frac{a_{k+1}}{a_k} = r \quad \text{for all positive integers } k$$

Such a sequence is called a **geometric progression** (or **geometric sequence**). If we rewrite the preceding equation as $a_{k+1} = a_k r$, we see that this is a recursive equation. Thus if we are given a_1 and the common ratio r, we can compute any term in a geometric progression. The progression begins

$$a_1, a_1 r, a_1 r^2, a_1 r^3, a_1 r^4, \ldots$$

The nth term in this geometric progression appears to be

(5)
$$a_n = a_1 r^{n-1}$$

and one can verify by mathematical induction that this is indeed the case.
   Let us look at some examples which are of a similar nature to those we did for arithmetic progressions.

EXAMPLE 1 If in a geometric progression $a_1 = -2$ and $r = 3$, find the first four terms and the eighth term.

SOLUTION

$$a_1 = -2$$
$$a_2 = a_1 r = (-2)3 = -6$$
$$a_3 = a_1 r^2 = (-2)3^2 = -18$$
$$a_4 = a_1 r^3 = (-2)3^3 = -54$$
$$a_8 = a_1 r^7 = (-2)3^7 = -4374$$

EXAMPLE 2 If $a_6 = -160$ and $a_{21} = 5{,}242{,}880$ are terms in a geometric progression, find a_1, r, and a_{32}. (Assume r is real.)

SOLUTION By equation (5),

$$a_6 = a_1 r^5 = -160 \quad \text{and} \quad a_{21} = a_1 r^{20} = 5{,}242{,}880$$

Dividing a_{21} by a_6, we obtain

$$\frac{a_1 r^{20}}{a_1 r^5} = \frac{5{,}242{,}880}{-160} \quad \text{or} \quad r^{15} = -32{,}768$$

Thus $r = -32{,}768^{1/15} = -2$. Substituting this back in the first equation yields $a_1(-2)^5 = -160$, so $a_1(-32) = -160$, or $a_1 = 5$. Finally, we obtain a_{32} using equation (5):

$$a_{32} = a_1 r^{31} = 5(-2)^{31} \approx -1.07374 \times 10^{10}$$

Suppose we are asked to find the nth partial sum S_n of a geometric progression. We could compute S_n by computing the first n terms and then computing their sum. However, that would be time-consuming, so again, as with arithmetic progressions, we seek an easier way.

The formula for S_n is

$$S_n = a_1 + a_1 r + a_1 r^2 + a_1 r^3 + \cdots + a_1 r^{n-2} + a_1 r^{n-1}$$

If $r = 1$, then $S_n = na_1$. Let us assume $r \neq 1$. Multiply both sides of the equation for S_n by r and subtract rS_n from S_n, obtaining

$$S_n = a_1 + a_1 r + a_1 r^2 + \cdots + a_1 r^{n-2} + a_1 r^{n-1}$$
$$rS_n = a_1 r + a_1 r^2 + a_1 r^3 + \cdots + a_1 r^{n-1} + a_1 r^n$$
$$\overline{S_n - rS_n = a_1 + 0 + 0 + \cdots + 0 - a_1 r^n}$$
$$(1 - r)S_n = a_1(1 - r^n)$$

Since we are assuming $r \neq 1$, we may divide by $1 - r$ to obtain

(6)

If $r \neq 1$ and $S_n = a_1 + a_1 r + a_1 r^2 + \cdots + a_1 r^{n-1}$, then

$$S_n = a_1 \frac{1 - r^n}{1 - r}$$

This is the formula we were seeking for the nth partial sum of a geometric progression.

EXAMPLE 3 Suppose a geometric progression has $a_1 = 5$ and $r = -2$. Find S_{30}.

SOLUTION By formula (6),

$$S_{30} = \frac{a_1(1 - r^{30})}{1 - r} = \frac{5[1 - (-2)^{30}]}{1 - (-2)}$$

$$= \frac{5(1 - 1{,}073{,}741{,}824)}{3}$$

$$= -1{,}789{,}569{,}705$$

EXAMPLE 4

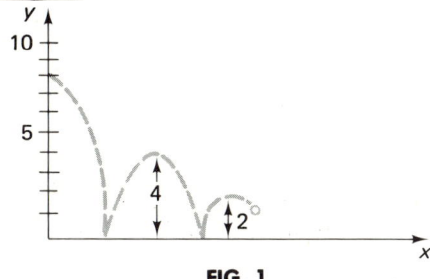

FIG. 1

A certain ball, when dropped from a height, rebounds to one-half of the original height. Suppose this ball is dropped from a point 8 feet above the ground. See Fig. 1. **(a)** How high does it go on the seventh bounce? **(b)** How far has it traveled, counting up and down distance only, when it hits the ground for the eighth time?

SOLUTION Let a_k be how high it rebounds after the kth bounce. Then $a_1 = \frac{1}{2}(8) = 4$ and $a_{k+1} = \frac{1}{2}a_k$, $k \geq 1$. Thus, $a_1, a_2, \ldots$ is a geometric progression and $a_k = (\frac{1}{2})^{k-1}a_1$. For **(a)**, after the seventh bounce, it rebounds

$$a_7 = \left(\frac{1}{2}\right)^6 a_1 = \frac{1}{2^6}(4) = \frac{1}{2^4} = \frac{1}{16} \text{ ft}$$

For **(b)**, on the first fall it travels 8 feet. Then it rebounds up $a_1 = 4$ feet and falls $a_1 = 4$ feet. For the second rebound, it bounces up $a_2 = 2$ feet and then falls down the 2 feet, etc. Therefore, when the ball hits the ground for the eighth time, it has traveled

$$8 + 2a_1 + 2a_2 + \cdots + 2a_7 = 8 + 2(4) + 2(4)(\tfrac{1}{2}) + \cdots + 2(4)(\tfrac{1}{2})^6$$

$$= 8 + 2(4)\frac{1 - (\frac{1}{2})^7}{1 - \frac{1}{2}} \quad \text{by (12)}$$

$$= 8 + 16\left(\frac{2^7 - 1}{2^7}\right) = 8 + \frac{127}{8} = 23.625 \text{ ft}$$

It is interesting and useful to watch what happens as more and more terms of a geometric progression are added together, i.e., to watch what happens to the sums S_n as n increases.

EXAMPLE 5 For the geometric progression $1, \frac{1}{4}, (\frac{1}{4})^2, (\frac{1}{4})^3, \ldots, (\frac{1}{4})^n, \ldots$, **(a)** use your calculator to find the following partial sums: $1, 1 + \frac{1}{4}, 1 + \frac{1}{4} + (\frac{1}{4})^2, 1 + \frac{1}{4} + (\frac{1}{4})^2 + (\frac{1}{4})^3, \ldots, 1 + \frac{1}{4} + (\frac{1}{4})^2 + \cdots + (\frac{1}{4})^{10}$. **(b)** What number do you think these are approaching? **(c)** Use (6) to find $1 + \frac{1}{4} + \cdots + (\frac{1}{4})^5$, $1 + \frac{1}{4} + \cdots + (\frac{1}{4})^8$, $1 + \frac{1}{4} + \cdots + (\frac{1}{4})^{10}$, $1 + \frac{1}{4} + \cdots + (\frac{1}{4})^{15}$, $1 + \frac{1}{4} + \cdots + (\frac{1}{4})^{20}$.

SOLUTION **(a)** You want to find eleven answers, $1, 1 + .25, 1 + .25 + .25^2, \ldots, 1 + .25 + .25^2 + \cdots + .25^{10}$. Each one may be obtained from the preceding answer by adding a power of .25. Put 1 in your calculator (your first answer); add .25 (getting your second answer); add $.25^2$ (getting your third answer), etc. With an RPN calculator, this is straightforward:

$$[1]; \quad [ENT][.25][+]; \quad [.25][ENT][2][y^x][+];$$
$$[.25][ENT][3][y^x][+]; \quad \text{etc.}$$

With an algebraic calculator with algebraic hierarchy, it is also straightforward:

$$[1]; \quad [+][.25][+]; \quad [.25][y^x][2][=][+];$$
$$[.25][y^x][3][=][+]; \quad \text{etc.}$$

With an algebraic calculator without algebraic hierarchy, then (1) if you have parentheses, do the preceding but with parentheses around each $.25^n$; (2) without parentheses, use the memory and $[M+]$, adding each y^n into memory and recalling each answer.

The answers are

1, 1.25, 1.3125, 1.328125, 1.3320312, 1.3330078, 1.3332519, 1.3333129, 1.3333282, 1.3333324, 1.3333327

(b) They seem to be approaching $1.33333\cdots = 4/3$.
(c) By (6),

$$1 + .25 + (.25)^2 + \cdots + (.25)^n = \frac{1 - (.25)^{n+1}}{1 - \frac{1}{4}} = \frac{4}{3}(1 - .25^{n+1})$$

If $n = 5$, we have $\frac{4}{3}(1 - .25^6) \approx 1.3330078$.
If $n = 8$, we have $\frac{4}{3}(1 - .25^9) \approx 1.3333282$.
If $n = 10$, we have $\frac{4}{3}(1 - .25^{11}) \approx 1.3333327$.
If $n = 15$, we have $\frac{4}{3}(1 - .25^{16}) \approx 1.3333333$ on a calculator.
If $n = 20$, we have $\frac{4}{3}(1 - .25^{21}) \approx 1.3333333$ on a calculator.

Example 5 is quite interesting and instructive in that the partial sums seem to be approaching a fixed number, 4/3, and that a calculator actually gives a fixed answer for n "large." By examining the formula for the nth partial sum,

$$S_n = 1 + \frac{1}{4} + (\frac{1}{4})^2 + \cdots + (\frac{1}{4})^{n-1} = \frac{4}{3}\left[1 - (\frac{1}{4})^n\right]$$

we see that the term $(\frac{1}{4})^n$ is arbitrarily close to 0 for n large. Then S_n is arbitrarily close to $\frac{4}{3}(1 - 0) = \frac{4}{3}$ for n large. When this happens, we say the **limit** of the partial sums is $4/3$ or that the **(infinite) geometric series**

$$1 + \tfrac{1}{4} + \left(\tfrac{1}{4}\right)^2 + \left(\tfrac{1}{4}\right)^3 + \cdots$$

converges, and it **adds** (or **sums**) to $4/3$. We write

$$1 + \tfrac{1}{4} + \left(\tfrac{1}{4}\right)^2 + \cdots = \tfrac{4}{3}$$

In general, for the geometric sequence $a_1, a_1 r, a_1 r^2, \ldots$, the infinite sum

$$a_1 + a_1 r + a_1 r^2 + \cdots$$

is called an **(infinite) geometric series**. If $|r| < 1$, then

$$S_n = a_1 + a_1 r + \cdots + a_1 r^{n-1} = \frac{a_1(1 - r^n)}{1 - r}$$

is arbitrarily close to $a_1/(1 - r)$ for n large. We then have the following:

---

(7)

If $|r| < 1$, the (infinite) geometric series

$$a_1 + a_1 r + a_1 r^2 + \cdots$$

converges, and it has sum

$$\frac{a_1}{1 - r}$$

If $|r| \geq 1$, then the partial sums $S_n = a_1 + a_1 r + \cdots + a_1 r^{n-1}$ are not arbitrarily close to a fixed number for n large. For these cases, we say that the sequence of partial sums **has no limit** and that the infinite series **diverges**.

EXAMPLE 6 For each of the following geometric series, determine if it converges or diverges. If it converges, find its sum. (a) $50 + 10 + 2 + \tfrac{2}{5} + \tfrac{2}{25} + \cdots$; (b) $\tfrac{1}{50} + \tfrac{1}{10} + \tfrac{1}{2} + \tfrac{5}{2} + \tfrac{25}{2} + \cdots$; (c) $5 - 5 + 5 - 5 + \cdots$.

SOLUTION (a) Here, the ratio is $r = a_2/a_1 = \tfrac{10}{50} = \tfrac{1}{5}$. Since $|r| < 1$, this series converges, and its sum is

$$\frac{a_1}{1 - r} = \frac{50}{1 - \tfrac{1}{5}} = \frac{250}{4} = 82.5$$

(b) For this series, the ratio is $r = a_2/a_1 = (1/10)/(1/50) = 5$. Since $|5| \geq 1$, this series diverges. In fact, if we examine the partial sums,

$$S_1 = \tfrac{1}{50} = .02, \qquad S_2 = .02 + \tfrac{1}{10} = .12$$

$$S_3 = .12 + \tfrac{1}{2} = .62, \qquad S_4 = .62 + \tfrac{5}{2} = 3.12$$

$$S_5 = 3.12 + \tfrac{25}{2} = 15.62, \qquad S_6 = 15.62 + \tfrac{125}{2} = 78.12, \qquad \text{etc.}$$

we can see that the sums are becoming arbitrarily large. In this case, we say that the series **diverges to plus infinity**.

(c) For this series, the ratio is $r = a_2/a_1 = (-5)/5 = -1$. Since $|-1| \geq 1$, this series diverges. If we examine the partial sums of this series,

$$S_1 = 5, \qquad S_2 = 5 - 5 = 0$$
$$S_3 = 0 + 5 = 5, \qquad S_4 = 5 - 5 = 0$$
$$S_5 = 0 + 5 = 5, \qquad S_6 = 5 - 5 = 0, \quad \text{etc.}$$

we can see that the sums do not diverge to plus infinity or minus infinity but that they oscillate back and forth between 5 and 0. Since the partial sums are not arbitrarily close to a *single* fixed number, the partial sums do not have a (unique) limit, and this is why we say that the series diverges.

We can use our result on the sum of a geometric series to express any infinite repeating decimal as a rational number (i.e., as a quotient of two integers).

EXAMPLE 7 Express $c = .14189189189\ldots$ as a rational number.

SOLUTION We write

$$c = .14189189189\ldots = .14 + (.00189 + .00000189 + .00000000189 + \cdots)$$

The part in parentheses is a geometric series with $a_1 = .00189$ and $r = .001$. Since $|r| < 1$, we can apply (7) to find its sum:

$$S = \frac{.00189}{1 - .001} = \frac{.00189}{.999} = \frac{189}{99,900} = \frac{21}{11,100} = \frac{7}{3700}$$

Then

$$c = .14 + S = \frac{14}{100} + \frac{7}{3700} = \frac{37 \cdot 14 + 7}{3700} = \frac{525}{3700} = \frac{21}{148}$$

EXERCISES

In Exercises 1-6, find the second, third, and ninth terms in a geometric progression with the given first term and common ratio.

1. $a_1 = 3, r = 2$
2. $a_1 = 81, r = \frac{1}{3}$
3. $a_1 = 4.8, r = .1$
4. $a_1 = \log 5, r = -2$
5. $a_1 = \frac{2}{9}, r = \frac{3}{2}$
6. $a_1 = 100, r = .2$

In Exercises 7-12, the first three terms of a geometric progression are given. Find the indicated term and partial sum.

7. $8, -4, 2$; find a_{15} and S_{15}.
8. $4, 4\sqrt{3}, 12$; find a_{16} and S_{16}.
9. $1, e, e^2$; find a_{20} and S_{20}.
10. $\sin \theta, 1, \csc \theta$; find a_{12} and S_{12}.
11. $729, -486, 324$; find a_{17} and S_{17}.
12. $-3, -12, -48$; find a_{10} and S_{10}.

In Exercises 13–16, the terms given are from a geometric progression. Find the common ratio and the first term. Assume the ratios are real.

13. $a_7 = 56$, $a_{12} = -1792$
14. $a_8 = 648$, $a_{14} = 30.233088$
15. $a_{12} = 63.21$, $a_{15} = 30.03$
16. $a_6 = \sin 2\theta$, $a_{10} = 2\cot\theta + 2\cot^3\theta$

In Exercises 17–20, express the given infinite repeating decimal as a rational number.

17. .165165165...
18. 5.2383838...
19. 632.01201201...
20. .014474747...

21.a. Use your calculator to find $1, 1 + \frac{1}{2}, 1 + \frac{1}{2} + (\frac{1}{2})^2, 1 + \frac{1}{2} + (\frac{1}{2})^2 + (\frac{1}{2})^3, \ldots, 1 + \frac{1}{2} + (\frac{1}{2})^2 + \cdots + (\frac{1}{2})^{10}$. What number do you think these are approaching?

  b. Use formula (6) to find $1 + \frac{1}{2} + \cdots + (1/2^{10})$, $1 + \frac{1}{2} + \cdots + (1/2^{20})$, $1 + \frac{1}{2} + \cdots + (1/2^{30})$, $1 + \frac{1}{2} + \cdots + (1/2^{50})$, $1 + \frac{1}{2} + \cdots + (1/2^{100})$. For which of these numbers, if any, would your calculator give an answer of 2? Is the sum ever really 2?

22. Replace $\frac{1}{2}$ in Exercise 21 with $-\frac{1}{2}$, and do a similar problem.

For the value of x given in Exercises 23–30, find the partial sums indicated in parts (a)–(c) and then answer part (d): (a) $S_6 = 1 + x + x^2 + \cdots + x^5$. (b) $S_{16} = 1 + x + x^2 + \cdots + x^{15}$. (c) $S_{25} = 1 + x + x^2 + \cdots + x^{24}$. (d) What, if anything, do the partial sums $S_{n+1} = 1 + x + x^2 + \cdots + x^n$ approach as n gets very large.

23. 2
24. 3/5
25. -3
26. $-.9$
27. 10
28. -1
29. 1
30. .1

In Exercises 31–34, find the sum of the geometric series.

31. $43.2 + 4.32 + .432 + \cdots$
32. $1 - \frac{1}{6} + \frac{1}{36} + \cdots$
33. $5.05 + 1.212 + .29088 + \cdots$
34. $1 + \cos^2 x + \cos^4 x + \cdots$

35. A ball is dropped from the top of a building 82 feet tall. On each rebound, the ball rises to seven-tenths of the height from which it last fell. Find the total distance the ball travels until the time it hits the ground for the tenth time. Use a geometric series to approximate the total distance the ball travels before coming to rest.

36. A vacuum pump removes one-fourth of the air in a container at each stroke. After 20 strokes, what percentage of the original amount of air remains in the container?

37. A piece of paper is .02 inch thick. For 10 times, it is folded in half so that its thickness is doubled each time. How thick is the result?

38. A superball is dropped from the top of the Gateway Arch in St. Louis, which is 195 meters high. Suppose the ball rebounds seven-tenths of the distance after each fall.
  a. How high does it bounce on its twelfth bounce?
  b. How far has it traveled when it hits the ground for the thirteenth time?
  c. How far does it travel altogether before it comes to rest?

39. A pendulum swings through an arc 90 centimeters long. On each successive swing, the pendulum covers an arc ten-elevenths of the preceding swing.
  a. How far does the pendulum go on its ninth swing?
  b. How far has it traveled altogether by the end of its ninth swing?
  c. What is the total distance covered by the time it comes to a complete rest?

40. If a body is falling near the surface of the earth, then, neglecting air resistance, the body accelerates at a constant rate of 32 feet per second. Thus each second the body falls, it falls 32

feet farther than it did the previous second. If a body starts from rest, it falls 16 feet during the first second.

a. How far does it fall during the tenth second?

b. How far has it fallen altogether during the first 10 seconds?

41. At the beginning of each year, Mrs. Jones deposited $1000 in a bank account which paid 6.5% interest, compounded annually. How much was in the account after 10 years?

42. $10,000 is placed in an account that pays 6%, compounded monthly. Suppose $100 is withdrawn at the end of each month.

a. How much is in the account after 5 years?

b. Will the account ever run out?

43. You deposit $20 in the savings account at the end of every month for 5 years. If the interest on the account is 6% per year, compounded monthly, how much money is in the account at the end of the 5 years?

44.a. Solve the problem posed by the following Old English children's rhyme:

> As I was going to St. Ives
> I met a man with seven wives;
> Every wife had seven sacks;
> Every sack had seven cats;
> Every cat had seven kits.
> Kits, cats, sacks and wives,
> How many were going to St. Ives?

b. Solve the problem in part a if the last line were

> How many were going away from St. Ives?

45. Prove Formula 1 of Sec. 5.5 by **(a)** first arguing that $A = P(1 + r)^n + P(1 + r)^{n-1} + \cdots + P(1 + r)$ and **(b)** then using (6).

46. Prove Formula 2 of Sec. 5.5 by **(a)** first arguing that $A(1 + r)^n = P(1 + r)^{n-1} + P(1 + r)^{n-2} + \cdots + P$ and **(b)** then using (6) and solving.

SECTION 12.3. ENUMERATION; PERMUTATIONS

Enumeration is a word for "the act of ascertaining the number of...," i.e., for counting. Some things are easy to enumerate. It is trivial to determine the number of letters in the alphabet or the number of stars on the U.S. flag. However, it is not quite so easy to determine the number of possible arrangements of the desks of the senators on the floor of the U.S. Senate. We shall learn some techniques of enumeration that will enable us to count some rather complicated things quickly and accurately.

Let us begin by looking at a simple example. Suppose you are going to label the quality of certain products by a two-place code. In the first position, you are going to put one of the three letters A, B, or C. In the second position, you are going to put one of the four numbers 1, 2, 3, or 4. How many possible labels are there? In this case, because there are only a few, it is feasible just to list all the possibilities:

$A1$, $A2$, $A3$, $A4$, $B1$, $B2$, $B3$, $B4$, $C1$, $C2$, $C3$, $C4$

Thus we see there are 12 possible labels.

However, we could have counted this as follows: There are three possible entries in the first position, and *for each* such entry, there are four possible entries in the second position. If you think very carefully about this statement, you should realize that it is really describing a *multiplication*: There are $3 \cdot 4 = 12$ possible labels.

This is an example of the following:

> **(8)   THE FUNDAMENTAL PRINCIPLE OF ENUMERATION**
>
> Suppose event A can be performed with m possible outcomes, and *for each* such outcome event B can be performed with n possible outcomes. Then the number of possible outcomes of performing event A and then event B is mn.

The key phrase is "for each." In almost all cases where this phrase is or could be used, you will be able to translate this phrase as multiplication.

The preceding process could continue. If *for each* outcome that event A and then event B can be performed, there are l possible outcomes for event C, then the number of possible outcomes of performing event A, then event B, and then event C is mnl, etc.

EXAMPLE 1 Suppose you have numbers carved on four blocks of wood. A 9 is carved on one block, an 8 on a second, a 6 on a third, and a 4 on a fourth. How many distinct three-digit numbers could you form?

SOLUTION You could put any of the four numbers in the first position. *For each* such choice, there are three numbers left, any of which could be used in the second position. *For each* such choice, there are two numbers left, any of which could be used in the third position. Thus there are $4 \cdot 3 \cdot 2 = 24$ possible three-digit numbers.

In Example 1, the order of the digits is important. Furthermore, once we have used a digit, we cannot use it again in the same three-digit number. In other words, once we have taken say the block with 9 on it from our pool of reserves, we do not *replace* it with another block with 9 on it in our pool of reserves. We describe the example by saying it is a *permutation of four objects taken three at a time without replacement*.

In general, a **permutation** is an ordering. The phrase "a permutation of n objects taken r at a time without replacement" indicates a filling of r positions, in order, from the n objects in a set, without replacement. We count the number of ways this could be done [which we denote by $P(n, r)$, but $_nP_r$ and P_r^n are also standard notations] as follows: There are n objects we could place in the first position. For each such placement there are $n - 1$ objects we could place in the second position. For each such placement, there are $n-2$ objects we could place in the third position. We continue this, until finally we place one of the $[n-(r-1)]$ possible objects in the rth position. From the Fundamental Principle of Enumeration (8), we conclude,

---

(9)

The number of permutations of n objects taken r at a time without replacement is $P(n,r) = n(n-1)(n-2) \cdots (n-r+1)$.

The words "without replacement" are often omitted. Note that $P(n,r)$ is the product of r consecutive descending integers, starting with n.

EXAMPLE 2
$$P(8,3) = 8 \cdot 7 \cdot 6 = 336$$
$$P(13,8) = 13 \cdot 12 \cdot 11 \cdot 10 \cdot 9 \cdot 8 \cdot 7 \cdot 6 = 51{,}891{,}840$$

If $r = n$, we just say *a permutation of n objects* instead of "a permutation of n objects n at a time." In this case, we are referring to an ordering of all the elements in a set. The number of permutations of n objects is

---

(10)

$$P(n,n) = n(n-1)(n-2) \cdots 2 \cdot 1$$

EXAMPLE 3 John, Joe, Jamie, and Janet are playing a round of golf. Any ties will be resolved by a playoff. How many ways can they finish?

SOLUTION The number of ways they can finish is the number of permutations of four people, so it is
$$P(4,4) = 4 \cdot 3 \cdot 2 \cdot 1 = 24$$

If n is a positive integer, the product of the first n consecutive integers $n(n-1)(n-2) \cdots 2 \cdot 1$ occurs so frequently in mathematics that we have special notation for it. We write an exclamation mark after n and say **n factorial**. Thus

---

(11)

$$n! = n(n-1)(n-2) \cdots 2 \cdot 1$$

For convenience, we define

---

(12)

$$0! = 1$$

We do not define $n!$ if n is negative or not an integer.

It would be very nice if all scientific calculators had a factorial button. Unfortunately, not all of them do, including some of the more common models. In addition, $70! \approx 1.1978571 \times 10^{100}$, so that for $n \geq 70$, $n!$ is outside of the range of most calculators. Consequently, it is important that we understand how to work with factorials.

EXAMPLE 4

a. $7! = 7 \cdot 6 \cdot 5 \cdot 4 \cdot 3 \cdot 2 \cdot 1 = 5040$

b. $\dfrac{10!}{0!} = \dfrac{10!}{1} = 10! = 10 \cdot 9 \cdot 8 \cdot 7 \cdot 6 \cdot 5 \cdot 4 \cdot 3 \cdot 2 \cdot 1 = 3{,}628{,}800$

c. $\dfrac{8!}{4!} = \dfrac{8 \cdot 7 \cdot 6 \cdot 5 \cdot 4 \cdot 3 \cdot 2 \cdot 1}{4 \cdot 3 \cdot 2 \cdot 1} = 8 \cdot 7 \cdot 6 \cdot 5 = 1680$

d. $\dfrac{12!}{11!} = \dfrac{12 \cdot 11 \cdot 10 \cdot 9 \cdot 8 \cdot 7 \cdot 6 \cdot 5 \cdot 4 \cdot 3 \cdot 2 \cdot 1}{11 \cdot 10 \cdot 9 \cdot 8 \cdot 7 \cdot 6 \cdot 5 \cdot 4 \cdot 3 \cdot 2 \cdot 1} = 12$

e. $\dfrac{100!}{97!} = \dfrac{100 \cdot 99 \cdot 98 \cdot 97 \cdot 96 \cdots 2 \cdot 1}{97 \cdot 96 \cdots 2 \cdot 1} = 100 \cdot 99 \cdot 98 = 970{,}200$

f. $8(7!) = 8(7 \cdot 6 \cdot 5 \cdot 4 \cdot 3 \cdot 2 \cdot 1) = 8!$

Note that part **(e)** could *not* be computed on a calculator with a [!] button by pressing [100][!][÷][97][!].

These examples indicate some general properties of factorials:

(13)
$$n[(n-1)!] = n!, \quad (n+1)(n!) = (n+1)!$$
$$n(n-1)[(n-2)!] = n!, \quad \text{etc.}$$
$$\frac{n!}{(n-1)!} = n, \quad \frac{(n+1)!}{(n-1)!} = (n+1)n$$
$$\frac{n!}{r!} = \frac{n(n-1)\cdots(r+1)r(r-1)\cdots 1}{r(r-1)\cdots 1} = n(n-1)\cdots(r+1)$$
$$\frac{n!}{(n-r)!} = \frac{n(n-1)\cdots(n-r+1)(n-r)(n-r-1)\cdots 1}{(n-r)(n-r-1)\cdots 1}$$
$$= n(n-1)\cdots(n-r+1)$$

Since the last expression is what we denoted earlier by $P(n,r)$ we can rewrite (9) and (10):

(14)
$$P(n,r) = \frac{n!}{(n-r)!}$$

(15)
$$P(n,n) = n!$$

Section 12.3 Enumeration; Permutations

So far we have discussed permutations without replacement. What happens if we want to fill r positions, in order, from a set of n distinct objects *with replacement*. The term "with replacement" tells us that after an object is chosen from our collection of possible objects and put in a position, that object is also put back (replaced) into the collection of possible objects so that it may also be used in subsequent positions. Thus we could place any of n objects in the first position. For each choice, there are still n objects in the collection (after replacement), and any of these may be placed in the second position. We continue this way until we place any of the n objects in the rth position. Therefore by (8),

(16)

The number of permutations of n objects r at a time with replacement is n^r.

EXAMPLE 5 How many three-digit numbers can you make using only the digits 4, 6, 8, and 9.

SOLUTION This is similar to Example 1, except that here replacement is allowed. There are four possibilities for the first place; for each such choice, there are four possibilities for the second place; and for each such choice, there are four possibilities for the third place. Thus there are $4^3 = 64$ such three-digit numbers.

The preceding principles apply in other situations.

EXAMPLE 6 The license plate of a certain state has two letters followed by four numbers. Find the number of possible license plates.

SOLUTION The same letter can be used in both of the first two positions, and similarly the same digit (0–9) can appear more than once in the last four positions. The number of letter combinations for the first two positions is $26 \cdot 26 = 26^2$. For each, we could have any of the ten digits 0–9 in the four number positions, a total of 10^4 four-digit numbers. Thus, according to (8), the total number of possible license plates is $26^2 \cdot 10^4 = 6{,}760{,}000$.

Some problems require addition.

EXAMPLE 7 How many three-letter words can be formed using the letters a, b, and c if repeats are allowed except that **(a)** the letter b must appear exactly once or **(b)** either the letter b does not appear or it appears exactly once? (The term *word* means any sequence of letters, whether or not they form a meaningful word in any language.)

Note: It may be helpful to first list all 27 three-letter words using a, b, and c on a separate piece of paper and check off the ones described as we go along.

SOLUTION (a) The letter b must appear in either the first, second, or third position. If it appears in the first position, then there are two choices for the second position and two choices for the third position. Therefore by (8) there are $1 \cdot 2 \cdot 2 = 4$ words with the letter b in exactly the first position. Similarly, there are 4 words with a b in exactly the second position and 4 words with a b in exactly the third position. The total number of words we want, then, is the sum of the numbers in each case, since no two of the cases overlap. Consequently there are $4 + 4 + 4 = 12$ words with exactly one b. (b) If no b appears, then there are two choices for each of the three positions, so there are $2^3 = 8$ such words. By part (a), there are 12 words with exactly one b. The total number of words with either no b or one b in them is the sum of the numbers in each case, since the cases do not overlap. Therefore there are $8 + 12 = 20$ words.

Finally, we should point out that when doing problems of enumeration, the calculator is of little or no help in the actual enumeration process. That is, a calculator does not help us find the answer in factorial form. The real advantage of a calculator is that after we have completed the enumeration and we have the answer in factorial form, we can use the calculator to compute this number. We certainly have a much better feeling for the magnitude of a number in decimal form than in factorial form. For instance, roughly how large is 52!/39!? By the calculator, we see that $52!/39! \approx 3.95424 \times 10^{21}$.

EXERCISES

In Exercises 1–8, compute the given number.

1. a. 0!   b. 1!   c. 2!   d. 3!   e. 4!
   f. 5!   g. 6!   h. 7!   i. 8!   j. 9!

2. $\dfrac{12!}{5!}$

3. $\dfrac{32!}{29!}$

4. $\dfrac{78!}{75!}$

5. $\dfrac{203!}{199!}$

6. $\dfrac{(11!)(2!)}{110(9!)}$

7. $\dfrac{(7!)(0!)(12!)}{42(5!)}$

8. a. $\dfrac{0 \cdot (5!)}{(0 \cdot 5)!}$ b. $\dfrac{(0 \cdot 5)!}{0 \cdot (5!)}$

In Exercises 9–18, write the number in factorial notation.

9. $7 \cdot 6 \cdot 5 \cdot 4 \cdot 3 \cdot 2 \cdot 1$
10. $52 \cdot 51 \cdots 2 \cdot 1$
11. $9 \cdot 8 \cdot 7$
12. $47 \cdot 46 \cdot 45 \cdot 44 \cdot 43$
13. $25(24!)$
14. $52 \cdot 51 \cdot 50(49!)$
15. $(n + 1)n!$
16. $(n + 2)[(n + 1)!]$
17. $7(7!) + 7!$
18. $n(n!) + n!$

In Exercises 19–22, compute the given number.

19. $P(8, 5)$
20. $P(11, 11)$
21. $P(80, 2)$
22. $P(198, 5)$

23. How many four-letter words can be formed using the English alphabet if **(a)** repetitions are allowed? **(b)** no letter may be used twice?

24. How many five-letter words can be formed using the Greek alphabet (which has 24 letters) if **(a)** repetitions are allowed? **(b)** no repetitions are allowed?

25. How many four-digit numbers can be formed if **(a)** repetitions are allowed? **(b)** no number may be used twice? (*Note:* A four-digit number may not begin with zero.)

26. How many 11-digit numbers can be formed if **(a)** repetitions are allowed? **(b)** no repetitions are allowed?

27. How many four-letter words can be formed using the English alphabet if repetitions are allowed except that **(a)** the letter q must appear exactly once? **(b)** the letter q may not be repeated (it appears either 0 or 1 time)?

28. How many five-letter words can be formed using the Greek alphabet if repetitions are allowed except that **(a)** the letter π must appear exactly once? **(b)** the letter π may not be repeated? (*Note:* The Greek alphabet contains 24 letters.)

29. How many four-digit numbers can be formed if repetitions are allowed except that **(a)** the digit 0 must appear exactly twice? **(b)** the digit 0 must appear exactly once? **(c)** the digit 0 may not appear more than twice (i.e., 0 appears either 2, 1, or 0 times)?

30. How many 11-digit numbers can be formed if repetitions are allowed except that **(a)** the digit 0 must appear exactly twice? **(b)** the digit 0 must appear exactly once? **(c)** the digit 0 may not appear more than once?

31. There are 1821 entrants in a photo contest. First through sixth places will be awarded prizes. How many different lists of winners are possible?

32. Forty-eight basketball teams enter the NCAA tournament. First, second, third, and fourth places are awarded prizes. How many different lists of winners are possible?

33. In how many ways can an instructor assign seats to the 25 students in his class if **(a)** there are 25 seats in the room? **(b)** there are 30 seats in the room?

34. A club containing 42 people is going to elect a president, a vice-president, a secretary, and a treasurer. How many different outcomes are possible?

35. A coin is tossed five times. How many outcomes are possible?

36. In how many ways can a 50-question true-false test be answered?

37. A test consists of 10 multiple-choice questions. In how many ways can the test be answered if **(a)** each question has four choices? **(b)** the first four questions have three choices and the remainder have five choices? **(c)** question number n has n choices?

38. Eight students are to be assigned to sit in a row of eight chairs.
  a. How many ways can this be done?
  b. If there are four boys and four girls and they are to sit boy, girl, boy, girl, etc., how many ways can this be done?
  c. If there are four boys and four girls and the girls are seated in the first four seats, how many ways can this be done?

39. There are 200 men and 200 women entered in a marathon race. How many different prize lists are possible if **(a)** only the people who place 1–6 irrespective of sex are given prizes? **(b)** the first-, second-, and third-place men and the first-, second-, and third-place women receive prizes?

40. How many four-digit numbers can be formed from the digits 2, 4, 6, 7, and 9 if **(a)** repetitions are not allowed? **(b)** repetitions are allowed? or **(c)** odd numbers must begin with even digits and even numbers must begin with odd digits (repetitions allowed)?

41. In how many ways can the manager of a nine-player softball team arrange the batting order if **(a)** the pitcher bats last and the best hitter bats fourth? **(b)** the four infielders bat in the first four positions? **(c)** either the pitcher bats last, or otherwise the catcher bats immediately after the pitcher?

42. How many seven-digit phone numbers can be formed using all the digits if **(a)** none of the first three digits may be zero? **(b)** none of the first three digits may be zero, and if the next to last digit is zero, then the last digit is also zero?

SECTION 12.4. DISTINGUISHABLE PERMUTATIONS; COMBINATIONS

In this section, we shall consider two related topics. One is the number of distinguishable permutations of a set of n objects when the objects are not all different. The other is the number of ways an unordered set of r objects can be chosen from a set of n distinct objects. The relationship between these two topics will emerge as we proceed.

We first consider permutations of objects which are not all different. Suppose we are given the word

$$aaabbc$$

If we interchange the second and third letters, we would again get

$$aaabbc$$

We cannot tell that anything happened just by looking at the result, so this is called an **indistinguishable permutation**. However, if we interchanged the third and fourth letters, we get

$$aababc$$

Now we can tell something happened, though not exactly what (for example, we might have interchanged the first two letters, also), so the result is called a **distinguishable permutation**. In this situation, we would not ask how many total permutations there are but just ask how many distinguishable permutations there are. So let P be the number of distinguishable permutations of $aaabbc$. Suppose we temporarily place marks above the letters so that we can tell them apart. Then the number of permutations of this new set is $P(6, 6) = 6!$. We shall count this another way. For each of the P distinguishable permutations, we could reorder the three a's (among the positions they occupy) in 3! different ways and the two b's in 2! ways (and the one c in 1! way). So with the added marks we get $P \times 3! \times 2! \times 1!$ permutations. By the preceding, there are 6! such permutations, so

$$P \, 3!2!1! = 6! \quad \text{or} \quad P = \frac{6!}{3!2!1!}$$

In this case, we had six objects but only three distinct types of objects, three of the first type, two of the second type, and one of the third type. Of course,

$$3 + 2 + 1 = 6$$

We determined that the number of distinguishable permutations is

$$\frac{6!}{3!2!1!}$$

For the general case, suppose you have n objects but only k distinct types of objects, r_1 of the first type, r_2 of the second type, ..., r_k of the kth type. Then

$$r_1 + r_2 + \cdots + r_k = n$$

Let $P(n; r_1, r_2, \ldots, r_k)$ denote the number of distinguishable permutations [so previously we had $P(6; 3, 2, 1)$]. To determine the formula, we proceed as previously. If we put marks on all the objects so that we can tell them apart, we would get $n!$ permutations. Counting this another way, for each of the $P(n; r_1, r_2, \ldots, r_k)$ distinguishable permutations, we could reorder the r_1 objects of the first type (among the positions they occupy) in $r_1!$ ways, reorder the r_2 objects of the second type in $r_2!$ ways, etc. As before, we observe that

$$P(n; r_1, r_2, \ldots, r_k) r_1! r_2! \cdots r_k! = n!$$

and conclude that

(17)
$$P(n; r_1, r_2, \ldots, r_k) = \frac{n!}{r_1! r_2! \cdots r_k!}$$

EXAMPLE 1 How many distinct 10-letter words can be formed using the letters in BOOKKEEPER?

SOLUTION If we interchange like letters, we get the same word, so we see we want to count only the distinguishable permutations of BOOKKEEPER. There are $n = 10$ objects but only $k = 6$ distinct types of objects. There are $r_1 = 1$ B, $r_2 = 2$ O's, $r_3 = 2$ K's, $r_4 = 3$ E's, $r_5 = 1$ P, and $r_6 = 1$ R. Thus

$$P(10; 1, 2, 2, 3, 1, 1) = \frac{10!}{1!2!2!3!1!1!} = 151{,}200$$

is the answer.

We now shift our attention to the subject of combinations. A **combination** of n objects taken r at a time is a choice of an *unordered* set of r objects from a set of n distinct objects. In contrast, recall that a permutation of n objects taken r at a time is a choice of an *ordered* set of r objects from a set of n distinct objects. The difference between combinations and permutations is illustrated by the following two problems:

1. Choose a first-, second-, and third-prize winner from among 20 pictures entered in a photography contest.
2. Choose 3 of the 20 pictures for display.

In the first problem, order is crucial. A list of winners such as first: #5, second: #8, and third: #12 is different from first: #12, second: #8, and third: #5. But in the second problem, displaying pictures #5, #8, and #12 is the same as displaying pictures #12, #8, and #5. The order of choice does not matter at all.

The first problem is a permutation of 20 objects taken 3 at a time, and the number of such is $P(20, 3) = 20 \cdot 19 \cdot 18 = 6840$. The second problem is a combination of 20 objects taken 3 at a time. We can count the number of such combinations as follows: In making a choice of 3 pictures, what we are really doing is dividing the set of 20 pictures into two sets, one set of 3 pictures to be displayed and one set of 17 pictures which will not be displayed. If you thought of this as having 3 letters D (for display) and 17 letters N (for not display), the number of different ways we can choose 3 pictures for display is precisely the number of different ways we can assign the 20 letters to the 20 pictures. Thus it is exactly the number of distinguishable permutations of 3 Ds and 17 Ns. This is

$$P(20; 3, 17) = \frac{20!}{3!17!} = \frac{20 \cdot 19 \cdot 18}{3 \cdot 2 \cdot 1} = 1140$$

There is another useful way we can obtain this number C of distinct choices of 3 pictures from 20. Given one of the C choices, we could, as a second step, assign the three prizes to the three pictures on display; this could be done in $P(3, 3) = 3!$ ways. Now for each of the C choices of 3 pictures, there are 3! ways to assign first, second, and third prizes. But this gives all the ways to assign prizes, namely, $P(20, 3) = 20 \cdot 19 \cdot 18$. Thus

$$C(3!) = 20 \cdot 19 \cdot 18 \quad \text{or} \quad C = \frac{20 \cdot 19 \cdot 18}{3!} = 1140$$

as before.

Choosing an unordered set of r elements from a set of n distinct elements is called a *combination of n objects taken r at a time*. The number of distinct combinations of n objects taken r at a time is sometimes denoted $C(n, r)$ (but $_nC_r$ and C_r^n are other standard notations). This is analogous to our notation for the number $P(n, r)$ of permutations of n objects taken r at a time. However, a more common notation in the case of combinations is $\binom{n}{r}$. Thus $C(n, r) = \binom{n}{r}$, and either is acceptable, although we shall tend to use the latter. We can compute the number $\binom{n}{r}$ in general in the two ways we did in the example: Either $\binom{n}{r} = P(n; r, n - r) = \frac{n!}{r!(n-r)!}$ or $\binom{n}{r} = \frac{P(n, r)}{r!} = \frac{n!}{r!(n-r)!}$. Thus

---

(18)

The number of combinations of n objects taken r at a time is

$$C(n, r) = \binom{n}{r} = \frac{n!}{r!(n-r)!} = \frac{n \cdot (n-1) \cdots (n-r+1)}{r!}$$

Again we emphasize that the difference between permutations and combinations is that in permutations order is important but in combinations order is not important.

We now give several examples to illustrate some of the properties of the number $\binom{n}{r}$.

EXAMPLE 2

$$\binom{42}{2} = \frac{42!}{2!\,40!} = \frac{42 \cdot 41}{2!} = 861$$

$$\binom{9}{8} = \frac{9!}{8!\,1!} = 9$$

$$\binom{n}{0} = \frac{n!}{0!\,(n-0)!} = \frac{n!}{n!} = 1 \quad \text{(recall that } 0! = 1\text{)}$$

$$\binom{n}{n} = \frac{n!}{n!\,(n-n)!} = \frac{n!}{n!\,0!} = 1$$

$$\binom{n}{1} = \frac{n!}{1!\,(n-1)!} = n$$

$$\binom{n}{2} = \frac{n!}{2!\,(n-2)!} = \frac{n(n-1)}{2}$$

It is helpful to remember when writing $\binom{n}{r}$ as $\dfrac{n \cdot (n-1) \cdots (n-r+1)}{r!}$ for specific values of n and r that the same number of factors appear in the numerator as in the denominator. For example,

$$\binom{28}{5} = \frac{28!}{5!\,23!} = \frac{\overbrace{28 \cdot 27 \cdot 26 \cdot 25 \cdot 24}^{5 \text{ factors}}}{\underbrace{5 \cdot 4 \cdot 3 \cdot 2 \cdot 1}_{5 \text{ factors}}} = 98{,}280$$

A property of the number $\binom{n}{r}$ is

(19)
$$\binom{n}{r} = \binom{n}{n-r}$$

since

$$\binom{n}{n-r} = \frac{n!}{(n-r)!\,[n-(n-r)]!} = \frac{n!}{(n-r)!\,r!} = \frac{n!}{r!\,(n-r)!} = \binom{n}{r}$$

EXAMPLE 3

$$\binom{36}{33} = \frac{36!}{33!3!} = \frac{36 \cdot 35 \cdot 34}{3 \cdot 2 \cdot 1} = 7140$$

$$\binom{36}{3} = \frac{36!}{3!33!} = \frac{36 \cdot 35 \cdot 34}{3 \cdot 2 \cdot 1} = 7140$$

In many problems involving enumeration, it is reasonably obvious that order is not important, and hence combinations are involved.

EXAMPLE 4 Four people at a party of 50 people will be chosen at random. **(a)** If each is given a door prize of $10, how many different possible outcomes are there? **(b)** If the first person chosen receives $20, the second person chosen receives $15, the third person chosen receives $10, and the fourth person chosen receives $5, how many different possible outcomes are there now?

SOLUTION **(a)** As long as a person wins, it does not matter whether the person was chosen first, second, third, or fourth. Thus order is not important, and so the number of different lists of winners is

$$\binom{50}{4} = \frac{50 \cdot 49 \cdot 48 \cdot 47}{4 \cdot 3 \cdot 2 \cdot 1} = 230{,}300$$

(b) Now it does matter whether a person is chosen first, second, third, or fourth, so the number of lists is

$$P(50, 4) = 50 \cdot 49 \cdot 48 \cdot 47 = 5{,}527{,}200$$

Sometimes both permutations and combinations can be involved.

EXAMPLE 5 A group of 46 people is going to elect a committee of five by electing a chairperson, a secretary, and three members at large. How many different ways can this be done (if one considers it to be important who the officers are)?

SOLUTION We can choose the chairperson and secretary in $P(46, 2) = 46 \cdot 45 = 2070$ ways (order is important). For each such choice, there are 44 people left from which to choose the three members at large. Since order is not important, they can be chosen $\binom{44}{3} = \frac{44 \cdot 43 \cdot 42}{3 \cdot 2 \cdot 1} = 13{,}244$ ways. Thus by (8), there are $2070(13{,}244) = 27{,}415{,}080$ ways we could choose the chairperson, secretary, and the three members at large. [Another way of looking at this problem is that it is a distinguishable permutation, and the answer is $P(46; 1, 1, 3, 41)$.]

Example 5 demonstrates that sometimes what is involved may depend on your point of view. For instance, if you feel that what the committee accomplishes depends only on which five people are on the committee and not on who the officers are, then from your point of view, there are only $\binom{46}{5} = \frac{46 \cdot 45 \cdot 44 \cdot 43 \cdot 42}{5 \cdot 4 \cdot 3 \cdot 2 \cdot 1} = 1{,}370{,}754$ distinct committees.

EXERCISES

In Exercises 1–10, find the given number.

1. $\binom{5}{2}$
2. $\binom{11}{3}$
3. $\binom{13}{9}$
4. $\binom{86}{0}$
5. $\binom{120}{120}$
6. $\binom{27}{7}$
7. $\binom{n}{n-1}$
8. $\binom{n}{3}$
9. $\binom{n+1}{n-1}$
10. $\binom{n+6}{n+1}$

In Exercises 11–14, solve for n.

11. $\binom{n}{1} = 37$
12. $\binom{n}{2} = 55$
13. $\binom{n}{3} = 2\binom{n}{2}$
14. $\binom{n+2}{4} = 6\binom{n}{2}$

15. a. Show $\binom{8}{2} + \binom{8}{3} = \binom{9}{3}$.
    b. Show that $\binom{n}{r} + \binom{n}{r+1} = \binom{n+1}{r+1}$ whenever $0 \leq r \leq n-1$.

16. In the word MOON,
    a. First put a mark over the second O, and then list all the permutations. How many are there?
    b. Now erase that mark. How many distinguishable permutations are there?

17. How many distinct arrangements are there of the letters in *Mississippi*?

18. How many distinct arrangements are there of the letters in *Massachusetts*?

19. a. How many ways can the product $a_1 a_2 b_1 b_2 b_3 c_1 c_2 c_3 c_4$ be expressed?
    b. How many ways can the product $a^2 b^3 c^4$ be expressed without exponents?

20. a. How many ways can the product $a_1 a_2 a_3 a_4 b c_1 c_2$ be expressed?
    b. How many ways can the product $a^4 b c^2$ be expressed without exponents?

21. A volleyball coach has 14 players ready to play. How many different teams of 6 could the coach put on the court?

22. A basketball coach has 14 players ready to play. If any player can play any position, how many different teams of 5 players could the coach put on the court?

23. A parking lot contains 200 1-year-old cars. Eight of them are to be chosen at random and tested to see how well their pollution devices are working. How many different sets of 8 cars are possible?

24. A local organization with 91 members can afford to send only 4 delegates to a national meeting. In how many ways can the 4 be chosen?

25. There are nine baseball teams in a conference.
    a. If each team plays every other team once, how many games are played?
    b. If each team plays every other team four times, how many games are played?

26. a. How many lines are determined by eight points, no three of which are collinear?
    b. How many triangles are determined by these same points? (You only count nondegenerate triangles having these points as vertices.)

27. a. A sorority with 27 members sends 3 members to a convention. How many ways can they pick 3 to go?

b. If they must state that 1 member is a delegate, 1 member is an alternate, and 1 member is a second alternate, how many ways can they now designate 3 to go?

28. A fraternity has 21 members.

a. How many ways can it select four officers?

b. How many ways can it select a president, vice-president, secretary, and treasurer?

29. A basketball coach has three centers, six forwards, and five guards. No guard can play forward or center, etc. How many different teams consisting of one center, two forwards, and two guards could the coach put on the floor?

30. A committee on product safety is to be formed by choosing 4 consumers from a list of 33, 3 manufacturers' representatives from a list of 24, and 2 government representatives from a list of 8. How many different committees are possible?

31. In a 25-question test, a student must answer 11 of the first 15 and 7 of the last 10. In how many ways can this be done?

32. If there are 61 Democrats and 39 Republicans in the U.S. Senate, in how many ways can a committee of 5 Democrats and 4 Republicans be chosen?

33. Twenty-eight people enter an essay contest. First prize is $100, second prize is $50, and third prize is $25. If the list of winners also includes four honorable mentions in alphabetical order, how many lists are possible?

34. In rating 23 brands of hair dryers, a consumer magazine picks a first, second, third, fourth, and fifth best brand and then 7 more as acceptable. In how many ways can it do this?

35. Fourteen men and 11 women go on a picnic. It takes 5 people to do the cooking.

a. In how many ways could the 5 be chosen arbitrarily?

b. In how many ways could they pick a chief cook, assistant cook, and 3 helpers?

c. In how many ways could they pick 3 men and 2 women?

d. In how many ways could they pick the 5 if at least 3 are to be men?

36. Find the total number of subsets of a set with **(a)** four elements, **(b)** five elements, and **(c)** n elements (guess).

SECTION 12.5. PROBABILITY

Suppose you are asked to draw two numbers between 1 and 10 when you arrive at a party. You choose 6 and 10. You are told that later a random number between 1 and 10 will be chosen, and if it is one of your numbers, you will win a door prize. How likely are you to win a door prize? You have two of the ten possible numbers, so it would seem that two out of ten times, i.e., two-tenths of the time, you should win. The last number, $\frac{2}{10}$, is called your **probability** of winning.

In general, suppose something is being done in which there are a finite number n of possible outcomes, all equally likely. Suppose for m of these outcomes, event E occurs, while for $n - m$ of the outcomes, event E fails to occur. Then

(20)

The *probability p that the event E occurs* is

$$p = \frac{m}{n} = \frac{\text{number of successful outcomes}}{\text{number of possible outcomes}}$$

(21)

The probability p_f that *E fails to occur* is

$$p_f = \frac{n-m}{n} = \frac{\text{number of unsuccessful outcomes}}{\text{number of possible outcomes}}$$

From the definitions, it is immediate that

(22)

$$0 \le p \le 1, \qquad 0 \le p_f \le 1, \qquad \text{and} \qquad p + p_f = 1$$

What do probabilities mean? If $p = 1$, then it is certain that event E will occur. If $p = 0$, it is certain that event E will not occur. However, if $0 < p < 1$, there is often more confusion as to what p denotes. Suppose, for instance, that $p = \frac{1}{5}$, and that after four outcomes, event E has not occurred. Can we be sure that E will occur on outcome 5? If all possibilities are equally likely and one outcome does not affect another, then the answer is no. All we can say is that *in the long run* event E will have occurred in approximately one-fifth of the outcomes. However, there could be extended streaks where event E does not occur at all, or event E could occur in several outcomes in a row.

Let us look at some examples of probability problems.

EXAMPLE 1 Determine the probability that when one die is thrown, the number appearing on top is even.

SOLUTION First, recall that dice is the plural of die and that one die is a cube. On each of the six faces of a die there is a different number of dots from 1 to 6.

This problem is to determine the probability that after a die is thrown, the face on top has an even number of dots. We want to find

$$p = \frac{\text{number of faces with an even number of dots}}{\text{total number of faces}}$$

Since there are six possible faces and three of these have an even number of dots, $p = \frac{3}{6} = \frac{1}{2}$.

EXAMPLE 2 If a committee of 3 is to be chosen from a group of 38 people of which you are a member, determine the probability that you will be on the committee.

SOLUTION We wish to determine

$$p = \frac{\text{number of committees of 3 which contain you}}{\text{total number of committees with 3 members}}$$

Since there are 38 people to choose from, there are $\binom{38}{3}$ possible committees. If you are chosen for the committee, there are 37 people left from which to choose the other two members. Thus the number of committees which include you is $1 \cdot \binom{37}{2} = 666$. Thus the probability that you will be on the committee is $\frac{666}{8436} \approx .0789473$.

EXAMPLE 3 There is a bin with five defective and four good items on a quality control person's desk. If three items are picked at random, what is the probability that at least one of them is defective?

SOLUTION We want to know the probability p that one, two, or all three of the items are defective. To compute p directly, we would have to compute three probabilities: the probability that exactly one, that exactly two, and that all three of the items chosen are defective. Then p is the sum of these three probabilities. However, it is easier to compute p_f, that is, the probability that none of the items is defective. Then $p = 1 - p_f$. Now

$$p_f = \frac{\text{number of ways to pick 3 good items}}{\text{number of ways to pick 3 items}}$$

Since there are $5 + 4 = 9$ items altogether, there are $\binom{9}{3}$ ways to pick three. Since there are four good items, there are $\binom{4}{3}$ ways to pick three good items. Thus $p_f = \binom{4}{3}/\binom{9}{3} = \frac{1}{21}$ so $p = 1 - \frac{1}{21} = \frac{20}{21} \approx .952381$.

EXERCISES

1. If a coin is flipped, what is the probability of obtaining heads?
2. If two coins are flipped, what is the probability of obtaining heads on both coins?
3. If one die is tossed, what is the probability of obtaining a 5 or a 6?
4. A five-card poker hand is dealt from a standard deck of cards. What is the probability that it contains all four aces?
5. A bridge hand (13 cards) is dealt. What is the probability that it contains
   a. Only red cards?
   b. Only hearts?
   c. No jacks, queens, kings, or aces?

d. All four tens?
   e. The ace of spades?
   f. One ace, two kings, no queens, and three jacks?
6. A card is drawn from a standard deck. What is the probability that it is
   a. An ace?
   b. Red?
   c. 10 or below?
   d. A jack, queen, king, or ace?
7. Mike is in a group of 34 students. Five of these people will be chosen to write up class notes. What is the probability that Mike is one of these students? What is the probability that he is not chosen?
8. A three-digit number is formed using any digit except 0. What is the probability that 4 does not appear in the number? What is the probability that the number ends in 44?
9. A box contains four red, seven white, and five blue balls. Two balls are chosen at random. What is the probability that
   a. They are both red?
   b. Neither one is red?
   c. At least one is white?
10. Four married couples go to a movie. The eight people are seated at random in a row of eight seats. What is the probability that each person is sitting next to his or her spouse?
11. A drug that was used to treat 1000 patients with a certain illness was found to be of benefit in 300 cases. A doctor uses the drug to treat 20 of his patients. Suppose we assume the past results determine the actual probabilities. What is the probability that
    a. None of the patients will benefit from the use of the drug?
    b. All the patients will benefit from the use of the drug?
12. In a primary election in a certain city, 5000 people register as Democrats and 4000 people register as Republicans. The day before the general election you choose 10 names from the phone book and make calls to remind them to vote. Assuming there are no Independents, what is the probability that you contact an equal number of Democrats and Republicans?

SECTION 12.6. THE BINOMIAL THEOREM

The expression $(a + b)^n$ occurs many places throughout mathematics and its applications, so it is important that we have a formula for its expansion. The binomial theorem gives us such an expansion. We first look at the first several cases:

$$(a + b)^0 = 1$$
$$(a + b)^1 = a + b$$
$$(a + b)^2 = a^2 + 2ab + b^2$$
$$(a + b)^3 = a^3 + 3a^2b + 3ab^2 + b^3$$
$$(a + b)^4 = a^4 + 4a^3b + 6a^2b^2 + 4ab^3 + b^4$$
$$(a + b)^5 = a^5 + 5a^4b + 10a^3b^2 + 10a^2b^3 + 5ab^4 + b^5$$

Think of the first term of $(a+b)^n$ as $a^n b^0$ and the last term as $a^0 b^n$. Then we can see that $(a+b)^n$ is a sum of $n+1$ terms. In each term, the sum of the powers of a and b is n, with the powers of a decreasing by 1 and the powers of b increasing by 1 from one term to the next. There is a symmetry to the coefficients, but where do they come from? To see where, we consider the case $n = 3$ by multiplying it out without combining terms until the last step:

$$\begin{aligned}(a+b)^3 &= (a+b)(a+b)(a+b) \\ &= (aa + ab + ba + bb)(a+b) \\ &= aaa + \underbrace{aab + aba + baa} + \underbrace{abb + bab + bba} + bbb \\ &= a^3 \;\;+\;\;\;\;\;\;\; 3a^2 b \;\;\;\;\;\;\;\;+\;\;\;\;\;\; 3ab^2 \;\;\;\;\;\;\;+ b^3\end{aligned}$$

By examining the next to last step, we see that really what we are doing is forming all possible products of three factors obtained by choosing either a or b from the first binomial $a+b$ and putting it in the first position, then choosing either a or b from the second binomial $a+b$ and putting it in the second position, and finally choosing either a or b from the third binomial and putting it in the third position. From Sec. 3, we know there are $2^3 = 8$ such products of three a's and b's. How many contain zero b's? We choose none of the b's, so from Sec. 4 we know the number is $\binom{3}{0} = 1$. How many contain exactly one b? This is the number of different ways we can choose one of the three positions to insert a b; that is, $\binom{3}{1} = 3$ (a is inserted in the other positions). How many contain exactly two b's? This is the number of different ways we can choose two positions to insert b's; that is, $\binom{3}{2} = 3$. Finally, the $\binom{3}{3} = 1$ term bbb contains three b's. Thus $(a+b)^3 = \binom{3}{0} a^3 + \binom{3}{1} a^2 b^1 + \binom{3}{2} a^1 b^2 + \binom{3}{3} b^3$.

The binomial theorem is just the corresponding formula for $(a+b)^n$.

---

(23)   **THE BINOMIAL THEOREM**

Let n be a positive integer. Then

$$(a+b)^n = \binom{n}{0} a^n + \binom{n}{1} a^{n-1} b^1 + \binom{n}{2} a^{n-2} b^2 + \cdots$$
$$+ \binom{n}{n-1} a^1 b^{n-1} + \binom{n}{n} b^n$$

Since the coefficient of the term $a^{n-r} b^r$ in the expansion of $(a+b)^n$ is $\binom{n}{r}$, the number $\binom{n}{r}$ is called a **binomial coefficient**.

Let us look at a few examples.

EXAMPLE 1   Find $(a+b)^8$.

SOLUTION

$$(a+b)^8 = \binom{8}{0}a^8 + \binom{8}{1}a^7b + \binom{8}{2}a^6b^2 + \binom{8}{3}a^5b^3 + \binom{8}{4}a^4b^4$$
$$+ \binom{8}{5}a^3b^5 + \binom{8}{6}a^2b^6 + \binom{8}{7}ab^7 + \binom{8}{8}b^8$$

To simplify this further, we should calculate the numbers $\binom{8}{r}$, $r = 0, 1, \ldots, 8$. In doing so, we use $\binom{8}{r} = \binom{8}{8-r}$:

$$\binom{8}{0} = \frac{8!}{0!8!} = 1 = \binom{8}{8}, \quad \binom{8}{1} = \frac{8!}{1!7!} = 8 = \binom{8}{7}$$

$$\binom{8}{2} = \frac{8!}{2!6!} = 28 = \binom{8}{6}, \quad \binom{8}{3} = \frac{8!}{3!5!} = 56 = \binom{8}{5}$$

$$\binom{8}{4} = \frac{8!}{4!4!} = \frac{8 \cdot 7 \cdot 6 \cdot 5}{4 \cdot 3 \cdot 2 \cdot 1} = 70$$

Thus

$$(a+b)^8 = a^8 + 8a^7b + 28a^6b^2 + 56a^5b^3 + 70a^4b^4$$
$$+ 56a^3b^5 + 28a^2b^6 + 8ab^7 + b^8$$

EXAMPLE 2 Expand $\left(2\sqrt{x} - \frac{1}{x}\right)^6$.

SOLUTION

$$\left(2\sqrt{x} - \frac{1}{x}\right)^6 = \left[(2x^{1/2}) + \left(-\frac{1}{x}\right)\right]^6 = \binom{6}{0}(2x^{1/2})^6$$
$$+ \binom{6}{1}(2x^{1/2})^5\left(-\frac{1}{x}\right) + \binom{6}{2}(2x^{1/2})^4\left(-\frac{1}{x}\right)^2 + \binom{6}{3}(2x^{1/2})^3\left(-\frac{1}{x}\right)^3$$
$$+ \binom{6}{4}(2x^{1/2})^2\left(-\frac{1}{x}\right)^4 + \binom{6}{5}(2x^{1/2})\left(-\frac{1}{x}\right)^5 + \binom{6}{6}\left(-\frac{1}{x}\right)^6$$

This simplifies to

$$64x^3 - 192x^{3/2} + 240 - \frac{160}{x^{3/2}} + \frac{60}{x^3} - \frac{12}{x^{5/2}} + \frac{1}{x^6}$$

Note that the third term contains no x.

In the expansion of $(a+b)^n$, the first term is $\binom{n}{0}a^{n-0}b^0$, the second term is $\binom{n}{1}a^{n-1}b^1$, etc., so that $\binom{n}{r}a^{n-r}b^r$ is the $(r+1)$st term in the expansion (and not the rth term).

EXAMPLE 3 Find the sixth term in the expansion of $\left(\frac{x}{2} + y\right)^9$.

SOLUTION The sixth term is

$$\binom{9}{5}\left(\frac{x}{2}\right)^{9-5} y^5 = \frac{9\cdot 8\cdot 7\cdot 6}{4\cdot 3\cdot 2\cdot 1} \frac{x^4}{2^4} y^5 = 126\frac{x^4 y^5}{16} = \frac{63}{8} x^4 y^5$$

EXAMPLE 4 Find the coefficient of x in the expansion of $\left(x + \dfrac{1}{x^2}\right)^{25}$.

SOLUTION The $(r+1)$th term in the expansion is

$$\binom{25}{r} x^{25-r}\left(\frac{1}{x^2}\right)^r = \binom{25}{r} x^{25-r}\left(\frac{1}{x^{2r}}\right) = \binom{25}{r} x^{25-3r}$$

Thus the expansion starts out as follows: $\binom{25}{0} x^{25} + \binom{25}{1} x^{22} + \binom{25}{2} x^{19} + \cdots$. We would like to determine the term involving x without writing out the expansion that far. All we need to do is find r so that $x^{25-3r} = x$; i.e., $25 - 3r = 1$, or $r = 8$. Thus the coefficient of x is $\binom{25}{8}$.

EXERCISES

In Exercises 1–12, use the binomial theorem to expand the given expression, and then simplify.

1. $(a + b)^7$
2. $(u - v)^6$
3. $(2x + y)^8$
4. $(x - 4y)^5$
5. $(3x + u^2)^6$
6. $\left(\dfrac{a}{2} - \dfrac{b}{3}\right)^4$
7. $(x - y)^{10}$
8. $(x^2 + x^{-1})^5$
9. $(x^2 - x^{-1})^5$
10. $(2x^3 - 1/2x^3)^6$
11. $\left(\dfrac{x}{y} + \dfrac{y}{x}\right)^4$
12. $(2 + y)^7$

In Exercises 13–22, write out only the terms requested.

13. Find the first three terms in the expansion of $(2x^3 - 7y)^{15}$.
14. Find the last three terms in the expansion of $(4x^2 - \tfrac{1}{2}y)^{11}$.
15. Find the last two terms in the expansion of $(x^2 y + u^{-1} v)^{12}$.
16. Find the first four terms in the expansion of $(c^2 d - 10 a^{-1})^{28}$.
17. Find the tenth term in the expansion of $(x^3 - 1/x)^{15}$.
18. Find the third term in the expansion of $(a^{2/3} + b^{1/3})^9$.
19. Find the sixteenth term in the expansion of $(3x^2 - 5y)^{16}$.
20. Find the second term in the expansion of $(3x^2 - \tfrac{1}{3}\sqrt{x}\,)^7$.
21. Find the middle term in the expansion of $(bx - y^2)^{14}$.
22. Find the two middle terms in the expansion of $(2a - \tfrac{1}{2}\sqrt{b}\,)^{15}$.
23. Find the term involving x^8 in $(2x^2 - 3y)^9$.

24. Find the term involving y^3 in $(\sqrt{x} - 2\sqrt{y})^{10}$.
25. Find the coefficient of x^3 in the expansion of $(x - 1/x)^9$.
26. Find the term containing x^{29} in the expansion of $(x^{-2} + x^3)^{12}$.
27. Find the term in the expansion of $(x^2 - 3x^{-2})^{10}$ which does not contain x.
28. Find the term in the expansion of $(2xy^4 + 3z/y)^{20}$ which does not contain y.
29. Show that $\binom{n}{0} + \binom{n}{1} + \binom{n}{2} + \cdots + \binom{n}{n} = 2^n$. [$Hint:$ $2^n = (1+1)^n$.]
30. Show that $\binom{n}{0} + 2\binom{n}{1} + 2^2\binom{n}{2} + \cdots + 2^n\binom{n}{n} = 3^n$.
31. Show that $\binom{n}{0} - \binom{n}{1} + \binom{n}{2} - \cdots + (-1)^n\binom{n}{n} = 0$.

SECTION 12.7. MATHEMATICAL INDUCTION

In this section, we introduce a method of proof known as **mathematical induction** (or simply **induction**). It is used to prove that certain kinds of statements and formulas about integers are always true.

For example, suppose we wish to know the sum of the first n odd positive integers. If we compute the first several cases directly, we get

If $n = 1$,	$1 = 1$	$= 1^2$
If $n = 2$,	$1 + 3 = 4$	$= 2^2$
If $n = 3$,	$1 + 3 + 5 = 9$	$= 3^2$
If $n = 4$,	$1 + 3 + 5 + 7 = 16$	$= 4^2$
If $n = 5$,	$1 + 3 + 5 + 7 + 9 = 25$	$= 5^2$
If $n = 6$,	$1 + 3 + 5 + 7 + 9 + 11 = 36$	$= 6^2$

These cases lead us to conjecture that the sum is always equal to the square of the number of terms in the sum. We now try to express this conjecture as a formula. When adding the first n odd positive integers, it is easy to see that the nth term is $2n - 1$. Thus the formula we want is

(24) $$1 + 3 + 5 + \cdots + (2n - 1) = n^2$$

The preceding direct computations show that this formula is true for the integers $n = 1, 2, \ldots, 6$. However, we do not yet know that this is true for $n = 7$ or 23 or 1429, for there are many formulas or statements which are true for the first several positive integers but not true for them all. Hence no amount of direct computation can show a formula or statement is true for all positive integers. A valid method for showing this is mathematical induction.

Intuitively, induction is the domino theory of mathematics. Suppose there is a row of dominos, all standing on end. Suppose you know two things about the dominos:

i. The first domino will fall. (Perhaps you are about to push it over.)
ii. No matter what domino you pick in the whole line, if that domino falls, then the next domino in the line will fall. (Perhaps you can look down the whole line and see that they are standing close enough for this to be true.)

Then you can conclude that all dominos are going to fall over. Be careful to note that ii does not assert that each one is going to fall; it just asserts that *if* any one falls, *then* the next one will fall. But by i we know the first one will fall, and by ii we know the first one's falling will push over the second; then, again by ii, the second's falling will push over the third, etc. And in this way we conclude that they all are going to fall. This is how mathematical induction works, and we now formally describe it.

Let P_n denote a formula or statement that depends on the positive integer n. For example, P_n might be formula (24).

---

(25)   **THE PRINCIPLE OF MATHEMATICAL INDUCTION**

Suppose for the statements P_n we can prove two things:

i. (Initial step) The statement is true when $n = 1$; that is, P_1 is true.
ii. (Induction step) For every positive integer k, if P_n is true for $n = k$, then P_n is also true for $n = k + 1$. That is, if P_k is true, then P_{k+1} is true.

Then P_n is true for all positive integers n.

The reasoning that this is valid is essentially the same as in the domino theory. In fact, it is really an axiom of the positive integers.

We first use induction to prove that formula (24) is true for all integers n and then make several comments.

EXAMPLE 1   Use induction to prove that

$$1 + 3 + 5 + \cdots + (2n - 1) = n^2$$

is true for any positive integer n.

SOLUTION   Let P_n be the statement

$$1 + 3 + 5 + \cdots + (2n - 1) = n^2$$

We have two things to do:

i. *Initial step:* We must show P_1 is true. But when $n = 1$, P_1 is simply $1 = 1^2$, which is true.

ii. *Induction step:* We have to show that for every positive integer k, if P_k is true, then P_{k+1} is true. Consequently we pick an arbitrary positive integer k and suppose P_k is true. We have to show that P_{k+1} is true; that is, we must show that

$$1 + 3 + 5 + \cdots + (2k - 1) + [2(k + 1) - 1] = (k + 1)^2$$

is true. Since P_k is assumed true, we know that

$$1 + 3 + 5 + \cdots + (2k - 1) = k^2$$

is true. We observe that if we add $2(k + 1) - 1$ to both sides of this equation, then the resulting equation would (still be true and) have the same left-hand side as P_{k+1}. We then simplify the right-hand side, hoping to obtain the same right-hand side as P_{k+1}:

$$1 + 3 + 5 + \cdots + (2k - 1) + [2(k + 1) - 1] = k^2 + [2(k + 1) - 1]$$
$$= k^2 + 2k + 2 - 1$$
$$= k^2 + 2k + 1$$
$$= (k + 1)^2$$

We now know P_{k+1} is true. By the principle of mathematical induction, the proof is complete.

We must make several comments.

1. Often we must experiment and try several cases in order to come up with a formula or statement we conjecture is true (as in Example 1). We then try to prove our conjecture by induction. But induction does us no good until we have an actual formula or statement to prove.
2. It is essential that we do *both* the initial and induction steps. If we leave out either one, the formula or statement has not been proved and hence may not be true for all n.

EXAMPLE 2 Recall that a positive integer p is called a *prime number* if it is greater than 1 and its only integer divisors are ± 1 and $\pm p$. Let P_n be the statement "$n^2 - n + 41$ is a prime number." Suppose you have verified directly that P_n is true for $n = 1, 2, 3, \ldots, 40$. (Go ahead and try it. It is true.) Explain why this does not constitute a proof that P_n is true for every positive integer n.

SOLUTION Direct verification of any finite number of cases of P_n can never show P_n is true for all integers n. In verifying that P_1 is true, you have done the initial step of induction. However, the induction step has not been done. In fact, it could never be done, since P_{41} is false, for $41^2 - 41 + 41 = 41^2$ is not a prime number.

EXAMPLE 3 Let P_n be the statement "$n = n + 1$." Explain what is wrong with the following "proof" that P_n is true for every positive integer n.

Pick an arbitrary $k \geq 1$ and suppose P_k is true. Then

$$k = k + 1$$

Adding 1 to both sides gives

$$k + 1 = (k + 1) + 1$$

Thus, P_{k+1} is true. By induction, P_n is true for all positive integers n.

SOLUTION We never verified that P_1 is true (and indeed it is not). That is, we never did the initial step. Indeed, the only thing we have shown is that *if* $k = k + 1$, *then* $k + 1 = k + 2$.

want to prove that P_{k+1} is true; i.e., we want to show that
$$0 < x^{k+1} < 1$$
Since P_k is assumed true, we know that
$$0 < x^k < 1$$
If we multiply this by x, the inequalities stay in the same direction (since $x > 0$), so we get
$$0 \cdot x < x^k \cdot x < 1 \cdot x \quad \text{or} \quad 0 < x^{k+1} < x$$
Since $x < 1$, we get
$$0 < x^{k+1} < x < 1 \quad \text{or} \quad 0 < x^{k+1} < 1$$
Thus, P_{k+1} is true, and the proof is complete.

Sometimes statements about integers are true for all integers greater than or equal to some fixed integer other than 1. For such cases, there is an alternate form of mathematical induction.

---

(26) THE (ALTERNATE) PRINCIPLE OF MATHEMATICAL INDUCTION

Suppose N is a fixed integer, and suppose for the statement P_n we can prove two things:

i'. (Initial step) The statement is true when $n = N$; that is, P_N is true.

ii'. (Induction step) For every integer $k \geq N$, if P_n is true for $n = k$, then P_n is also true for $n = k + 1$. That is, if P_k is true, then P_{k+1} is true.

Then P_n is true for all integers $n \geq N$.

EXERCISES

In Exercises 1–17, use induction to prove that the given formula is true for every positive integer n.

1. $1 + 2 + 3 + \cdots + n = \dfrac{n(n+1)}{2}$
2. $2 + 4 + 6 + \cdots + 2n = n(n+1)$
3. $1 + 4 + 7 + \cdots + (3n - 2) = \dfrac{n(3n-1)}{2}$
4. $1 + 5 + 9 + \cdots + (4n - 3) = n(2n - 1)$
5. $3 + 9 + 15 + \cdots + (6n - 3) = 3n^2$
6. $2 + 7 + 12 + \cdots + (5n - 3) = \dfrac{n(5n-1)}{2}$
7. $1^3 + 2^3 + 3^3 + \cdots + n^3 = \dfrac{n^2(n+1)^2}{4}$

We give several more examples of correct applications of induction.

EXAMPLE 4 Prove that for every positive integer n

$$1^2 + 2^2 + 3^2 + \cdots + n^2 = \frac{n(n+1)(2n+1)}{6}$$

SOLUTION Let P_n be the given statement.

i. *Initial step:* Let $n = 1$. Then

$$\frac{n(n+1)(2n+1)}{6} = \frac{1(1+1)(2(1)+1)}{6} = \frac{1(2)(3)}{6} = 1 = 1^2$$

ii. *Induction step:* Pick an arbitrary positive integer k, and assume P_k is true. We want to prove that P_{k+1} is true; i.e., we want to show that

$$1^2 + 2^2 + \cdots + k^2 + (k+1)^2 = \frac{(k+1)[(k+1)+1][2(k+1)+1]}{6}$$

Since P_k is assumed true, we know that

$$1^2 + 2^2 + \cdots + k^2 = \frac{k(k+1)(2k+1)}{6}$$

Adding $(k+1)^2$ to both sides and simplifying, we obtain

$$1^2 + 2^2 + \cdots + k^2 + (k+1)^2 = \frac{k(k+1)(2k+1)}{6} + (k+1)^2$$

$$= \frac{k(k+1)(2k+1) + 6(k+1)^2}{6}$$

(*Note:* The algebra is easier if we factor out a $k+1$ instead of multiplying the numerator out.)

$$= (k+1)\left[\frac{k(2k+1) + 6(k+1)}{6}\right]$$

$$= (k+1)\left[\frac{2k^2 + 7k + 6}{6}\right]$$

$$= \frac{(k+1)(k+2)(2k+3)}{6}$$

$$= \frac{(k+1)[(k+1)+1][2(k+1)+1]}{6}$$

Thus, P_{k+1} is true. By induction, the proof is complete.

EXAMPLE 5 Prove: If x is any real number such that $0 < x < 1$, then for every positive integer n, $0 < x^n < 1$.

SOLUTION Let P_n be the statement $0 < x^n < 1$.

i. *Initial step:* $0 < x < 1$, by hypothesis, so P_1 is true.

ii. *Induction step:* Pick an arbitrary integer $k \geq 1$, and assume P_k is true. We

8. $1^2 + 3^2 + 5^2 + \cdots + (2n-1)^2 = \dfrac{n(4n^2-1)}{3}$

9. $\dfrac{1}{1\cdot 2} + \dfrac{1}{2\cdot 3} + \dfrac{1}{3\cdot 4} + \cdots + \dfrac{1}{n(n+1)} = \dfrac{n}{n+1}$

10. $\dfrac{1}{1\cdot 2\cdot 3} + \dfrac{1}{2\cdot 3\cdot 4} + \dfrac{1}{3\cdot 4\cdot 5} + \cdots + \dfrac{1}{n(n+1)(n+2)} = \dfrac{n(n+3)}{4(n+1)(n+2)}$

11. $\dfrac{1}{1\cdot 3} + \dfrac{1}{3\cdot 5} + \dfrac{1}{5\cdot 7} + \cdots + \dfrac{1}{(2n-1)(2n+1)} = \dfrac{n}{2n+1}$

12. $1\cdot 2 + 2\cdot 3 + 3\cdot 4 + \cdots + n(n+1) = \tfrac{1}{3} n(n+1)(n+2)$

13. $\left(1 + \dfrac{1}{1}\right)\left(1 + \dfrac{1}{2}\right)\left(1 + \dfrac{1}{3}\right) \cdots \left(1 + \dfrac{1}{n}\right) = n+1$

14. $\dfrac{1}{2} + \dfrac{2}{2^2} + \dfrac{3}{2^3} + \cdots + \dfrac{n}{2^n} = 2 - (n+2)2^{-n}$

15. $1 + 2n \le 3^n$

16. $1 + 2 + 3 + \cdots + n < \tfrac{1}{8}(2n+1)^2$

17. If $0 < a < b$, then $(a/b)^n < a/b$ for $n \ge 2$.

18. If n is an integer ≥ 2 and $a > -1$, prove that $(1+a)^n > 1 + na$. (Where does the proof break down if $a = -1$? If $a < -1$?)

19. If n is an integer ≥ 2, prove that

$$\left(1 - \dfrac{1}{2^2}\right)\left(1 - \dfrac{1}{3^2}\right)\left(1 - \dfrac{1}{4^2}\right) \cdots \left(1 - \dfrac{1}{n^2}\right) = \dfrac{n+1}{2n}$$

20. If n is an integer ≥ 2, prove that

$$\dfrac{1}{\sqrt{1}} + \dfrac{1}{\sqrt{2}} + \dfrac{1}{\sqrt{3}} + \cdots + \dfrac{1}{\sqrt{n}} > \sqrt{n}$$

21. If n is an integer ≥ 4, prove that

$$1 + \dfrac{1}{1!} + \dfrac{1}{2!} + \dfrac{1}{3!} + \cdots + \dfrac{1}{n!} < 3 - \dfrac{1}{n}$$

22. If n is an integer ≥ 3, prove that $\left(\dfrac{n+1}{n}\right)^n < n$

23. There is a positive integer N such that if $n \ge N$, $2^n n!/n^n \le 1$. Find N, and then prove the inequality.

24. Find a simple expression for

$$(1-x)(1+x)(1+x^2)(1+x^4) \cdots (1+x^{2^n}), \quad n \ge 0$$

Then prove your answer by induction.

25. Prove: For $n \ge 2$,

$$\dfrac{1}{n+1} + \dfrac{1}{n+2} + \cdots + \dfrac{1}{2n} > \dfrac{1}{2}$$

26. Prove: $1\cdot 2 + 2\cdot 3 + 3\cdot 4 + \cdots + n(n+1) > n^3/3$.

27. Prove that 3 divides $n^3 - n + 3$.

28. Prove that 9 divides $10^{n+1} + 3\cdot 10^n + 5$.

29. Prove that 64 divides $9^n - 8n - 1$.

30. Find a formula for the sequence defined by $a_1 = 2$, $a_{n+1} = 2a_n - 1$. Prove your formula by induction.

31. Let $b_1 = \sqrt{2}$, $b_2 = \sqrt{2 + \sqrt{2}}$, $b_3 = \sqrt{2 + \sqrt{2 + \sqrt{2}}}$, ..., and in general $b_{n+1} = \sqrt{2 + b_n}$, $n \geq 1$. Prove that $b_n < 2$, for $n \geq 1$.

32. Show that n lines in a plane, no two of which are parallel and no three of which are concurrent (meet in a single point), separate that plane into $n(n + 1)/2$ separate regions.

33. Prove that the number of distinct subsets of a finite set of n elements is 2^n.

34. Prove that if n people stand in a line at a ticket window, and if the first person in line is a woman and the last person in line is a man, then somewhere in the line there is a man directly behind a woman.

35. Suppose n people enter a chess tournament in which each person is to play one game against each of the others. Show that the total number of games that will be played in the tournament is $n(n - 1)/2$.

36. Find the error in the following proof: *Theorem:* Everyone is of the same sex. *Proof:* Let P_n be the statement "If P is a set of n people, then all the people in P are of the same sex."

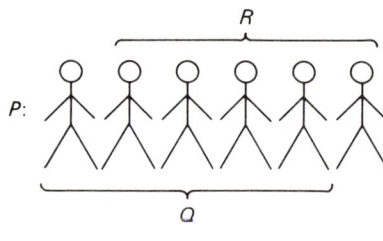

FIG. 2

i. *Initial step:* Clearly, P_1 is true.

ii. *Induction step:* Pick an arbitrary $k \geq 1$, and assume P_k is true; i.e., assume if k people are grouped together, then they are all of the same sex. We wish to prove P_{k+1}, so pick an arbitrary set of $k + 1$ people P. (See Fig. 2 for the case $k = 5$.) Then P is the union of two overlapping sets Q and R of k people each. Since P_k is true, everyone in Q is of the same sex, and everyone in R is of the same sex. Since Q and R overlap, everyone in P is of the same sex. Since P was arbitrary, P_{k+1} is true. By induction, we are done.

REVIEW EXERCISES

In Exercises 1 and 2, find the first, fifth, and seventieth terms.

1. $a_n = \dfrac{1}{2n}$ **2.** $a_n = \dfrac{2^n}{n}$

In Exercises 3–7, find the sums.

3. S_6 if $a_n = 1/2^n$, $n \geq 1$
4. S_{22} if $a_n = 3 + 2(n - 1)$, $n \geq 1$
5. S_{47} if $a_1 = 5$, $a_2 = 2$, ... is an arithmetic progression
6. S_{44} if $a_n = 5 \cdot 2^{n-1}$, $n \geq 1$

7. S_{18} if $a_1 = 6$, $a_2 = 2, \ldots$ is a geometric progression

8. How many integers between 100 and 200 are divisible by 7? What is their sum?

9. A ball is dropped from 30 feet above the pavement. After each fall the ball rebounds to two-thirds of that fall. How high does it go after the ninth bounce? How far has it traveled when it hits the ground for the tenth time? How far does it travel altogether before it comes to rest?

10. Suppose at the end of each month you deposit $100 in a savings account that pays 6% interest, compounded monthly. How much do you have after 5 years? If you then start withdrawing $100 a month, how much is left in the account after the next 5 years?

In Exercises 11–19, compute the given number.

11. $12!$

12. $\dfrac{15!}{9!}$

13. $\dfrac{83!}{81!}$

14. $P(9, 3)$

15. $P(164, 4)$

16. $\binom{7}{5}$

17. $\binom{164}{4}$

18. $\binom{208}{206}$

19. $\binom{15}{0}$

20. A social security number is of the form ***–**–****, where each * may be any digit from 0 to 9. How many distinct social security numbers are possible?

21. There are 23 freshmen and 17 sophomores entered in an essay contest. How many different prize lists are possible if (a) all entrants compete for first, second, third, and fourth and (b) a first and second are awarded to the two best freshmen and a first and second are also awarded to the two best sophomores?

22. How many four-letter words are possible if (a) repetitions are allowed, (b) repetitions are not allowed, and (c) repetitions are not allowed, the word contains exactly one vowel, and it is the second letter?

23. A bin contains 186 widgets. An inspector is going to choose 8 of them to see if they are defective. How many different sets of 8 may the inspector choose?

24. At a university, there are 18 professors in the psychology department. The students are going to "rate their professors" and choose the best, next best, third best, and fourth best, followed by a group of 9 professors merely described as being acceptable. How many ways can this be done?

25. Seventeen students are going to be split into three committees: 5 students on the advisory committee, 3 students on the grievance committee, and 7 students on the judiciary committee (no student can be on two committees). How many ways can this be done?

26. A store has 18 toasters in stock, 2 of which are defective. Three toasters are then sold. What is the probability (a) none of the 3 are defective, (b) one of the 3 is defective, and (c) 2 of the 3 are defective?

27. Use the binomial theorem to expand $(2x - x^{-2})^6$, and then simplify.

28. Find the eighth term in the expansion of $(u^5 + u^{-1})^{18}$.

29. Find the third term in the expansion of $(4y^3 - \frac{1}{16}z)^8$.

30. Find the coefficient of x^{-5} in the expansion of $(2x^{-3} + \frac{1}{4}x)^{11}$.

In Exercises 31–37, use induction to prove the formula.

31. $4 + 8 + 12 + \cdots + 4n = 2n(n+1)$

32. $2 + 10 + 18 + \cdots + (8n - 6) = 2n(2n - 1)$

33. $n + 1 \leq 2^n$, $n \geq 1$

34. $n^2 < 4^n$, $n \geq 0$

35. If $0 < a < 1$, then $(1 - a)^n \geq 1 - na$, $n \geq 1$
36. 2 divides $n^2 + n$, $n \geq 0$
37. $2^{n-1} \leq n!$, $n \geq 0$
38. Consider the statement P_n: $n = n + 3$.

   a. If P_k is true, is P_{k+1} true?
   b. Is P_1 true?
   c. Is P_n true for every positive integer n?

Reference Material

APPENDIX A

A.1. PREFIXES

The following prefixes, in combination with the basic unit names, provide multiples and submultiples. For example, 1 kilometer = 1000 meters, 1 kilogram = 1000 grams, etc.

tetra 10^{12} kilo 10^3 deci 10^{-1} micro 10^{-6} femto 10^{-15}
giga 10^9 hecto 10^2 centi 10^{-2} nano 10^{-9} alto 10^{-18}
mega 10^6 deca 10 milli 10^{-3} pico 10^{-12}

A.2. CONVERSION FACTORS

Weight: 1 kilogram (kg) ≈ 2.2046226 pounds (lb) (avdp)
          1 lb (advp) = .453592037 kg (exactly)
                    = 16 ounces (oz)

Length: 1 inch (in.) = 2.54 centimeters (cm) (exactly)
          1 foot (ft) = .3048 meters (m) (exactly)
          1 mile (mi) = 1.609344 kilometers (km) (exactly)
                    = 5280 ft (ft)
                    = 8 furlongs (fur)

Volume: 1 liter (*l*) ≈ 1.0566882 quarts (qt)
          1 qt = .946352946 *l* (exactly)
             = 32 oz
             = 2 pints (pt)
          1 gallon (gal) = 3.785411784 *l* (exactly)
             = 4 qt

Velocity of light ≈ 299,792.5 km/sec
                  ≈ 186,282 mi/sec

Velocity of sound ≈ 1088 ft/sec, in air at sea level at 32°F

A.3. THE GREEK ALPHABET

Letters		Names	Letters		Names	Letters		Names	Letters		Names
A	α	Alpha	H	η	Eta	N	ν	Nu	T	τ	Tau
B	β	Beta	Θ	θ	Theta	Ξ	ξ	Xi	Υ	υ	Upsilon
Γ	γ	Gamma	I	ι	Iota	O	o	Omicron	Φ	ϕ	Phi
Δ	δ	Delta	K	κ	Kappa	Π	π	Pi	X	χ	Chi
E	ϵ	Epsilon	Λ	λ	Lambda	P	ρ	Rho	Ψ	ψ	Psi
Z	ζ	Zeta	M	μ	Mu	Σ	σ	Sigma	Ω	ω	Omega

APPENDIX B

Using a Calculator

SECTION B.1. BASIC INFORMATION

When you turn on your calculator or start a new problem, always *clear the machine*. Although most machines clear automatically when turned on, you never know when that first surge will leave some unwanted numbers in the wrong place. It is best to get into the habit and be on the safe side.

When you are not using the machine, turn it off to save the battery.

Sometimes it takes a few moments for a calculator to perform an operation. The display is blank or lightly flashing when this is going on. **You must give your calculator the time it needs to complete a computation before pressing the next button**, or an error may occur.

Entering Numbers

To enter the number 231.72, press [2][3][1][·][7][2]. To enter -231.72, you do the same thing, except you also press the *change sign* key, which usually looks like [+/−] but may be [SC] or [CHS]. This may be done *at any time* during the entry *after the first digit has been entered*. The entry of large and small numbers is discussed in Sec. 1.3.

The Basic Operations [+], [−], [×], [÷]

Your calculator has one of two kinds of logic: either *algebraic* or *reverse Polish notation* (RPN). If it has algebraic, it has an [=]; if it has RPN, it has an *enter* button, usually either [ENTER↑], [ENT], or [EN]. To find 2 + 3, you press

Alg.: [2][+][3][=];   RPN: [2][ENT][3][+]

and similarly for −, ×, and ÷.

If you have an algebraic machine, you *must* know if it has standard or algebraic hierarchy. To illustrate what this means, consider the expression $2 + 3x$. You know that if you evaluate this expression at $x = 4$, you obtain 14. In particular you do *not* obtain 20 since $2 + 3 \times 4$ means $2 + (3 \times 4)$ and not $(2 + 3) \times 4$. This illustrates that when we write such expressions without parentheses, there is a **hierarchy** of operations; we perform all multiplications and divisions first and then additions and subtractions. An algebraic hierarchy machine has the electronics built into it to keep straight which operations take precedence over others; i.e., it keeps straight the hierarchy of the operations and will correctly compute the above as 14. On the algebraic hierarchy machine, if you want $(2 + 3) \times 4$, you press [2][+][3][=][×][4][=]. However, a standard algebraic machine just computes numbers as they go along. So if you want $(2 + 3) \times 4$, you need only press [2][+][3][×][4][=] (you may but do not have to press [=] after the [3]). If you want $2 + 3 \times 4$, you must either use parentheses [2][+][(][3][×][4][)][=] or change the order in which you enter the data, e.g., [3][×][4][+][2][=]. Either way is not that hard to get used to, but *you must keep in mind what your calculator is doing as you are doing computations*.

On an RPN calculator, if you want $2 + 3 \times 4$, press [2][ENT][3][ENT][4][×][+]; if you want $(2 + 3) \times 4$, press [2][ENT][3][+][ENT][4][×]. An RPN calculator has a stack of (usually four) registers. In the first computation above, you are using the *bottom three* registers. When you then press [×], the 3 and 4 are multiplied, and 12 is put in the bottom, and the 2 is moved down one. When the [+] is then pressed, the bottom two are added. The thing you must keep track of with an RPN calculator is that when the stack is full and you enter another number, the top number is lost (out the top). Four registers are sufficient for most computations if you perform operations whenever possible and keep track of what is going on.

Error Sign

If you try to compute something that is not allowable, for example, divide by zero or take the square root of a negative number or compute a number that is outside of the calculator's range, the machine will start to compute it and then given an error sign. These signs vary from calculator to calculator. The sign may be a flashing "E," a symbol such as "Γ" or "F" on the left-hand side of the display, the word "ERROR," a row of dots, etc. Try to compute $1 \div 0$ or $\sqrt{-1}$ on your machine to see what its error sign is, so you will recognize what happened the next time you see it.

EXERCISES

Use your calculator to compute the following after *estimating first*.

		Estimate	Answer
1.	$571.81 + 219.96$	$\approx 600 + 200 = 800$	791.77
2.	$.08412 - .10908$	$\approx .08 - .11 = -.03$	$-.02496$

3.	$4.821 \times (-621.5)$	$\approx 5 \times (-600) = -3000$	-2996.25
4.	$.9198 \div (.0185)$	$\approx 1 \div .02 = 100 \div 2$ $= 50$	49.7189
5.	$8.11 + 2.81 \times 3.10$	$\approx 8 + 3(3) = 17$	16.821
6.	$42.1 - 8.11 \div (-4.92)$	$\approx 42 - 8 \div (-4) = 44$	43.7484
7.	$121.9 \times (-8.21) \div (.00212)$	$\approx 100(-8)/.002$ $= 100,000(-8)/2$ $= -400,000$	$-472,075$
8.	$48.11 - 19.19 - 14.21$	$\approx 50 - 20 - 10 = 20$	14.71
9.	$48.11 - (19.19 - 14.21)$	$\approx 50 - (20 - 10) = 40$	43.13

SECTION B.2. MEMORY AND PARENTHESES

Calculators vary as to the number of memories and how you may use them. Consult your owner's manual to see how to use the memories on your calculator. Typically, calculators have an *add into memory* and/or *store in memory* and *recall from memory* buttons. The add into memory button, commonly labeled [M+] or [SUM], adds whatever is in the display to the contents of the memory. Before using this the first time in a computation, it is essential that the memory contain zero. Check your instructions as to how to clear your memory. The store into memory button, commonly labeled [M in] or [STO], replaces the contents of memory by whatever is in the display. Whatever was in the memory before is lost. The recall from memory button, commonly labeled [MR] or [RCL], puts the contents of the memory into the display.

Warning: The [MR] button will usually *not* set the memory to zero.

Calculators also vary as to how many levels of parentheses they have and even whether or not they have parentheses. Parentheses are used in a very natural way. The following examples will illustrate how parentheses and memory are used to do more complicated problems. Since we wish to illustrate only the capabilities of various calculators, all the examples will use very simple numbers. However, first consult your owner's manual to determine the capabilities of your calculator.

EXAMPLE 1 Use a calculator to compute $2 \cdot 8 + 4 \cdot 5 + 3 \cdot 3 = 45$.

SOLUTION (a) Algebraic calculator with hierarchy: The hierarchy takes care of everything, so simply press

$$[2][\times][8][+][4][\times][5][+][3][\times][3][=]$$

(b) Algebraic, with parentheses: Press

$$[2][\times][8][+][(][4][\times][5][)][+][(][3][\times][3][)][=]$$

(c) Algebraic, with [M +]: Press
$$[2][\times][8][=][M+][4][\times][5][=][M+][3][\times][3]$$
$$[=][M+][MR]$$

(d) RPN: Press
$$[2][ENT][8][\times][4][ENT][5][\times][+][3][ENT][3][\times][+]$$

EXAMPLE 2 Use a calculator to compute
$$(2 + 8)(4 + 5)(3 + 3) = 540$$

SOLUTION (a) Algebraic, with parentheses: Press
$$[2][+][8][=][\times][(][4][+][5][)][\times][(][3][+][3][)][=]$$

(b) Algebraic, with [STO] (*do not* use the [M +] button): Press
$$[2][+][8][=][STO][4][+][5][=][\times][MR][=][STO][3][+][3]$$
$$[=][\times][MR][=]$$

(c) RPN: Similar to part (d) of the solution to Example 1, except interchange [+] and [×].

EXAMPLE 3 Use a calculator to compute $(2 + 3)/(3 + 5) = .625$.

SOLUTION (a) Algebraic, with parentheses: Press
$$[2][+][3][=][\div][(][3][+][5][)][=]$$

(b) Algebraic, with memory (either [STO] or [M +]): Press
$$[3][+][5][=][STO][2][+][3][=][\div][MR][=]$$

(c) RPN: Press
$$[2][ENT][3][+][3][ENT][5][+][\div]$$

EXERCISES

Use your calculator to compute the following after estimating first.

1. $23.41 \times 48.19 + 213.1 \times 9.181 + 7.191 \times 417.8$
   Est.: $\approx 20 \times 50 + 200 \times 9 + 7 \times 400 \approx 1000 + 1800 + 2800 \approx 6000$
   Ans.: 6089.00
2. $.2133 \times .987 + .581 \times .612 + .018 \div .2113$
   Est.: $\approx .2(1) + .6(.6) + .02/.2 = .2 + .36 + .1 \approx .7$
   Ans.: .651286
3. $(54.11 + 23.1)(76.001 - 18.91)(52.13 + 4.009)$
   Est.: $\approx 70(60)(60) \approx 4000(60) = 240{,}000$
   Ans.: 247,460

4. $[(.1181 + .0932)(.7919 - .5818)]/(.6912 + .1123)$
   Est.: $[.2(.2)]/.8 = .1/2 = .05$
   Ans.:   .0552509

5. $\dfrac{8.127 + 2.916}{3.911 - 1.121}$
   Est.: $\approx \dfrac{8 + 3}{4 - 1} = \dfrac{11}{3} \approx 3.7$
   Ans.:   3.95806

6. $\dfrac{88.91 + 15.01}{23.04 - 8.12}$
   Est.: $\approx \dfrac{100}{15} = \dfrac{20}{3} \approx 7$
   Ans.:   6.96515

7. $\dfrac{.0129 - .1281}{.918 + .2001}$
   Est.: $\approx \dfrac{-.1}{1.1} \approx -.1$
   Ans.: $-.103032$

SECTION B.3. THE SQUARE, SQUARE ROOT, AND RECIPROCAL BUTTONS

Most scientific calculators have the square, square root, and reciprocal buttons as separate buttons. The square button is usually labeled x^2. The square root button is labeled $\sqrt{}$ or $\sqrt{x}$; the reciprocal button is usually labeled $1/x$ or x^{-1} (remember that x^{-1} means $1/x$). Their use is the same for both algebraic and RPN machines. For example, to find 3^2, $\sqrt{2}$, or $1/2$, press $[3][x^2]$ (getting 9), $[2][\sqrt{}]$ (getting $1.414\ldots$), or $[2][1/x]$ (getting .5).

It is important to understand how to use one of these buttons when the operation is part of a larger expression. For example, consider the two expressions

1. $7 + \sqrt{9}$
2. $\sqrt{7 + 9}$

You can easily do these in your head and see that expression 1 is equal to 10 and that expression 2 is equal to 4. The question is, How do you use your calculator to compute these numbers (so you can compute similar expressions with numbers which are not so simple)? You press

1. Alg.: $[7][+][9][\sqrt{}][=]$
   RPN: $[7][ENT][9][\sqrt{}][+]$

2. Alg.: $[7][+][9][=][\sqrt{}]$
   RPN: $[7][ENT][9][+][\sqrt{}]$

The point to understand is that the $[\sqrt{}]$ button works *only* on the number which is in the display, and it leaves alone anything else which may be in the machine. So to compute the square root of an expression, make sure the value of that expression *only* is sitting in the display.

Thus the difference between computing expressions 1 and 2 is very slight: simply interchanging the last two operations. However, the consequences of such small differences can be extreme. This simple example illustrates the following point which cannot be overemphasized: *Your calculator will not do any thinking for you.* It will simply compute exactly whatever corresponds to the button you push. Consequently, *you must keep track of what you are doing when using a calculator*. That is, you must know the algebra which is going on and the capabilities of the machine.

The preceding discussion for the [√] button applies to the [1/x] and [x^2] buttons in the corresponding way.

To compute $8 + \frac{1}{2}$ ($= 8.5$), press

    Alg.:   [8][+][2][1/x][=]
    RPN: [8][ENT][2][1/x][+]

To compute $1/(8 + 2)$ ($= \frac{1}{10} = .1$), press

    Alg.:   [8][+][2][=][1/x]
    RPN: [8][ENT][2][+][1/x]

To compute $2(3^2)$ ($= 18$), press

    Alg.:   [2][×][3][x^2][=]
    RPN: [2][ENT][3][x^2][×]

To compute $(2 \cdot 3)^2$ ($= 36$), press

    Alg.:   [2][×][3][=][x^2]
    RPN: [2][ENT][3][×][x^2]

We now give an example which combines these properties.

EXAMPLE 1   Compute

$$\left(4.3\sqrt{5.128} - \frac{1}{.1249}\right)^2$$

SOLUTION   Press

    Alg.:   [4.3][×][5.128][√][=][−][.1249][1/x][=][x^2]
    RPN: [4.3][ENT][5.128][√][×][.1249][1/x][−][x^2]

The answer is ≈ 2.99630.

EXERCISES

Compute the given number.

1. $2 + \sqrt{3}$
    Ans.: ≈ 3.73205

2. $2 - \sqrt{3}$
    Ans.: $\approx .267949$

3. $3.1 + 4.2^2$
    Ans.: $= 20.74$

4. $8.1 - \dfrac{1}{7.3}$
    Ans.: ≈ 7.96301

5. $(3.1 - \sqrt{4.2})^2$
    Ans.: ≈ 1.10378

6. $\sqrt{4.81 - \dfrac{1}{3.41}}$
    Ans.: ≈ 2.12526

7. $\left(5.7\sqrt{19.81} - \dfrac{1}{3.1}\right)^2$
    Ans.: ≈ 627.363

8. $\left(1 + \dfrac{.04}{6}\right)^2$
    Ans.: ≈ 1.01338

9. $\sqrt{1 + \dfrac{.08}{365}}$
    Ans.: ≈ 1.00011

10. $4.2 + \dfrac{1}{2 - 1/1.8}$
    Ans.: ≈ 4.89231

11. $\sqrt{8.91 + \sqrt{3.41 + (2.81)^2}}$
Ans.: ≈ 3.50321

12. $\left(2.9 - 9.1 \div \sqrt{89.7} - \dfrac{1}{3.4}\right)^2$
Ans.: ≈ 2.70621

13. $\sqrt{\left(\dfrac{1}{14.97} - 3.1\sqrt{8.24}\right)^2 - 5.8^2}$
Ans.: ≈ 6.66048

ANSWERS TO ODD-NUMBERED EXERCISES

CHAPTER 1

Section 1.1

1. Commutative   3. Additive identity   5. Distributive   7. Closure   9. Additive Inverse
11. Associative   13. Rational, $\frac{3}{1}$ 15. Rational, $\frac{183}{10}$ 17. Irrational   19. Irrational
21. Rational, $\frac{2}{3}$ 23. Rational, $\frac{5}{6}$ 25. a, f, g   27. $.5\bar{0}$ 29. $.8\bar{3}$ 31. $.\overline{09}$ 33. $.\overline{857142}$
35. 5/3   37. 1/6   39. 10/3

Section 1.2

1. 763,552   3. .000000121164   5. $-8/125$ 7. 14   9. 9/8   11. 16×10^{10}
13. $(1/15) \times 10^{-8}$ 15. x^{18} 17. $32s^{15}t^{20}$ 19. $-3k^{15}l^5$ 21. 1   23. $y + x/y$
25. $1/(2x - 3y)$ 27. d^{18}/c^{12} 29. 72/17   31. 1   33. $y + x$ 35. $3a^2$ 37. $6b^8/a^4$
39. $\frac{-3}{4} x^{11} y^4$ 41. $3b^4/256$ 43. $8xy^2 - 12/y^2$ 45. y^6/a^{12} 47. $b/(a^2 + b^2)$
49. $-972 x^{14}/u^5 y^{13} z^5$ 51. 1   53. x^{m^2}

Section 1.3

1. a. 3   b. 3   c. 5   d. 4   e. 4   f. 7   g. Probably 4   3. 4.761×10^3 5. 2.71×10^5
7. 6.71×10^{13} 9. 1×10^1 11. 410.2   13. .4   15. 8,149,000,000
17. -10^4; -1.15193×10^4 19. 10^{16}; 1.25302×10^{16} 21. -800; -729.99
23. 5×10^{16}; 7.52960×10^{16} 25. 10^4; 1.08178×10^4 27. -10^{-9}; -1.01287×10^{-9}
29. 127.563; $-272,841$; 2.12855×10^{-37}; -1.67593×10^{44}
31. -60.2566; $-27,806.8$; -5.72653×10^{-30}; -2.98421×10^{35}
33. $2,380.95   35. 1.584×10^9 in   37. 5.86570×10^{12} mi   39. 8.07161×10^{17} mi

Section 1.4

1. $-48; 120{,}000; 97{,}378.2$ 3. $-.247685; 500; -401.592$ 5. $5x^5 - x^4 - 2x^3 + x^2 + 12$
7. $2x^3 - 3x^2 - 4x - 1$ 9. $4a^4b - 4a^3b^2 + 4a^2b^3 - 4ab^4$ 11. $x - 3$ 13. $y^3 + x^2y$
15. $x^{11}y^{-6} + x^8y^{-4}$ 17. $x^3 + x^2y + 4y^3$ 19. $x^4 - y^4$ 21. $.04c^2 + .12cd + .09d^2$
23. $9x^4 - 12 + 4x^{-4}$ 25. $36m^2 - 49n^2$ 27. $x^2 + 13x + 36$
29. $4z^2 + 4z - 35$ 31. $24c^4 - 25c^2 + 6$ 33. $42p^2 - 11pq - 3q^2$ 35. $63u^2 + 151uv^5 + 90v^{10}$
37. $-112xy$ 39. $x^{2m}y^{2n} - 2x^my^nz^p + z^{2p}$ 41. $x^3 + 8$ 43. $343 - a^3b^3$

Section 1.5

1. $3x(x-1)^2$ 3. $11(z-1)(z^2+z+1)$ 5. $(a-6)(a^2+2)$ 7. $\frac{1}{3}(c+5d)^2$
9. $(n+3)(n+4)$ 11. $t(2s^2+t)(s^2+3t)$ 13. $(x+1)(x^2-x+1)(x-1)(x^2+x+1)$
15. $z(z+2)^2$ 17. $-4ab$ 19. $(a+b)(x+y)$ 21. $(x+5)(x^2-3)$
23. $(x+4)(x+2)(x-2)$ 25. $(3x-1)(x^2-6)$ 27. $(x+24)(x+1)$
29. $yz(2y+z)(3y+2z)$ 31. $5(x-3y)(x+y)$ 33. $(3z-2)^2$ 35. $5(b-\frac{1}{3})(b^2+\frac{1}{3}b+\frac{1}{9})$
37. $4(2c-d)(c-2d)$ 39. $4(c+5d)(c^2-5cd+25d^2)$ 41. $a[(a-b)^2-(a-b)b+b^2]$
43. $(2a+3b)(3a+2b)$ 45. $3(d-2c)(d^2+2cd+4c^2)(d+2c)(d^2-2cd+4c^2)$
47. $(x+2)(x-1)(x^2+x+1)$ 49. $(x+2)(x-2)^2(x^2+2x+4)$

Section 1.6

1. $2; 1; -; -; -2; -.01; -10^3; x \neq 0, 1$ 3. $2; -; -; \frac{1}{2}; \frac{2}{3}; .00990099; .999001; x \neq -1, 0$
5. $(x+1)/(x+4)$ 7. $(x^2-2x-1)/x(x-1)$ 9. $(3x+4)/(2x-1)$ 11. $(a^2+b^2)/(a^2-b^2)$
13. $(-x-11)/(2x+1)(3x-2)$ 15. $x-2$ 17. $z(2z-3)/3(4z-1)(z+7)$ 19. 0
21. $3/x$ 23. $(x-3)/(x+2)$ 25. $(x^2+x+1)/(x+1)$ 27. $x-y$ 29. $(x^2-2x+4)/x$
31. $-y/(y+1)^2$ 33. $1/(x+1)$ 35. $(y+x)/(y-x)$ 37. 1

Section 1.7

1. 81 3. 10 5. 3 7. 6 9. $5^{3/8} \approx 1.82858$ 11. $6^{7/9} \approx 4.02929$
13. $y^{1/6}$ 15. $1/y^{1/4}$ 17. $12xy^2$ 19. x^2 21. $a^{4/3}/b$ 23. $x^{1/6}$ 25. $x^{2/3}$
27. $1/x^{3/2}y^{1/3}$ 29. $x^{2/5} - 1$ 31. $y^{4/7}/9x^{1/2}$ 33. $\frac{5}{2}x^2y$ 35. $a^{1/2}$ 37. $m - n$
39. $-1/y^{11/2}$ 41. $\sqrt{6}/3$ 43. $\sqrt[3]{36}/4$ 45. $\sqrt{x}/x$ 47. $\sqrt[3]{xy}/y$ 49. $\sqrt{2} - 1$
51. $-\frac{1}{4}(\sqrt{5}+3)$ 53. $5 - 2\sqrt{6}$ 55. $(1 - x^{1/3} + x^{2/3})/(1+x)$ 57. $-1/(5+2\sqrt{6})$
59. $1/(\sqrt{2+x}+\sqrt{2})$ 61. ± 2 63. -2 65. $\pm 8^{3/2} \approx \pm 22.6274$ 67. 32

Review

1. a. Commutative   b. Distributive   3. a. $.25\bar{0}$ b. $1.1\bar{6}$ c. $.\overline{27}$ d. $.\overline{461538}$
5. $5x^3 - 8x^2 + 4x - 7$ 7. $-14b$ 9. $-2y^2 + 5y$ 11. $c^4 + 2c^3d - 2cd^3 - d^4$
13. $9s^4 - 24 + 16s^{-4}$ 15. $16x^4 - 8x^2y^2 + y^4$ 17. $8x(y-2z)$ 19. $(2a+7b)(3a-5b)$
21. $(3s^2 + 4t^3)^2$ 23. $(x-3)(x^2-2)$ 25. $(s-t)(s+t)(s^2+st+t^2)(s^2-st+t^2)$
27. $108a^9b^8$ 29. $a/16b^3$ 31. $54n$ 33. s^9t^6 35. $xy/(x+y)$ 37. $.5b^3/a$ 39. $7^{4/5}$
41. $x^{17/12}y^{11/6}$ 43. $a+b$ 45. $(2x-3)/(3x+4)$ 47. $(x+3)/(2x-5)$
49. $(z^2+z+1)/(z+1)$ 51. $-2/(5x-4)$ 53. $\sqrt{22}/11$ 55. $8(\sqrt{x}+2)/(x-4)$

CHAPTER 2

Section 2.1

1. $-9/4$ 3. $0, -4$ 5. $-1/2$ 7. $-1/8$ 9. $3, -4$ 11. $15/2$
13. $3/2, -2/3$ 15. all reals   17. -1 19. 16 21. $\varnothing$ 23. -5 25. $\varnothing$
27. $12/19$ 29. 3 31. $8, -5$ 33. 2 35. $0, .196602$ 37. 67.8548 39. $.302370$

41. .243605   **43.** $C = \frac{5}{9}(F - 32)$ **45.** $n = (S - a + d)/d$ **47.** $x = 1/(m + 1), -2/(m + 1)$
49. $w = (P - 2l)/2$ **51.** $m = F/(v_1 - v_2)$ **53.** $y = 4/k^2, -1/k^2$
55. a. Third equation; 3.   **b.** Third equation; 0, 2.

Section 2.2

1. 45.8333%   **3.** .133333%   **5.** 6, 18   **7.** 5, 13   **9.** $14,200   **11.** $42.95
13. 152,000   **15.** $30,000, $14,000   **17.** $7,500, $1,500   **19.** 82   **21.** 6, 12, 15   **23.** 7 in.
25. 2 in.   **27.** $45\frac{5}{11} \approx 45.4545$ mph   **29.** $16\frac{2}{3} \approx 16.6667$ kph   **31.** 2 mph   **33.** 48 mph
35. 20 min   **37.** $1\frac{1}{9}$ hr ≈ 1.11111 hr   **39.** 60 min, 90 min   **41.** $3\frac{3}{7}$ hr ≈ 3.42857 hr
43. 12 hr, 20 hr   **45.** 30 mi

Section 2.3

1. $1/2, -2$ **3.** ± 3 **5.** $2x^2 - 11x - 90 = 0$ **7.** $x^2 - 36.2x + 327.61 = 0$
9. $x^2 + \sqrt{2}x - 4 = 0$ **11.** $3/2, -1$ **13.** ± 3 **15.** $(1 \pm \sqrt{33})/8 \approx .843070, -.593070$
17. $1.11118, -1.38270$ **19.** $3/4, -2$ **21.** $23/2, 4$ **23.** $r = \pm\sqrt{\frac{3V}{\pi h}}$
25. $r = (-2\pi h \pm \sqrt{4\pi^2 h^2 + 4\pi A})/2\pi$ **27.** $x = (3 - k \pm \sqrt{k^2 - 2k + 5})/(2 - 2k)$
29. No real solution   **31.** Two real solutions   **33.** $-3/4, k = -8$ **37.** .858572 ft   **39.** 1 m
41. $h = 4$ in., $b = 7$ in.   **43.** $-3 + 3\sqrt{2} \approx 1.24264$ in.   **45.** $90\sqrt{2} \approx 127.279$ ft   **47.** 30, 40 kph
49. 9, 16 mph   **51.** 2 in.   **53. a.** 6.30840 sec   **b.** 5.36999 sec   **c.** 7.41080 sec
55. a. 160 ft   **b.** 1 sec, 6 sec, 7 sec   **c.** $s \le 196$. The highest point the ball reaches is 196 ft.
59. $10^{10}, 10^{-10}$ **61.** $-10^5, 10^{-8}$

Section 2.4

1. 0, 1   **3.** 1   **5.** 3   **7.** -103 **9.** $-3^{-1/3} \approx -.693361$ **11.** $\emptyset$ **13.** 5   **15.** -2
17. $6, -19/16$ **19.** 2   **21.** -3 **23.** $-1, -1/2$ **25.** 8   **27.** 4
29. $\pm 2, \pm 3$ **31.** ± 1 **33.** 1, 81   **35.** $27, -1/27$ **37.** $64, 1/64$ **39.** .150130
41. $8, -27$ **43.** $1, -3/2$ **45.** $3, 1/2$ **47.** 0, 3   **49.** $1, -2$ **51.** $[(1 + \sqrt{5})/2]^4 \approx 6.85410$
53. $\pm 3/2$ **55.** $\pm\sqrt{(3 + \sqrt{21})/6} \approx \pm 1.12417$ **57.** $m = T^2 g/4\pi^2$ **59.** $x = (a^{2/3} - 4^{2/3})^{3/2}$
61. $g = 4\pi^2 m/T^2$ **63.** $x = (\sqrt{a} - \sqrt[4]{y})^2$ **65.** 60.5269 ft   **67.** 572.142 ft   **69.** 423.357 ft

Section 2.5

1. a. $-1 < 1$ **b.** $-7 < -5$ **c.** $-8 < -4$ **d.** $2 > 1$ **e.** $0 = 0$
13. a. $(-2, 3)$ **b.** $(-2, 3]$
   c. $[-2, 3)$ **d.** $[-2, 3]$
15. a. $(-7, -3)$ **b.** $\mathbb{R}$
   c. $\emptyset$ **d.** $(-\infty, -7) \cup (-3, \infty)$
17. a. $x < 5$ **b.** $x \le -2$ **c.** $-3 \le x$ **d.** $0 < x$
19. $(-2, \infty)$ **21.** $(5/2, \infty)$
23. $[.481499, \infty)$ **25.** $[.684375, \infty)$

27. $(-\infty, 10]$ **29.** $(-\infty, 9/11]$

31. $[-11/2, 13/2]$ **33.** $[34/3, 12)$

35. $(-\infty, -11/2] \cup [13/2, \infty)$ **37.** $(-\infty, 34/3] \cup (12, \infty)$

39. $\emptyset$ **41.** $(-5/3, 0]$ **43.** $(-\infty, -5/3] \cup [-3/2, \infty)$

45. $[8, \infty)$ **47.** $[-16, \infty)$

49. $k < 1/9, k \neq 0$ **51.** $[93\frac{1}{3}, 100]$, so really $[94, 100]$ **53.** A must go at least 275 mi. No.
55. a. $15\frac{5}{9} \leq C \leq 26\frac{2}{3}$ **b.** $-40 \leq F \leq 32$ **57. a.** $96.1538 \lesssim t \lesssim 128.205$ **b.** $7.8 \leq v \leq 10.4$
59. $\emptyset, \emptyset, \emptyset, \{a\}$ **61.** $\{-1\}, \emptyset, \{-7, -5, -2, -1, 0, 1\}, \{0, 1, 3\}, \{0, 1, 3\}$
63. $[1, \infty), (3, \infty), \emptyset, (-\infty, 2) \cup (3, \infty), (3, \infty) \cup [1, 2), (3, \infty) \cup [1, 2).$

Section 2.6

1. $7 - \sqrt{7}$ **3.** $2^{\sqrt{2}} - 2$ **5.** 5   **7.** 6   **9.** $3|x(z-1)|$ **11.** $2|p|^3/9r^4$

13. ± 3 **15.** $(-\infty, -\frac{1}{2}] \cup [\frac{1}{2}, \infty)$

17. $(-4.5, -1.5)$ **19.** -4 **21.** $\mathbb{R}$

23. $\pm 1/3$ **25.** $(-6, 6)$

27. $(-\infty, -\frac{1}{4}] \cup [\frac{5}{4}, \infty)$ **29.** $.948, -.801333$

31. $(-.7, .9)$ **33.** $[-\frac{8}{7}, 1]$ **35.** $y = \pm x$

37. a. e.g., $x = -1, y = 2$
     b. $x < 0$ and $y > 0$ or $x > 0$ and $y < 0$
     c. $|x + y| \leq |x| + |y|$
39. $-a, a$

Section 2.7

1. $(-\infty, -1) \cup (2, \infty)$ **3.** $(1, 5)$

5. $(-\infty, -1/3] \cup [1/2, \infty)$ **7.** $(-\infty, -3/4] \cup [4/3, \infty)$

9. $(-\infty, -3] \cup [3, \infty)$ **11.** $[-1, 1]$

13. $(-1/2, 3]$ **15.** $(-1/2, 2)$

17. $(-\infty, 0) \cup (1, \infty)$ **19.** $[0, 1]$

21. $(-\infty, 0) \cup (0, 1]$ **23.** $(-1, 2) \cup (3, \infty)$

25. $(-\infty, -2) \cup (0, 2)$ **27.** $(-1, 2) \cup [7/2, \infty)$

29. $[-3, -5/2) \cup (5/2, 3]$ **31.** $(-\infty, 0) \cup \{2\}$

33. $(-\infty, -2) \cup (-1, 1)$ 　　**35.** $(-\infty, -3) \cup \{-2\} \cup [1, 2)$
37. $[1/2, 3]$ 　**39. a.** $(-2, 2)$ 　**b.** $(-\infty, -2) \cup (2, \infty)$

Review

1. $2, 4$ 　**3.** -1.13081 　**5.** $(-3 \pm \sqrt{5})/2 \approx -.381966, -2.61803$
7. $r = (-2\pi h \pm \sqrt{4\pi^2 h^2 + \frac{20}{3}\pi A})/\frac{10}{3}\pi$ 　**9.** ± 1 　**11.** 24 　**13.** $\pm 1, \pm 5$ 　**15.** $8, -27$
17. $3 - \sqrt{6}$ 　**19.** $\frac{27}{2}|x|^3 y^4$ 　**21.** One real solution 　**23.** No real solution
27. $(-\infty, -3]$ 　　**29.** $(-\infty, \infty)$
31. $(-2, 4)$ 　　**33.** $(-\infty, -1) \cup (6, \infty)$
35. $[-7/2, -2) \cup (1, \infty)$
37. $(-\infty, -\sqrt{5}] \cup [-\sqrt{3}, \sqrt{3}] \cup [\sqrt{5}, \infty)$
39. $219.231, 269.231$ km/hr 　**41.** $\$16{,}850, \$1600.75$ 　**43.** 90 　**45.** 336 mi $\leq d \leq 588$ mi

Section 3.1

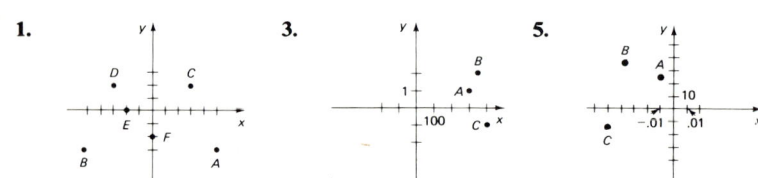

7. a. 13; 　**b.** $(-1, -1/2)$ 　**9. a.** 7; 　**b.** $(7, 3/2)$ 　**11. a.** $2\sqrt{a^2 + b^2}$; 　**b.** (b, a)
13. a. $s^{3/2} + s^{-3/2}$; 　**b.** $(\frac{1}{2}(s^{3/2} + s^{-3/2}), \frac{1}{2})$ 　**15.** Not collinear 　**17.** Collinear
19. $(5, 2), (5, 6); (-3, 2), (-3, 6); (-1, 4), (3, 4)$ 　**21.** $(0, 1), (0, 9)$ 　**23.** $(-2, -7), (-2, 17)$
25. $(0, -3/4)$ 　**27. a.** $(a/2, b/2)$

Section 3.2

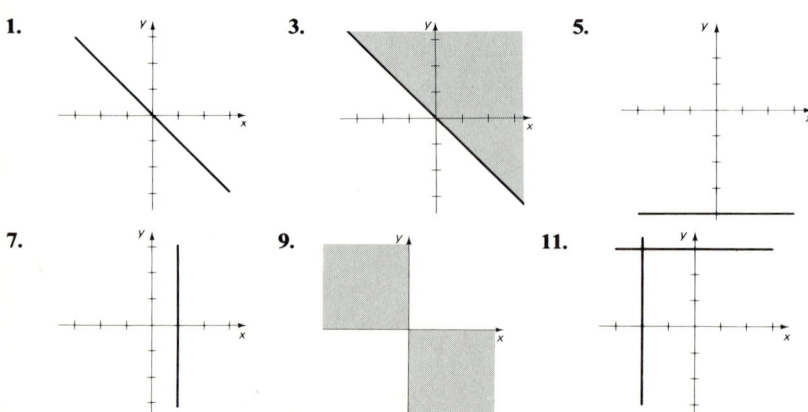

13.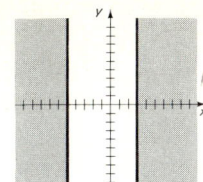
15.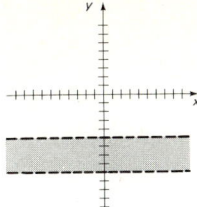
17.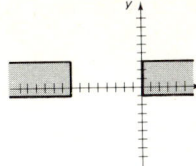

19.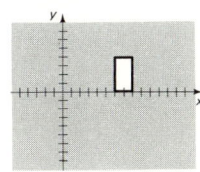
21.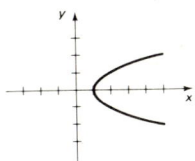
23.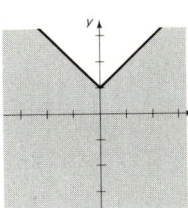

25.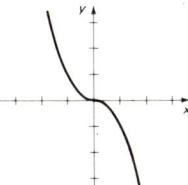
27.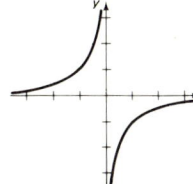
29.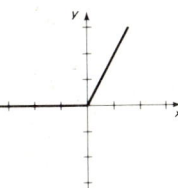

31.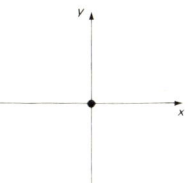
33.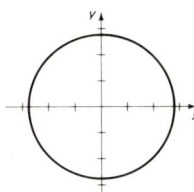
$C = (0,0), r = 2\sqrt{2}$
35.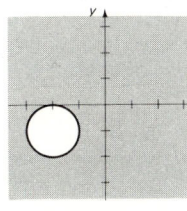
$C = (-2, -1), r = 1$

37.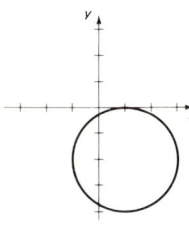
$C = (1, -2), r = 2$
39.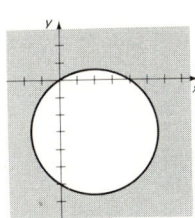
$C = (2, -3), r = \sqrt{13}$
41.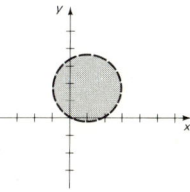
$C = (1, 5/3), r = \sqrt{34}/3$

43.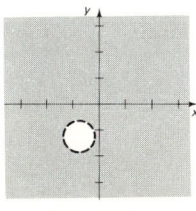
$C = (-\frac{3}{4}, -\frac{5}{4}), r = 1/2\sqrt{2}$

45. y-axis   47. None   49. x-axis   51. x-axis   53. Origin

55. x-axis, y-axis, origin **57.** x-axis **59.** Origin **61.** None **63.** y-axis

65. **a.** 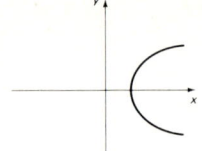 **b.**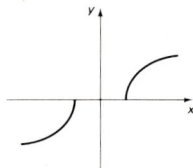

c. 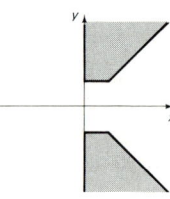 **d.**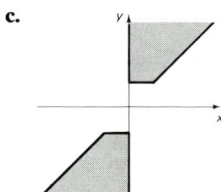

67. **a.** 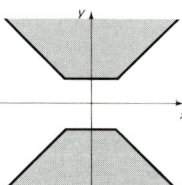 **b.**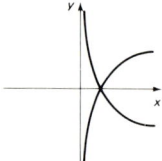

c. 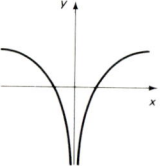 **d.**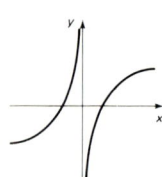

69. **a.** 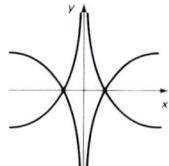 **b.**

c. **d.**

71. $(x + 1)^2 + (y + 3)^2 = 5^2$ **73.** Inside

Section 3.3

1. 34, 6, 1, 9   **3.** 3, 2, 1, 0   **5.** **a.** 3   **b.** -18 **c.** $7/a + 3$,   **d.** $7a + 24$ **e.** 7
7. **a.** 2   **b.** 10   **c.** 5/2   **d.** $5a^2$ **e.** 10   **9.** **a.** $-.8$ **b.** 1.8   **c.** 1   **d.** 0   **f.** -1

11. a. 1.4   b. 2.1   c. .9   d. 1.8   f. 1.2   13. $[4, \infty)$ 15. $(-\infty, 0]$ 17. all $x \neq \pm 3$
19. all $x \neq -3, -1, 2$ 21. $(-\infty, -3/2] \cup [3/2, \infty)$ 23. $(-2, 1] \cup (6, \infty)$
25. $(-\infty, -3/2) \cup (-3/2, 1]$ 27. all $x \neq \pm 1$ 29. b, d   31. b, d   33. $r = C/2\pi$
35. $A = 6V^{2/3}$ 37. a. $V = (10 - 2x)(12 - 2x)x$,   b. $0 < x < 5$
39. a. $T = \sqrt{1 + x^2}/3$ b. $T = \sqrt{5}/3$ hr   41. $d = \begin{cases} 15t, \ 0 \leq t \leq 2 \\ \sqrt{400(t-2)^2 + 225t^2}, \ 2 < t \end{cases}$
43. $C = \begin{cases} 20n, \ 0 \leq n < 25 \\ (22.40 - .10n)n, \ 25 \leq n \leq 110 \end{cases}$

Section 3.4

1. $y = [2(x - \tfrac{3}{2})]^2$ 3. $y = 1/2(x + 2)$ 5. $y = -3[-\tfrac{1}{2}(x - 4)]^3 = \tfrac{3}{8}(x - 4)^3$
7. $y = 5|-3(x - \tfrac{1}{3})| - 1 = 15|x - \tfrac{1}{3}| - 1$ 9. $y = -\sqrt{2x - 4}$ 11.

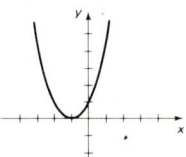

13, 15.   17.   19.

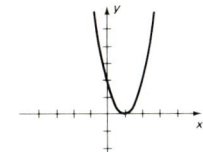   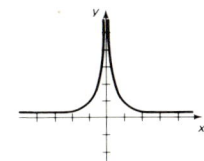   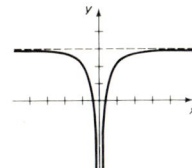

21.   23.   25.

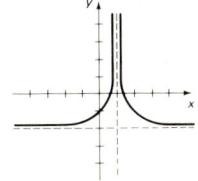   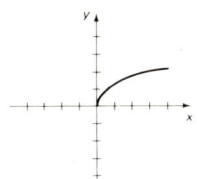   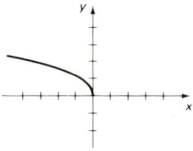

27.   29.   31.   33.

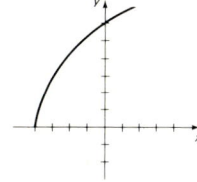   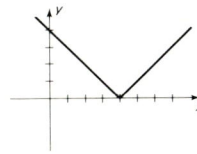   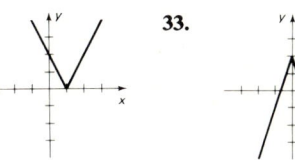

35.   37.   39.   41.

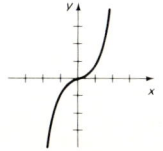   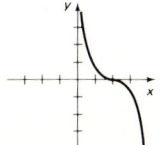   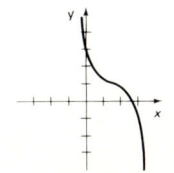   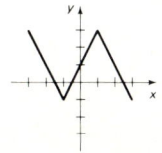

43.  45. 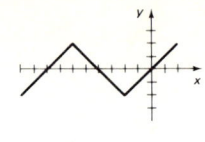 47.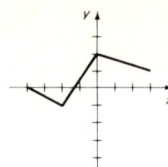

49. 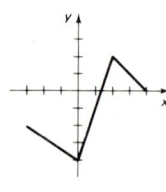 51. 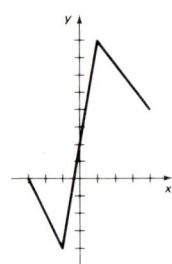 53.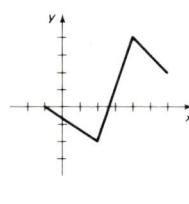

Section 3.5

1. **a.** -1 **b.** $y = -x + 3$ 3. **a.** $1/6$ **b.** $y = \frac{1}{6}x - \frac{9}{2}$ 5. **a.** Undefined **b.** $x = 2$ 7. Yes
9. No 11. $(0, 6), (8, 0)$ 13. $(3, -4), (9, 0)$ 15. $(-1, 0), (5, -2)$
17. $y = -2x + 1$ 19. $y = -\frac{5}{6}x + \frac{31}{6}$ 21. $y = \frac{2}{3}x + \frac{4}{3}$ 23. **a.** $x = 4$, **b.** $y = -2$
25. $y = -\frac{3}{2}x - \frac{7}{2}$ 27. $y = \frac{2}{3}x - \frac{13}{3}$ 29. 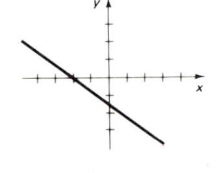 $m = -3/4, b = -3/2$ 31. 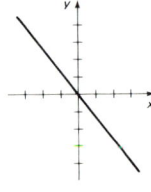 $m = -\frac{4}{3}, b = 0$

33. 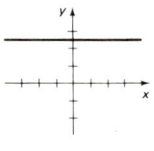 $m = 0, b = 5/2$ 35. 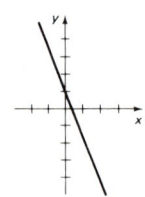 $m \approx -2.79338, b \approx 1.07550$ 37. 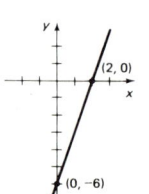 39.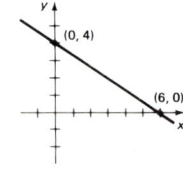

41. 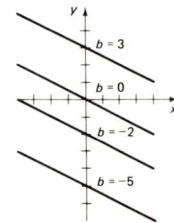 43. $g(x) = -\frac{1}{2}x + \frac{5}{2}$ 45. $F(x) = -\frac{3}{2}x + 4$ 47. $F = \frac{9}{5}C + 32$

49. **a.** $v = -32t + 90$ **b.** $90/32 = 2.8125$ sec **c.** -38 ft/sec 51. **a.** $T = -h/240 + 20$ **b.** $35/3°C$
55. $x = 2, y = 9/2$ or $x = 6, y = 3/2$

Section 3.6

1. $y = 7x/3$ 3. $m = .108/n$ 5. $w = 4z^3/3$ 7. $c = 36/d^{1/3}$ 9. $u = .5vw^3$
11. $A = \frac{2}{3}(B + C)$ 13. $f(x) = 4x/9$ 15. $f(x) = 4x^{3/2}/27$ 17. $y = kz^6$ 19. $x = ky^{-1}$
21. **a.** 8/3 in.   **b.** 9/2 lbs   23. **a.** 375 ft-lbs   **b.** $20\sqrt{10/3} \approx 36.5148$ ft/sec   25. 3137.5 cu ft
27. 50 ohms   29. **a.** 36.3361 km,   **b.** 47.3373 m

Section 3.7

1. $8x - 1$; $8x - 11$ 3. $12x^2 + 1$; $6x^2 + 2$ 5. $x + 1, x \geq -2$; $\sqrt{x^2 + 1}$
7. $8x - 1$; $2\sqrt[3]{x^3 - 1}$ 9. $x/(1 - 3x), x \neq 0, 1/3$; $x - 3, x \neq 3$ 11. $1/x^4, x \neq 0$; $1/x^4, x \neq 0$
13. $f(x) = x^2, g(x) = x + 1$ 15. $f(x) = x + 1, g(x) = x^2$ 17. $f(x) = \sqrt[3]{x}, g(x) = 1/x$
19. $f(x) = 1/x, g(x) = \sqrt[3]{x}$ 21. 1.25, 2.25   23. .2, −1.8   25. .7, .7   27. .84, −1.16
29. Yes   31. No   33. No   35. Yes   37. Yes   39.

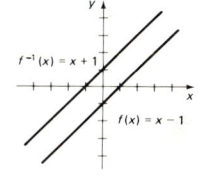

41. 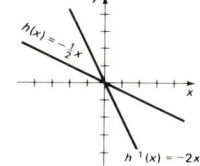   43. 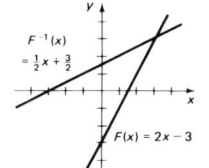   45.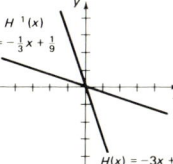

47. 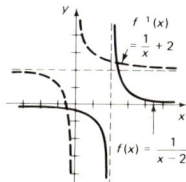   49. 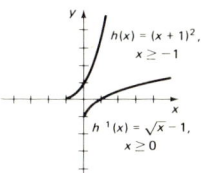   51.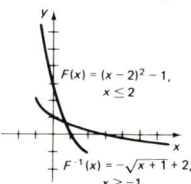

53. $f^{-1}(x) = x - 1$ 55. $f^{-1}(x) = \frac{1}{2}x - \frac{3}{2}$ 57. $f^{-1}(x) = \frac{x}{x - 1}, x \neq 1$
59. $f^{-1}(x) = \sqrt{x + 1}, x \geq 0$

67. 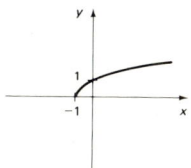   69. 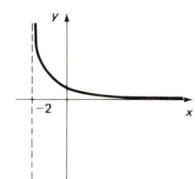   71.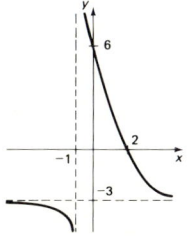

$D_f = [0, \infty) = R_{f^{-1}}$
$R_f = [-1, \infty) = D_{f^{-1}}$

$D_h = (0, \infty) = R_{h^{-1}}$
$R_h = (-2, \infty) = D_{h^{-1}}$

$D_F = \{x \neq -3\} = R_{F^{-1}}$
$R_F = \{x \neq -1\} = D_{F^{-1}}$

73. **a.** $c = 1, d$ arbitrary   **b.** $c = 1, d$ arbitrary   **c.** If $a \neq 1, d = b$. If $a = 1, b, d$ both arbitrary.

Review

1. 29 **3. a.** Both axes **b.** Second and fourth quadrants including axes **5.** $(x-3)^2+(y+4)^2=7^2$

7. $y=-\frac{1}{2}x+\frac{5}{2}$ **9.**

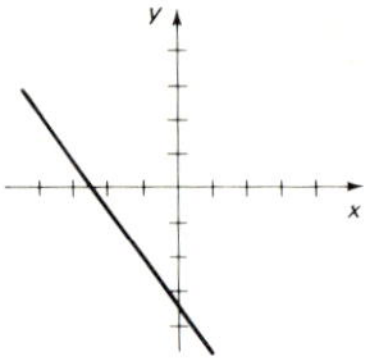

11.

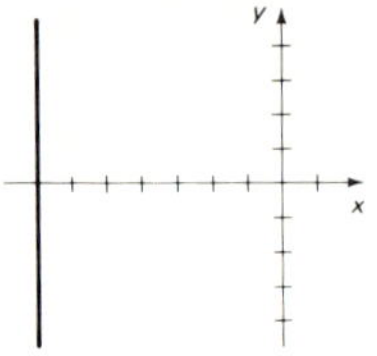

13.

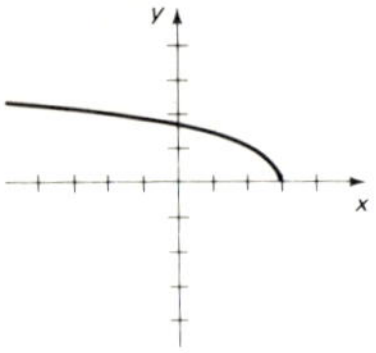

15.

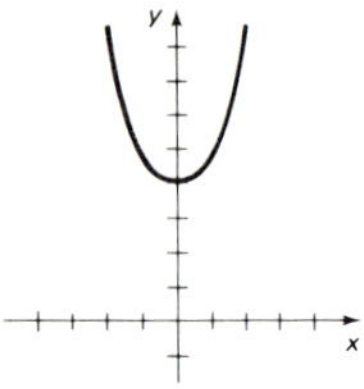

17.

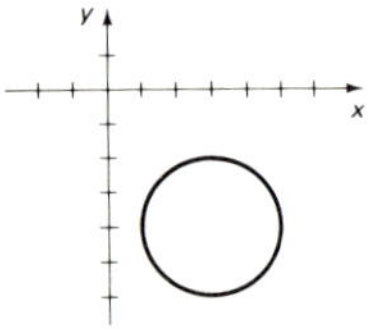

19.

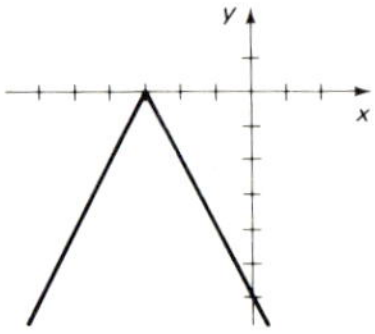

21.

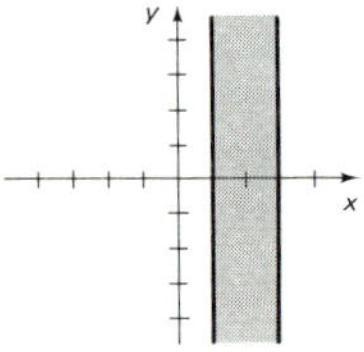

23.

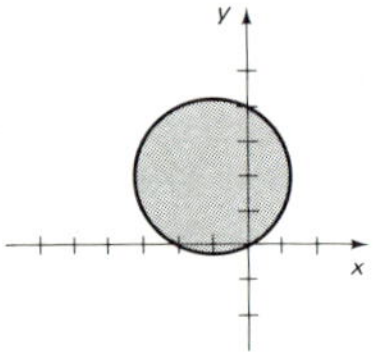

25.

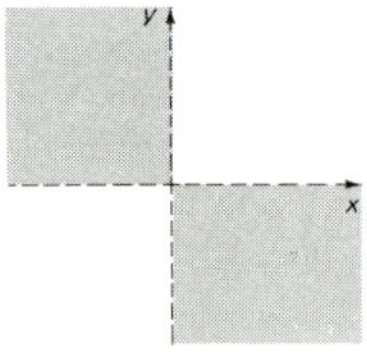

27.

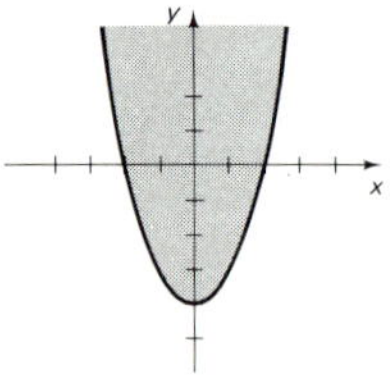

Hill & Hestenes 3R.21

29. a. -1 **b.** $1/2$ **c.** 0 **d.** $a^2/(a^2-2),\ a \neq \pm\sqrt{2}$ **e.** $a^2/(a-2)^2,\ a \neq 2$ **f.** $(1+h)/(h-1),\ h \neq 1$ **g.** $1/(1-2a),\ a \neq 1/2$ **h.** $(a-2)/a,\ a \neq 0$ **31.** $4x^2-12x+8,\ 2x^2-5$

33.

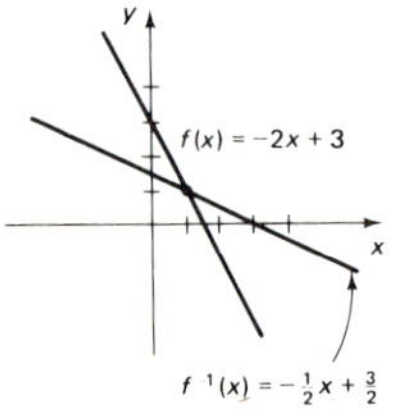

35.

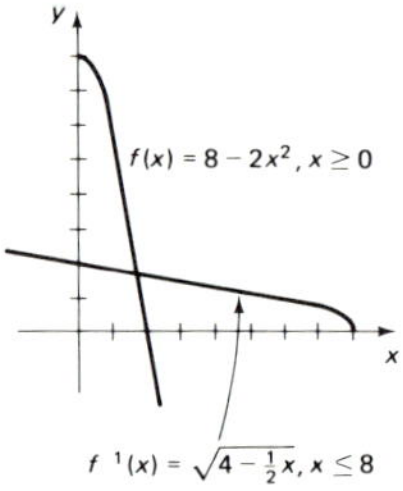

37. x-axis, y-axis, origin **39.** x-axis, y-axis, origin **41.** $(31/4, -2)$ **43.** $C = 2\sqrt{\pi A}$ **45.** 262.5

Section 4.1

1. a.

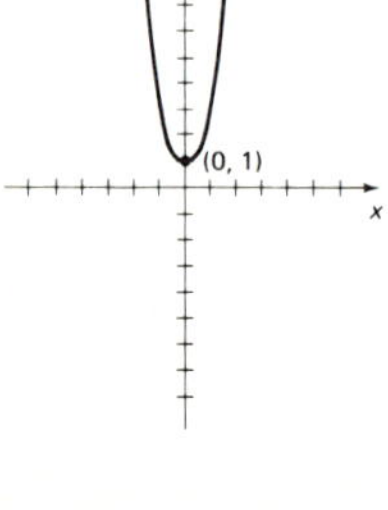

b.

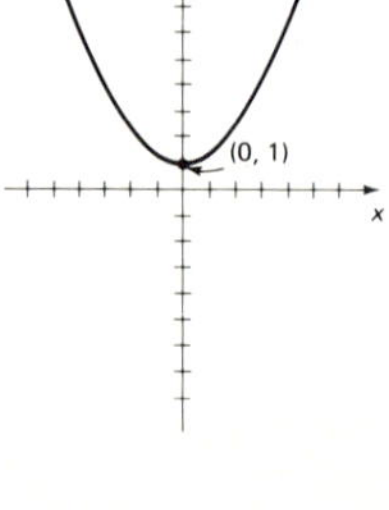

c.

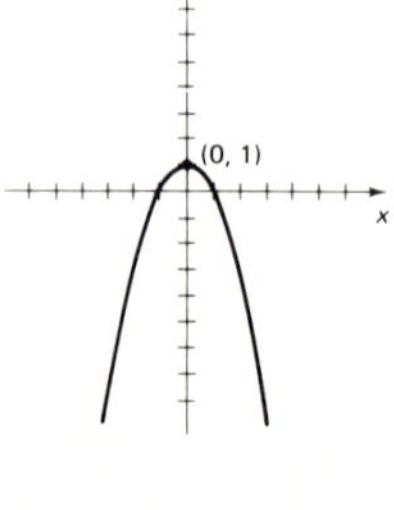

d.

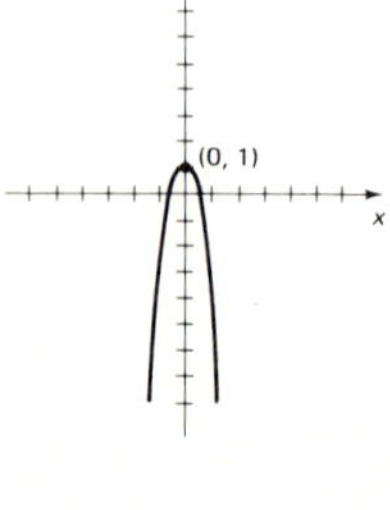

3. a. 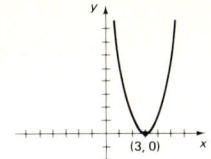 b.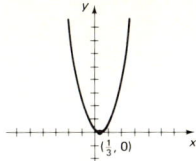

c. 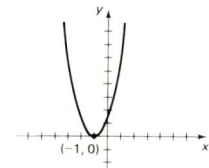 d.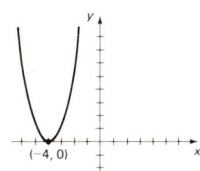

5. 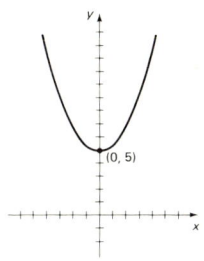 7. 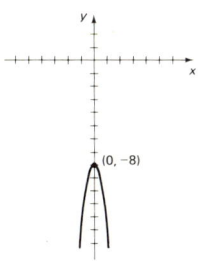 9.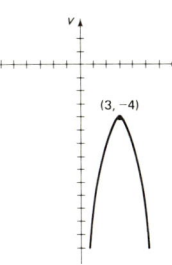

11. 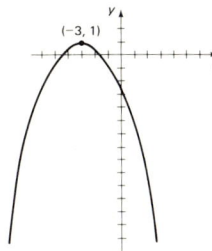   13. x: $(1, 0)$, $(4, 0)$; y: $(0, 4)$ 15. x: $(1/3, 0)$, $(-4, 0)$; y: $(0, 4)$

17. x: $(-3/2, 0)$; y: $(0, 9)$ 19. x: $(0, 0)$, $(8, 0)$; y: $(0, 0)$

21. 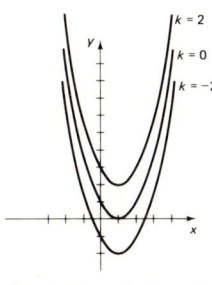   23. 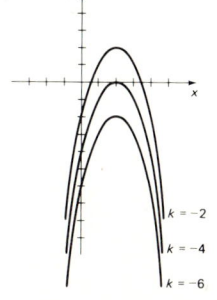   25.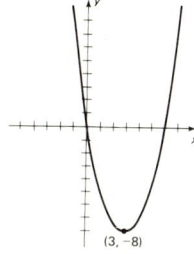

$k < 0$; $k = 0$; $k > 0$ $k > -4$; $k = -4$; $k < -4$ $f(x) = (x - 3)^2 - 8$

27.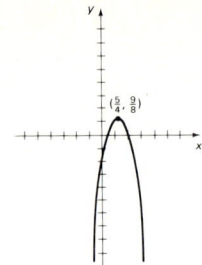
$h(x) = -2(x - 5/4)^2 + 9/8$

29.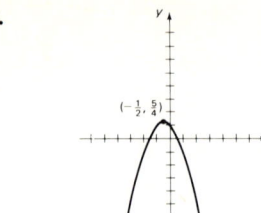
$m(x) = -(x + 1/2)^2 + 5/4$

31.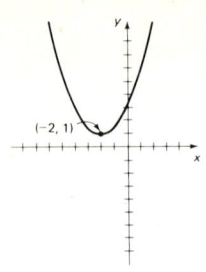
$F(x) = \frac{1}{2}(x + 2)^2 + 1$

33.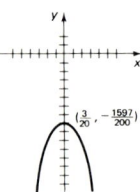
$H(x) = -\frac{2}{3}(x - \frac{3}{20})^2 - \frac{1597}{200}$

35. $b = h = 15, A = 225/2$

37. $x = 2$ in., $y = 3/2$ in., $A = 3$ sq in.

39. $x = 40$ m, $y = 120$ m, $A = 4800$ sq m

41. **a.** $11/8$ sec **b.** $241/4$ ft **c.** 3.31552 sec

43. **a.** 800 m **b.** 10 sec **c.** 22.7775 sec

Section 4.2

1. a. 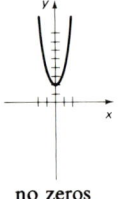 no zeros **b.** 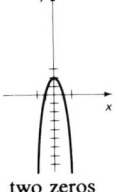 two zeros **c.** 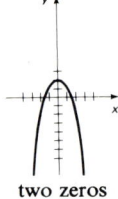 two zeros **d.** 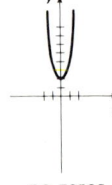 no zeros

3.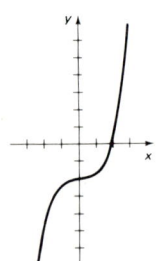

5.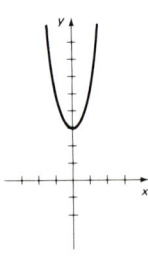

7.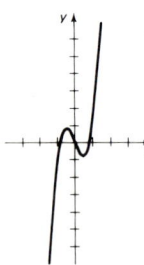

9.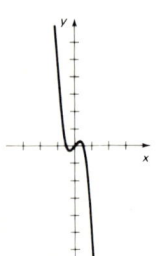

11.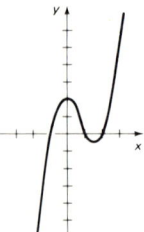

13.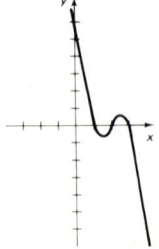

15. 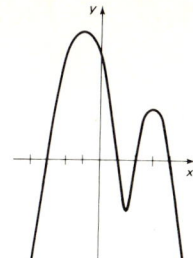 **17.**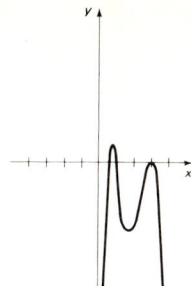

Section 4.3

1. 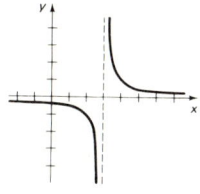 **3.** 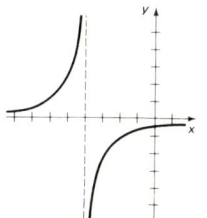 **5.** 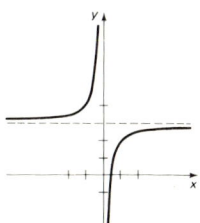 **7.**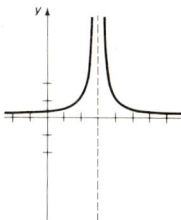

9. 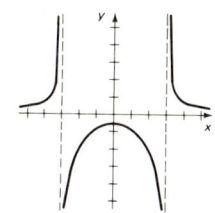 **11.** 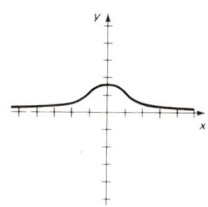 **13.** 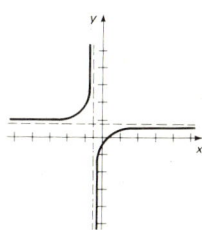 **15.**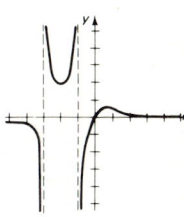

17. 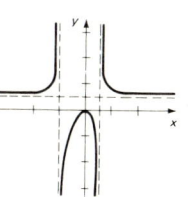 **19.** 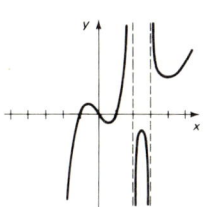 **21.** 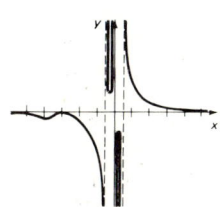 **23.**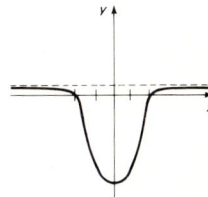

25. 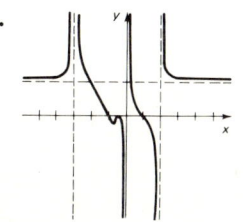 **27.**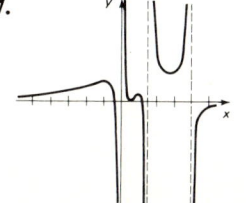

Section 4.4

1.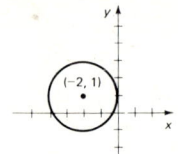

$(x + 2)^2 + (y - 1)^2 = 2^2$

3.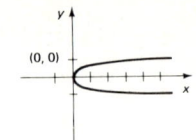

$x = 4y^2$

5.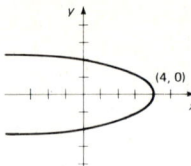

$x = -y^2 + 4$

7.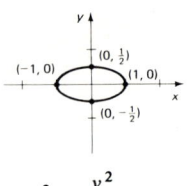

$x^2 + \dfrac{y^2}{\left(\frac{1}{2}\right)^2} = 1$

9.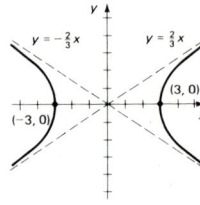

$\dfrac{x^2}{3^2} - \dfrac{y^2}{2^2} = 1$

11.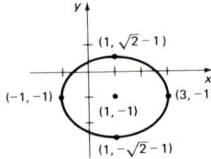

$\dfrac{(x - 1)^2}{2^2} + \dfrac{(y + 1)^2}{(\sqrt{2})^2} = 1$

13.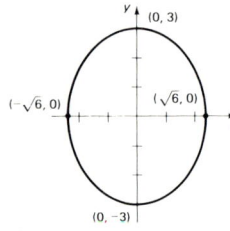

$\dfrac{x^2}{(\sqrt{6})^2} + \dfrac{y^2}{3^2} = 1$

15.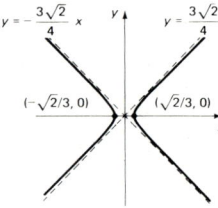

$\dfrac{x^2}{(\sqrt{2}/3)^2} - \dfrac{y^2}{(1/2)^2} = 1$

17.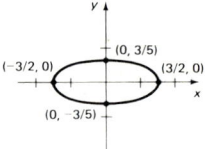

$\dfrac{x^2}{\left(\frac{3}{2}\right)^2} + \dfrac{y^2}{\left(\frac{3}{5}\right)^2} = 1$

19.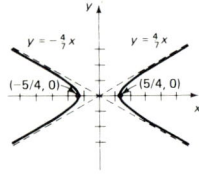

$\dfrac{x^2}{\left(\frac{5}{4}\right)^2} - \dfrac{y^2}{\left(\frac{5}{7}\right)^2} = 1$

21.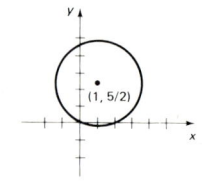

$(x - 1)^2 + (y - \tfrac{5}{2})^2 = (\sqrt{29}/2)^2$

23.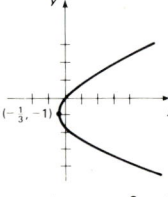

$x = \tfrac{1}{3}(y + 1)^2 - \tfrac{1}{3}$

25.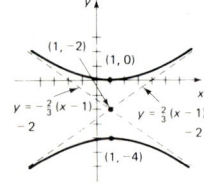

$\dfrac{(y + 2)^2}{2^2} - \dfrac{(x - 1)^2}{3^2} = 1$

27.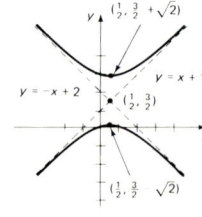

$\dfrac{(y - \frac{3}{2})^2}{(\sqrt{2})^2} - \dfrac{(x - \frac{1}{2})^2}{(\sqrt{2})^2} = 1$

29.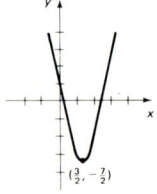

$y = 2(x - \tfrac{3}{2})^2 - \tfrac{7}{2}$

Review

1. 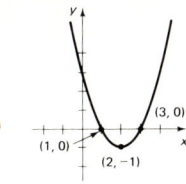 **3.** 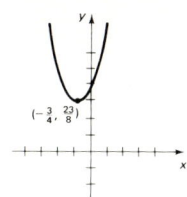 **5.** $x = 3m, y = m$ **7.**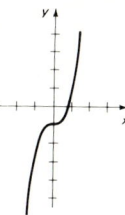

9. 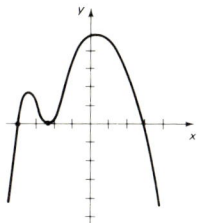 **11.** 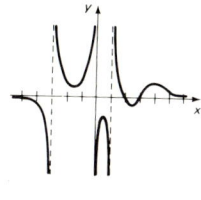 **13.**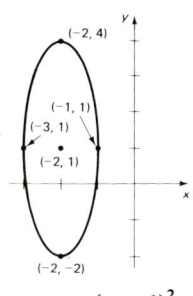

$$(x + 2)^2 + \frac{(y - 1)^2}{3^2} = 1$$

15. 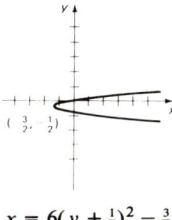 **17.**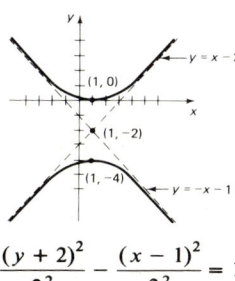

$x = 6(y + \tfrac{1}{2})^2 - \tfrac{3}{2}$

$$\frac{(y + 2)^2}{2^2} - \frac{(x - 1)^2}{2^2} = 1$$

CHAPTER 5

Section 5.1

1. 1330.45   **3.** .562341   **5.** .0214936   **7.** 1.08380   **9.** 1.48985   **11.** 480.540
13. 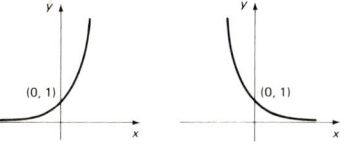   **15. a.** 1.7 **b.** 7 **c.** .8 **d.** 10   **17.** 2.2; 2.23607

19. .9; .870551   **21.** 1.1; 1.12246   **23.** 6; 6.24025

25, 27. 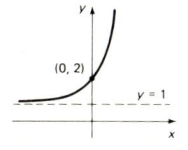   **29.** 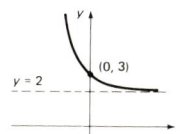   **31.**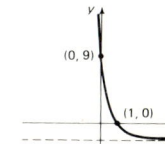

Answers to Odd-Numbered Exercises **529**

33. 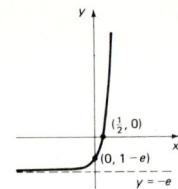 **35.** 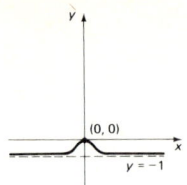 **37.** 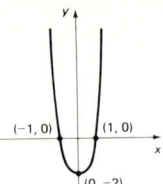 **39. a.**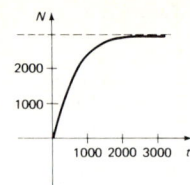

b. 543, 1896, 2504
c. 215 hrs

Section 5.2

1. 3   **3.** 0   **5.** 1   **7.** −2   **9.** 3/2   **11.** −12   **13.** −2   **15.** 1.26186
17. 2.77124   **19.** 5.28540   **21.** 1.35621   **23.** −2.09590   **25.** −.748070   **27.** −.827087
29. 15   **31.** 4   **33.** 7/2   **35.** 2   **37.** $3^9 = 19{,}683$ **39.** $\sqrt{2}$ **41.** 1/2   **43.** 729
45. 11   **47.** 4

Section 5.3

1. 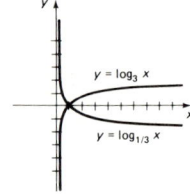   **3.** 1.26186; −1.26186   **5.** −.203114; .203114   **7.** −2.72683; 2.72683

9. 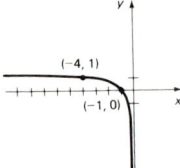   **11.** 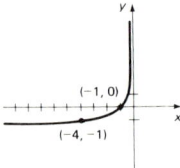   **13.** 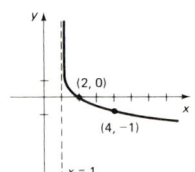   **15.**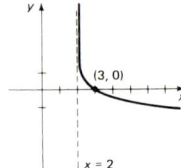

17. 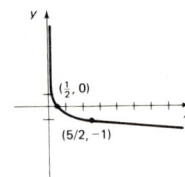   **19.** 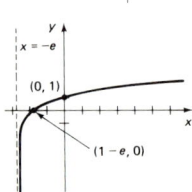   **21.** 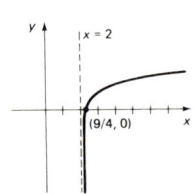   **23.**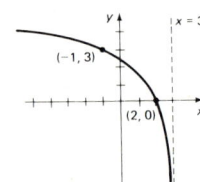

25. 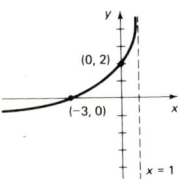   **27.** 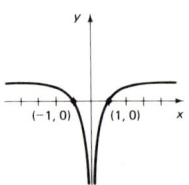   **29.**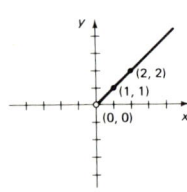

31. a. −.287682; .792481; 2.07944; .333025   **b.** No, no   **c.** Both = $\log_y x$ **33.** 1.43   **35.** .25
37. −.82   **39.** −1.07   **41.** 5   **43.** 20   **45.** x **47.** −1/2   **49.** 36
51. $(23/4)\log_4 x$ **53.** $\log_{12}(x^2 - x + 1)$ **55.** 1/2   **57.** 3.16993   **59.** .319508   **61.** 6.56931
63. .474517   **65.** 16; .25; 1×10^{16}; 1×10^{46}

530   ANSWERS TO ODD-NUMBERED EXERCISES

Section 5.4

1. 3; 6.19174; 7.24465; 7.37431; 7.38906; 7.38906
3. -19; 1; 2.03704×10^{-10}; 1.68297×10^{-9}; 2.06111×10^{-9}; 2.06115×10^{-9}
5. **a.** $10,794.60 **b.** $11,040.20 **c.** $11,126.70 **d.** $11,127.70
7. **a.** 8.16334% **b.** 8.00321% **c.** 7.87294% **d.** 7.84723%
9. **b.** 11. **a.** $5882.80 **b.** $5663.77 **c.** $5594.44 **d.** $5588.62 13. **a.** $72,900 **b.** 15.2755 years
15. 20.5403%; 9.90308 years 17. **a.** 1614 ants **b.** September 13
19. **a.** $P = 50e^{-.0330070t}$ **b.** 69.7605 days **c.** .000292942 grams 21. **a.** 26.6643 watts **b.** 1247.66 days
23. **a.** 2.20300; 2.39999; 2.40000 **b.** 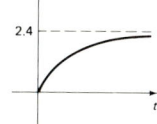 **c.** 1.43341; never 25. 31.6228 times stronger

Section 5.5

1. **a.** $62.08 **b.** $4469.76 **c.** $669.76
3. 20 years: **a.** $179.95 **b.** $43,188 **c.** $23,188
 25 years: **a.** $167.84 **b.** $50,352 **c.** $30,352
 30 years: **a.** $160.92 **b.** $57,931.20 **c.** $37,931.20
5. **a.** Think that you lend the bank money and it pays you back. **b.** $59.96 **c.** yes, $50 **d.** 11.5813 years
7. **a.** $353.39 **b.** $43,249.40 9. **a.** $4,492,760 **b.** 25.3073 years 11. **a.** $18.08 **b.** $16.61
13. **a.** $205.58 **b.** $199.19

Section 5.6

1. 4 3. 2.52372 5. 17/2 7. $\pm\sqrt{10} \approx \pm 3.16228$ 9. 1; -3 11. 8.43168
13. 10 15. 4^{-16}; 4 17. $-\log_4 3 \approx -.792481$ 19. 255; $-3/4$ 21. 11/9 23. 1; 100
25. 1; 10,000 27. 0; $\log_3 8 \approx 1.89279$ 29. $\log_4(8 \pm \sqrt{63}) \approx \pm 1.99717$ 31. ± 2
33. $\log_2 \sqrt{3} \approx .792481$ 35. $-.684177$ 37. 1; $10^{\sqrt{3}} \approx 53.9574$; $10^{-\sqrt{3}} \approx .0185331$

Section 5.7

1. $440.26 3. $338.35 5. $613.35 7. $308.65 9. $689 11. $284.04 13. 9.6%
15. 10.8%

Review

1. .00478656 3. 2; 1.62066 5. 1/4; 4 7. 5 9. 2 11. ± 7

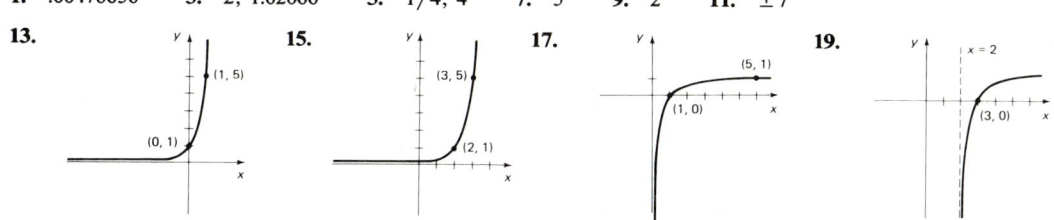

13. (0, 1), (1, 5) 15. (2, 1), (3, 5) 17. (1, 0), (5, 1) 19. $x = 2$, (3, 0)

21. 2 23. $x^2 - 4x$ 25. $-.25$; .86; 2.54 27. .284065 29. .387904 31. 1/2
33. $\log_4 7 \approx 1.40268$ 35. **a.** $7454.51 **b.** $7595.92 **c.** $7609.35 **d.** $7609.81
37. $4340.37; 3.49279 years 39. 32.9877 grams; 108.048 days 41. **a.** $250.28 **b.** $16,016.80
 c. $4,016.80 43. $510.89

CHAPTER 6

Section 6.1

1. $(0,1), (1,2)$ 3. $(-7,3), (-1/4, -3/2)$ 5. No solution 7. $(3,1)$
9. $(2,3), (9, 2/3)$ 11. $(3,0), (-9/5, 12/5)$ 13. $\left(\dfrac{1-\sqrt{39}}{2}, \dfrac{1+\sqrt{39}}{2}\right), \left(\dfrac{1+\sqrt{39}}{2}, \dfrac{1-\sqrt{39}}{2}\right)$
15. $(1, \sqrt{3}), (1, -\sqrt{3}), (-1, \sqrt{3}), (-1, -\sqrt{3})$ 17. No solution 19. $(\sqrt{2}, 2), (-\sqrt{2}, 2)$ 25. No solution
21. **a.** $k < -1$; **b.** $k = -1$; **c.** $k > -1$ 23. $(x, y) = (2, 10)$
27. Two solutions 29. One solution 31. $((\log_2 9) \log_2 3, \log_2 9)$ 33. $.641$

Section 6.2

1. $(1, 2)$, independent 3. Inconsistent 5. $(-1/3, -4)$, independent
7. $(-1/2, 22/3)$, independent 9. Inconsistent 11. $((1 + 2s)/3, s)$, dependent
13. $(1/2, 1/3)$, independent 15. $(2.3, -3.1)$, independent 17. Inconsistent
19. $((2 + 4s)/3, s)$, dependent 21. $(\tfrac{6}{5}, 3)$ 23. $(2, 4)$ 25. $(1, 2), (1, -2), (-1, 2), (-1, -2)$
27. $(1, 3), (1, -3), (-1, 3), (-1, -3)$ 29. No solution 31. **a.** -3 **b.** inconsistent
33. **a.** 3 **b.** independent 35. $(-4, -7)$ 37. 22 nickels, 18 dimes
39. Stream 1.5 kph, woman 4.5 kph 41. Children 330, adults 210

Section 6.3

1. $(-2, 1, 3)$, independent 3. $(4, 1/2, -1/3)$, independent 5. Inconsistent
7. $(\tfrac{8}{5} - \tfrac{3}{5}s, \tfrac{6}{5} - \tfrac{1}{5}s, s)$, dependent 9. $(3, -1, -5)$, independent 11. $(1, -1, 2)$, independent
13. Inconsistent 15. $(3s, 2s + 1, s)$, dependent 17. $(1, -1, 2, -1/2)$, independent
19. $(s/7, 3s/7, s)$, dependent 21. $(0, 0, 0)$, independent 23. $(-2, -1, 3)$, independent
25. $(1/3, 2, -1/2)$, independent 27. $(\tfrac{3}{4} + \tfrac{1}{4}s, -\tfrac{1}{2} - \tfrac{1}{2}s, s)$, dependent 29. $(1, 2, 3)$
31. $f(x) = -2x^2 - 3x + 1$ 33. 8 nickels, 16 dimes, 16 quarters
35. 36 beetles, 30 spiders, 12 centipedes 37. A 24 hrs, B 4.8 hrs, C 12 hrs 39. $34l, 22l, 44l$
41. Candy 55 kg, nuts 100 kg, raisins 45 kg

Section 6.4

1. $(2, -1)$ 3. $(3, -1, -2)$ 5. $(-1, 5, -2)$ 7. $((9 - s)/18, (9 + 5s)/6, s)$ 9. Inconsistent
11. $(-1, -2, 3)$ 13. $(1, -1, 2, 2)$ 15. $(2, -1, -2, 3, 1)$

Section 6.5

1. -1 3. 2 5. -31 7. 24 9. -10 11. $(-1, 2)$ 13. $(1/2, -5/8)$
15. $(1, 2, -1)$ 17. $(-1/3, 1/2, 1)$ 19. 5 21. -6 25. $x^3 - 9x$ 27. $-z^4 - 4z^2 + 2z$

Section 6.6

1. $\begin{bmatrix} 6 & 4 & 0 \\ -3 & 3 & -1 \end{bmatrix}$ 3. $\begin{bmatrix} 4 & 1 \\ 7 & 1 \end{bmatrix}$ 5. $\begin{bmatrix} 4 & 1 & -1 & -3 \\ 2 & 1 & -1 & -6 \end{bmatrix}$

7. $AB = \begin{bmatrix} 2 & -3 \\ -11 & -1 \end{bmatrix}$, $BA = \begin{bmatrix} 1 & 7 \\ 5 & 0 \end{bmatrix}$ 9. $AB = \begin{bmatrix} -6 & -1 \\ 0 & 1 \end{bmatrix}$, $BA = \begin{bmatrix} 6 & 2 & -9 \\ 0 & 1 & 0 \\ 8 & 4 & -12 \end{bmatrix}$

11. $AB = \begin{bmatrix} 5 & -2 & 1 \\ 3 & 0 & -3 \end{bmatrix}$, BA undefined 13. $AB = \begin{bmatrix} 1 & -26 \\ 0 & 27 \end{bmatrix} = BA$ 15. Both undefined

19. $\begin{bmatrix} 7 & -1 \\ 6 & -9 \end{bmatrix} \begin{bmatrix} x \\ y \end{bmatrix} = \begin{bmatrix} -9 \\ 2 \end{bmatrix}$ 21. $\begin{bmatrix} 5 & 1 & -9 \\ 3 & 5 & 8 \end{bmatrix} \begin{bmatrix} x \\ y \\ z \end{bmatrix} = \begin{bmatrix} 5 \\ -8 \end{bmatrix}$

23. $\begin{bmatrix} -6 & -19 \\ 23 & 16 \end{bmatrix} \begin{bmatrix} s \\ t \end{bmatrix} = \begin{bmatrix} 2 \\ 3 \end{bmatrix}$ 25. $\begin{bmatrix} 6 & 23 & -16 \\ 22 & 5 & -2 \end{bmatrix} \begin{bmatrix} p \\ q \\ r \end{bmatrix} = \begin{bmatrix} 10 \\ -3 \end{bmatrix}$

27. $\begin{bmatrix} 0 & 0 & -4 \\ 5 & -6 & 1 \\ 0 & 1 & -2 \end{bmatrix} \begin{bmatrix} p \\ q \\ r \end{bmatrix} = \begin{bmatrix} 2 \\ -5 \\ 1 \end{bmatrix}$ 29. $\begin{bmatrix} \text{Sales} \\ \text{Costs} \end{bmatrix} \begin{matrix} \text{Dec.} & \text{Apr.} & \text{Aug.} \\ 8750 & 5750 & 7350 \\ 6210 & 4080 & 5220 \end{matrix}$

Section 6.7

1. 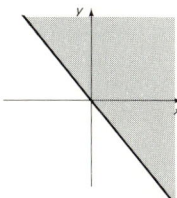 3. 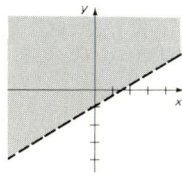 5. 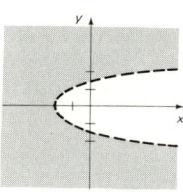 7.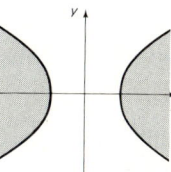

9. 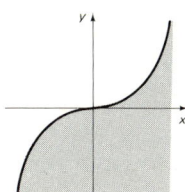 11. 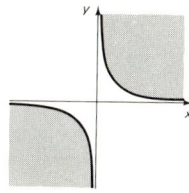 13. 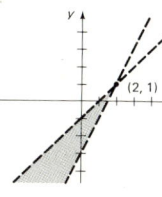 15.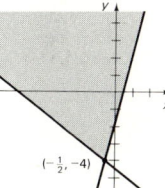

17. 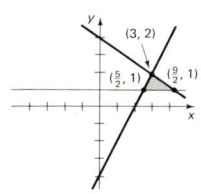 19. 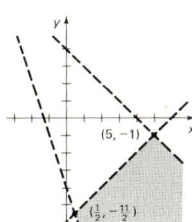 21.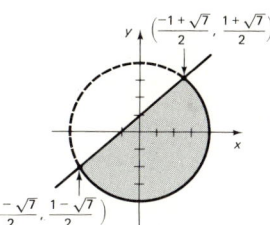

23. 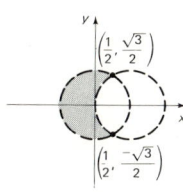 25. 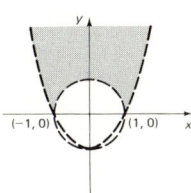 27. 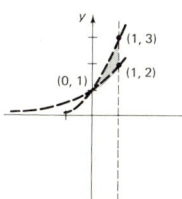 29.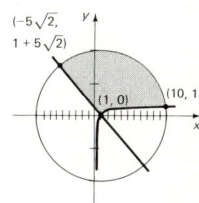

31. x = no. cartons A, y = no. cartons B
$x \geq 0$
$y \geq 0$
$50x + 40y \leq 5000$
$\frac{3}{2}x + 2y \leq 200$

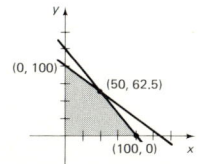

33. x = no. \$4 tickets, y = no. \$5 tickets
$x \geq 200$
$y \geq 100$
$x + y \leq 500$
$4x + 5y \geq 2000$

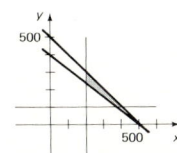

Answers to Odd-Numbered Exercises 533

35. x = no. of A
y = no. of B
$.30x + .45y \leq 500$
$x \geq 400$
$y \geq 500$

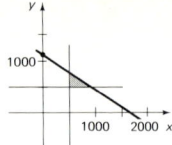

Section 6.8

1. 6 at $(2, 0)$, -7 at $(0, 3)$ **3.** 9 at $(1, 2)$, -1 at $(0, 0)$ **5.** 2 at $(1, 1)$, -80 at $(10, 6)$
7. 39 at $(10, 4)$, 0 at $(1, 1)$ **9.** 60 of A, 50 of B; $415 **11.** 60 of A, 25 of B; $1225
13. A: 0 hours J_1, 40 hours J_2 B: 30 hours J_1, 10 hours J_2; $1740 **15.** 4,000 deluxe; 4,000 standard; $520
17. 2 cups of A, 3 cups of B; 65¢ **19.** 50 of A and C, 100 of B; $29,500

Review

1. $(-1, 0), (3, 8)$ **3.** $(-1, 2), (2, -1)$ **5.** $(41/3, 14/3)$, independent
7. $(-.323582, .772214)$, independent **9.** $(2, -1, 3)$, independent **11.** $(-1/3, 1/2, 1)$, independent
13. Inconsistent **15.** 22 **17.** $A + B = \begin{bmatrix} 4 & -2 & -2 \\ 6 & 2 & 2 \end{bmatrix}$ **19.**

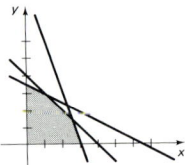

21. 12, 18 **23.** x = no. of A; y = no. of B
$600x + 200y \leq 3600$
$80x + 160y \leq 1040$
$3x + 3y \leq 24$
$x \geq 0, y \geq 0$

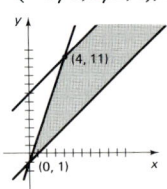

CHAPTER 7

Section 7.1

1. $480°, 8\pi/3$ **3.** $540°, 3\pi$ **5.** $-120°, -2\pi/3$ **7.** $-320°, -16\pi/9$ **9.** $14.42°$
11. $-24.25°$ **13.** $231.549°$ **15.** $84.1019°$ **17.** $-132.694°$ **19.** $1227.08°$
21. $9°28'4.8''$ **23.** $-75°36'48.6''$ **25.** $431°58'48''$ **27.** $3\pi/4 \approx 2.35619$ **29.** $5\pi/12 \approx 1.30900$
31. $2\pi/3 \approx 2.09440$ **33.** $\pi \approx 3.14159$ **35.** -1.49051 **37.** 5.43214 **39.** $-.806342$
41. 15.0348 **43.** $210°$ **45.** $225°$ **47.** $1620°$ **49.** $450°$ **51.** $171.887°$ **53.** $45.8366°$
55. $469.825°$ **57.** $19.9962°$ **59.** $143°14'22.0''$ **61.** $39°4'32.6''$ **63.** $-91°40'23.7''$
65. $-5°12'50.1''$ **67.** $114.887°, 2.00516$ **69.** 47.7149 m **71. a.** $120°, 2\pi/3$ **b.** 39.7935 in.
73. a. $6°, .104720$ **b.** $.0785398$ in. **75.** 2950 ft **77.** 3500 mi **79.** 391.111 rad

For the remainder of the book, when a question asks for all trigonometric functions of θ, the answers are given in the order $\sin \theta, \cos \theta, \tan \theta, \csc \theta, \sec \theta, \cot \theta$.

Section 7.2

1. $5, 6, 10, 4/5, 4/5$ **3.** $39, 5, 13, 5/12, 5/12$ **5.** $\sqrt{13}, 24/\sqrt{13}, 36/\sqrt{13}, 2/3, 2/3$
7. $(-3/5, 4/5)$ **9.** $(4, -8)$ **11.** $4/5, 3/5, 4/3, 5/4, 5/3, 3/4$
13. $7/25, 24/25, 7/24, 25/7, 25/24, 24/7$ **15.** $1/\sqrt{2}, 1/\sqrt{2}, 1, \sqrt{2}, \sqrt{2}, 1$
17. $\sqrt{11}/6, 5/6, \sqrt{11}/5, 6/\sqrt{11}, 6/5, 5/\sqrt{11}$ **19.** $12/13, 5/13, 12/5, 13/12, 13/5, 5/12$

21. $1/\sqrt{2}, 1/\sqrt{2}, 1, \sqrt{2}, \sqrt{2}, 1$ **23.** $\sqrt{3}/2, 1/2, \sqrt{3}, 2/\sqrt{3}, 2, 1/\sqrt{3}$
25. $2/3, \sqrt{5}/3, 2/\sqrt{5}, 3/2, 3/\sqrt{5}, \sqrt{5}/2$ **27.** $1 + \cot^2\theta = \csc^2\theta, \tan^2\theta + 1 = \sec^2\theta$
29. 5.89744 ft **31.** 56 ft **33.** 6.4 ft

Section 7.3

		$\sin\theta$	$\cos\theta$	$\tan\theta$	$\csc\theta$	$\sec\theta$	$\cot\theta$
1. 3.	$30°, \pi/6$	$1/2$	$\sqrt{3}/2$	$1/\sqrt{3}$	2	$2/\sqrt{3}$	$\sqrt{3}$
	$60°, \pi/3$	$\sqrt{3}/2$	$1/2$	$\sqrt{3}$	$2/\sqrt{3}$	2	$1/\sqrt{3}$

5. $\sqrt{3}$ **7.** 1.53987 **9.** .826590 **11.** 1.11740 **13.** $1/\sqrt{2}$
15. $\sqrt{3}/2$ **17.** .841471 **19.** .647859 **21.** .726543 **23.** 1.12207
25. 2.36522 **27.** 1.23607 **29.** .977076 **31.** 2.04950 **33.** $4/\sqrt{2}$
35. .966789 **37.** .250220 **39.** -1.76948 **41.** 36 ft **43.** 235,075 sq yd
45. $1/\sqrt{2}$ **47.** 25.1725 m

Section 7.4

1. $30°, \pi/6$ **3.** $60°, \pi/3$ **5.** $60°, \pi/3$ **7.** $45°, \pi/4$ **9.** $45°, \pi/4$
11. $51.7575°, .903339$ **13.** $84.2608°, 1.47063$ **15.** $53.1301°, .927296$ **17.** $36.8699°, .643501$
19. $23.5782°, .411517$ **21.** $11.5370°, .201358$ **23.** $11.3099°, .197396$ **25.** $38.6598°, .674741$
27. $71.5651°, 1.24905$ **29.** $89.4271°, 1.56080$ **31.** $.6, 36.8699°$ **33.** $4.71094, 78.0156°$
35. $.181731, 79.5294°$ **37.** $26.3878°$ **39.** $37.6699°$ **41.** $.380506$ **43.** $72.1101°$

Section 7.5

1. 3.32257 ft, 12.8374 ft **3.** 12.2257 m, 26.5174 m **5.** 1.56847 in., 3.24036 in.
7. 14.2808 ft, 12.9826 ft **9.** 6.13562 cm, 49.0850 cm **11.** 80.4562 in., 257.872 in.
13. 137.589 ft, 166.071 ft **15.** $24.0965°, 65.9035°$ **17.** $36.2931°, 53.7069°$ **19.** $23.8830°, 66.1170°$
21. 11.5175 ft **23.** 13.33 mi, 400.0 mph **25.** 598.794 m **27.** 47 ft **29.** $53.35°$
31. 10.7907, 12.4600 **33.** 1136 ft **35.** 140 ft **37.** 3.6 mi **39.** 9221 yd **41.** 63.4738 ft
43. 188.369 mph

Review

1. $62.305°$ **3.** $4817.87°$ **5.** $-187.595°$ **7.** $\pi/3$ **9.** 1.46608 **11.** -8.22609
13. $120°$ **15.** $131.780°$ **17.** $-2.69290°$ **19.** $1/4, \sqrt{15}/4, 1/\sqrt{15}, 4, 4/\sqrt{15}, \sqrt{15}$
21. $\sqrt{15}$ **23.** .0174524 **25.** 1.13897 **27.** 1.12464 **29.** .249845 **31.** .603374
33. -14.5107 **35.** $400\sqrt{3}$ ft **37.** $30°, \pi/6$ **39.** $28.6854°, .500654$ **41.** $45°, \pi/4$
43. $60°, \pi/3$ **45.** 3.98178, 4.88215 **47.** 1.20366 **49.** 31.1868 ft **51.** 352.564 ft

CHAPTER 8

Section 8.1

1. $3/5, -4/5, -3/4, 5/3, -5/4, -4/3$ **3.** $3/\sqrt{13}, 2/\sqrt{13}, 3/2, \sqrt{13}/3, \sqrt{13}/2, 2/3$
5. $-3/\sqrt{10}, -1/\sqrt{10}, 3, -\sqrt{10}/3, -\sqrt{10}, 1/3$
7. $-2/\sqrt{13}, 3/\sqrt{13}, -2/3, -\sqrt{13}/2, \sqrt{13}/3, -3/2$
9. $\pm 3\sqrt{5}/7, -2/7, \mp 3\sqrt{5}/2, \pm 7/3\sqrt{5}, -7/2, \mp 2/3\sqrt{5}$
11. $\pm\sqrt{3}/2, .5, \pm\sqrt{3}, \pm 2/\sqrt{3}, 2, \pm 1/\sqrt{3}$

13. $\pm 3/\sqrt{10}, \pm 1/\sqrt{10}, 3, \pm\sqrt{10}/3, \pm\sqrt{10}, 1/3$
15. $\pm 4/\sqrt{17}, \mp 1/\sqrt{17}, -4, \pm\sqrt{17}/4, \mp\sqrt{17}, -1/4$ 17. $Q_3, -, -, +$ 19. $Q_4, -, +, -$
21. $Q_2, +, -, -$ 23. $Q_4, -, +, -$ 25. $\pm 24/25, -7/25, \mp 24/7, \pm 25/24, -25/7, \mp 7/24$
27. $\pm 5/13, \pm 12/13, 5/12, \pm 13/5, \pm 13/12, 12/5$ 29. $\pm 12/13, 5/13, \pm 12/5, \pm 13/12, 13/5, \pm 5/12$
31. $-\sqrt{5}/3, -2/3, \sqrt{5}/2, -3/\sqrt{5}, -3/2, 2/\sqrt{5}$ 33. $-2/\sqrt{5}, -1/\sqrt{5}, 2, -\sqrt{5}/2, -\sqrt{5}, 1/2$
35. $-3/5, 4/5, -3/4, -5/3, 5/4, -4/3$ 37. $-1/4, \sqrt{15}/4, -1/\sqrt{15}, -4, 4/\sqrt{15}, -\sqrt{15}$
39. $-2/5, -\sqrt{21}/5, 2/\sqrt{21}, -5/2, -5/\sqrt{21}, \sqrt{21}/2$ 41. $\sqrt{3}/2, 1/2, \sqrt{3}, 2/\sqrt{3}, 2, 1/\sqrt{3}$
43. $3/5, -4/5, -3/4, 5/3, -5/4, -4/3$ 45. $-72/7, 10.7143$ 47. $27/4, 27.8310$

Section 8.2

1. $.554118, -.832438, -.665657, 1.80467, -1.20129, -1.50228$
3. $-.777146, -.629320, 1.23490, -1.28676, -1.58902, .809784$
5. $-.515501, -.856889, .601597, -1.93986, -1.16701, 1.66224$ 7. $(-15.8873, 3.64452)$
9. $(.934994, -5.0631)$ 11. $(-.184, -.556722)$ 13. $(2.33825, -1.6802)$ 15. $0, 1, 0, —, 1, —$
17. $-1, 0, —, -1, —, 0$ 19. -1 21. Undefined   23. 0 25. 7.35498 27. 12.7445
29. $.432189$ 31. $-.285093$ 33. $.0125298$ 35. $.669926$ 37. -1.09842 39. 567.886

Section 8.3

1. $5/13, \pm 12/13, \pm 5/12, 13/5, \pm 13/12, \pm 12/5$ 3. $\pm 4/5, \mp 3/5, -4/3, \pm 5/4, \mp 5/3, -3/4$
5. $-24/25, \pm 7/25, \mp 24/7, -25/24, \pm 25/7, \mp 7/24$ 7. $1/3, -2\sqrt{2}/3, -1/2\sqrt{2}, 3, -3/2\sqrt{2}, -2\sqrt{2}$
9. $\sqrt{3}/2, -1/2, -\sqrt{3}, 2/\sqrt{3}, -2, -1/\sqrt{3}$ 11. $-1/\sqrt{2}, 1/\sqrt{2}, -1, -\sqrt{2}, \sqrt{2}, -1$
13. $2/3, -\sqrt{5}/3, -2/\sqrt{5}, 3/2, -3/\sqrt{5}, -\sqrt{5}/2$ 15. $-\sqrt{3}/2, 1/2, -\sqrt{3}, -2/\sqrt{3}, 2, -1/\sqrt{3}$
17. $-2/\sqrt{5}, -1/\sqrt{5}, 2, -\sqrt{5}/2, -\sqrt{5}, 1/2$ 19. $-2\sqrt{2}/3, 1/3, -2\sqrt{2}, -3/2\sqrt{2}, 3, -1/2\sqrt{2}$
21. $.998342$ 23. $.997053$ 25. $-.994210$ 27. $-.335839$ 29. 0 31. $-.977566$
33. 13.1836 35. -1

Section 8.4

1. $Q_3, 57°$ 3. $Q_2, 77°$ 5. $Q_1, 4°$ 7. $Q_1, 28°45'$ 9. $Q_2, 2\pi/9$ 11. $Q_3, \pi/5$
13. $Q_2, 1.14159$ 15. $Q_2, .707963$

17.

$\pi/6$	$\pi/4$	$\pi/3$	$2\pi/3$	$3\pi/4$	$5\pi/6$	$7\pi/6$
$1/2$	$1/\sqrt{2}$	$\sqrt{3}/2$	$\sqrt{3}/2$	$1/\sqrt{2}$	$1/2$	$-1/2$
$\sqrt{3}/2$	$1/\sqrt{2}$	$1/2$	$-1/2$	$-1/\sqrt{2}$	$-\sqrt{3}/2$	$-\sqrt{3}/2$
$1/\sqrt{3}$	1	$\sqrt{3}$	$-\sqrt{3}$	-1	$-1/\sqrt{3}$	$1/\sqrt{3}$

$5\pi/4$	$4\pi/3$	$5\pi/3$	$7\pi/4$	$11\pi/6$
$-1/\sqrt{2}$	$-\sqrt{3}/2$	$-\sqrt{3}/2$	$-1/\sqrt{2}$	$-1/2$
$-1/\sqrt{2}$	$-1/2$	$1/2$	$1/\sqrt{2}$	$\sqrt{3}/2$
1	$\sqrt{3}$	$-\sqrt{3}$	-1	$-1/\sqrt{3}$

$-\pi/6$	$-\pi/4$	$-\pi/3$	$-2\pi/3$	$-3\pi/4$	$-5\pi/6$	$-7\pi/6$
$-1/2$	$-1/\sqrt{2}$	$-\sqrt{3}/2$	$-\sqrt{3}/2$	$-1/\sqrt{2}$	$-1/2$	$1/2$
$\sqrt{3}/2$	$1/\sqrt{2}$	$1/2$	$-1/2$	$-1/\sqrt{2}$	$-\sqrt{3}/2$	$-\sqrt{3}/2$
$-1/\sqrt{3}$	-1	$-\sqrt{3}$	$\sqrt{3}$	1	$1/\sqrt{3}$	$-1/\sqrt{3}$

$-5\pi/4$	$-4\pi/3$	$-5\pi/3$	$-7\pi/4$	$-11\pi/6$
$1/\sqrt{2}$	$\sqrt{3}/2$	$\sqrt{3}/2$	$1/\sqrt{2}$	$1/2$
$-1/\sqrt{2}$	$-1/2$	$1/2$	$1/\sqrt{2}$	$\sqrt{3}/2$
-1	$-\sqrt{3}$	$\sqrt{3}$	1	$1/\sqrt{3}$

19. −.958924 **21.** −.268266 **23.** .491800 **25.** 1.32135 **27.** −.998630 **29.** −.5
31. −.902886 **33.** −1.11260

Section 8.5

1, 3, 5. Graphs in text. **7.** 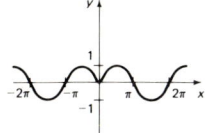 **9.** 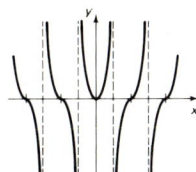 **11.**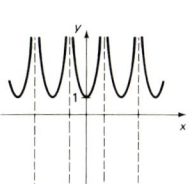

13. 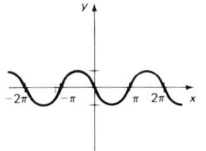 **15.** 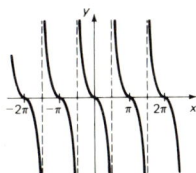 **17.**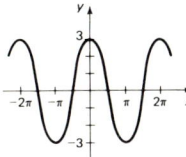

19. 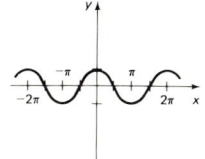 **21.** 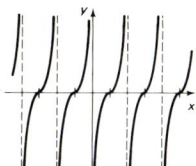 **23.**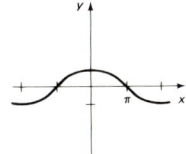

25. 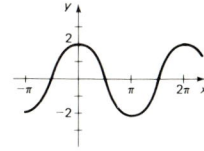 **27.** 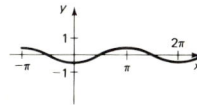 **29.**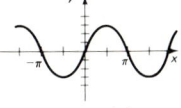

31. 2 **33.** 2 **35.** 3

Section 8.6

1. $2, 2\pi$ 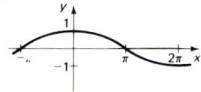 **3.** $\frac{1}{2}, 2\pi$ 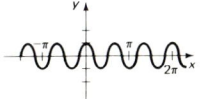 **5.** $3, 2\pi$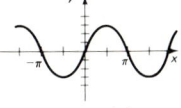

7. $1, 4\pi$ **9.** $1, 2\pi/3$

11. $1, 2\pi, \pi/3$ 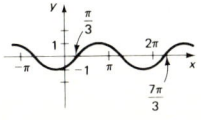 **13.** $1, 2\pi, -1$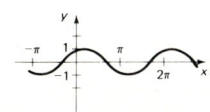

15. $1, 2\pi, \pi$ 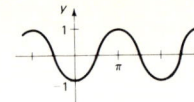   **17.** $1, 2\pi/3, \pi/3$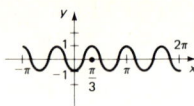

19. $1, 2\pi/3, 2$ 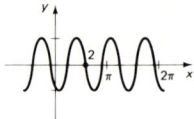   **21.** $3, 2, -1/2$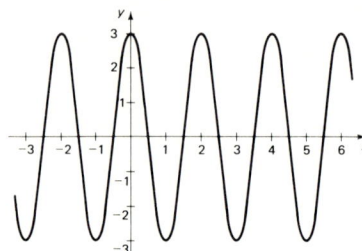

23. $4, 2\pi/3, 1$ 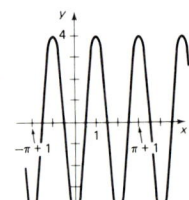   **25.** $1/3, 2\pi, 2\pi$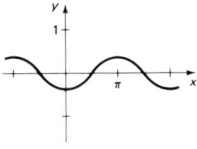

27. 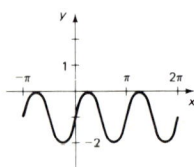   **29.** 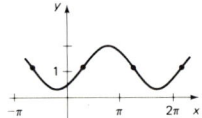   **31.**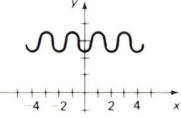

33. 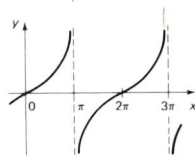   **35.** 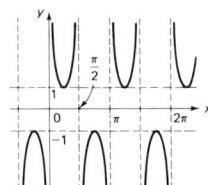   **37.**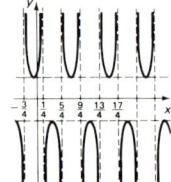

39.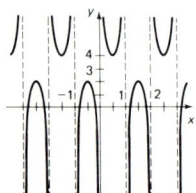

Section 8.7

1. .961411, .961411, 55.0848°   **3.** 1.69784, 1.69784, 97.2789°
5. $-1.36948, -1.36948, -78.4654°$ **7.** .0820593, .0820593, 4.70165°
9. .989813   **11.** .700551   **13.** .745356   **15.** 60.8582   **17.** -112.098 **19.** .00824613
21. $\pi/6$ **23.** $2\pi/3$ **25.** $-\pi/4$ **27.** $\pi/4$ **29.** $\pi/4$ **31.** $\pi/4$ **33.** $5\pi/6$
35. $5\pi/6$ **37.** C   **39.** C   **41.** A   **43.** A   **45.** C   **47.** C   **49.** .2   **51.** $-\pi/3$
53. 38°   **55.** $5\pi/8$ **57.** $-78°$ **59.** 57°   **61.** 34°   **63.** -479 **65.** $-1, \pi$

67. $1/\sqrt{1-u^2}$ **69.** $\sqrt{u^2-1}, u \geq 1; -\sqrt{u^2-1}, u \leq -1$

71. 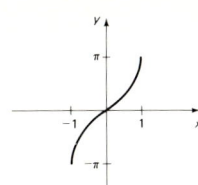   **73.** 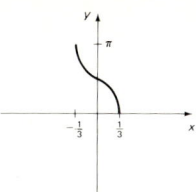   **75.**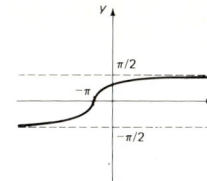

77. 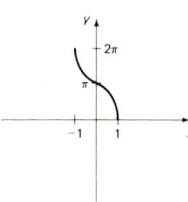   **79.** 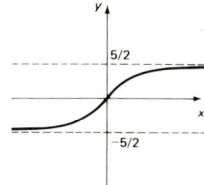   **81.**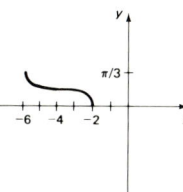

83. 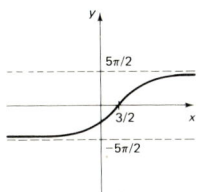   **85.** 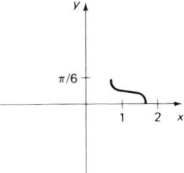   **87.** 1.61683, .834441

Section 8.8

1. 2.12 ft   **3.** 880 hz   **5.** 18.5 ft   **7.** 1.25 ft, .625 ft   **9.** 4.29×10^{14} hz
11. 107 m

Review

1. $4/\sqrt{17}, -1/\sqrt{17}, -4, \sqrt{17}/4, -\sqrt{17}, -1/4$
3. $\pm\sqrt{21}/5, 2/5, \pm\sqrt{21}/2, \pm 5/\sqrt{21}, 5/2, \pm 2/\sqrt{21}$
5. $\pm 3/\sqrt{10}, \mp 1/\sqrt{10}, -3, \pm\sqrt{10}/3, \mp\sqrt{10}, -1/3$
7. $\pm 12/13, \mp 5/13, -12/5, \pm 13/12, \mp 13/5, -5/12$ **9.** 4/5, 3/5, 4/3, 5/4, 5/3, 3/4
11. $5/\sqrt{26}, -1/\sqrt{26}, -5, \sqrt{26}/5, -\sqrt{26}, -1/5$
13. $-\sqrt{15}/4, -1/4, \sqrt{15}, -4/\sqrt{15}, -4, 1/\sqrt{15}$
15. (−1.36, 1.38576)   **17.** 0   **19.** .0936614   **21.** 2.5   **23.** 0   **25.** 5.22829
27. 1.09637, −.889555   **29.** 5°, −.996195
31. $3, 2\pi, \pi/3$ **33.** $2, 4\pi, \pi/2$ **35.**

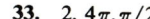
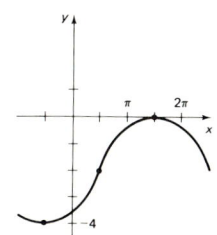
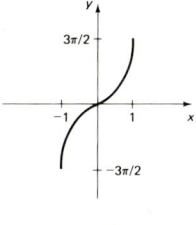

37. $-\pi/6$ **39.** $-62°$

CHAPTER 9

Section 9.1
41. Not an identity, e.g. $\pi/6$ 43. Identity   45. Not an identity, e.g., $\pi/6$
47. Not an identity, e.g., $\pi/4$

Section 9.2
1. $\pi/4 + 2n\pi, 3\pi/4 + 2n\pi$ 3. $-\pi/6 + n\pi$ 5. $\pm\pi/3 + 2n\pi$ 7. $\pm\pi/3 + 2n\pi$
9. $\emptyset$ 11. $n\pi$ 13. $\pm\pi/3 + 2n\pi, \pi + 2n\pi$
15. $\pi/6 + 2n\pi, 5\pi/6 + 2n\pi, -.339837 + 2n\pi, -2.80176 + 2n\pi$ 17. $2n\pi$ 19. $\emptyset$
21. $-\pi/4 + n\pi$ 23. $\pi + 2n\pi, \pi/2 + 2n\pi$ 25. $\pm\pi/3 + n\pi, n\pi$
27. $-.695004 + 2n\pi, -2.44659 + 2n\pi, .401053 + 2n\pi, 2.74054 + 2n\pi$

Section 9.3
1. $(\sqrt{3} - 1)/2\sqrt{2}$ 3. $(\sqrt{3} - 1)/2\sqrt{2}$ 5. $-2 - \sqrt{3}$ 7. $\sqrt{2}(\sqrt{3} + 1)$
9. $1/\sqrt{2}$ 11. $.5$ 13. 1 15. $33/65, -63/65, 56/65, -16/65, 33/56, 63/16$
17. $(\sqrt{6} + 1)/2\sqrt{3}, (\sqrt{6} - 1)/2\sqrt{3}, (\sqrt{3} - \sqrt{2})/2\sqrt{3}, -(\sqrt{3} + \sqrt{2})/2\sqrt{3}, (1 + \sqrt{6})/(\sqrt{3} - \sqrt{2}),$
$(1 - \sqrt{6})/(\sqrt{3} + \sqrt{2})$ 45. $\sin 4\theta + \sin 2\theta$ 47. $-3(\sin 7\theta + \sin 3\theta)/2$
49. $(\cos 6t - \cos 18t)/2$ 51. $2 \sin 4\theta \cos \theta$ 53. $10 \cos \frac{7}{2}\theta \cos \frac{1}{2}\theta$ 55. $2 \sin \frac{3}{2}x \cos \frac{1}{2}x$

Section 9.4
1. $\sqrt{3}/2, -1/2, -\sqrt{3}$ 3. $-120/169, -119/169, 120/119$ 5. $-24/25, 7/25, -24/7$
7. $\sqrt{2 - \sqrt{3}}/2$ 9. $-\sqrt{2 - \sqrt{2}}/2$ 11. $3/4$ 13. $2/\sqrt{7}$
27. $7\pi/12, 7\pi/12 + \pi, 11\pi/12, 11\pi/12 + \pi$ 29. $\pi/2, 3\pi/2, 7\pi/6, 11\pi/6$ 31. $3\pi/2, \pi/6, 5\pi/6$
33. $\pi, \pi/3, 5\pi/3$ 35. $0, \pi, 2\pi$

Section 9.5
1. $\gamma = 82.61°, a = 241.987, b = 359.974$ 3. $\alpha = .991593, b = 115.167, c = 58.2914$
5. $\beta = 1.97159, a = .0320852, b = .0400342$ 7. $\alpha = 44.61°, a = 4591.96, c = 5532.98$
9. $\gamma = 75.37°, a = 6.89226, c = 23.8889$ 11. No triangle
13. $\beta = 33.3195°, \alpha = 90.2305°, a = 38.6484$
15. $\gamma = 1.11622, \beta = 1.52968, b = 18.9056$; or $\gamma = 2.02538, \beta = .620516, b = 11.0020$
17. No triangle   19. $\beta = .352465, \alpha = 2.47193, a = 121.594$; or $\beta = 2.78913, \alpha = .0352643, a = 6.90654$
21. 168.511 ft   23. 127.899 ft   25. 108.454 ft, 34.9642 ft, (No)   27. 153.007 mi, 279.024 mi
29. 762.510 nautical mi   31. 263.507 km

Section 9.6
1. $c = 9.28588, \alpha = 93.7047°, \beta = 26.2953°$ 3. $a = .0747981, \gamma = 22.6739°, \beta = 35.3261°$
5. $b = 466.491, \alpha = 2.17246, \gamma = .584133$ 7. $a = 165.205, \beta = 12.2629°, \gamma = 153.327°$
9. $b = 8.89065, \alpha = 27.5157°, \gamma = 112.484°$ 11. No triangle
13. $\alpha = 50.2808°, \beta = 98.7143°, \gamma = 31.0049°$
15. $\alpha = 122.271°, \beta = 17.3422°, \gamma = 40.3866°$ 17. 10.1214 hrs   19. 36.3966 ft, 50.1332°
21. .687161 km   23. 104.031°   25. 33.1472 km

Section 9.7
1. 72.7328, 80.1039°   3. 5.72242, 260.293°   5. 34.9042, 3.54805°   7. 0
9. **a.** 7 mph, S   **b.** 1 mph, S   **c.** 5 mph, 143.130°   **d.** 2.83363 mph, 131.529°   **e.** 6.47847 mph, 160.886°

11. a. 7.21110 mph, 56.3099° to AB **b.** 1.5 mi **c.** 15 min **d.** 15 min **13. a.** 41.8103° upstream **b.** 8.94427 min **c.** No **15.** 322.360°, 405.449 mph **17.** 128.176°, 398.765 kmph **19.** 38.4204 km, 4.54758° **21.** 17.2516 lbs **23.** 24.8346° **25.** 13.2501 lbs

Section 9.8

1.–9.

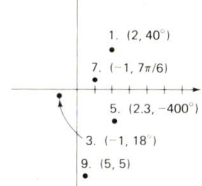

11. $(4\sqrt{2}, 135°)$ **13.** $(6.92315, -79.1770°)$ **15.** $(6, 180°)$ **17.** $(\sqrt{10}, -71.5651°)$
19. $(-1, \sqrt{3})$ **21.** $(-5.86889, 1.24747)$ **23.** $(0, 0)$ **25.** $(.832294, -1.81859)$
27. $r\cos\theta = 7$ **29.** $\theta = \tan^{-1} 2$ **23.** $r^2(\cos^2\theta + 4\sin^2\theta) = 4$ **25.** $x^2 + y^2 = 9$
35. $y = -x$ **37.** $y = 7$ **39.** $(x^2 + y^2)x = 3y - x$

41. 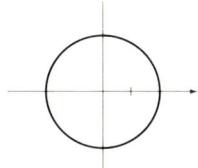 **43.**  **45.**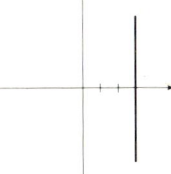

47. 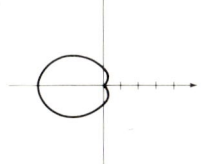 **49.** 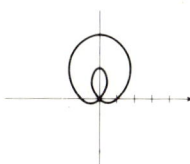 **51.**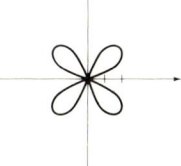

Review

11. $-\pi/6 + 2n\pi, -5\pi/6 + 2n\pi$ **13.** $n\pi, \pm 1.23096 + 2n\pi$ **15.** $2n\pi, \pm 2\pi/3 + 2n\pi$
17. $16/65, -63/65, -16/63$ **19.** $\gamma = 70.5°, a = 47.1915, b = 114.276$
21. $\alpha = 1.13648, \beta = 1.37680, b = 30.2326$ or $\alpha = 2.00511, \beta = .508161, b = 14.9914$
23. $a = .797409, \beta = 1.83834, \gamma = .323256$ **25.** 134.444 mi or 19.6074 mi **27.** 9.02936°
29. 123.796, 31.5351° **31.** 138.879 mph, 233.797°

CHAPTER 10

Section 10.1

1. $3 - i$ **3.** 5 **5.** $7 + 7i$ **7.** $15 - 7i$ **9.** 0 **11.** $6 - i$ **13.** $10 - 2i$
15. $3 + 3i$ **17.** $-3 - 2i$ **19.** $-8 - 5i$ **21.** $-4 - 11i$ **23.** $11 - 10i$ **25.** $2 - 9i$
27. $1 + 3i$ **29.** $9 + 15i$ **31.** $-5 - 12i$ **33.** i **35.** 1 **37.** $3i$ **39.** $\frac{8}{13} - \frac{1}{13}i$
41. $\frac{1}{17} + \frac{13}{17}i$ **43.** $\frac{1}{5} - \frac{2}{5}i$ **45.** $3 - 4i$ **47.** $-i$ **49.** $x = -2, y = 7/3$ **51.** $x = -4, y = 7$
53. $x = 3, y = 3^{1/9}$ **55.** $x = 3/2, y = 1/3$ **57.** $x = \pi/2 + 2n\pi, y = \pm\pi/3 + 2n\pi$

Section 10.2

1. $2i$ **3.** $-2\sqrt{5}\,i$ **5.** $11i$ **7.** -90 **9.** 30 **11.** $13/2$ **13.** $12 - 5i$
15. $-20 - 20\sqrt{3}\,i$ **17.** $6 + 10i$ **19.** $-16\sqrt{2}\,i$ **25.** $-\frac{1}{3}i$ **27.** $1 + 2i$ **29.** $-\frac{5}{4}i$

31. $-3 + i$ 33. $-\frac{6}{25} + \frac{17}{25}i$ 35. $-\frac{11}{5} + \frac{2}{5}i$ 37. $1 \pm i$ 39. $\frac{3}{4} \pm \frac{1}{4}\sqrt{7}\,i$
41. $-\frac{2}{3} \pm \frac{1}{3}\sqrt{2}\,i$ 43. $1, -\frac{1}{2} \pm \frac{1}{2}\sqrt{3}\,i$ 45. $5, -\frac{5}{2} \pm \frac{5}{2}\sqrt{3}\,i$ 47. $\pm 2, \pm 2i$ 49. $\pm 2, \pm 1 \pm \sqrt{3}\,i$
51. $\pm i, \pm \frac{1}{2}\sqrt{2}\,i$ 53. $-\frac{1}{2} \pm \frac{1}{2}\sqrt{3}\,i, 0, 0$ 55. $i, -2i$ 57. $-\frac{1}{2} + \frac{1}{2}(2 + \sqrt{3})i, -\frac{1}{2} + \frac{1}{2}(2 - \sqrt{3})i$
59. $-\frac{1}{2} + \frac{1}{2}i, -\frac{1}{2} + \frac{1}{2}i$ 61. $.833333 \pm 1.28019i$ 63. $.381532 \pm .339104i$ 65. $x^2 + 4 = 0$
67. $x^2 + 4x + 13 = 0$ 69. $x^2 + 3 = 0$ 71. $x^3 + (-1 + 2i)x^2 + (3 - 2i)x - 3$
73. $x^3 - x^2 + 2$ 75. $x^3 + (-4 - i)x^2 + (5 + 4i)x - 5i$

Section 10.3

1. $|z_n| = \sqrt{2}$, all n
$1 + i = \sqrt{2}(\cos \pi/4 + i\sin \pi/4) = \sqrt{2}\operatorname{cis} \pi/4$
$-1 + i = \sqrt{2}(\cos 3\pi/4 + i\sin 3\pi/4) = \sqrt{2}\operatorname{cis} 3\pi/4$
$-1 - i = \sqrt{2}(\cos 5\pi/4 + i\sin 5\pi/4) = \sqrt{2}\operatorname{cis} 5\pi/4$
$1 - i = \sqrt{2}(\cos 7\pi/4 + i\sin 7\pi/4) = \sqrt{2}\operatorname{cis} 7\pi/4$

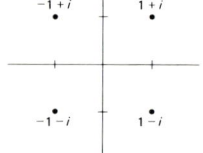

3. $|z_n| = 2$, all n
$\sqrt{3} + i = 2(\cos \pi/6 + i\sin \pi/6) = 2\operatorname{cis} \pi/6$
$-\sqrt{3} + i = 2(\cos 5\pi/6 + i\sin 5\pi/6) = 2\operatorname{cis} 5\pi/6$
$-\sqrt{3} - i = 2(\cos 7\pi/6 + i\sin 7\pi/6) = 2\operatorname{cis} 7\pi/6$
$\sqrt{3} - i = 2(\cos 11\pi/6 + i\sin 11\pi/6) = 2\operatorname{cis} 11\pi/6$

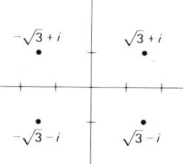

5. $|z_1| = 2\sqrt{10}, |z_2| = \sqrt{10}, |z_3| = 3\sqrt{10}, |z_4| = 2\sqrt{10}$
$6 + 2i = 2\sqrt{10}(\cos 18.4349° + i\sin 18.4349°) = 2\sqrt{10}\operatorname{cis} 18.4349°$
$-3 + i = \sqrt{10}(\cos 161.565° + i\sin 161.565°) = \sqrt{10}\operatorname{cis} 161.565°$
$-9 + 3i = 3\sqrt{10}(\cos 161.565° + i\sin 161.565°) = 3\sqrt{10}\operatorname{cis} 161.565°$
$6 - 2i = 2\sqrt{10}[\cos(-18.4349°) + i\sin(-18.4349°)] = 2\sqrt{10}\operatorname{cis}(-18.4349°)$

7. $-3\sqrt{3} - 3i$ 9. $-.393159 + 2.44864i$ 11. $19.2\operatorname{cis} 419°$ 13. $740.479\operatorname{cis} 10.12$
15. $.5\operatorname{cis}(-70°)$ 17. $(5/3)\operatorname{cis}(\pi/5)$ 19. $2.14\operatorname{cis} 177°$ 21. $1.31785\operatorname{cis}(-2\pi/15)$
23. **a.** $16\sqrt{3} + 16i$ **b.** $-1/4$

Section 10.4

1. $3^6\operatorname{cis} 612°$ 3. $2^{-8}\operatorname{cis}(-1704°)$ 5. $3.7^5\operatorname{cis}(-5\pi/3)$ 7. $\frac{1}{2}\sqrt{2} - \frac{1}{2}\sqrt{2}\,i$
9. $154.686 - 606.795i$ 11. $-3245.02 - 2574.04i$

13. $2\operatorname{cis}(-\pi/9) \approx 1.87939 - .684040i$
$2\operatorname{cis} 5\pi/9 \approx -.347296 + 1.96962i$
$2\operatorname{cis} 11\pi/9 \approx -1.53209 - 1.28558i$

15. $2\operatorname{cis} \pi/12 \approx 1.93185 + .517638i$
$2\operatorname{cis} 5\pi/12 \approx .517638 + 1.93185i$
$2\operatorname{cis} 9\pi/12 \approx -1.41421 + 1.41421i$
$2\operatorname{cis} 13\pi/12 \approx -1.93185 - .517638i$
$2\operatorname{cis} 17\pi/12 \approx -.517638 - 1.93185i$
$2\operatorname{cis} 21\pi/12 \approx 1.41421 - 1.41421i$

17. $1.29239\operatorname{cis} 29.2620° \approx 1.12747 + .631726i$
$1.29239\operatorname{cis} 101.262° \approx -.252399 + 1.26751i$
$1.29239\operatorname{cis} 173.262° \approx -1.28347 + .151636i$
$1.29239\operatorname{cis} 245.262° \approx -.540827 - 1.17379i$
$1.29239\operatorname{cis} 317.262° \approx .949216 - .877078i$

19. a. 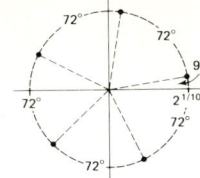 **b.**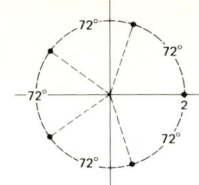

21. $-4.67945 + 3.67945i,$
$-.320551 - 6.79449i$

23. $.618034 - .618034i,$
$-1.61803 + 1.61803i$

25. $-1.64952 - 1.30384i,$
$1.14952 + .303845i$

Review

1. $7 + 5i$ **3.** $-4 + 16i$ **5.** -1 **7.** $-\frac{1}{2} - \frac{1}{2}i$ **9.** $-81\sqrt{3}\,i$
11. $2\sqrt{2} + 2\sqrt{2}\,i$ **13.** $1.70197 - 5.75355i$ **15.** $29{,}524.5 + 29{,}524.5\sqrt{3}\,i$ **17.** $x = .5, y = 3.5$
19. $-\frac{17}{26} - \frac{7}{26}i$ **21.** $2, -1 \pm \sqrt{3}\,i$ **23.** $[-2 + (3 + \sqrt{5})i]/6, [-2 + (3 - \sqrt{5})i]/6$
25. $\sqrt[10]{17}\ \text{cis}\ (-14.0362°)$
$\sqrt[10]{17}\ \text{cis}\ (57.9638°)$
$\sqrt[10]{17}\ \text{cis}\ (129.964°)$
$\sqrt[10]{17}\ \text{cis}\ (201.964°)$
$\sqrt[10]{17}\ \text{cis}\ (273.964°)$

CHAPTER 11

Section 11.1

1. $-x^4 + x^2 + 5x - 12$ **3.** $-3x^{448} + x^{329} - x^{325} - x^{116} + x^{112} + 3x$
5. $2x^3 - 6ix^2 - 2x - 1 + 3i$ **7. a.** 11   **b.** 7   **c.** 21   **d.** 3   **e.** $0 \le \deg r(x) \le 3$ or $r(x) = 0$
9. a. 16   **b.** $0 \le \deg\{f(x) + g(x)\} \le 8$ or $f(x) + g(x) = 0$ **c.** 24   **d.** 0
e. $0 \le \deg r(x) \le 7$ or $r(x) = 0$ **11.** $2x^2 + x - 7, -28x + 38$ **13.** $3x^2 - \frac{9}{2}x + \frac{27}{4}, -\frac{73}{4}x^2 - \frac{9}{2}x + \frac{31}{4}$
15. $0, x^2 - x + 1$ **17.** -22 **19.** -244 **21.** 5
23. $q(x)$ is different; $f(x) = (3x - 1)q(x) + r(x) = (x - \frac{1}{3})[3q(x)] + r(x)$ **25.** 6   **27.** $f(2) = 0$
29. $f(-2) = 0$ **31.** $f(a) = 0$ **33.** -4 **35.** $(1 \pm \sqrt{3})/2$

Section 11.2

1. $3x^3 + 4x^2 + 9x + 19; 34$ **3.** $3x + 4; 12.9$ **5.** $-6x^3 + 49x^2 - 393x + 3144; -25151$
7. $-11t^3 + 44t^2 - 169t + 678; -2712$ **9.** $21t^3 + 4t^2 + 14t + 3; 17$ **11.** $x^2 + 4x - 5.1; -15.9$
13. $x^2 + 2cx + 4c^2; 38c^3$ **15.** $x^3 - 3bx^2 + 8b^2x - 27b^3; 0$ **17.** 5, 755, 345
19. $-137.28, 640.23, -51.112$ **21.** $2.83530, 2.11986, .940029$
23. $((((-6)x + 18)x + 1)x + 0)x - 47$
25. $(((((((1)t + 0)t + 0)t + 6)t - 11)t + 0)t + 0)t - 14)t + 1$ **27.** $11{,}851{,}139;\ 1{,}351{,}973;\ 54502.7$
29. $12.2676;\ 21.7444;\ 12.5539$

Section 11.3

1. $(x-1)(x+2)(x-5)$ 3. $(x-i)(x-3)[x-(-2+5i)]$ 5. $(x+1)^3(x+3)^2$
7. $(x-i)^3(x-6)(x+6)$ 9. a. $x^3-(4+i)x^2+(5+5i)x-6-6i$
  b. $x^5-5x^4+12x^3-26x^2+32x-24$ 11. a. $x^3-9ix^2-27x+27i$ b. $x^6+27x^4+243x^2+729$
13. $1, 3, -4$ with multiplicities 2, 1, 1 respectively
15. $-1+\sqrt{3}i, -1-\sqrt{3}i, \sqrt{2}$ with multiplicities 4, 4, 3 respectively
17. $1-i, -2, 1 \pm \sqrt{3}i$ with multiplicities 3, 1, 1, 1 respectively   19. $(-5 \pm \sqrt{3}i)/2$
21. $-2+2i, (-1 \pm \sqrt{3}i)/2$ 23. $x^4+3x^3-48x^2-92x+336; x^n-a_{n-1}x^{n-1}+a_{n-2}x^{n-2}-a_{n-3}x^{n-3}+\cdots+(-1)^n a_0$ (i.e., change the sign of every other term)

Section 11.4

1. $\pm 1, \pm 3, \pm 1/2, \pm 3/2$ 3. ± 1
5. $-1, -2, -4, -1/2, -1/3, -2/3, -4/3, -1/4, -1/6, -1/12$
7. $1, 2, 3, 6, 1/2, 3/2, 1/3, 2/3, 1/6, 1/9, 2/9, 1/18$
9. 0 is a root, $\pm 1, \pm 2, \pm 4, \pm 5, \pm 10, \pm 20, \pm 1/2, \pm 5/2, \pm 1/4, \pm 5/4, \pm 1/8, \pm 5/8$
11. f(real number) is always >0.   13. $\pm 1, -5$ 15. $-\frac{1}{2}$
17. $\frac{2}{3}, \pm i\sqrt{3}$ 19. No rational zeros   21. $-2, -2, \frac{3}{2}, \pm i$ 23. $-1, \frac{1}{7}, -\frac{11}{2}, \pm i$
25. $\frac{1}{3}, \frac{1}{5}, \frac{1}{7}, \frac{5}{2}$ 27. $-1, -1, -2, -2, -\frac{1}{3}, -\frac{1}{3}$ 29. No rational zeros   31. $\frac{1}{2}$
33. $w = 3/2$ ft, $h = \frac{1}{2}$ ft, $l = 4$ ft   35. $3/2$ cm, 1.20871 cm

Review

1. 118   3. $f(2) = 0$ 5. $5x^5-10x^4+13x^3-42x^2+72x-144; 283$
7. $2x^3+3x^2+(-1+4i)x+(1+2i); -1+2i$ 9. $(((((-2)x+6)x+1)x-5)x-11)x+8$
11. $x^4-6x^3-x^2+54x-72$ 13. a. $x^4-(5+2i)x^3+(10+8i)x^2-(10+12i)x+4+8i$
  b. $x^5-6x^4+19x^3-36x^2+38x-20$ 15. $1/3, 2/3, -3/2$

CHAPTER 12

Section 12.1

1. $1, 2, 3, 4, 5, 11$ 3. $7, 9, 11, 13, 15, 27$ 5. $2, \sqrt{2}+\frac{1}{2}, \sqrt{3}+\frac{1}{3}, 2+\frac{1}{4}, \sqrt{5}+\frac{1}{5}, \sqrt{11}+\frac{1}{11}$
7. $8, 8, 8, 8, 8, 8$ 9. $1/2, \sqrt{3}/2, 1, \sqrt{3}/2, 1/2, -1/2$ 11. $3, 1, 3, 1, 3$
13. $-4, -2, -1, -1/2, -1/4$ 15. $2, 4, -64, 16777216, -1.32923 \times 10^{36}$
17. $2, -3, -6, 18, -108$ 19. a. $1, 1, 2, 3, 5, 8$ b. 75025 cells   21. $4, 6, 8, 10, 202$
23. $6.3, 6.5, 6.7, 6.9, 26.1$ 25. $11, 14, 17, 20, 308$ 27. $3-2i, 5-i, 7, 9+i, 201+97i$ 29. $139, 3290$
31. $1006; 84,180$ 33. $3315; 57,171$ 35. $31x-49y, 486x-621y$ 37. $-44, 3, 46, 31$
39. $7, 4, 195, 4848$ 41. $1.1, .54, 27.56, 716.5$ 43. 2450   45. $128; 70,336$ 47. $18; 31; 639$
49. 784 ft, 76 ft   51. $2, \$73.50$

Section 12.2

1. $6, 12, 768$ 3. $.48, .048, 4.8 \times 10^{-8}$ 5. $1/3, 1/2, 729/128$ 7. $1/2048, 10923/2048$
9. $e^{19}, (1-e^{20})/(1-e)$ 11. $1.10986, 437.4 [1+(2/3)^{17}] \approx 437.844$ 13. $-2, 7/8$
15. $.780291, 968.198$ 17. $55/333$ 19. $210460/333$ 21. a. Last sum $\approx 1.99902, 2$
  b. $2(1-2^{-11}), 2(1-2^{-21}), 2(1-2^{-31}), 2(1-2^{-51}), 2(1-2^{-101})$, last two or three, no
23. $63; 65,535; 33,554,431; +\infty$ 25. $-182; -10,761,680; 2.11822 \times 10^{11}$; nothing
27. $111,111; 1.11111 \times 10^{15}; 1.11111 \times 10^{24}, +\infty$ 29. $6; 16; 25; +\infty$ 31. 48   33. 6.64474
35. 449.225 ft, 464.667 ft   37. 20.48 in.   39. 41.9857 cm, 570.143 cm, 990 cm
41. $\$14,371.56$ 43. $\$1395.40$

Section 12.3

1. 1; 1; 2; 6; 24; 120; 720; 5040; 40,320; 362,880   3. 29,760   5. 1,648,441,200
7. $12! = 479,001,600$ 9. 7!   11. $9!/6!$ 13. 25!   15. $(n+1)!$ 17. 8!   19. 6720
21. 6320   23. a. $26^4 = 456,976$;   b. $P(26,4) = 358,800$ 25. a. $9(10^3) = 9000$,
b. $9 \cdot 9 \cdot 8 \cdot 7 = 4536$ 27. a. $4(25^3) = 62,500$,   b. $25^4 + 4(25^3) = 453,125$ 29. a. $3(9^2) = 243$,
b. $3(9^3) = 2187$,   c. $3(9^2) + 3(9^3) + 9^4 = 8991$ 31. $P(1821,6) \approx 3.61641 \times 10^{19}$
33. a. $25! \approx 1.55112 \times 10^{25}$,   b. $P(30,25) \approx 2.21044 \times 10^{30}$ 35. $2^5 = 32$ 37. a. $4^{10} = 1,048,576$;
b. $3^4 5^6 = 1,265,625$;   c. $10! = 3,628,800$ 39. a. $P(400,6) \approx 3.94456 \times 10^{15}$,
b. $P(200,3)^2 \approx 6.21007 \times 10^{13}$ 41. a. $7! = 5040$,   b. $4!5! = 2880$,   c. $8! + 8(7!) = 80,640$

Section 12.4

1. 10   3. 715   5. 1   7. n 9. $n(n+1)/2$ 11. 37   13. 8
17. $P(11; 1, 4, 4, 2) = 11!/1!4!4!2! = 34,650$ 19. a. $9! = 362,880$,   b. $P(9; 2, 3, 4) = 9!/2!3!4! = 1260$
21. $\binom{14}{6} = 3003$ 23. $\binom{200}{8} \approx 5.50990 \times 10^{13}$ 25. a. $\binom{9}{2} = 36$,   b. $4\binom{9}{2} = 144$
27. a. $\binom{27}{3} = 2925$ b. $27 \cdot 26 \cdot 25 = 17,550$ 29. $\binom{3}{1}\binom{6}{2}\binom{5}{2} = 450$ 31. $\binom{15}{11}\binom{10}{7} = 163,800$
33. $28 \cdot 27 \cdot 26 \binom{25}{4} = 248,648,400$ 35. a. $\binom{25}{5} = 53,130$ b. $25 \cdot 24 \binom{23}{3} = 1,062,600$
c. $\binom{14}{3}\binom{11}{2} = 20,020$ d. $\binom{14}{3}\binom{11}{2} + \binom{14}{4}\binom{11}{1} + \binom{14}{5} = 33,033$

Section 12.5

1. 1/2   3. 1/3   5. a. $\binom{26}{13}/\binom{52}{13} \approx 1.63785 \times 10^{-5}$
b. $\binom{13}{13}/\binom{52}{13} \approx 1.57476 \times 10^{-12}$ c. $\binom{36}{13}/\binom{52}{13} \approx 3.63896 \times 10^{-3}$
d. $\binom{4}{4}\binom{48}{9}/\binom{52}{13} \approx 2.64105 \times 10^{-3}$ e. $\binom{1}{1}\binom{51}{12}/\binom{52}{13} = .25$
f. $\binom{4}{1}\binom{4}{2}\binom{4}{3}\binom{36}{7}/\binom{52}{13} \approx 1.26198 \times 10^{-3}$ 7. $\binom{1}{1}\binom{33}{4}/\binom{34}{5} = \frac{5}{34}$, $1 - \frac{5}{34} = \frac{29}{34}$
9. a. $\binom{4}{2}/\binom{16}{2} = \frac{1}{20}$ b. $\binom{12}{2}/\binom{16}{2} = \frac{11}{20}$ c. $1 - \binom{9}{2}/\binom{16}{2} = \frac{7}{10}$
11. a. $(.7)^{20} \approx 7.97923 \times 10^{-4}$ b. $(.3)^{20} \approx 3.48678 \times 10^{-11}$

Section 12.6

1. $a^7 + 7a^6 b + 21a^5 b^2 + 35a^4 b^3 + 35a^3 b^4 + 21a^2 b^5 + 7ab^6 + b^7$
3. $256x^8 + 1024x^7 y + 1792x^6 y^2 + 1792x^5 y^3 + 1120x^4 y^4 + 448x^3 y^5 + 112x^2 y^6 + 16xy^7 + y^8$
5. $729x^6 + 1458x^5 u^2 + 1215x^4 u^4 + 540x^3 u^6 + 135x^2 u^8 + 18xu^{10} + u^{12}$
7. $x^{10} - 10x^9 y + 45x^8 y^2 - 120x^7 y^3 + 210x^6 y^4 - 252x^5 y^5 + 210x^4 y^6 - 120x^3 y^7 + 45x^2 y^8 - 10xy^9 + y^{10}$
9. $x^{10} - 5x^7 + 10x^4 - 10x + 5x^{-2} - x^{-5}$ 11. $x^4 y^{-4} + 4x^2 y^{-2} + 6 + 4x^{-2} y^2 + x^{-4} y^4$
13. $32,768x^{45} - 1,720,320x^{42} y + 42,147,840x^{39} y^2$ 15. $12x^2 yu^{-11} v^{11} + u^{-12} v^{12}$
17. $\binom{15}{9}(x^3)^6(-1/x)^9 = -5005x^9$ 19. $\binom{16}{15} 3x^2(-5y)^{15} \approx -(1.46484 \times 10^{12})x^2 y^{15}$
21. $\binom{14}{7}(bx)^7(-y^2)^7 = -3432 b^7 x^7 y^{14}$ 23. $\binom{9}{5}(2x^2)^4(-3y)^5 = -489,888 x^8 y^5$
25. $\binom{9}{3} = 84$ 27. $\binom{10}{5}(-3)^5 = -61,236$

Section 12.7

23. 6

Review

1. $1/2, 1/10, 1/140$ **3.** $63/64$ **5.** -3008 **7.** $9(1 - 3^{-18})$
9. .780369 ft, 146.879 ft, 150 ft **11.** 479,001,600 **13.** 6806 **15.** 697,224,024 **17.** 29,051,001
19. 1 **21. a.** $P(40, 4) = 2{,}193{,}360$ **b.** $23 \cdot 22 \cdot 17 \cdot 16 = 137{,}632$ **23.** $\binom{186}{8} \approx 3.05006 \times 10^{13}$
25. $P(17; 5, 3, 7, 2) = 49{,}008{,}960$ **27.** $64x^6 - 192x^3 + 240 - 160x^{-3} + 60x^{-6} - 12x^{-9} + x^{-12}$
29. $448y^{18}z^2$

INDEX

INDEX

A

Abscissa, 102
Absolute value, 88
   of a complex number, 429
Addition formulas, 381–2
Algebraic expression, 25
Algebraic logic, 5
Amplitude, 354
Angle, 294
   between quadrant, 328
   of depression, 312
   of elevation, 312
   initial side, 294
   negative, 295
   positive, 295
   quadrantal, 328
   reference, 342
   right, 295
   special, 309, 345
   terminal side, 294
   vertex, 294
Approximation, 3
Arccosine function, 315
Arcsine function, 315
Arctangent function, 317

Argument of a complex number, 429
Arithmetic progression, 463
Arithmetic sequence, 463
Associative law:
   for matrices, 271
   for numbers, 10
Asymptote, 116
Axis, 101
   imaginary, 428
   polar, 408
   real, 428

B

Backsubstitution, 247, 253
Base, 14
   of an exponential, 201
   of a logarithm, 207
Between quadrant angle, 328
Binomial coefficient, 491
Binomial Theorem, 491

C

Center (of a circle), 116
Circle, 116, 188, 191
Coefficient:
   binomial, 491
   matrix, 274
Combination, 482
Common difference (of an arithmetic
       progression), 463
Commutative property:
   for matrices, 271, 273
   for numbers, 9
Complement method, 280
Completing the square, 64, 118, 175, 190, 194,
   196
Complex number, 417
   absolute value, 429
   argument, 429
   conjugate, 419
   imaginary part, 418
   polar form, 429
   pure imaginary, 418
   real part, 418
   standard form, 417
   trigonometric form, 429
Complex plane, 428
Component (of a vector), 405
Composite function, 158
Compound continuously, 223
Compound interest, 221–2
Conditional equation, 49
   trigonometric, 374, 378–81
Conic (section), 188
Conjugate, complex, 419
Consistent, 249
Constant:
   of proportionality, 152, 153
   of variation, 152, 153
Contraction, 134–6
Converges (series), 471
Coordinate:
   axes, 101
   Cartesian, 101
   line, 79
   plane, 101
   polar, 408–9
   rectangular, 101
Cosecant function, 304, 329
Cosine function, 304, 329
Cotangent function, 304, 329
Cramer's Rule, 265, 267

D

Decreasing function, 163
Degree, 295
   of a polynomial, 180
De Moivre's Theorem, 434
Dependent, 249
Dependent variable, 121
Depreciation, 224
Determinant, 264, 266, 268
Dimensions (of a matrix), 270
Discriminant, 67
Distance:
   on a line, 89
   in the plane, 104
Distinguishable permutation, 481
Distributive property:
   for matrices, 273
   for numbers, 10
Diverges (series), 471
Dividend (polynomials), 443
Divisible, 442
Division algorithm, 443, 444
Divisor (polynomials), 443
Domain, 121
Double-angle formulas, 388–9

E

Electromagnetic spectrum, 370
Elementary trigonometric identities, 306, 337
Ellipse, 188, 193
Enumeration, fundamental principle of, 475
Equality:
   of matrices, 270
   of polynomials, 441
   of vectors, 401

Equation:
  algebraic, 49
  conditional, 49, 374
  identity, 49, 374
  quadratic, 63
Equivalent equations, 55
Equivalent systems, 241, 248–9
Estimate, 3
  graphical, 125, 204, 215
  rough, 203, 204, 210
Expansion, 134–6
Exponent, 14
Exponential decay, 225
Exponential function, 201
Exponential growth, 224
Extraneous solutions, 52

F

Factorial, 476
Factors, 442
Factor theorem, 445
Fibonacci sequence, 463
Frequency (of a motion), 367
Function, 121, 124 (*See also* Trigonometric functions)
  composite, 158
  decreasing, 163
  domain, 121
  exponential, 201
  image, 121
  increasing, 163
  inverse, 160
  linear, 143
  logarithmic, 207
  periodic, 340
  polynomial, 180
  quadratic, 173
  range, 121
  rational, 184
  of two variables, 285
  value, 121
Fundamental frequency, 369
Fundamental Principle of Enumeration, 475
Fundamental Theorem of Algebra, 452

G

Gaussian elimination, 247
General term of a sequence, 462
Geometric progression, 467
Geometric sequence, 467
Geometric series, infinite, 471
Gradian (grad.), 295
Graph:
  of a function, 123
  of an inequality, 277
  of a relation, 109
  of a set of real numbers, 82
  of a system, 279
Greater than, 79

H

Half-angle formulas, 388–90
Half-life, 225
Half plane, 277
Harmonics, 369
Heading (for navigation), 403
Homogeneous, 258
Horizontal line test, 162, 359
Hyperbola, 188, 196
Hypotenuse, 302

I

Identity. (*See also* Trigonometric identity)
  additive, 10
  equation, 49
  matrix, 273
  multiplicative, 10
Image, 121
Imaginary axis, 428
Imaginary part (of a complex number), 418
Inconsistent, 250
Increasing function, 163

Independent, 249
Independent variable, 121
Indistinguishable permutation, 481
Induction, mathematical, 495, 498
Inequalities, 79–97, 276–282
   graph, 279
Infinite geometric series, 471
Infinite sequence, 462
Infinity, 181, 186
Initial point (of a vector), 401
Initial side (of an angle), 294
Integers, 11
Intercepts:
   of a line, 147
   of a parabola, 176
Interest:
   compound, 221–2
   simple, 56
Interpolation:
   linear, 236–7
   rough, 203, 210, 236
Intersection, 85
Interval:
   closed, 83
   finite, 82
   half-open, 83
   infinite, 82
   open, 83
Inverse:
   additive, 10
   function, 160
   multiplicative, 10
   trigonometric functions, 314, 359–61
Irrational number, 12

L

Law of Cosines, 397–8
Law of Sines, 392
Laws of exponents, 14
Length of a vector, 401
Less than, 79
Line, equation of:
   intercept, 151
   point-slope, 146
   slope-intercept, 147

Linear equation, 50
Linear function, 143
Linear interpolation, 235–7
Linear programming, 283
Logarithm:
   common (log), 211
   function, 207
   natural (ln), 211
Logistic function, 228

M

Magnitude of a vector, 401
Mathematical induction, 495, 498
Matrix, 260, 270
   associated with a system, 260, 274
   coefficient, 274
   square, 264
   triangular form, 261
Midpoint formula, 107
Minute, 296
Multiplicity (of a zero), 453

N

Navigator's compass, 394
Negative angle, 295
Nested form of a polynomial, 450
Normal distribution curve, 205
Number line, 79

O

Ordered pair, 101
Ordinate, 102
Origin:
   on the line, 79

Origin (*cont.*)
   in the plane, 101
   in polar coordinates, 408–9
Overtone, 369

P

Parabola, 174, 188
Parallel (lines), 148
Partial sum, 464
Period:
   of a function, 340
   of a motion, 367
   of the trigonometric functions, 350, 352
Periodic function, 340
Permutation, 475
   distinguishable, 481
   indistinguishable, 481
   with replacement, 478
   without replacement, 475
Perpendicular (lines), 150
Phase shift, 356
Polar axis, 408
Polar coordinates, 408–9
Polar form of a complex number, 429
Pole, 408
Polynomial, 180
Polynomial function, 180
Positive angle, 295
Probability, 487–8
Product formulas, 388
Progression:
   arithmetic, 463
   geometric, 467
Proportional:
   directly, 152
   inversely, 153
Pure imaginary, 418
Pythagorean Theorem, 302, 303

Q

Quadrant, 101
Quadrantal angle, 328

Quadratic equation, 63
Quadratic formula, 65
Quadratic function, 173
Quotient (of polynomials), 443

R

Radian (rad.), 297
Radical, 39
Radius, 116
Range, 121
Rational expression, 33
Rational function, 184
Rationalize, 43, 45
Rational number, 11
Rational root theorem, 456
Real axis, 428
Real number, 9
Real part (of a complex number), 418
Recursive, 462
Reference angle, 342
Reflect (a graph), 114, 135, 136, 162
Reflection, 134–6
Relation, 109
Remainder theorem, 444
Remainder (polynomials), 443
Resolving (a vector), 405
Reverse (an inequality), 81
Richter scale, 228
Right angle, 295
Right triangle, 302
Root, 64, 180 (*See also* Zero)
   double, 64
   of multiplicity two, 64
Root, nth:
   of a complex number, 436–7
   of a real number, 39
RPN logic, 5

S

Scientific notation, 20
Secant function, 329, 304

Second, 296
Sequence:
   arithmetic, 463
   geometric, 467
   infinite, 462
Set, 49
Significant figure, 22
Similar triangles, 303
Simple harmonic motion, 367
Simple zero, 453
Sine function, 304, 329
Sinking fund, 231
Slope, 143
Slope-intercept form, 147
Solution:
   of an equation, 49
   of an inequality, 81, 227
   of a system, 241
Solution set:
   of an equation, 49
   of an inequality, 82
Solve a triangle, 391
Sound, 368
Special angle, 309, 345
Square root, 39, 422–3
Standard position, 328
Sum formulas, 388
Symmetric:
   about the origin, 113
   about the x-axis, 112
   about the y-axis, 112
Synthetic division, 448

T

Tangent function, 304, 329
Term of a sequence, 462
Terminal point (of a vector), 401
Terminal side (of an angle), 294
Transitivity, 80
Translation:
   horizontal, 130
   vertical, 130
Triangles:
   right, 302
   similar, 303

Triangular form:
   of a matrix, 261
   of a system, 247, 253
Trigonometric form of a complex number, 429
Trigonometric functions, 304, 329
   inverse, 314, 359–61
Trigonometric identity, 374–7
   addition formulas, 381–2
   double-angle formulas, 388–9
   elementary, 306, 337
   half-angle formulas, 388–90
   product formulas, 388
   sum formulas, 388

U

Union, 83
Unknown, 49

V

Value (of a function), 121
Variable, 25
   dependent, 121
   independent, 121
Varies:
   directly, 152
   inversely, 153
   jointly, 155
Vector, 401
Vector sum, 402–3
Vertex:
   of an angle, 294
   of a graph, 279
   of a hyperbola, 195
   of a parabola, 174
Vertical line test, 121

W

Well-defined, 329

X

x-axis, 101
x-coordinate, 102
x-intercept, 147

Y

y-axis, 101

y-coordinate, 102
y-intercept, 147

Z

Zero, 180
Zero matrix, 271
Zero vector, 403